COLLINS TREE GUIDE

OWEN JOHNSON

ILLUSTRATED BY
DAVID MORE

WILLIAM COLLINS

HarperCollins Publishers
1 London Bridge Street
London SE1 9GF

WilliamCollinsBooks.com

Collins is a registered trademark of
HarperCollins Publishers

First published in 2004
This paperback edition published 2006

Text © 2004 Owen Johnson
Illustrations © 2004 David More

15

10 9

The author asserts his moral right to be identified as the author
of this work. All rights reserved.

All rights reserved. No parts of this publication may be reproduced,
stored in a retrieval system or transmitted, in any form or by any
means, electronic, mechanical, photocopying, recording or
otherwise, without the prior permission of the publishers.

A catalogue record for this book is available from the British
Library.

ISBN 978 0 00 720771 8

Designed and edited by D & N Publishing, Hungerford, Berkshire.

Printed and bound in Hong Kong by Printing Express

CONTENTS

How to Use this Book	4
Introduction	5
Keys	9
Winter Shoots	9
Conifers	12
Leaf Shapes	14
Species Descriptions	20
Tree Ferns and Ginkgo	20
Yews and Nutmeg Trees	22
Conifers	26
Cypresses	30
Junipers	54
Silver Firs	66
Cedars	90
Larches	94
Spruces	100
Hemlocks	116
Pines	122
Broadleaves (Dicotyledons)	150
Poplars	150
Willows	164
Hickories	174
Walnuts	178
Birches	182
Alders	190
Southern Beeches	200
Beeches	204
Oaks	214
Elms	240
Magnolias	258
Sorbus	292
Hawthorns	284
Apples	306
Pears	316
Cherries	322
Hollies	362
Maples	368
Horse Chestnuts	392
Limes	400
Eucalypts	414
Ashes	436
Monocotyledons	450
Glossary	452
Index	454

HOW TO USE THIS BOOK

Identifying a Tree

Use the pictorial keys to leaf-shape (p.14) and to winter twigs (p.9), or the key to conifers (p.12), or simply thumb through the illustrations until you find a match for your tree. The text will then point you to any other trees with similar characteristics.

Larger genera are introduced by 'Things to Look For': a quick-reference list of particular diagnostic features.

Species Descriptions

Symbols Used

🗝 Key Species (for major genera): the commonest or most distinctive forms, to which all the others are compared.

🌲 The tree is (more or less) evergreen.

🍃 The tree carries its leaves and leaf-buds in opposite pairs (or 3s) on extension shoots. (All trees without this icon carry them one by one.)

☠ The tree is in part poisonous. (The toxicity of many rarer trees in this book is unknown.)

Most species descriptions include the following details:

- Preferred common name.
- Current scientific name.
- Alternative/illegitimate common and scientific names in parentheses at the start of the first line of the description.
- Basic natural distribution, including status in Britain and Ireland (native, naturalized).
- The date of the first known introduction of a species to Britain, the first occurrence of a hybrid, or the first distribution of a cultivar.
- Frequency in Britain and Ireland: Dominant (the principal tree species in an area); Abundant (locally ubiquitous and likely to be found in most places); Frequent (present in small numbers in most areas); Occasional (individuals in many large gardens or towns but absent from the majority of parks or woods); Rare (likely to be absent from many counties and found in only a few parks and gardens in others).
- Scientific family name (ending -ae, and usually -aceae), included in the species description if the tree is the first one to be treated in a new family and the family is not widespread enough to warrant an introductory paragraph. Families are covered consecutively through the book, ordered by probable evolutionary relationships.
- The tree's APPEARANCE (the most diagnostic features are *italicized*): Shape The height given is that of the tallest specimen currently recorded in Britain/Ireland. **Bark** Of a mature trunk unless stated. **Shoots** Colour and appearance by the first winter (most start green). **Buds** Colour and appearance (of leaf-buds, unless stated) during winter. **Leaves** Note that examples can always be found whose size exceeds the suggested size parameters. (The 'bottom' of the leaf is the end nearest the stalk.) **Flowers** and **Fruit** (or **Cones**).
- COMPARE Cross-refers to trees elsewhere in the book with which confusion is possible.
- VARIANTS Subspecies, varieties and cultivars that are widely grown or readily separable.
- OTHER TREES Some rarer allies (often not illustrated).

Illustrations

The illustrations show leaves, flowers and fruits for major species and cultivars. Although 'typical' specimens have been chosen, each tree's potential variety cannot be illustrated, but *is* described. Many individual trees have also been portrayed but, as a tree's shape is particularly dependent on its environment, these must not be assumed to be characteristic.

INTRODUCTION

This book is intended as a portable companion to anyone looking at trees in the countryside, parks and gardens of Britain and of non-Mediterranean Europe.

Its predecessor, the late Alan Mitchell's *A Field Guide to the Trees of Britain and Northern Europe* (1974), was the first such guide to all trees thought likely to be found outside specialist collections. Over the last 30 years two developments have helped necessitate an update. Nurseries now supply, and gardeners and local authorities plant, a continually increasing range of trees, while the active recording and measuring of them – a pursuit largely inspired by Alan Mitchell's own life-work – continues to reveal a cornucopia of rarities in unlikely places. This guide now treats nearly 1600 taxa, and, to keep it handy, more use has been made of illustrations. We have tried to include those trees which, however rarely, may be found in the general run of gardens, at the expense of others which attract interest when seen but are confined to larger collections where they will, with luck, be correctly labelled. Conifer species, unfashionable through the second half of the 20th century and often short-lived in much of Britain, are the exception in occurring generally in less diversity than they did 30 years ago, but Alan Mitchell's thorough coverage has been retained, partly in the hope of interesting more people in these fascinating plants. All the major tree species native to any part of Europe are featured, along with microspecies native within Britain and Ireland. Fruiting selections, and introductions hardy only in the very mildest areas, have been excluded.

Deciding which species to include in a tree guide is difficult not only because of the range of species and cultivars – at least 6000 – in Europe's collections as a whole, but also because the concept 'tree' is a vague one. 'Bird' is a precise scientific term, but 'trees' occur among groups of plants all across the evolutionary spectrum from ferns to palms. The beech family (Fagaceae) is a group of plants across which similarities of flower-structure suggest a close relationship; all are trees. But the rose family (Rosaceae) contains a jumble of herbs, shrubs and trees (such as cherries), while the figwort family (Scrophulariaceae) consists of herbs such as foxglove and just one genus of tree: the foxglove trees or *Paulownia*. For the purposes of this book, a 'tree' is any plant which commonly grows to 3 m on a single stem at least 20cm thick. The rules have been bent to include some common natives (Dogwood, Alder Buckthorn) which very seldom do this, and to exclude many *Rhododendron*, *Pieris* etc., which eventually can, but of which old examples are rare.

Our aim has been to keep to the spirit of Alan Mitchell's guide, and I have been more than happy to take the opportunity to borrow several descriptions that could not be bettered. Emphasis has continued, where possible, to be placed on those features (bark, habit, shoots or leaves) which best assist field-identification year-round, rather than the flowers and fruit which display more absolute distinguishing characteristics but are often unavailable. Many exotic trees have also been widely cultivated to date in a limited range of clones, and this has allowed us to treat as diagnostic some features of bark and habit which may become less reliable in the wild (or among the wild-collected specimens in major arboreta). Botanical terminology has been kept to a bare minimum.

Looking at Trees

Tree-watching is an engrossing but under-subscribed hobby. Although trees are big organisms, they are easily overlooked, especially perhaps by eyes tuned to the movements of tiny birds or used to scanning the ground for rare herbs. Their impact on our surroundings is as overwhelming as that of landforms or weather, and is likely to be taken for granted in the same way: only when we look at old photographs or confront the effects of a freak storm do we realize with a jolt how much of a difference the growth, the loss and the variety of trees can make. It was discovered only in 1989 (by the eminent field botanist Dr Francis Rose) that one of the trees growing – in towering, wild groves – on the scarps of the South Downs is the very rare Broad-leaved Lime; contemporaneously, ancient True Services, identified on cliffs near Cardiff, added a new species to the list of British natives. An entirely new species – *Zelkova sicula* – was discovered growing as a population of 200 trees in south-east Sicily in 1991. Everywhere, too, are long-forgotten arboreta, where knowledge of the site's origins has long been lost but the rare trees persist, standing incongruously in suburban verges, in the waste ground behind an industrial estate, or in an overgrown spinney. Most cemeteries and town parks have at least one tree rarer than passers-by will in general realize, and the definition of a tree in this book as 'rare' should not tempt you to assume a misidentification.

Learning to recognize trees is an open-ended process. The first steps are hardest: once you can confidently distinguish a hornbeam from a beech, you can begin to notice similar trees which are clearly something different, and you will also be able to guess whether the tree is a rarer hornbeam or represents a new group. Human brains habitually skip the unfamiliar: it can be salutary to spend time studying every tree in, say, a churchyard –

even the small ones in the corner shrubberies – as the process may lead you to identify a species you have never consciously spotted before but which, once learnt, will crop up everywhere.

Winter is the hardest time to identify trees, as broadleaves with neither flowers nor foliage offer few clues, and rarities are bound to be missed. Autumn is a good study period, as foliage tints highlight differences and leaves, out of reach in summer, can be picked off the ground. (Leaves which decay slowly, such as the Wild Service's, can blow large distances during winter and examples may have to be tracked down by following trails of foliage.) Spring, when winter buds have burst but leaves are not expanded, is a tricky time but the only window for identifying some flowering trees.

Binoculars are useful tools (leaves may be just too high up for details of toothing to be seen from the ground; most silver firs carry cones only near their tops). A hand-lens (×10) can also reveal minute tell-tale hairs or glands on leaves or shoots.

Labelled arboreta are the best places to learn the less common trees, though it must be borne in mind that labels on the trees in many town parks, for example, are frequently incorrect.

Emphasis has been placed in this book on the most concrete or fail-safe ways of differentiating trees: are the buds, for instance, alternately spaced or in opposite pairs; do the leaves have hairs underneath? This yes-or-no methodology lends itself to description and depiction, but quite quickly you will instead be able to stand at a cemetery gate or look out of a train window and identify every tree you can see simply from subtle variations in colour, texture and habit.

Until this stage is attained, more hard-and-fast distinguishing features have to be looked for, but these can vary from genus to genus. A full checklist would include:

- **Shape:** Do the branches weep, or rise steeply, or spread in plates? Does the tree look as if it will become large?
- **Bark:** Is it rough or smooth (at what age/diameter)? What colour is it? (In clean or moist air the bark may be covered by moss, lichens or green and orange algae.) Is there a graft? (a horizontal discontinuity, showing that a nursery has 'worked' a cutting of a clone, or of a species hard to grow from seed, on to an easily propagated rootstock. A graft suggests a rarity or a nameable cultivar, though by no means all such trees will be grafted. Sometimes a loop of wire grown into the trunk can create a graft-like scar.)
- **Shoots:** Are they hairy, or grooved, or warty? What colour are they? (The hairs may be tiny and transparent. The easiest way to spot them is to hold a specimen against a dark background next to a light-source; they will catch the light and halo the shoot.) Is there a distinct smell if you scratch the wood? Do the leaves/leaf-buds come in opposite pairs, or one by one?
- **Buds:** What shape are they? Are they hairy? Have they any protective scales, and of what colour?
- **Leaves:** How big are they? (Resist the temptation to choose the biggest leaf you can find: measure a well-grown, average one.) What shape? Are the margins lobed or toothed? Are there any hairs – above, below, along the veins, on the stalk, or in tufts under the main vein-joints? The underside of the leaf holds more clues than the top: what colour is it? How conspicuous are the veins?
- **Flowers and fruit:** their structural range and evolutionary stability make these the most useful parts of the tree for identification and the basis for the differentiation of scientific families and genera, though they will often not be available in the field.

Further Reading

The learning process does not end when you are familiar with the trees in this book. Works which can take you further include:
Trees of Britain & Europe by Keith Rushforth (HarperCollins, 1999). A fully descriptive guide to 1200 species.
Trees and Shrubs Hardy in the British Isles by W. J. Bean (Murray; 8th Edition 1973–1989).
Manual of Cultivated Broad-leaved Trees and Shrubs and *Manual of Cultivated Conifers* by Gerd Krussmann (English translations 1984ff).

British and Irish Native Trees

Because of extinctions caused by successive Ice Ages, and the brief window between the end of the last Ice Age and the area's isolation by the flooding of the North Sea, the native tree flora of Britain and Ireland is singularly impoverished. Excluding hybrids, varieties and unashamed bushes, the 63 species and microspecies that seem most likely to have got here without human agency are:

Field Maple *Acer campestre* (England)
Common Alder *Alnus glutinosa*
Strawberry Tree *Arbutus unedo* (SW Ireland, local)
Silver Birch *Betula pendula*
Downy Birch *Betula pubescens*
Box *Buxus sempervirens* (SE England, local)
Hornbeam *Carpinus betulus* (SE England)
Dogwood *Cornus sanguinea* (England, Wales and Ireland)
Hazel *Corylus avellana*
Midland Thorn *Crataegus laevigata* (S England)
Hawthorn *Crataegus monogyna*
Spindle *Euonymus europaeus*
Beech *Fagus sylvatica* (S England, S Wales)
Alder Buckthorn *Frangula alnus* (England, Wales and Ireland)
Common Ash *Fraxinus excelsior*
Sea Buckthorn *Hippophae rhamnoides* (E coasts of England and of S Scotland)
Holly *Ilex aquifolium*
Juniper *Juniperus communis* (local)
Wild Crab *Malus sylvestris*
Scots Pine *Pinus sylvestris* (Scottish Highlands; naturalized elsewhere)
White Poplar *Populus alba*
Black Poplar *Populus nigra* ssp. *betulifolia* (England and Wales; local)
Aspen *Populus tremula*
Wild Cherry *Prunus avium*
Bird Cherry *Prunus padus* (N Britain)
Blackthorn *Prunus spinosa*
Plymouth Pear *Pyrus cordata* (SW England; very rare)
Wild Pear *Pyrus pyraster* (England; very rare)
Sessile Oak *Quercus petraea*
English Oak *Quercus robur*
Purging Buckthorn *Rhamnus cathartica* (England, Wales and Ireland)
White Willow *Salix alba*
Goat Willow *Salix caprea*
Grey Sallow *Salix cinerea* ssp. *oleifolia*
Crack Willow *Salix fragilis*
Bay Willow *Salix pentandra* (N Britain and Ireland only)
Purple Osier *Salix purpurea* (local)
Almond Willow *Salix triandra* (local)
Common Osier *Salix viminalis*
Elder *Sambucus nigra*
Sorbus anglica (SW England, Wales and County Kerry; local)
Whitebeam *Sorbus aria* (S England; Ireland)
Arran Whitebeam *Sorbus arranensis* (N Arran)
Rowan *Sorbus aucuparia*
Bristol Service *Sorbus bristoliensis* (Avon Gorge)
French Hales *Sorbus devoniensis* (SW England and SE Ireland, local)
True Service *Sorbus domestica* (Wales; very rare)
Sorbus eminens (Wye Valley and Avon Gorge)
Sorbus hibernica (central Ireland; local)
Sorbus lancastriensis (S Cumbria)
Sorbus porrigentiformis (N Devon, Mendips and S Wales; local)
Arran Service *Sorbus pseudofennica* (N Arran)
Cliff Whitebeam *Sorbus rupicola* (limestone uplands; local)
Exmoor Service *Sorbus subcuneata* (Exmoor)
Wild Service *Sorbus torminalis* (England and Wales)
Sorbus vexans (N Devon)
Sorbus wilmottiana (Avon Gorge)
Common Yew *Taxus baccata*
Small-leaved Lime *Tilia cordata* (England and Wales; local)
Broad-leaved Lime *Tilia platyphyllos* (England and Wales; very local)
Wych Elm *Ulmus glabra*
European White Elm *Ulmus laevis* (very rare)
Field Elm *Ulmus minor* (in England, in at least some clones)

Well-naturalized, locally at least, are:
Norway Maple *Acer platanoides*
Sycamore *Acer pseudoplatanus*
Horse Chestnut *Aesculus hippocastanum*
Snowy Mespil *Amelanchier lamarckii*
Sweet Chestnut *Castanea sativa*
Common Apple *Malus domestica*
Medlar *Mespilus germanica*
Grey Poplar *Populus canescens*
Myrobalan Plum *Prunus cerasifera*
Sour Cherry *Prunus cerasus*
Common Plum *Prunus domestica*
Bullace *Prunus insititia*
Cherry Laurel *Prunus laurocerasus*
Portugal Laurel *Prunus lusitanica*
St Lucie Cherry *Prunus mahaleb*
Black Cherry *Prunus serotina*
Common Pear *Pyrus communis*
Turkey Oak *Quercus cerris*
Holm Oak *Quercus ilex*
False Acacia *Robinia pseudoacacia*
Sorbus croceocarpa
Tamarisk *Tamarisk gallica* and other species

BOTANICAL NAMES

Plants' scientific names are always worth learning, as common names can vary from region to region. 'Sycamore' in the United States is a species of *Platanus* ('plane'), while in Scotland 'plane' can be *Acer pseudoplatanus* ('Sycamore'); 'Siberian Crab', used in this guide for *Malus baccata*, is used by some gardeners for *M.* × *robusta*.

Customarily italicized, a scientific name has two main parts: the generic name ('surname'), followed by the specific name ('forename'): *Quercus* (never *quercus*) is the name for the group of related trees commonly called 'oaks'. *Quercus robur* (never *Robur*) is one distinct species, the English Oak. The fuller form of the name is *Quercus robur* L., the 'L.' denoting the botanical authority who published this name: the 18th-century Swedish botanist, Carl von Linné, who wrote as 'Linnaeus'. The convention antedates Darwin's theories of evolution: any nomenclature is ultimately a convenient but artificial way for us to parcel up the constantly evolving continuum of the genes of all life.

Obvious sexual hybrids between species are denoted with an '×': *Quercus* × *rosacea* is the hybrid of *Quercus robur* and *Quercus petraea* (Sessile Oak), but this entity can just as legitimately be called *Quercus petraea* × *robur*. Fertile, stable hybrids (e.g. *Sorbus hybrida*) generally lose their '×'. If two trees customarily allocated to different genera hybridize, a new generic name is needed and the '×' precedes this: × *Crataemespilus grandiflora* is the hybrid of *Crataegus laevigata* (Midland Hawthorn) and *Mespilus germanica* (Medlar). Sometimes hybrids occur when genetic material, rather than mixing sexually, fuses in a natural grafting process, creating a 'chimaera'; for these trees, a '+' is used. *Aesculus* + *dallimorei* (Dallimore's Chestnut) is a fusion of tissues of Common Horse Chestnut (*Aesculus hippocastanum*) and Yellow Buckeye (*Aesculus flava*), with cells of both entities co-existing in a single plant. + *Laburnocytisus adamii* is an intergeneric chimaera, Common Laburnum (*Laburnum anagyroides*) having fused with a purple broom (*Cytisus purpureus*). Some genera (eg. *Sorbus*) are partly divided not into species but into microspecies, whose individuals are self-fertile and go on producing seedlings with very little genetic diversity.

A third botanical name is sometimes used to distinguish different forms of a species – either subspecies (ssp.) or variety (var.); variety suggests a difference in appearance rather than habitat. Corsican Pine is *Pinus nigra* ssp. *laricio*, Austrian Pine *Pinus nigra* ssp. *nigra*. A third category, forma ('f.'; plural formae), may be used to describe a population of 'sports': Sycamores with purple-backed leaves – often found in the wild – are *Acer pseudoplatanus* f. *purpureum*.

A 'cultivar' is, strictly, one particular clone or asexually propagated individual plant, and its name is never italicized because the rules of scientific nomenclature no longer obtain. Instead, the first letter is capitalized and the word is enclosed in single quotation marks *or* preceded by 'cv': *Malus domestica* 'Granny Smith' or *Malus domestica* cv. Granny Smith. Nurseries naming their cultivars traditionally used pseudo-scientific names (eg. 'Pendula' for weeping trees); since 1959 names have been required to be in the vernacular (eg. *Gleditsia triacanthos* 'Sunburst'). In practice, cultivar names often serve to cover 'variants' (groups of clones with similar characteristics): *Populus nigra* 'Italica' is used for Lombardy Poplars of various inherited shapes.

The allocation of families and genera is an art with no written laws, but the rules for naming species are precise: the correct name is the earliest one to have been published in a reputable journal which unambiguously describes an understood species. A plant's specific name must not be the same as its generic name, though an animal's can be (eg. *Troglodytes troglodytes*, the Wren). Names change for three reasons: a new description, published earlier, may be rediscovered; an existing name may be deemed to embrace trees from another species; or new ideas on a tree's family relationship may come into fashion, prompting a different generic name (plus the adoption of the next-oldest species name if the previous one is already in use in the new genus for a different tree).

Spellings can be important: *Acer maximowiczii* is a rare Snake-bark Maple; *Acer maximowiczianum* is the very dissimilar Nikko Maple. *Acer pensylvanicum* ('from Pennsylvania') is spelled like this because of a misprint in Linnaeus' original description, though his *Stewartia* (honouring John Stuart, Earl of Bute) is often 'corrected' to *Stuartia*. Specific names should inflect (like Latin adjectives) to agree with the gender of the genus name – but as botanists are seldom classical scholars there is frequently an unresolved confusion between, for example, '*europaeus*' and '*europaea*'.

WINTER SHOOTS OF COMMON OR EASILY RECOGNIZED TREES

Bark and habit tend to offer more clues to the identity of mature trees in winter, but cannot be used to identify saplings. Most twigs on old trees are slow-growing, with a gnarled surface and a congested rather than alternate/opposite distribution of buds. Fast-growing 'extension' shoots are shown here, but their shapes have been standardized.

Buds in opposite pairs

Euodia (p.356) — Woolly scale-less buds

Common Dogwood (p.424) — Black buds without scales; shoots crimson in sun lime-green in shade

Foxglove Tree (p.444) — Tiny buds. No end bud

Indian Bean (p.444) — Tiny buds. No end bud

Elder (p.448) — Buds like pineapples; shoots wander

Common Ash (p.436) — Black buds

Golden Ash (p.436)

Manna Ash (p.438) — Pale woolly buds

White Ash (p.440) — Dark woolly buds

Purging Buckthorn (p.398) — Sharp brown buds. Many pairs displaced

Katsura (p.274) — Some bud-pairs displaced

Dawn Redwood (p.64) — Buds underneath shoots

Smooth Japanese Maple (p.384) — Two end-buds

Japanese Maple 'Sango-Kaku' (p.384)

Spindle (p.448) — Dark green shoot

Box Elder (p.390) — Sometimes bloomed white

Field Maple (p.368) — Small grey downy buds

Silver Maple (p.378)

Cappadocian Maple (p.374) — Strong shoots crimson

Norway Maple (p.372) — Big sharp red-brown buds

Sycamore (p.370) — Green buds

Amur Cork Tree (p.356) — Blunt green buds

Moosewood 'Erythrocladum' (p.382) — Snake-bark maples have stalked side-buds

Horse Chestnut (p.392) — Sticky buds

Red Horse Chestnut (p.394) — Often 2 end-buds

Indian Horse Chestnut (p.394) — Often 2 end-buds

10 WINTER SHOOTS OF COMMON OR EASILY RECOGNIZED TREES

Buds alternate

Caucasian Wingnut (p.172) Stalked buds without scales

Alder Buckthorn (p.398) Tiny woolly rufous buds without scales. White streaks on shoot.

Stag's-horn Sumach (p.360) Velvety shoots; buds without scales

Common Larch (p.94) Buds broader than high

Japanese Larch (p.96) Some white bloom on shoot

Swamp Cypress (p.64)

Tulip Tree (p.272) Buds like paddles

London Plane (p.280) Horn-shaped buds

Snowbell Tree (p.434) Buds like tiny mittens

Walnut (p.178) Mitre-shaped buds

Black Walnut (p.178) Hairy grey mitre-shaped buds

Willows have a smooth, curving shoot and one bud-scale:

Grey Sallow (p.168) Finely hairy shoot

Goat Willow (p.168) Shoots red in sun, grey/green in shade

Violet Willow (p.168) Some purple bloom on shoot

Weeping Willow (p.168) Bright yellow shoot

Crack Willow (p.166) Dull orange shoot

Coral-bark Willow (p.164) Bright orange shoot

White Willow (p.164) Finely hairy shoot

Silver Willow (p.164) White-hairy shoot

Bat Willow (p.164) Purplish finely hairy shoot

Common Osier (p.170) Close-set blunt silky-hairy buds

Oaks have a cluster of buds at each shoot-tip:

English Oak (p.216)

Hungarian Oak (p.226) Hairy shoot; many, loose bud-scales

Red Oak (p.234) Small sharp buds, tipped with hairs

Turkey Oak (p.218) All buds with big whiskers

Most alders have side-buds on stalks:

Common Alder (p.190) Often mauve buds; club-shaped

Grey Alder (p.192) Hairy shoot

Italian Alder (p.192) Variable bloom on shoot

Only 2 or 3 bud-scales:

Sweet Chestnut (p.212) Buttress behind each bud; no bigger end bud

Small-leaved Lime (p.400) Buds like boxing gloves; no bigger end bud

Broad-leaved Lime (p.400) Fine hairs; 3 hairy bud-scales

Crimean Lime (p.402) Shoot green in shade, orange-pink in sun

Silver Lime (p.404) Grey wool

Laburnum (p.350) Silky-white buds

White Poplar (p.150) Silvery hairs; small sharp buds

Grey Poplar (p.150) Silvery hairs rub off

Birches have big long sharp buds on slender shoots:

Downy Birch (p.182) Usually hairy shoot

Silver Birch (p.182) Rough white warts on shoot

Paper-bark Birch (p.184) Rough white warts on shoot

Hornbeam (p.194) Long, sharp, appressed buds

Snowy Mespil (p.320) Buds like miniature beech-buds

Common Beech (p.204) Very long, spreading-buds

WINTER SHOOTS OF COMMON OR EASILY RECOGNIZED TREES

From tiny and round to long and sharp buds:

Tree of Heaven (p.358)	Tapering twisted shoots; big pale leaf-scars
Honey-Locust (p.348)	Thorns in 3s on some trees
False Acacia (p.354)	Thorns in pairs
Myrobalan Plum (p.340)	Thin green shoots
Pissard's Plum (p.340)	Dark purple shoots
Blackthorn (p.340)	Side-shoots can end in spines. Shoots purple in sun, green in shade.
Hawthorn (p.284)	Individual thorns along twigs
Broad-leaved Cockspur Thorn (p.288)	Long straight individual thorns
Wild Service (p.296)	Buds like peas
Hazel (p.198)	Roughly hairy twig; very blunt buds
Turkish Hazel (p.198)	Redder buds
Roblé Beech (p.200)	Very thin hairy shoots; small sharp buds
Rauli (p.200)	Hairy shoots with green warts; big conic buds
Caucasian Elm (p.252)	Shoots with white hairs
Sea Buckthorn (p.410)	Close-set conic orange buds
English Elm (p.244)	Small short buds
Judas Tree (p.348)	Leaf-scar all round bud
Golden Rain Tree (p.398)	Prominent leaf-scars with dark rims
Black Cherry (p.344)	
White Mulberry (p.258)	
Persian Ironwood (p.278)	Sharp blackish buds
Medlar (p.290)	Prominent lenticels on shoot
Plum (p.340)	Short, sharp reddish buds
Wild Crab (p.306)	May be spiny
Orchard Apple (p.306)	Hairy buds and twigs
Siberian Crab (p.310)	
Purple Crab (p.314)	Purple shoots
Common Pear (p.316)	May be spiny
Willow-leaved Pear (p.320)	White-downy
Maidenhair Tree (p.20)	Many buds on spurs
Wych Elm (p.242)	Shoots stiffly hairy. Short sharp purplish buds
Black Mulberry (p.258)	Shoots with some stiff hairs
Dove Tree (p.412)	Shoots with pale freckles
Wild Cherry (p.322)	Flower-buds sharp and clustered
Japanese Cherry ('Kanzan') (p.326)	
Bird Cherry (p.342)	Deep brown shoot with fawn lenticels
Aspen (p.152)	Painfully sharp buds. Suckers have woolly shoots
Sweet Gum (p.278)	Green/red conic buds
Black Poplar (p.152)	Buds appressed
Hybrid Black Poplar ('Regenerata') (p.156)	
Balsam Poplar (p.160)	Big sticky buds strong balsam scent
Japanese Rowan (p.302)	Slender red rather sticky buds
Swedish Whitebeam (p.298)	Green and brown bud-scales
Whitebeam (p.292)	Long sharp buds; green and brown scales
Rowan (p.300)	Purplish bud-scales fringed with silvery hairs
Sargent's Rowan (p.302)	Very sticky bud
Shag-bark Hickory (p.176)	Hickories have huge end-buds
Saucer Magnolia (p.270)	Flower-buds silky, very large
Campbell's Magnolia (p.264)	Beak-shaped blue-bloomed leaf-buds

KEY TO CONIFERS

Broad leaves

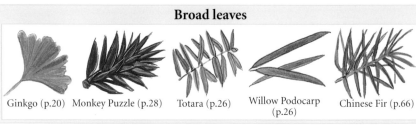

Ginkgo (p.20)   Monkey Puzzle (p.28)   Totara (p.26)   Willow Podocarp (p.26)   Chinese Fir (p.66)

Leaves very short or scale-like

Fully flattened ferny sprays

leaves fine and short:

False Cypresses (p.36)

leaves thicker, glossier:

Thujas (p.32)   Hiba (p.32)

leaves longer:

Nootka Cypress (p.44) Side-leaves with spreading points

Northern Incense Cedar (p.30) 'Stretched' appearance

leaves spreading bluntly:

Southern Incense Cedar (p.30)

3D sprays

spreading points make shoots rough:

Patagonian Cypress (p.52)

Giant Sequoia (p.62)

Summit Cedar (p.58)

Leylandii (p.46)

Mexican Cypress (p.48)

shoots smooth:

True cypresses (p.48)
Carry rounded woody cones

Junipers (p.56)
May carry berries

Smooth Tasmanian Cedar (p.58)

Mixed scale-leaves and long narrow needles

scale-leaves up main shoots; needles on side-shoots:

Coast Redwood (p.62)

random patches of short needles:

Chinese Juniper and Pencil Cedar (p.56)

Narrow needles

IN ROSETTES (ON OLDER TWIGS) OF 10–50

Evergreen

Deciduous

Cedars (p.90)   Larches (p.94)   Golden Larch (p.98)

IN BUNDLES OF 2–8

Pines (p.122)

IN WHORLS LIKE UMBRELLA-SPOKES

Umbrella Pine (p.66)

IN WHORLS OF 3

Junipers (p.54)

ONE BY ONE ALONG THE SHOOTS

shoot (visible) soon woody, greyish/brownish

leaves soft, deciduous
side-shoots and buds in opposite pairs:

Dawn Redwood (p.64)

side-shoots and buds alternate:

Swamp Cypress (p.64)

leaves hard, evergreen
shoot with 'pegs' for each leaf-stalk:

Spruces (p.100)

leaf-stalks with round 'sucker' bases:

Silver Firs (p.66)

leaf-stalks pressed parallel with shoot:

Hemlocks (p.116)
Mountain Hemlock (p.118)

otherwise:

Douglas Firs (p.120)
(Tiny sucker at base)

shoot green for more than a year, or hidden by leaf-bases

leaves bent into ranks either side of shoots

leaves slightly yellowish green underneath:

Yews (p.22)

leaves jumbled around shoots

Prince Albert's Yew (p.26)

Totara (p.26)

leaves white-banded or white-green underneath
leaves sharply spined:

Nutmeg trees (p.24)

leaves very leathery:

Cephalotaxus (p.24)

leaves rather soft:

Chilean Plum Yew (p.26)

leaves evenly distributed all round shoots

leaves broad-based

Japanese Red Cedar (p.60)

Summit Cedar (p.58)

King William Pine (p.58)

Norfolk Island Pine (p.28)

leaves not broad-based

Junipers (p.54)

Sawara Cypress cultivars (p.42)

Japanese Red Cedar 'Elegans' (p.60)

¹⁴ LEAF SHAPES OF SOME BROADLEAVED TREES

This pictorial key focuses in particular on trees whose family allegiance and place in the book is not obvious. Many maples have the same radially lobed leaf design as Sycamore, and many oaks have lateral lobes like English Oak's, so these are not shown separately. The descriptions of all the trees whose leaves are included here will also point you towards any lookalikes.

The contrast in size between leaves has been reduced in these drawings. Each section progresses from very small to very large leaves.

- Triangular/heart-shaped leaves (ie. broadest very near the base): p.14
- Lobed leaves: p.15 (Lobes are promontories on the leaf-margin more than 1 cm deep *or* fewer than 20 in number.)
- Oval/boat-shaped leaves: evergreen: p.16 (Evergreen leaves are leathery/glossy and persist under the tree year-round.)
- Oval/boat-shaped leaves: deciduous, serrated: p.17
- Oval/boat-shaped leaves: deciduous, not serrated: p.18
- Oddities: p.18
- Compound leaves: p.19 (Separate leaflets make a repeated pattern; the central leaf-stalk has no buds and will not develop into a twig.)

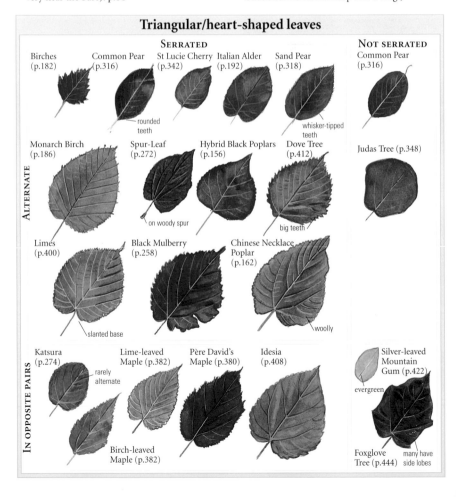

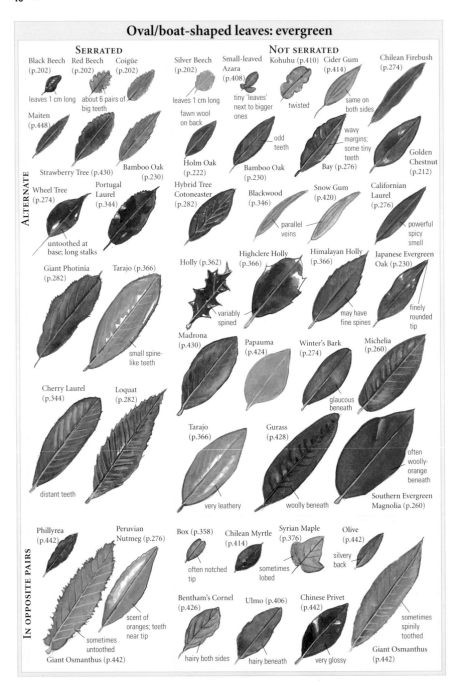

LEAF SHAPES OF SOME BROADLEAVED TREES

Oval/boat-shaped leaves: deciduous, not serrated

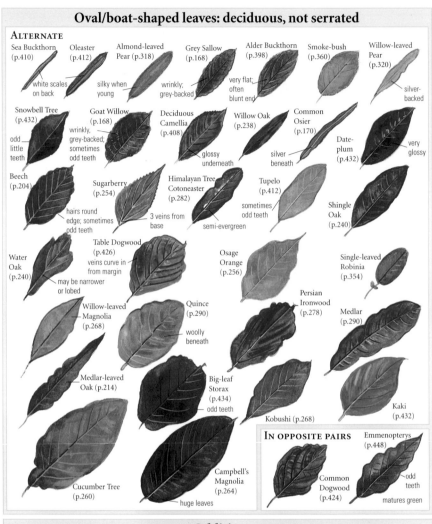

Oddities

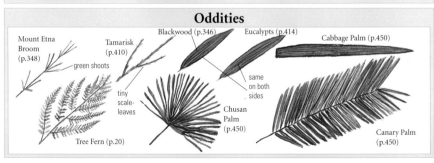

TREE FERNS & MAIDENHAIR TREE

The Common Tree Fern and Maidenhair Tree are two of the most ancient and singular trees we can grow. Over 300 species of tree fern in the Cyathaceae family survive in cool, humid parts of the tropics. The trunk (to 25 m long) is an accumulation of fibres on top of which the living plant sits; its roots draw moisture out of the trunk and do not reach the ground. The Ginkgoaceae family, dominant in the Mesozoic Era, has only one surviving species.

Common Tree Fern — *Dicksonia antarctica*

The hardiest species (SE Australia; Tasmania). Restricted to mild, moist areas in the UK where, after 150 years, plants are 5–6 m tall and seed freely. Trees sourced from the highest elevations are now planted more widely in small gardens and some town parks, often as mature transplants from rainforests under development.

APPEARANCE Leaves To 2 m; green beneath.
Other trees Silver Tree Fern, *Cyathea dealbata* (New Zealand), has *white-backed* leaves on a slender (20 cm) stem; confined to the mildest areas.

Maidenhair Tree — *Ginkgo biloba*

Brought from temple gardens in Chekiang Province, China, in 1758; now endangered in the wild. Quite frequent in areas with warmer summers.

APPEARANCE Shape Haphazard, and spiky with short 'spur' shoots; spire-shaped with few small, horizontal limbs from a slightly leaning trunk, or on boles narrowly diverging from low down. To 28 m. **Bark** Grey-brown, corky then craggy, its exposed growth-rings sometimes creating a beautiful lace-like silver patterning; oldest specimens – still healthy – developing the 'chi-chi' or 'breasts' which characterize ancient Chinese trees. **Shoots** Shiny, grey. **Buds** Fat but sharp, green/red. **Leaves** *Fan-shaped, 9 × 7 cm*; alternately up strong shoots and clustered on short 'spurs'. Fresh green; rich gold in late autumn. The marginal notches are variable and tend to be deeper on young plants. **Flowers** Trees either male or female. Male catkins yellow, 2–4 cm, in bunches in late spring; scarcely seen in Britain. Female flowers green 4 mm knobs, paired on 4 cm stalks; plum-like fruits ripen yellow in autumn (given enough summer warmth) and smell unpleasant. Grafts from male trees are consequently preferred as garden plants. The white nut in the fruit is, however, a delicacy in China.
COMPARE In winter, Golden Larch (p.98): a more symmetrical, broad-conic crown and square-cracked bark. Some specimens of Common Pear (p.316): similar spiky crown, dense with 'spur' shoots, but blackish bark.
VARIANTS 'Variegata' is very rare, as is the low, weeping 'Pendula'. Straight, narrow forms have been selected, especially in USA where they are invaluable as heat-tolerant street trees ('Sentry'; 'Fastigiata'). The best clones have tight crowns on vertical twigs and recall Lombardy Poplar, but very slender trees are part of the natural variation.

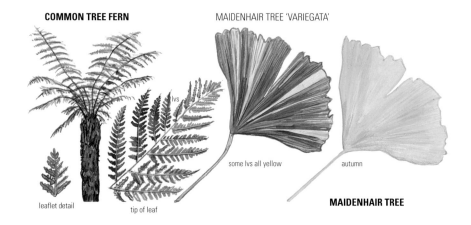

COMMON TREE FERN — leaflet detail, lvs, tip of leaf

MAIDENHAIR TREE 'VARIEGATA' — some lvs all yellow

MAIDENHAIR TREE — autumn

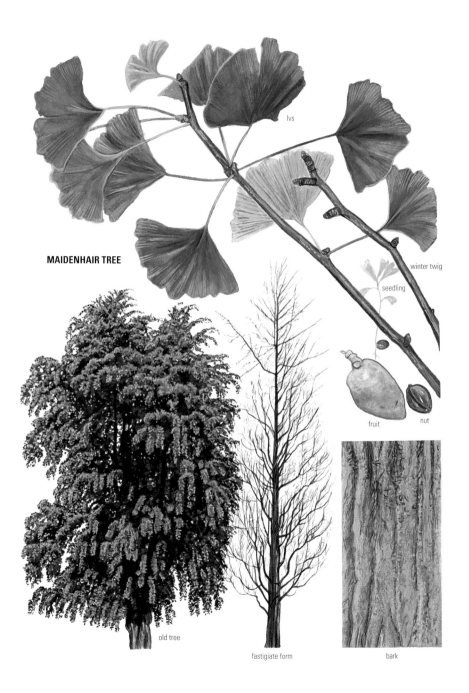

YEWS

Yews (6 similar species) are ancient trees with berry-like fruits – red 'arils'. (Family: Taxaceae.)

Things to Look for: Yews to Podocarps

- Bark
- Leaves: How big? What colours are they underneath? Are they spine-tipped? Do they curve (up, down)?

Common Yew   *Taxus baccata*

Europe, N to southern Scandinavia and E to Iran, and including Britain and Ireland; N Africa. Grows singly in woods and hedgerows on mineral soils, with some dense groves on limestone scarps, but long eradicated from many parts because it is toxic to livestock; abundantly planted in parks, shrubberies and gardens. One of the world's longest-lived trees: over 2000 years old in many English, Welsh and N French churchyards.
APPEARANCE Shape Often soon broad, but generally with a pointed tip; to 25 m. Heavy, blackish presence. **Bark** Shallowly scaling; grey, purplish and reddish – can be exquisitely colourful. **Shoots** *Green for 3 years*, with 2 mm *round* green buds. **Leaves** 30 × 3 mm, downward-arching, in one flattened rank on either side of the side-shoots but spiralling all round strong/erect growths; abruptly sharp-pointed but not rigid; dull *yellowish* grey-green beneath. Highly toxic: do not crush them to release scent. **Flowers** Nearly always dioecious. Male flowers whiten the crown before shedding clouds of pollen in early spring. **Fruit** Arils ripen in early autumn, cerise. The flesh wrapping the lethal seed is sweet.
COMPARE Chilean Plum Yew and Prince Albert's Yew (p.26). Nutmegs and *Cephalotaxus* (p.24): bigger/stiffer leaves. Coast Redwood (p.62): disparate scale-leaves clasping its shoots. Among trees with shoots green for some years, the yews' slightly yellowish under-leaves are diagnostic.
VARIANTS 'Adpressa' (1828 or 1838) has *half-length*, full-width leaves; dark blue-grey and matt *en masse*. Female; rare. 'Adpressa Variegata' ('Adpressa Aurea'; 1866) has yellow-edged, half-length leaves.
Golden Yew, f. *aurea*, has yellow-edged younger leaves; frequent in gardens and, rarely, found wild.
Westfelton Yew, 'Dovastoniana', arose in Shropshire by 1776 (where the original still thrives): its foliage *cascades from wide very low limbs*. Female; very occasional. 'Dovastonii Aurea' has yellow-edged younger leaves; rare and low.
'Lutea' ('Fructu-Luteo'; 1817) has *yellow* fruit; very rare but very striking.
Irish Yew, 'Fastigiata' ('Stricta', 'Hibernica'), is abundant; *erect* (narrowest in high-rainfall areas; cf. Hicks' Yew, below); blackish leaves all round *all* shoots (cf. Fastigiate Plum Yew, p.24, and Totara, p.26); normally female. One of 2 original seedlings (transplanted by 1780) still grows at Florencecourt, Co. Fermanagh. 'Fastigiata Aurea' (male: yellow-edged leaves) is frequent; the female equivalent 'Standishii' (1908) is very bright and dense but slow and occasional.
OTHER TREES Hicks' Yew, *T.* × *media* 'Hicksii', a rare clone of a hybrid with Japanese Yew (1900; to 9 m so far), resembles Irish Yew. A female plant, with large fruits; leaves *yellower* beneath, tipped by a short, quite soft *spine*; crown blobby and fuzzy, less crisply turreted. (Fastigiate Plum Yew, p.24, has bigger leaves, whiter beneath.)
Japanese Yew, *T. cuspidata* (Japan, Korea, Manchuria; 1855), is in some collections; a much hardier tree than Common Yew. The bud is larger (4 mm), and *brown*; its leaves are stiffer (tipped by *1 mm spines* – cf. Plum Yew, p.24), and a darker *brownish gold beneath*; most *arch almost to vertical*; females carry massed, *clustered* fruits.
Chinese Yew, *T. mairei*, has been grown (usually as *T. celebica*) in a few collections since 1908 – a sad bush. Slender, *yellowish*, sparse leaves curl backwards on either side of a shoot which turns *brown after 1 year*; its fruits *stay green*.

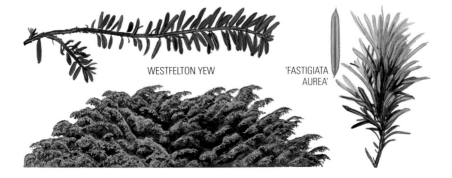

WESTFELTON YEW

'FASTIGIATA AUREA'

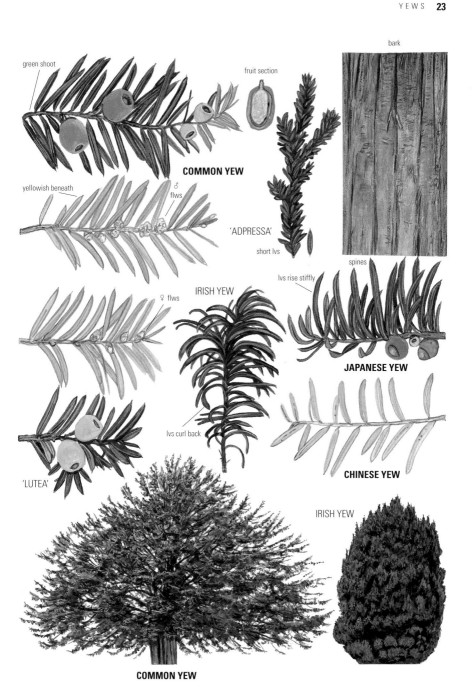

NUTMEGS & PLUM YEWS

Californian Nutmeg *Torreya californica*

A close relative of the yews which, being Californian, grows bigger. A scarce tree in the wild. 1851. Some large gardens in warmer, humid parts.
Appearance Shape A broad, open, fresh-green *spire*, to 22 m, with *whorls* of level branches; usually leans with age. **Bark** Pale red/grey; flat stringy ridges. **Buds** Finely *pointed*. **Leaves** Neatly ranked; about 40 × 3 mm, *rigid and rather fiercely spined; olive-green*, with usually narrow *whitish bands in 2 slight grooves* underneath. Rich, sage-like aroma. **Fruit** 4 cm (female trees; ripening in 2 years), somewhat resembling those of the unrelated culinary nutmeg.
Compare *Cephalotaxus* species (below): bushier, with softer, glossier-green leaves and scalier bark. Totara/Chilean Totara (p.26): stumpy, jumbled leaves. Santa Lucia Fir (p.88): 1 cm buds and twigs red-brown within 4 months.

Japanese Nutmeg *Torreya nucifera*

(Naya) 1764. Largely confined to collections.
Appearance Shape A smaller, often slenderer tree (to 13 m). **Bark** Darker red, more criss-cross fibrous ridges than Californian Nutmeg's. **Buds** *Blunt*, on green shoots which become *rich rufous* by their third year. **Leaves** Only about 30 × 2.5 mm and glossier green, with 2 whitish bands on their *flat* undersides; more closely parallel, and *curving downwards* – Californian Nutmeg's can curve slightly up (but cf. Chinese Plum Yew). **Fruit** Tasty nuts (only try them if you are sure this is not a kind of yew) contain an oil used in Japanese cuisine.
Other trees *T. grandis* (S and E China, 1855) is a spreading plant in a few collections: flimsier, *scentless* leaves on shoots *grey* by their third year.

Plum Yews (7 species) are true conifers (the flowers have 2 ovules), though their leaves and fruit resemble those of the nutmegs. (Family: Cephalotaxaceae.)

Plum Yew *Cephalotaxus harringtonia*

(Cow-tail Pine; incorporating var. *drupacea*) Japan, Korea, N China. 1829. Rare.
Appearance Shape Straggling, but often on a straight bole; to 9 m. **Leaves** Smaller and duller than Chinese Plum Yew's (5–8 cm); in many trees (probably selected in Japan and once distinguished as the type of the species) they rise *stiffly above the shoot* like a pigeon's wings in display-flight. Variably pale green/white-banded beneath; abruptly but softly spined.
Compare Japanese Yew (p.22): shorter leaves more yellow-brown beneath.
Variants Fastigiate Plum Yew, 'Fastigiata' (1861; probably from Korea), has the *habit of Irish Yew* (p.22), with back-curling leaves all round its little-branched, erect shoots. Its leaves, however, are *much bigger* (60 × 4 mm) and the plant itself *reaches only 7 m*. Very occasional in gardens and churchyards – like a giant mutant spurge.

Chinese Plum Yew *Cephalotaxus fortunei*

E and central China. 1848. In a few larger gardens.
Appearance Shape Like a thin, *sprawling* Common Yew (to 10 m) but with giant, eye-catching foliage. **Bark** Soon shaggy with long, shredding, red-brown scales. **Shoots** Green for 3 years. **Buds** Round. **Leaves** Often *8 cm,* slightly arched downwards in a neat rank either side of the shoot; *soft/flexible*, fine-pointed but unspined; a *glossy* plastic-green with 2 whitish bands across the flat underside. **Flowers** Dioecious. **Fruit** In bunches of 3–5, 20 mm long; shiny grey-blue then red-brown.
Compare Willow Podocarp (p.26): the only fully hardy tree-conifer with even longer, glossy leaves.

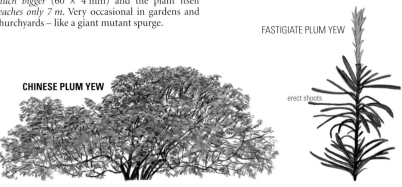

FASTIGIATE PLUM YEW

erect shoots

CHINESE PLUM YEW

NUTMEGS & PLUM YEWS

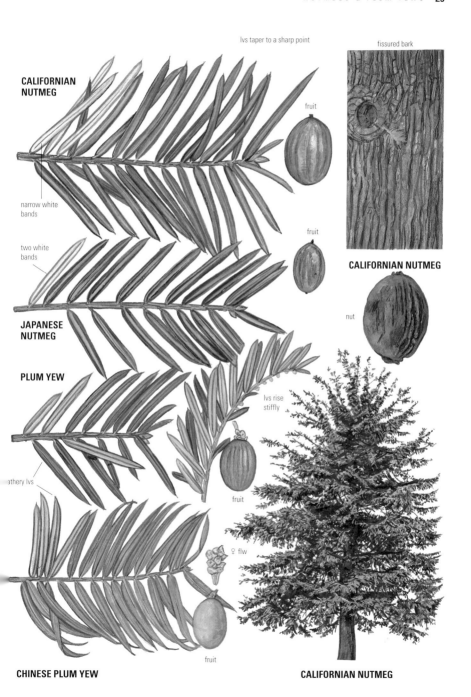

PODOCARPS & PRINCE ALBERT'S YEW

The Podocarpaceae family comprises about 130 large, generally tropical conifers.

Chilean Plum Yew — *Prumnopitys andina*

(*Podocarpus andinus*) S central Andes. 1860. Occasional in gardens and parks in milder areas.
APPEARANCE Shape Often many-stemmed. Densely upswept (sparser in dry districts) and spiky; then spreading, to 20 m (but not weeping); boldly *blue-*green. **Bark** *Smooth, black-grey.* **Shoots** Green for 2 years. **Leaves** Typically 25 × 2 mm; rather *crowded/jumbled above the shoots.* Soft and flexible; 2 broad greyish bands beneath. **Flowers** Usually dioecious; the bunched edible green 'plums' are rarely seen in Britain.
COMPARE Common Yew (p.22). Prince Albert's Yew (below): harder leaves, whiter beneath, red bark. Chilean Totara: broader leaves and peeling bark.

Totara — *Podocarpus totara*

A giant conifer from New Zealand. Straggling, sparse-leaved, yellowish plants and the odd sturdy tree survive in some big gardens in mild areas.
APPEARANCE Bark Brown; stringy, shaggily peeling ridges. **Leaves** 25 × 4 mm, *stiff,* spined, yellow-green beneath. Typically sparse *all round the twigs*; sometimes twisted into roughly flat ranks.
COMPARE Nutmeg trees (p.24): slenderer, with elegantly ranked leaves. Irish Yew (p.22); juvenile Monkey Puzzle (p.28). Chinese Fir (p.66): bluer leaves whose bases clasp the shoot.
OTHER TREES Chilean Totara, *P. nubigenus* (central S Andes, 1847), is in some collections: a happier tree with denser, bright, slightly bluish leaves, *whitish-banded beneath.*
Huon Pine, *Lagarostrobus franklinii* (*Dacrydium franklinii*), from S and W Tasmania and believed by some to be the world's longest-lived tree, is related though very dissimilar, and survives as a thin bush in a few mild gardens. (Its fine, smooth, thread-like foliage, in dangling '3D' sprays, suggests a cypress.)

Willow Podocarp — *Podocarpus salignus*

Chile. 1853. A desirable garden plant, very occasional in mild, moist areas.
APPEARANCE Shape A lax, glistening haystack, often on many stems, to 20 m: bamboo-like, but sparse in dry or shady sites. **Bark** Rufous, vertically stringy and stripping. **Leaves** Big (*to 12 × 1 cm*), thick, shiny, in flattish ranks; yellowish beneath.
COMPARE Kauri Pine (p.28): opposite leaves. Blackwood (p.346): bigger, parallel-veined 'phyllodes'. (The E Asiatic *P. macrophyllus* and *P. nagi* have even broader leaves; scarce bushes in Britain.)

Prince Albert's Yew — *Saxegothaea conspicua*

Central S Andes. 1847. Rare; in mild areas.
APPEARANCE Shape Dense and dark (to 15 m), rather weeping; sometimes bushy. **Bark** *Yew-like*: flaking in reds and purples. **Shoots** Becoming purplish in sun (green in shade for 3–4 years). **Leaves** 20 × 2 mm, closely set in *irregular* ranks. Curved, hard, sharp; *bright white bands beneath.* **Fruit** A 1 cm 'cone'.
COMPARE Common Yew (p.22): leaves yellow-green beneath. Similar bark and leaves (seen from above) make this an easily missed tree. Chilean Plum Yew (above): softer leaves, blue-grey beneath, and smooth grey bark. Hemlocks (p.116): brown twigs within 4 months.

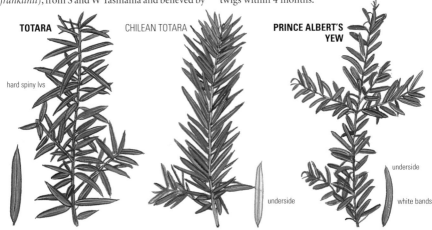

TOTARA — hard spiny lvs

CHILEAN TOTARA — underside

PRINCE ALBERT'S YEW — underside, white bands

PODOCARPS & PRINCE ALBERT'S YEW   27

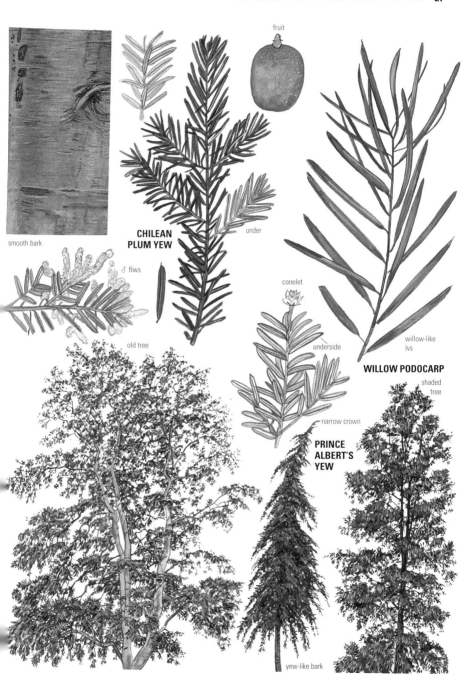

MONKEY PUZZLES

Araucariaceae is a small 'living fossil' family.

Monkey Puzzle — *Araucaria araucana*

(Chile Pine; *A. imbricata*) When the Scottish plant-hunter Archibald Menzies was given unfamiliar nuts for dessert during a banquet in Valparaiso in 1792, he is supposed to have slipped 5 into his pocket, germinating them on the voyage home. The 'monkey-puzzler' became a sensation and there is now believed to be more genetic diversity in European gardens than in the vulnerable remaining Andean stands (all on the slopes of dormant volcanoes). Occasional in gardens of all sizes, living longest in high-rainfall areas; sometimes seeding. Its instant recognizability makes it seem commoner than it is.

APPEARANCE Shape Trees vary in breadth, but have a straight trunk (very rarely forked) designed to carry the foliage quickly out of browsing dinosaurs' reach. To 30 m. The branch-whorls, whose scars are conspicuous up the trunk, are not produced annually: growth can stop for the winter at any stage and the average whorl represents some 18 months' growing. **Leaves** 4×2 cm; leathery, spined. Felled or blown trees coppice, and the regrowth can be sinuous, with smaller leaves (cf. Totara, p.26). **Flowers** Nearly always dioecious. **Female cones** 15 cm across; they ripen in 2 years and – fortunately – break up on the tree. The nuts (rather like Brazil nuts) are best roasted.

OTHER TREES Parana Pine, *A. angustifolia*, an important forestry tree from SE Brazil/Argentina, is scarcely hardy in Britain but grows in one or two collections, with slenderer, softer, sparser leaves.

Norfolk Island Pine — *Araucaria heterophylla*

(*A. excelsa*) Confined to Norfolk Island in the S Pacific (where it reaches 60 m) but widely planted throughout the tropics and in office atria. Common around the Mediterranean but in N Europe has survived outdoors only at Tresco Abbey on the Scilly Isles, where it grew to 30 m.

APPEARANCE Shape A perfect pagoda to great heights. **Bark** Becoming scaly. **Shoots** Arranged in one up-tilted plane. **Juvenile leaves** Rather like Japanese Red Cedar's (p.60) but needles are rounder, stiffer and darker, with a plastic-like texture enhancing the sense that the tree is too good to be real. **Adult leaves** Much shorter; seen only on old trees.

Bunya-bunya Pine — *Araucaria bidwillii*

Coasts of Queensland, Australia. Scarcely seen outdoors in N Europe.

APPEARANCE Shape Loosely conic. **Leaves** Broad (20×8 mm; cf. Totara, p.26), and *well spaced in opposite pairs*. **Cones** (On female trees only) can weigh 5 kg, and contain edible seeds.

OTHER TREES Kauri Pine, *Agathis australis*, a giant tree of the same family from New Zealand (where it reaches 90 m), scarcely survives in Britain but has grown as a thin plant in one or two collections since 1823. Trunk straight and grey; leaves (in pairs 2.5 cm apart) much longer (80×17 mm) and bright green, almost suggesting a broadleaved tree; they emerge bronzy-pink.

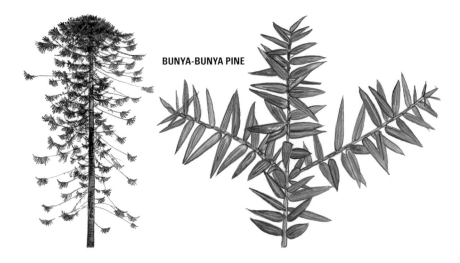

BUNYA-BUNYA PINE

MONKEY PUZZLES

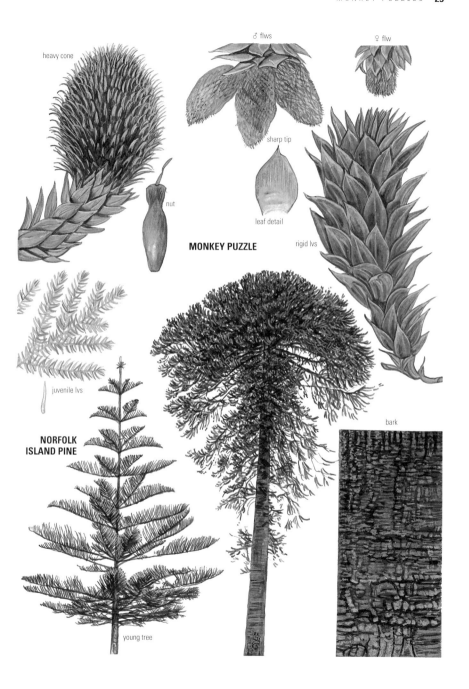

CYPRESSES & INCENSE CEDARS

The family Cupressaceae embraces most of the trees known as 'cypresses'. The minute, scale-like adult leaves make ferny sprays; juvenile leaves, persistent in some false cypress cultivars and in junipers, are short, paired, spreading needles – finer and straighter than those of Japanese Red Cedar (p.60). Cypresses can carry '2D' (fern-like) or '3D' (plume-like) sprays (usually juvenile) – but keep the type's distinctive aroma.

Things to Look for: All Cypresses

- Shape: How weeping?
- Leaves: Are foliage sprays 2D or 3D? Aroma? How glossy? White bands/spots (especially beneath)? Tip (how spreading? how sharp?)
- Cones or berries: How big? What colour?

Key Species

Incense Cedar (below): 2D sprays (long side-scales). **Western Red Cedar** (p.32): 2D sprays (broad, smooth). **Lawson Cypress** (p.36): 2D sprays (finer, almost smooth). **'Squarrosa' Cypress** (p.42): almost 3D sprays of soft juvenile leaves. **Leylandii** (p.46): almost 3D sprays (long-pointed scales). **Monterey Cypress** (p.48): fully 3D sprays (very smooth). **Common Juniper** (p.54): 3D sprays of sharp juvenile leaves. **Chinese Juniper** (p.56): fully 3D sprays (round smooth threads plus some sharp juvenile leaves).

Incense Cedar   *Calocedrus decurrens*

(*Libocedrus decurrens*) Mid-Oregon to S California. 1853. Occasional in larger older gardens, and increasingly fashionable. Resistant to honey-fungus and *Phytophthora* root-rot.
APPEARANCE Shape In Britain a *dense exclamation-mark* (to 40 m). Wild trees, curiously, are open and level-branched. Irish specimens are broader, with erect branching, and Continental ones often narrow, but with more spreading branches. **Bark** Rich red-brown; long, curling or spongy plates. **Leaves** In rising sprays. Side ones with tiny sharp points (fractionally incurved so the sprays are not prickly); longer than most cypresses', giving the broad (3 mm) shoots a *stretched* appearance. Rich matt green; *no white marks beneath*. Shoe-polish aroma. **Cones** 25 mm, with 6 scales, yellow then rufous; like lines of mantelpiece vases.
COMPARE False cypresses (pp.36–44): finer sprays of shorter leaves. Nootka Cypress (p.44): equally long leaves, hanging foliage. Lawson Cypress 'Erecta' (p.36): similar at a distance (more egg-shaped). Oriental Thuja (p.34) and Western Red Cedar 'Fastigiata' (p.32) also have erect sprays.
VARIANTS 'Aureovariegata' (1904), slower growing but shaped as the type, has *blotches of yellow foliage*.

Chilean Incense Cedar   *Austrocedrus chilensis*

(*Libocedrus chilensis*) S central Andes. 1847. In some collections in milder areas.
APPEARANCE Shape Columnar, but *only to 15 m*; a soft sage-green. **Bark** Greyer than Incense Cedar's; *small scales*. **Leaves** Fungous scent; *broad white marks beneath, and often on top* (cf. only Patagonian Cypress, p.52). The side-leaves' incurving tips project, but are *blunt* (cf. Hinoki Cypress, p.40). **Cones** Only 4 scales (rarely seen in N Europe).

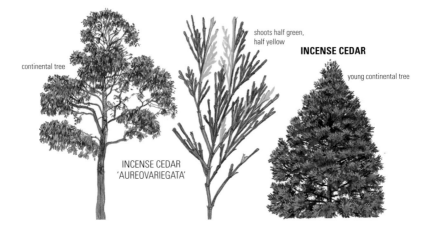

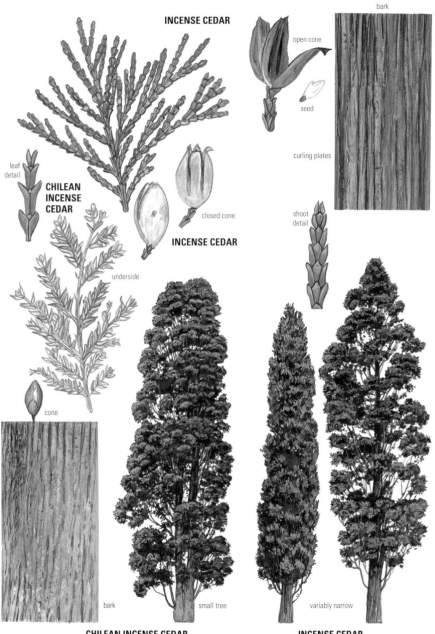

HIBA & WESTERN RED CEDAR

Hiba — *Thujopsis dolabrata*

An important forestry tree in Japan. 1853. Occasional: most frequent in wet areas; sometimes in small gardens for its lavish foliage.
APPEARANCE Shape A broad spire, to 25 m, or a ring of layered stems. **Bark** Red/grey; stringy strips. **Leaves** In 2D sprays *5 mm wide*, lizard-like; convex and *very glossy green* above, a big *dead white* curved streak beneath.
COMPARE Chilean Incense Cedar (p.30); Patagonian Cypress (p.52). Korean Thuja (p.34): much slenderer shoots.
VARIANTS Many were planted as 'Variegata', with a sprinkling of cream shoots; soon few spots remain. 'Aurea' (1866) is very rare but quite a bright gold.

While in the same family as cypresses, trees in the Thuja *genus have 2D sprays of neat, rather flattened, richly and sweetly aromatic foliage-threads; the minute leaf-points are incurved so that the shoot feels smooth. The oval cones stand above the branches. Sometimes called Arborvitae ('tree of life'); the foliage of Eastern White Cedar (rich in vitamin C) was infused to combat scurvy (the leaves in bulk are poisonous/allergenic). (Family: Cupressaceae.)*

Western Red Cedar — *Thuja plicata*

(*T. lobbii*; *T. gigantea*) A giant tree from NW North America, growing on windswept hillsides where its tops are repeatedly blown out but the basal trunk can survive for several millennia. 1853. Abundant in parks, gardens and hedges (though occasional in dry/polluted parts); some forestry plantations; self-seeding in old quarries, neglected mortar, etc.
APPEARANCE Shape When growing well, a tidy spire with a billowing base and erect leader, like a rocket taking off; sometimes forked or layering. *Bigger* than any similar cypress: 45 m so far. Broader, sparse or broken-topped in dry conditions. **Bark** Dark red-brown; soft, stringy ridges (often greyer and harder in plantations). **Leaves** In luxuriant, rather *glossy* drooping sprays, the sheen accentuating their neat design; a fresh deep olive-green (slightly bronzing in winter), with dull yellowish/white-green streaks beneath. *Strong, sweet pineapple/Philadelphus scent* fills the air on warm days.
COMPARE Eastern White Cedar and Japanese Thuja (p.34). Lawson Cypress (p.36): finer, duller shoots, rather sourly parsley-scented. Nootka Cypress (p.44) and Leylandii (p.46): shoots the same width, but dull and dark with spreading leaf-points and a sourer smell. Hiba (above): much bigger, glossier leaves.
VARIANTS 'Zebrina' (1868) is a bright *pale yellow* abundant garden tree, the colour formed by *zebra-stripes* of white-gold, green-gold and green across each spray. More than one clone: sometimes a tall spire (to 28 m); more often ('Irish Gold') slow and dumpy but very bright.
'Aurea' (still rare) has big, dishevelled, uniformly dark gold sprays (cf. Eastern White Cedar 'Lutea', p.34). The very rare 'Semperaurescens' is greener, but neat and glossy. 'Wintergold' (rather rare) has scraggily rising, gold-tipped sprays.
'Fastigiata' (1867), occasional in hedges and parks, has dark sprays held *well above the horizontal*, and a *narrow*, tight spire shape (though the lower branches frequently layer).

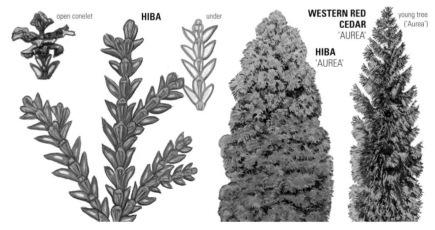

open conelet — **HIBA** — under — **WESTERN RED CEDAR** 'AUREA' — young tree ('Aurea') — **HIBA** 'AUREA'

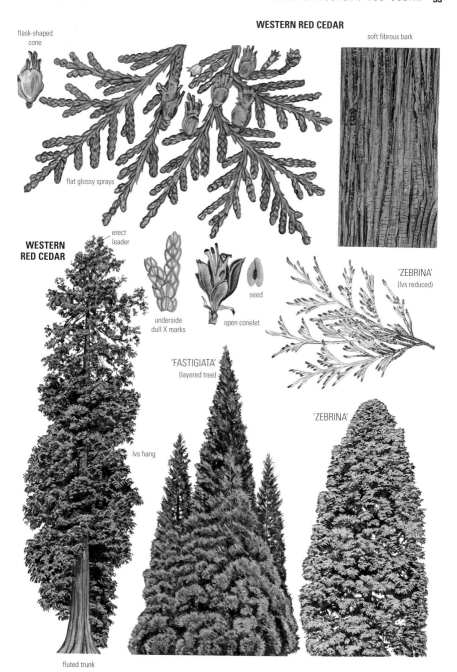

THUJAS & ORIENTAL THUJA

Eastern White Cedar — *Thuja occidentalis*

(Northern White Cedar) Western Red Cedar's east-coast counterpart, a tougher, hardier but much smaller plant. Probably the first American tree in Britain (1536?), but now very occasional.

APPEARANCE Shape Much like an ailing, stunted or bushy Western Red Cedar's; seldom 15 m. **Bark** Often greyer, its ridges more narrowly criss-crossed (but sometimes shaggy). **Leaves** In marginally finer, lace-like sprays, with a sharper smell (apples); often yellowish and wilted-looking. Under-leaves *uniformly yellowish: no white marks*. **Cones** Typically abundant, yellowing the whole crown.
VARIANTS Numerous: tough, and often happier. 'Holmstrup' (by 1951) is very abundant. A dense, *rounded spire* (eventually to 9 m); *brilliant green* foliage in close, often *vertical* rounded plates (cf. Oriental Thuja 'Elegantissima', below). 'Spiralis' (1923; much rarer) makes a narrow column (to 15 m) of dark, flattened, short curled sprays which spiral up turret-like branches. (A darker, taller, much rougher tree than the Hinoki Cypress 'Nana Gracilis', p.40.) 'Lutea' (*c*.1870, rare) makes a good, uniform yellow spire of small sprays (no whitish marks under its leaves; cf. Western Red Cedar 'Aurea', p.32). 'Ellwangeriana Aurea' (1895) is an old-gold blob (to 7 m) with largely vertical, edge-on sprays. Rare – a giant, less orange version of the ubiquitous bun-shaped 'Rheingold'.

Japanese Thuja — *Thuja standishii*

Japan. 1860. Very occasional.
APPEARANCE Shape Easily taken for a dumpy (to 24 m) yet imperturbably healthy Western Red Cedar, but subtly and prettily distinctive. **Bark** Develops harder (even shiny) ridges, some *crimson-pink*. **Leaves** Matt, the sprays *rougher* above; fine, brighter grey-white streaks underneath; greyer at first – each short, drooping, rather distant spray *neatly tiling the crown*.
OTHER TREES Hybrids with Western Red Cedar (intermediate) are growing fast in some collections.

Korean Thuja — *Thuja koraiensis*

Korea, N China. 1918. In a few collections. A wretched slow plant, grown for its foliage.
APPEARANCE Shape Slender, even bowed. **Leaves** Much finer than Hiba's (p.32); threads *almost entirely silver* beneath, as if sprayed.

Oriental Thuja — *Platycladus orientalis*

(Biota; *Thuja orientalis*) China – where known only from gardens. 1752. An oddball conifer, appreciating hot, dry sites and a rich/chalky soil; very hardy, yet thriving in the tropics. Quite frequent in lowland parks, old gardens, and especially country churchyards; rarely naturalizing.
APPEARANCE Shape A *balloon*, to 15 m: slightly sinuous stems from a short, grey-brown trunk. **Leaves** *Scentless*; in rather vertical 2D plates, *the same colour each side*; shoots duller and finer than thujas'.
COMPARE Lawson Cypress 'Erecta' (p.36): a bigger, parsley-scented tree with minutely pointed side-leaves. Eastern White Cedar 'Holmstrup' (above).
VARIANTS 'Elegantissima' has foliage-plates brightly yellow-tipped in summer, duller gold-green in winter. Frequent in rockeries and churchyards, to 10 m. Crisper habit than Eastern White Cedar 'Ellwangeriana Aurea' – like a giant coral.

healthier than type trees
EASTERN WHITE CEDAR 'AUREA'

'SPIRALIS'
dense short curled sprays

KOREAN THUJA
small tree

ORIENTAL THUJA 'ELEGANTISSIMA'
broadens with age

LAWSON CYPRESS

Lawson Cypress
Chamaecyparis lawsoniana

(Port Orford Cedar) A few stands in the Klamath and Siskiyou Mountains (Oregon/California), now threatened by fungal attack. 1854. Very abundant in the UK, in parks, gardens, hedges. Used in forestry to 'nurse' beech plantations; naturalizing on sandy banks etc.
APPEARANCE Shape Densely columnar (to 42 m so far); open and broken in drier sites; bole often much forked. *Drooping leader* and nodding sprays, hanging like rags on some old trees. **Bark** Reddish and purplish, *spongy*; harder-plated with age but *never spirally ridged*. **Leaves** Matt threads 1.8 mm wide; side-leaves *minutely* free-pointed but not prickly, and grey-white beneath along the joints; particularly prominent *translucent gland* in the centre of the top-leaves. Rather sour, *parsley* scent. **Cones** Abundant, pea-sized, ripening in 1 year.
COMPARE Nootka Cypress (p.44): longer, longer-pointed side-leaves. Sawara Cypress (p.42) and Taiwan Cypress (p.44): side-leaves with more spreading points. White Cypress (p.44): small sprays of finer foliage. Hinoki Cypress (p.40): blunt side-leaves. Thujas (pp.32–5): broader, smoother shoots.
VARIANTS A uniform tree in the wild, but after its arrival in Europe it immediately began to throw more 'sports' than any other species. Some are now the mainstay of small gardens. Those with aberrant leaves *still have the parsley scent*. With age, foliage colours tend to fade and habits become less pronounced.

Variegated forms include 'Albospica' (shoots brilliantly white-blotched/tipped), 'Versicolor' (gold-and-white speckling) and 'Albomaculata' (tall-growing; big ivory-yellow splodges). None are common.

Forms with erect foliage include 'Erecta Viridis' ('Erecta'), which grew from the first consignment of seed raised in Britain. Frequent, to 35 m: tightly *egg-shaped* (pointed at first; much broken with age); vivid green. Sprays rise steeply (their tips may nod); leaves slightly bigger than the type's but much shorter than Incense Cedar's (p.30); cf. Oriental Thuja (p.34). 'Kilmacurragh' (by 1951) makes a *narrow column*; still rare. 'Allumii' (1890; abundant in small gardens, but to 30 m) has largely edge-on, steely-grey sprays; 'Fraseri' (c.1893) makes a tighter, dull-green spire (cf. 'Youngii', below, and 'Green Spire'). 'Columnaris' (c.1940) is much narrower than 'Allumii', though with looser sprays at all angles. Fast-growing (to 22 m so far); its habit *abruptly degenerates* at 8 m. 'Grayswood Pillar' (by 1960) is marginally bluer. Both (?) are abundant. 'Green Spire' ('Green Pillar'; by 1947) has edge-on, *yellowish-green* sprays; *narrowly* spire-shaped; rare.

Seedlings can be very pendulous. Distinct weeping clones include 'Filiformis' (1878), spectacular but rare, with *thread-like streamers*, and 'Pendula Vera' (very rare), whose *smaller branches hang* as well as the sprays.

Forms with aberrant foliage include 'Intertexta' (1869), with big, distinctively *distantly branching*, drooping, lacy sprays of *thick shoots*. To 30 m; often forking; occasional in big gardens. 'Filifera' has distantly branching sprays of very *slender shoots*; very rare. 'Youngii' (by 1874) has *level, upcurling* sprays of *thick foliage*, bright green. A spire, to 25 m; rare.

LAWSON CYPRESS 'ERECTA VIRIDIS'

'GREEN SPIRE'

'GRAYSWOOD PILLAR'

'INTERTEXTA'

often forks

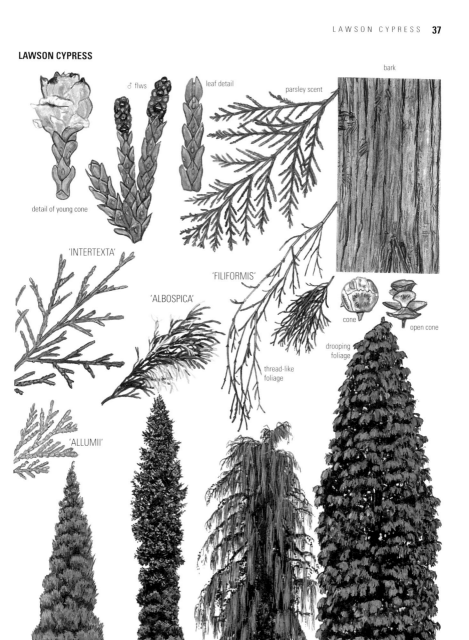

LAWSON CYPRESS CULTIVARS

Variants with aberrant foliage (continued)
'Wisselii' (1888) has *dark, dense blue-grey turrets*, blushing deep crimson with male flowers in spring. To 30 m; frequent. (Its '3D' sprays are like those of the true cypresses, none of which share its habit; cf. Hinoki Cypress 'Lycopodioides', p.40.)

Among forms with semi-juvenile foliage, 'Ellwoodii' (1929) is abundant, growing steadily to 12 m. A tight, blackish, spiky-topped column; leaves with 2 mm spreading points. 'Ellwood's Gold' (black-grey sprays yellow-edged in summer) and 'Ellwood's White' (many creamy-yellow blotches) are also abundant, but dwarfer. 'Fletcheri' (1913; abundant and reaching 17 m) is slightly *paler grey and fluffier*, its leaves with 3 mm points. 'Pottenii' (1900) has *small* sprays, *steeply rising then nodding*, and a *dense*, columnar shape. Abundant. Pampered trees live in chicken-wire strait-jackets. 'New Silver' is similar but silvery-blue; very rare.

Yellow-leaved forms colour most brightly in full sun: their northern sides are a dull, pale green. 'Lane' (1938) is the most abundant in small gardens: *horizontal*, short, flat sprays with *nodding* bright yellow ends, very *densely and neatly* arranged in a *stumpy* column. To 20 m in woodland but commonly semi-dwarf; trees sold as 'Smithii' (since 1898; occasional?) are more vigorous, with longer, less densely or regularly carried sprays. 'Lutea' (*c.*1870) is abundant as a tall, older tree (to 22 m). Sprays *hang like the type's* but are shorter; rich gold in sun. Narrowly, irregularly columnar; low branches may mutate into something like 'Lane'. 'Winston Churchill' (1945) is abundant; more *spire-shaped* than 'Lutea' but irregular, to 15 m so far: rich gold sprays, some hanging but many held rather stiffly at a *jumble of angles*. 'Lemon Queen' is a brilliant and tidier more recent clone. 'Grayswood Gold' has *very slender, erect*, pale yellow sprays and a neat *egg shape*. Spectacular but rather rare. 'Stewartii' (1920) is a tall *rounded-conic* tree (to 25 m), with long, *slender* sprays regularly *ascending then nodding* at the yellow tip; much *less densely* carried than in 'Lane' and not so bright. Frequent. 'Hillieri' (by 1920) is brilliant but scarce; *narrowly* columnar with short, densely arranged sprays, *some horizontal* with nodding tips, *some jumbled*. 'Golden King' (by 1931) is vigorous and rare: rather *distant big loose spiky sprays* of dark gold foliage, hanging near their tips. 'Westermannii' (*c.*1880) is tall and open, with very hanging, *pale dull yellow-green* sprays; rare.

Some forms have yellow young shoots which fade to grey – a striking though rather bilious combination. 'Naberi' (1929; rare) is *tightly conic*: foliage pale turquoise in shade, dark yellow in sun – bold, banded combinations recall Western Red Cedar 'Zebrina' (p.32). 'Silver Queen' (*c.*1883), occasional as an old tree with the habit of the type, is much duller. 'Elegantissima' (by 1920) has very pale, long-pendulous sprays; rare?

Seedlings of the type are often glaucous. Distinct grey/blue clones include 'Pembury Blue' (very frequent in small gardens), stumpily columnar, with *densely carried*, short sprays, intensely turquoise-grey. 'Triomf van Boskoop' (*c.*1890) is *tall* and frequent: *big* drooping blue-grey sprays and a shapely columnar-conic crown on an often *single bole*. (And see 'Allumii' etc, p.36.)

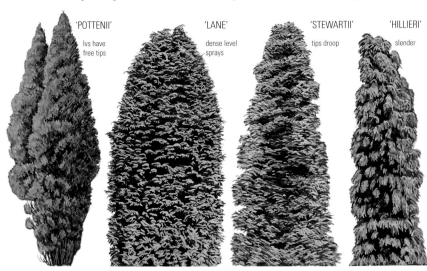

'POTTENII' — lvs have free tips

'LANE' — dense level sprays

'STEWARTII' — tips droop

'HILLIERI' — slender

LAWSON CYPRESS CULTIVARS

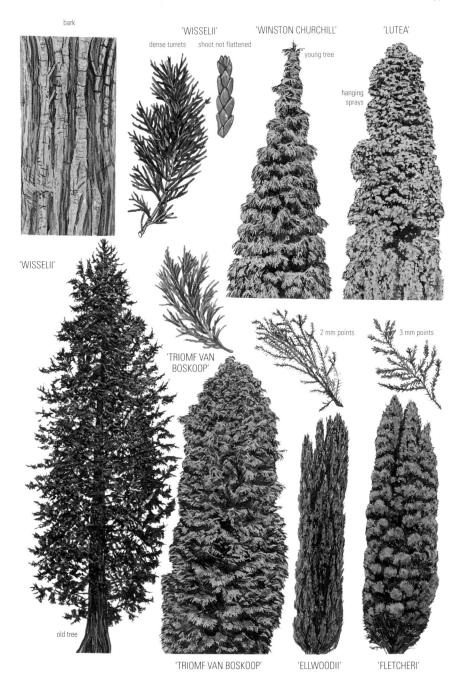

HINOKI CYPRESS

Hinoki Cypress *Chamaecyparis obtusa*

Mountains of central and S Japan – one of the country's major timber conifers (the 'Japanese Gateway' at Kew Gardens is made from it). The Japanese name means 'Tree of Fire' – wild stands ignite readily, perhaps even from the friction of branches rubbing in the wind. Introduced when Japan was opened to westerners in 1861; occasional in larger gardens. Almost always a happy, healthy tree in Britain and Ireland, as are its several popular cultivars.

APPEARANCE Shape Variably dense on a generally straight, single trunk to 25 m (sometimes multi-stemmed). The rich green foliage scarcely weeps and is often gathered into many horizontally projecting, densely sculpted branch-systems – *'Japanese-looking'* – between which the trunk remains visible. **Bark** Reddish, soft and rather stringy, or with harder vertical plates. **Leaves** In very neatly fanned sprays, matt yet rich green; side-leaves effectively *blunt*; each leaf with a crisp white base underneath. *Sweet*, Eucalyptus-like scent when crushed.

COMPARE Thujas (pp.32–4): broader shoots, none with such matt, horizontally fanned sprays. Chilean Incense Cedar (p.30): long-projecting, blunt side-leaves. True cypresses and junipers (pp.48–58): smooth '3D' shoots. The other false cypresses have minutely pointed side-leaves.

VARIANTS All have blunt scales and the same sweet smell. 'Crippsii' (1901) is one of the brightest and commonest of yellow cypresses, to 20 m so far. The pale gold younger growth tipping each independent, horizontal branch-system makes these particularly distinct and distinguishes the tree at a distance.

'Aurea', an older clone with much duller, more pendulous sprays (even in sun) is now rare.

'Tetragona Aurea' has *congested, blunt bright yellow turrets* and is very frequent. It was introduced in 1876 from Japan, where the green variety from which it must have mutated had long been lost to cultivation. Like many cypresses it is sold as a dwarf but grows a steady 10 cm a year, so that the oldest, 140 years old and still mostly thriving, are now 15 m tall. The dense '3D' foliage suggests a true cypress or juniper, but the yellow Monterey Cypresses (p.48) have longer, wispier, pointed turrets and Golden Chinese Juniper (p.56) is rather neatly conic, with odd patches of long, juvenile leaves.

'Nana Gracilis' (1874), also slow and frequent in rock gardens, has *brilliant dark green* foliage in crisp, incurled, ascending *coral-like fans*. Irregularly globular, on a straight stem; to 12 m so far (cf. Eastern White Cedar 'Holmstrup', p.34).

'Chabo-yadori' (by 1970) is a slow-growing, straight-stemmed, gauntly conic, rich-green clone with more densely horizontal branch systems than the type. Rather rare.

'Lycopodioides' (1861, from Japan) has dull, greyish, *tall blunt turrets* of curious, moss-like foliage. A gaunt tree in some large gardens, to 20 m. Most like the Lawson Cypress 'Wisselii' (which is taller-growing and more luxuriant, with many darker, crisper, sharp turrets; p.38).

'Filicoides' (1861, from Japan) has similar but moss-green foliage, the sparse, narrow, congested sprays *horizontal, with weeping tips*. A rare, rather eerie tree.

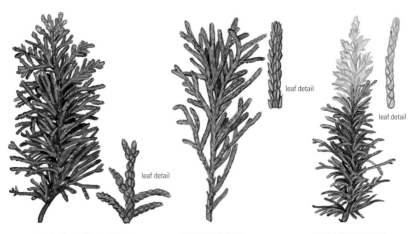

HINOKI CYPRESS 'FILICOIDES'   'LYCOPODIOIDES'   'TETRAGONA AUREA'

SAWARA CYPRESS

Sawara Cypress *Chamaecyparis pisifera*

Another giant cypress from the mountains of Japan. 1861. As frequent as Hinoki Cypress (p.40), but looking very different.
APPEARANCE Shape Rather *patchily sparse and open*, on a usually single but slightly sinuous trunk; particularly wretched in dry areas. **Bark** Softer and stringier than the other common false cypresses', red/grey. **Leaves** In rather meagre, mid-green sprays, slightly drooping; upper, lower and side-leaves *the same length*, with fine *spreading tips*; leaf-bases white underneath. Acrid, resinous aroma.
COMPARE Taiwan Cypress (p.44): very similar, but *no white markings* under its leaves. Lawson Cypress (p.36): more luxuriant sprays, the minute points of the side-leaves scarcely spreading. Nootka Cypress (p.44): longer, straighter side-leaves, not white beneath. White Cypress (p.44): even finer sprays.
VARIANTS are numerous and highly diverse:
'Aurea' (1861, from Japan) has foliage designed like the type's; *yellow* new growth fades to green through the summer. Rare, but easily overlooked.
'Filifera' (1861, from Japan) has *thread-like hanging shoots* alternating with small bunched sprays. Often many-stemmed and layering; to 20 m. Occasional; apt to revert.
'Filifera Aurea' (1889) is *bright yellow*, a tangled haystack of a tree with the thread-like shoots of 'Filifera'; abundant in small gardens (cf. Monterey Cypress 'Coneybeari', p.48).

'Plumosa' (1861, from Japan) has prickly semi-juvenile foliage, the leaves with 3 mm points, in ascending and softly curling, slightly 3D sprays. Generally a leaning column on forking, easily shattered trunks. The foliage, unlike the type's, is *very dense*, with dead leaves retained, so ideal for nesting birds. Frequent, especially in cemeteries, and healthier-looking than the species. To 25 m – but there are equally common semi-dwarf variants, such as 'Plumosa Compressa'.
'Plumosa Aurea' is also frequent: yellow-green, duller with age and reverting piecemeal to 'Plumosa'; the semi-dwarf 'Plumosa Aurea Compacta' is probably commoner.
'Squarrosa' (1843 – from Japan via Java) has effectively 3D sprays of long (6 mm), soft, paired juvenile leaves, bluish (from broad grey bands underneath) and contrasting prettily with its *bright rufous bark*. On a straight trunk/trunks, to 25 m; the massed, fuzzy sprays in horizontal branch-systems somewhat recall Hinoki Cypress. Frequent. (Common Juniper, p.54, and Meyer's Juniper, p.56, have rising sprays of much harsher needles, and are bushy; the Japanese Red Cedar 'Elegans', p.60, has soft needles carried alternately and *twice as long*, and a sprawling habit.)-
'Boulevard' (1934), very abundant as a slender *semi-dwarf* (to 6 m), has slightly longer leaves in laxer, drooping pale sprays, the tips bronzing in winter. Live growth is sadly patchy with age.
'Squarrosa Sulphurea' is a frequent variant: pale *creamy young foliage* fades to the same blue-grey.

SAWARA CYPRESS 'FILIFERA AUREA'

'PLUMOSA'

leaf detail

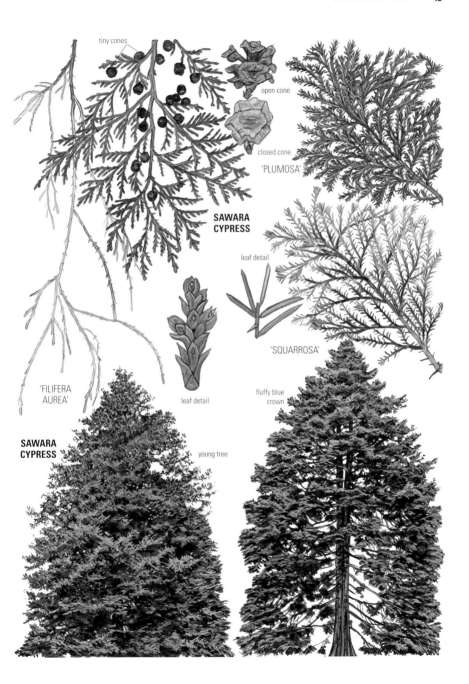

FALSE & NOOTKA CYPRESSES

Taiwan Cypress *Chamaecyparis formosensis*

(Formosan Cypress) A giant, threatened tree, perhaps to 3000 years old; in a few big gardens since 1910, but reaching only 18 m in the UK.
Appearance Shape The branches typically curve up in big 'U's. **Leaves** Very like Sawara Cypress (p.42), but *lacking white marks* beneath. Foliage tends to bronze in winter. Aroma of *aniseed/seaweed*.

White Cypress *Chamaecyparis thyoides*

The false cypress of the E coasts of the USA – a tough, hardy swamp-dweller unable to exploit good conditions and reaching only 15 m in the UK. 1736. Rare.
Appearance Shape Often *very slender*; densely and fluffily columnar. Dark grey ('Glauca') or yellowish. **Bark** Dark brownish; long peeling strips. **Leaves** *Smaller* than those of other false cypresses (1.2 mm threads), in smaller, shorter, fan-like sprays which have a sad, moth-eaten appearance and a gingery scent. The side-leaves have tiny spreading points and white edges underneath.
Compare Sawara Cypress (p.42: redder bark, open crown, bigger foliage). Pencil Cedar (p.56) and Mexican Cypress (p.48): similar in habit and in their fine foliage, but with 3D sprays.
Variants 'Variegata' (1831; yellow-splashed foliage) is confined to collections.

Nootka Cypress *Xanthocyparis nootkatensis*

(*Chamaecyparis/Cupressus nootkatensis*) Perhaps the longest-lived tree in the cloud forests of NW North America, but unlikely to make a giant in Britain. 1854. Rather occasional, and happiest in high-rainfall areas. The 2D, sour-smelling sprays suggest a false cypress, but the cones take 2 years to mature.
Appearance Shape Often neatly conic, to 30 m, though frequently on several trunks; a *solid surface* of close, *steeply hanging, dull, dark sprays*, with the interior clean and open: *look up and you see the sky*. **Bark** Stringy grey-red ridges. **Leaves** Threads broad, like Western Red Cedar's (p.32), but matt; the straight, 3 mm side-leaves' slightly spreading points make the sprays *harsh if rubbed backwards; no white marks* on the yellowish undersides (cf. Incense Cedar, p.30, and Taiwan Cypress, above). Heavy, oily scent (like Stinking Mayweed). **Cones** 9 mm, purple-green then red-brown in their second year.
Compare: Leylandii (p.46): *less pendulous*, imperfectly flattened sprays; columnar crown *dense inside*. Sawara Cypress (p.42) and Taiwan Cypress (above): sharp-tipped side-leaves but much finer, in meagre, less weeping sprays. Lawson Cypresses (p.36): large leaves of vigorous young trees faintly silver-marked beneath, and never so long or sharp.
Variants Afghan Hound Tree, 'Pendula' (1884), carries *curtains* of foliage from few, *gaunt* branches; very occasional.
'Lutea' ('Aurea'; 1891) has *yellow new growths*, greening in winter: rare but can be overlooked.
'Argenteovariegata' (1873) has foliage with some big *cream blotches*. Deservedly very rare.
Other trees Kashmir Cypress, *Cupressus cashmeriana*, is the most desired true cypress with comparable 2D sprays of long-pointed leaves (cones 18 mm, ripening in 2 years) – tender and very rare outdoors. The *clear turquoise foliage is spectacularly weeping*, though odd trees surviving in colder regions can be a sorry sight.

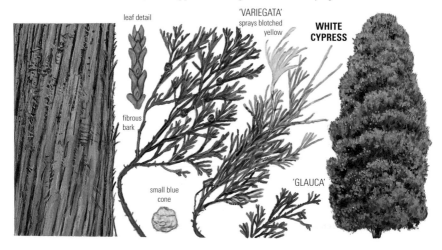

leaf detail

'VARIEGATA' sprays blotched yellow

WHITE CYPRESS

fibrous bark

small blue cone

'GLAUCA'

FALSE AND NOOTKA CYPRESSES 45

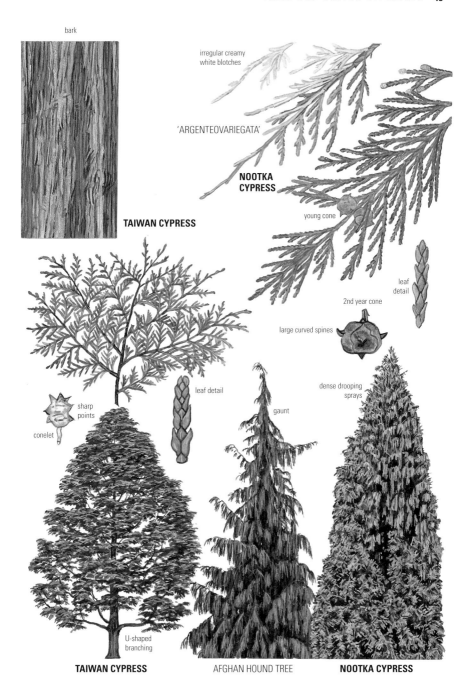

HYBRID CYPRESSES

Leylandii × *Cupressocyparis leylandii*

(*Cupressus* × *leylandii*) Nootka Cypress has hybridized several times in Britain or Ireland to make trees of huge hybrid vigour. The Leylandii – at once the *most planted* and most hated garden tree – is a cross with Monterey Cypress which seems to have first occurred (unnoticed) at Rostrevor (Co. Down) around 1870. The clone 'Rostrevor' remains rare; the easily propagated, ubiquitous 'Haggerston Grey' arose at Leighton Park, Powys, in 1888, when a Monterey pollinated a Nootka Cypress. Very occasional in forestry. Birds, at least, love it.

APPEARANCE Shape Very *densely columnar*, generally clothed to the base and tapering to an open, slanted tip; dark sooty-green surface, with little texture. Grows rapidly to 30 m in any soil, then blows down. **Bark** Dull red-grey; vertical/criss-crossing shallow stringy ridges. **Leaves** In rather '3D' plumes – often flattened in the ultimate 2 divisions, which are *at right-angles* to the third division; the 3 mm scales have sharp straight tips and no white markings underneath. **Flowers** Often *absent*, though 'Leighton Green' (below) can have massed yellow male flowers in summer and brown 2 cm cones.

COMPARE Mexican Cypress (p.48): finer, fluffy foliage. West Himalayan Cypress (p.52).

VARIANTS 'Leighton Green' occurred – also, strangely, at Leighton Park – in 1911 when a Nootka pollinated a Monterey Cypress. Sprays *flatter* (cf. Nootka Cypress, p.44; broader; dense columnar crown distinguishes). A brighter-green tree but harder to propagate, so now rather rare.

'Naylor's Blue', a sister-tree to 'Leighton Green', was propagated only when a whirlwind demolished it in 1954; rare. *Dark softly blue-grey* foliage, in rather 3D plumes; more raggedly open-crowned.

'Stapehill' ('Stapehill 20'; Ferndown, Dorset, 1940) sheds its brown inner foliage, so the *crown is open and wispy* and the rather 3D sprays appear motheaten: often looks moribund, but disappoints. Occasional, because it was claimed to do well on chalk.

'Castlewellan Gold' arose at Castlewellan (Co. Down) in 1962 when a Nootka pollinated a Golden Monterey Cypress. As the first yellowish Leylandii its ubiquity was assured. *Squat-conic*; 3D plumes gold-tipped in spring/summer then *dull olive*.

'Robinson's Gold' occurred – again in Co. Down – in the same year (at Bangor Castle, where there is no obvious father). As dull, but *slender-columnar*, with *2D sprays*; harder to propagate, and rare.

'Golconda', a 1977 sport of 'Haggerston Grey' (with its habit), is one of the *brightest* yellow cypresses, but scarce to date. The Dutch 'Gold Rider' is tidier.

'Harlequin' (1975; occasional to date) has many 1–10 cm pale yellow patches; 'Silver Dust' (1960) has rather fewer 1–5 cm white shoots.

OTHER TREES Alice Holt Cypress, × *C. notabilis* (*Cupressus* × *notabilis*) is a hybrid of Nootka and Smooth Arizona Cypresses (again from Leighton Park; 1956), confined to bigger gardens. Openly conic, with a *purple-grey*, hard, blistering bark and beautiful, rather *turquoise-grey*, lax, weeping 3D sprays structured like Leylandii's (cf. Kashmir Cypress, p.44).

Ovens Cypress, × *C. ovensii* (*Cupressus* × *ovensii*) is a Nootka Cypress/Mexican Cypress cross (Westonbirt Arboretum, 1961) in a few collections: openly columnar crown of 2D foliage like that of Nootka Cypress, but lemon-scented.

'SILVER DUST'

creamy-white shoots

'CASTLEWELLAN GOLD'

dumpy shape

brightest in spring

HYBRID CYPRESSES

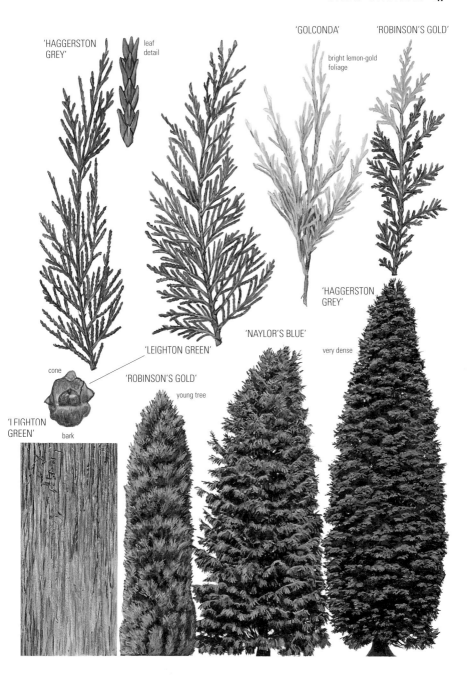

TRUE CYPRESSES

'True Cypresses' (20 species) have larger cones, which generally take 2 years to mature, than those in the Chamaecyparis *genus. The foliage of the commoner species is held in 3D sprays, and the leaves are often blunt. (Family: Cupressaceae.)*

Monterey Cypress

Cupressus macrocarpa

Confined in California to two windswept cliff-tops, but since 1838 much the commonest and biggest true cypress near S/SW coasts of Britain and Ireland: it is very salt-tolerant. Less frequent inland. Sometimes seeding. Seldom in plantations and now rare in hedges.
APPEARANCE **Shape** Typically barrel-like, on many steep limbs from a *short bole*; near the sea often very broad, or shaped like Cedar of Lebanon; often dying back (*Corinium* canker). *To 40 m: vigorous, wispy, sharp plumes* from the crown help distinguish it from other cypresses/junipers. **Bark** Grey-brown; stringy criss-cross ridges. **Leaves** In dark green 3D sprays, with appressed scales whose tiny points can hardly be snagged by running a nail down the shoot; no white markings. Rich, lemony aroma. **Cones** Shiny, 3 cm, the scales not sharply ridged; *seldom conspicuous* except on dying trees.
COMPARE Italian Cypress and Rough Arizona Cypress (p.50); West Himalayan Cypress, Gowen Cypress (p.52); Pencil Cedar (p.56). All have smooth, 3D shoots but are slower/narrower trees.
VARIANTS 'Lutea' (1892) has laxly spiky *lime-green* plumes and is frequent as an old tree, especially on coasts where it copes especially well.
'Goldcrest' (1946) is the most abundant of the newer, *acid yellow* clones. *Narrowly columnar* when young, often *slanted*; broader with age with sharp dense turrets towards tip, often *twisted as if wind-blown*. Young pot-plants carry many soft 5 mm juvenile needles.
'Donard Gold' (1935; rare?) is slightly duller and broader, with more spreading turrets.
'Coneybeari' makes a giant broad bush with dull yellow *thread-like hanging sprays*. Rare (cf. Sawara Cypress 'Filifera Aurea', p.42).
'Fastigiata' is *very narrowly* and neatly but sinuously columnar with an open, pointed tip; rare (cf. Italian Cypress, p.50).

Mexican Cypress

Cupressus lindleyi

(Often equated with the Cedar of Goa, *C. lusitanica*, a clone – of Indian origin? – cultivated in Portugal by 1634.) Central and S Mexico; Guatemala. Occasional in milder areas.
APPEARANCE **Shape** Rather *densely columnar* until old, but a *fluffy*, ragged outline, on a generally *pole-like* stem with stringy coppery-grey bark; to 30 m. **Leaves** *Sharp, spreading (but very fine) tips* (unlike other true cypresses' with 3D sprays). Variably grey (though without white markings); almost scentless. **Cones** Only 15 mm wide; a prickle on each scale.
COMPARE Leylandii (p.46): harder, closely set, dull dark sprays, dull bark; seldom a good stem. West Himalayan Cypress (p.52).
VARIANTS 'Glauca Pendula' has blue-grey, *tumbling plumes*; columnar or sprawling. Spectacular; perhaps now the most commonly planted form.
OTHER TREES Bentham's Cypress, *C. benthamii* (NE Mexico, c.1838; often treated as a variety), has bright, shining green foliage in *flatter, fern-like* sprays from *sinuous* branchlets; the leaf-tips *scarcely spread*. Confined to collections in mild areas. (Fluffier-crowned on a stronger stem than false cypresses, or West Himalayan Cypress.)

MONTEREY CYPRESS 'GOLDCREST'   'DONARD GOLD'   'LUTEA'

TRUE CYPRESSES **49**

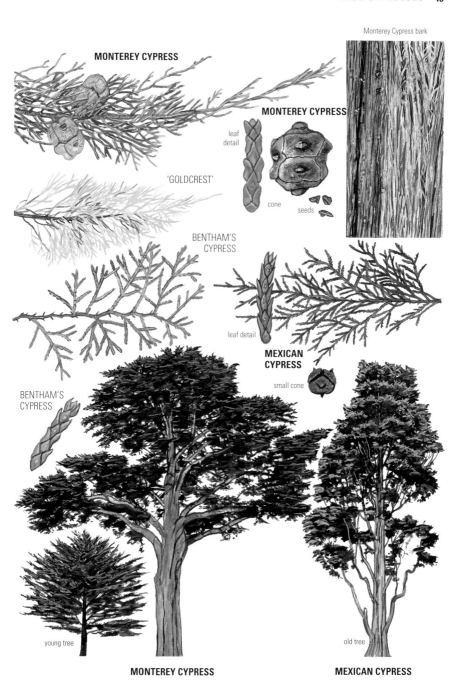

TRUE CYPRESSES

Italian Cypress *Cupressus sempervirens*

E Mediterranean to Iran, but long grown further W and N: the definitive Mediterranean landscape tree. Rather occasional (but hardy to NE Scotland): small gardens, parks and churchyards in warmer parts.
APPEARANCE Shape The variably densely columnar f. *stricta*, with steep branching and no persistent trunk, is the familiar form, to 24 m. Var. *horizontalis* is often unrecognized: spreading branches and a lumpy, irregular crown as broad as high. Can be blackish or greyish green. **Bark** Grey-brown stringy ridges, like many true cypresses. **Leaves** Dense, 3D sprays of dark leaves with closely appressed tips and no white markings. Sweet, resinous scent may be very faint. **Cones** 3 cm, dull *grey*, each scale with a *blunt knob*; abundantly *studding the dark crown* for a year.
COMPARE Monterey Cypress (p.48). The rare Monterey Cypress 'Fastigiata' and the spreading form of Italian Cypress make identification tricky. The more numerous, grey cones of Italian Cypress, with knobbly scales, are helpful, as are the more vigorous, plumy growths of Monterey Cypress. Monterey Cypress shoots are faintly *club-ended*; in Italian Cypress they *taper*. In the absence of juvenile foliage, Pencil Cedar (p.56) can be very like Italian Cypress' spreading variety, but its foliage smells soapy and it has *no cones*. Chinese Juniper (p.56) has thicker foliage threads. Rough Arizona Cypress (below) has smaller cones.
VARIANTS 'Swane's Golden', slow-growing, is very rare as yet. 'Green Pencil' is an absurdly tight recent selection of f. *stricta*. Only the as yet very rare Tibetan Cypress, *C. gigantea*, is as narrow, along with Skyrocket Juniper (p.58: no cones).

Rough Arizona Cypress *Cupressa arizonica*

Arizona, SW Texas and N Mexico. 1882. Rare.
APPEARANCE Shape Narrowly columnar. **Bark** Stringy, grey-brown. **Shoots** 3D sprays of smooth shoots. **Leaves** Greyish (rarely with central white spot). **Cones** 2 cm; each scale has a central prickle.
COMPARE Small cones, light horizontal branching, and fluffy edges to grey crown distinguish from Italian Cypress (above). (Gowen Cypress, p.52, has *dark* green leaves.)

Smooth Arizona Cypress *Cupressus arizonica* var. *glabra*

(*C. glabra*) Central Arizona. 1907. Now very frequent in parks and small gardens in several selections.
APPEARANCE Shape Conic, to 22 m, with a narrow tip or egg-shaped in age; edges somewhat spiky with dense, rugged, round-ended plumes. **Bark** Beautiful: purples and reds, flaking in smooth circular or snakeskin scales. Old trees often conspicuously grafted on the type. **Leaves** In bright pale grey 3D sprays, the appressed leaves *often with a central white spot of dried resin*. **Male flowers** Abundant and yellow all winter. **Cones** 2 cm, retained for many years.
COMPARE Guadalupe Cypress and allies (p.52).
VARIANTS 'Pyramidalis' (1928) is probably the common clone today: tidily shaped, with white spots on *nearly every leaf*; 'Hodgins' can be narrower.
'Aurea' has *yellow* young foliage slowly turning pale grey – rather rare but striking.

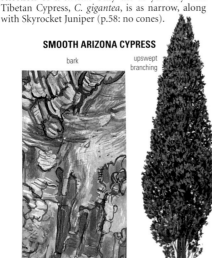

SMOOTH ARIZONA CYPRESS
bark
upswept branching

'PYRAMIDALIS'
bluer crown

ROUGH ARIZONA CYPRESS

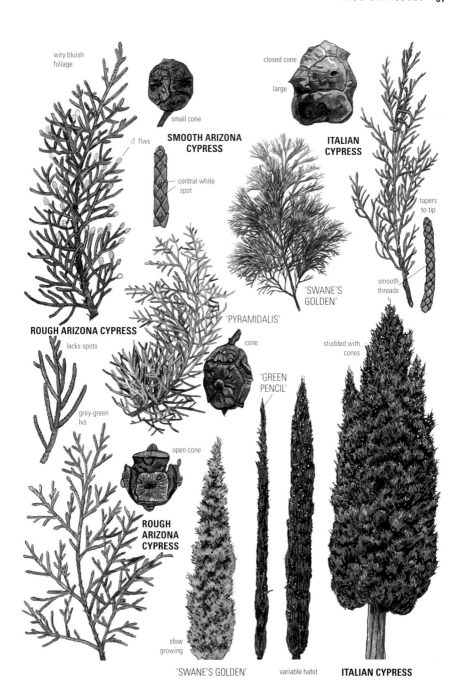

GUADALUPE TO PATAGONIAN CYPRESS

Guadalupe Cypress
Cupressus guadalupensis

Guadalupe Island, Mexico. 1880. One of several New World true cypresses with tiny and vulnerable wild populations, confined to a few botanic gardens in mild areas of Britain and Ireland but attracting much interest.
APPEARANCE Bark Glorious; redder and smoother than Smooth Arizona Cypress's (p.50). **Leaves** *Sea-green*; threads fine, wiry, smooth, in 3D sprays with long, unbranched lengths. Little aroma.
COMPARE Smooth Arizona Cypress (p.50): the only commonly planted tree in this group.
OTHER TREES Equally rare are Cuyamaca Cypress (*C. stephensonii*), from San Diego County, California, with a purplish bark (and white-spotted leaves), and Tecate Cypress (*C. forbesii*), which extends into Orange County and has paler green foliage. Gowen Cypress, *C. goveniana* var. *goveniana*, confined to two groves near Monterey, California (1848), is the most likely of these cypresses to be seen in the UK as a relatively hardy old tree, to 20 m: usually densely columnar, with long, narrow *blackish* spiky sprays whose shoots branch *almost at right-angles* (gorse-like effect); a few of its smoothly appressed *lemon-scented* leaves have white resin-spots. The dull dark grey bark shreds finely. *Small* (15 mm) cones, each scale with a forwards prickle, cluster at the bases of shoots.

West Himalayan Cypress
Cupressus torulosa

(Bhutan Cypress). W Himalayas (very similar species extend E to China). 1824: old gardens, rare.
APPEARANCE Shape An often densely columnar cypress with almost flattened, slender sprays of *hanging*, rather *distantly branched*, yellowish-green fluffy foliage. **Bark** Greyish- or chocolate-brown, contrasting with foliage; *very stringy*. **Leaves** With minute, incurved points (much finer than those of Mexican Cypress, p.48), and no white marks. Grassy aroma. **Cones** Tiny (14 mm), each scale with a rounded knob.
COMPARE Bentham's Cypress (p.48): less weeping sprays; faint, resinous scent. Taiwan Cypress (p.44).
VARIANTS Var. *corneyana* (*C. corneyana*), in a few collections, has duller, more 3D foliage, the shoots typically with twisted tips.

Patagonian Cypress
Fitzroya cupressoides

(Alerce; *F. patagonica*) A giant, endangered timber tree from the rain-drenched S central Andes, living for at least 3000 years. 1849. Rare: milder areas. Nearly all in Britain and Ireland are cuttings from one plant, and female. It is consequently hard to raise genetically diverse seedlings from wild-sourced trees without the risk of cross-pollination by this one clone.
APPEARANCE Shape Heavy, irregular branches; *weeping* foliage, in 3D sprays. To 20 m. **Bark** Reddish; lifting strips. **Leaves** Dark blue-green, in 3s (cf. some junipers), *bluntly hooked*; white bands either side. **Flowers** Dioecious. **Cones** 6 mm, massed (9 scales, in 3 whorls, to further the triplicate theme).
COMPARE Chilean Incense Cedar (p.30): flattened shoots. Summit Cedar (p.58).

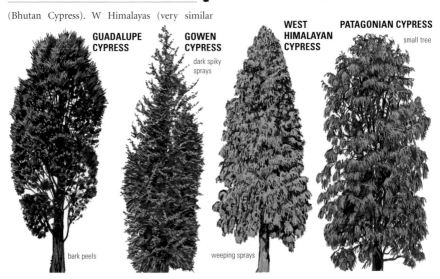

GUADALUPE CYPRESS — bark peels
GOWEN CYPRESS — dark spiky sprays
WEST HIMALAYAN CYPRESS — weeping sprays
PATAGONIAN CYPRESS — small tree

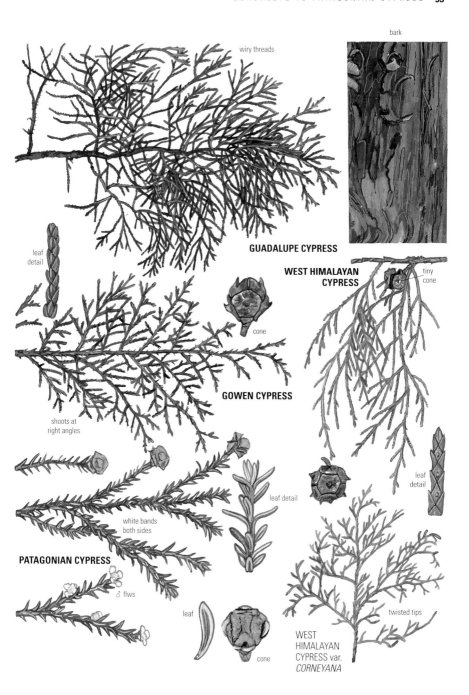

JUNIPERS

Many junipers (50 species) are bushy. They are the only 'cypresses' with berries (formed by the fusion of fleshy cone-scales). Most retain juvenile leaves – straight, sharp, often in 3s, in 3D sprays; their narrow bases do not hide the quickly brown shoots. (Family: Cupressaceae.)

Common Juniper   *Juniperus communis*

The only conifer native both to Eurasia (including Britain and Ireland) and to North America, N to the Arctic Circle; also in N Africa. In Britain, a scarce plant of chalk scarps, open moorland and Scottish pine woods, dwindling through changes in land management (it needs a sudden cessation in grazing to colonize grassland, before taller competition shades it out or makes it vulnerable to fungal infections). Rarely planted.
APPEARANCE **Shape** A softly greyish bush, to 8 m. Dense and varied, like a topiary menagerie. **Bark** Grey-brown; very shaggy and stringy. **Leaves** All juvenile: to 1 cm, with *1 white band* on the cupped inner side; in 3s up shoots red-brown after 1 year. Scent of apples/lemons. **Flowers** Dioecious. **Fruit** Berries, 7 mm, ripen black in 3 years, with 1–3 seeds: the flavouring for gin.
COMPARE Syrian Juniper (below): much longer leaves. Temple Juniper (below) and Meyer's Juniper (p.56): weeping habit. Taiwania (p.60): needles whose broad bases *clasp* its shoots. Sawara Cypress 'Squarrosa' (p.42): softer leaves (in *pairs*). VARIANTS Irish Juniper, 'Stricta' ('Hibernica'; 1838), is an occasional, tight, grey selection (to 8 m), with upright branching *and shoots*; Swedish Juniper (f. *suecica*) differs in *nodding* shoot-tips. 'Oblonga Pendula' (a few collections) has hanging shoots and leaves twice as long and *scarcely white* on the inner side (cf. Temple Juniper).

OTHER TREES Prickly Juniper, *J. oxycedrus* (mountains and dunes of S Europe to Iran; in some N European collections), has 10–25 mm leaves, often with *2 grey bands* on the inner side. Its berries, *6–15 mm and usually brownish*, ripen in 2 years.

Syrian Juniper   *Juniperus drupacea*

(*Arceuthos drupaceae*) S Greek mountains to N Syria. 1854. Rare.
APPEARANCE **Shape** In the UK a sad bush, or a splendid, dense glistening *column, to 20 m*. **Leaves** All juvenile: long (to 2 cm), *rigid; bright* green with 2 white bands on the cupped inner side; their bases, unlike other junipers', *run down the green shoot*.
FLOWERS Only male trees known in the UK (berries in the wild brown/blue-black, to 25 mm).

Temple Juniper   *Juniperus rigida*

Japan; Korea; N China (often in temple gardens). 1861. Rare.
APPEARANCE **Shape** Singularly non-rigid. Often gaunt and asymmetrical, to 15 m; the dull, *yellowish*, rather sparse foliage hangs, dramatically at best. **Bark** Dull brown; shaggy strips. **Leaves** All juvenile: sharp but *soft*, in *distant* whorls of 3; to 2 cm, with 2 narrow whitish bands flanking *a groove* on the inner side; grassy aroma. **Fruit** Abundant on female trees, 8 mm, ripening purple.
COMPARE Drooping Juniper (p.56): denser leaves.

TEMPLE JUNIPER — drooping sprays

COMMON JUNIPER

conic tops

SWEDISH JUNIPER — nodding tips

IRISH JUNIPER — very narrow

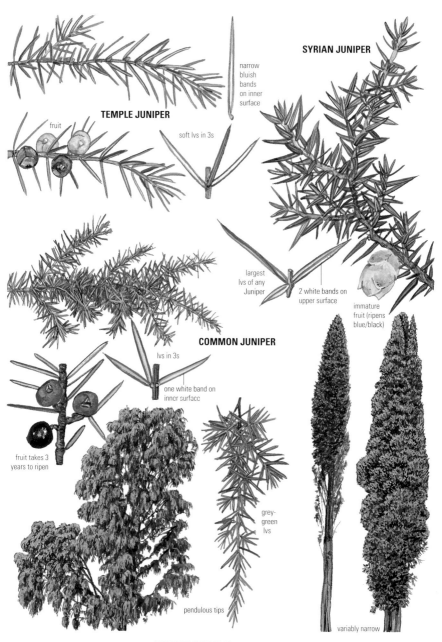

JUNIPERS

Drooping Juniper *Juniperus recurva*

(Himalayan Juniper; Coffin Juniper) Afghanistan to W China. 1830. Very occasional; larger gardens. **Appearance Shape** Densely nodding and *weeping* sprays; greyish, with much rufous dead foliage; to 15 m. **Bark** Shaggy; grey-brown. **Leaves** All juvenile, rustling in *crowded, forwards* whorls of 3; 5–8 mm (but to 10 mm in the widely grown brighter green Cox's Juniper, var. *coxii*); 1 silver band on the inner side.
Compare Temple Juniper (p.54): long lax leaves.

Meyer's Juniper
Juniperus squamata 'Meyeri'

An old Chinese selection of an undistinguished Asiatic bush. 1914. Abundant; to 8 m.
Appearance Shape Dark blue, arching, gaunt sprays *like pampas grass plumes*. **Bark** Very shaggy; brown. **Leaves** Dense and juvenile: 8 mm, spiny; 2 brilliant blue-white bands on the inner side.

Chinese Juniper *Juniperus chinensis*

China, Japan. 1804. Occasional as an older tree. Schizophrenic: *tufts of juvenile foliage* can persist often at the *base* of adult, scale-leaved, 3D sprays. **Appearance Shape** A tight column, an egg-shaped, or rugged; to 20 m, on fluted/fused stems. Dingy out of flower. **Bark** Grey-brown; stringy, twisting strips. **Juvenile leaves** To 1 cm, sharp, rigid, grey-blue inside; densely in 2s/3s. **Adult leaves** Cypress-like, their *pale blunt margins* making a faint pattern; in *quite thick* (1.8 mm), smooth threads; catty smell. **Flowers** Dioecious: male trees yellowish through winter with flower. **Fruit** Females carry 6–8 mm berries (whose separate scales remain visible), ripening from whitish to dark brown in *2 years*.
Compare Pencil Cedar (below): differences emphasized. Without juvenile foliage/berries, a lumpier tree than Monterey Cypress (p.48) and Italian Cypress (p.50), *never with woody cones* or open plumes of vigorous growth.
Variants Golden Chinese Juniper, 'Aurea' (1855), is occasional, to 14 m: bright and usually neatly conic. Male (so no berries) but much less spiky than Monterey Cypress 'Goldcrest' (p.48).
'Kaizuka' lacks juvenile foliage; a female plant, berrying freely. Nightmarish and broad: long, *gaunt, wandering, twisted turrets*. Rare.
'Keteleerii' makes a *dense greyish spire*, again with only adult foliage but plenty of berries. Very rare.

Pencil Cedar *Juniperus virginiana*

(Eastern Red Cedar) Ontario to Florida. 1664. Very occasional.
Appearance Shape A dark tree, to 22 m, columnar or spreading; easily confused with Chinese Juniper (differences emphasized). **Juvenile leaves** *Always in pairs, often at the shoot-tips*, but sometimes absent. **Adult leaves** With minute appressed *points*, in *fine* (1.2 mm) threads. *Soapy* scent. **Flowers** Dioecious. **Fruit** Berries blue-grey, only 4–6 mm; sweet and ripening in *1 year*.
Variants 'Canaertii' (1868; rare) makes a dense, *bright* green column. No juvenile foliage, but a female plant, with showy purple berries (half the size of those of the Chinese Juniper 'Keteleerii').

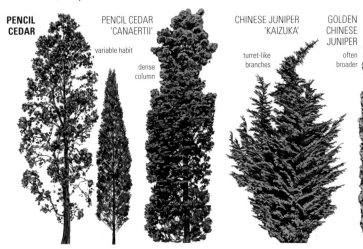

PENCIL CEDAR

PENCIL CEDAR 'CANAERTII'
variable habit
dense column

CHINESE JUNIPER 'KAIZUKA'
turret-like branches

GOLDEN CHINESE JUNIPER
often broader

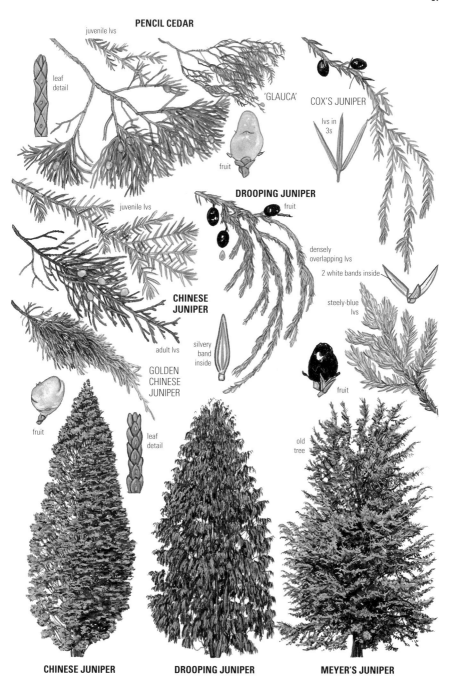

JUNIPERS & TASMANIAN CEDARS

VARIANTS OF PENCIL CEDAR (CONTINUED)
'Glauca', rather rare and female, has soft grey, mostly adult foliage and a rough, upswept habit (cf. White Cypress, p.44). A similar but much commoner and semi-dwarf plant, neatly but openly conic and with pale grey adult foliage, is 'Springbank', a male cultivar of Rocky Mountain Juniper.
OTHER TREES Rocky Mountain Juniper, *J. scopulorum*, takes the place of Pencil Cedar in the W USA: a very similar tree whose berries ripen in their *second year* (cf. Chinese Juniper, p.56). In a few N European collections. Its cultivar 'Skyrocket' is now quite frequent: *very narrowly columnar* (to 9 × 0.5 m) with distinctively *coarse, stringy* upright glaucous sprays (cf. Italian Cypress, p.50).
Spanish Juniper, *J. thurifera* (Spain, the French Alps and N Africa, 1752; in some N European collections as a narrow column, to 14 m), has very fine, *strongly musty-smelling* foliage often *intermediate between adult and juvenile states* (2 mm free tips to scale-leaves; cf. Mexican Cypress, p.48); berries grey-bloomed then purple, 8 mm. Stinking Juniper, *J. foetidissima* (Albania to Syria and the Crimea, 1910), scarcely grown in the UK, is very similar.
Grecian Juniper, *J. excelsa* (S Balkans to Turkmenistan, 1806; in some N European collections often as a dense spire, to 16 m), has *even slenderer* foliage threads (less than 1 mm wide), which may also remain intermediate between juvenile and adult states. Most plants carry both male and female flowers; 1 cm berries ripen purple-brown.
Phoenician Juniper, *J. phoenicea* (Mediterranean Europe, N Africa, the Canaries, 1683; in the odd N European collection), has brighter green adult foliage, the threads sometimes as broad as Chinese Juniper's; mature plants retain little or no juvenile foliage and carry both male and female flowers; the 8–14 mm *red-brown* berries have up to 9 seeds.

Taxodiaceae is a family of 17 primitive conifers, often gigantic and with spongy red bark. They survive in scattered montane populations in both hemispheres.

Summit Cedar *Athrotaxis laxifolia*

The least rare (in big gardens) of 3 species from the mountains of W Tasmania. It is intermediate, but shows no obvious signs of hybridization.
APPEARANCE Shape Sturdily open-conic, to 20 m. **Bark** Red-brown: hard, shaggy ridges. **Cones** 2 cm; abundant. **Leaves** Similar to those of Japanese Red Cedar (p.60), but much shorter (4 mm free), and more broadly tapered, flatter and harder, with *hooked* tips; green on *both sides* (some small, whitish zones). Young growths are *yellow*.
COMPARE Patagonian Cypress (p.52): blunter leaves strongly white-marked on each side. Giant Sequoia (p.62).

King William Pine *Athrotaxis selaginoides*

(King Billy Pine) W Tasmania, where most of the large old trees have been felled for their timber. c.1857. Very rare; tenderer than Summit Cedar.
APPEARANCE Shape A stumpy, broad-conic tree, easier to mistake for Japanese Red Cedar (p.60). **Leaves** *Longer* than Summit Cedar's (8 mm free) - but flatter, harder and more curved than Japanese Red Cedar's, tapering from a thicker base, and *brighter white* on the inner sides; shining green outside.
COMPARE Junipers (pp.54–8): straight slender needles in 3s and never a thick, straight bole. Norfolk Island Pine (p.28); Taiwania (p.60).

Smooth Tasmanian Cedar
Athrotaxis cupressoides

W Tasmania; the rarest in N Europe and tenderest of the 3 species.
APPEARANCE Shape Usually conic but very open. **Bark** Finely shredding. **Leaves** Foliage with appressed scales, like a true cypress (the tips just free), but the Samphire-like distant *whorls of 3 thick shoots* are highly distinctive.

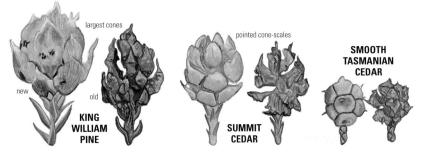

largest cones — new — old — **KING WILLIAM PINE** — pointed cone-scales — **SUMMIT CEDAR** — **SMOOTH TASMANIAN CEDAR**

JUNIPERS & TASMANIAN CEDARS

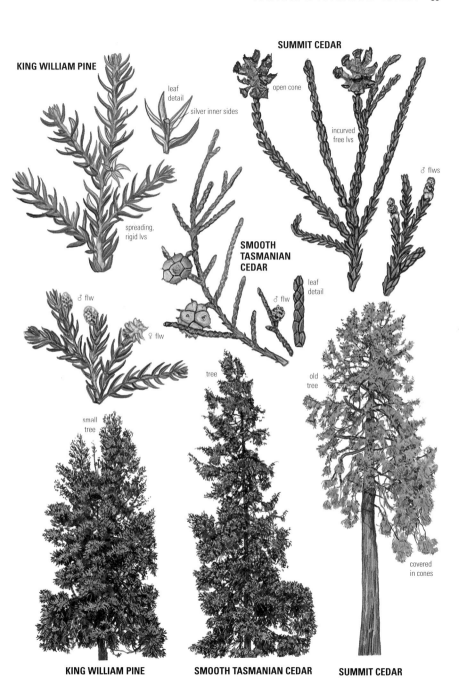

JAPANESE RED CEDAR & TAIWANIA

Japanese Red Cedar *Cryptomeria japonica*

A giant tree in the mountains of Japan. Rather occasional; rare in dry/cold areas; in a few forestry plantations.
APPEARANCE Shape Conic on a straight stem to 40 m (occasionally layered or low-forking); rather *patchy*, rich, hard green foliage. **Bark** Orange-brown; the soft, stringy ridges much flatter and wider than Giant Sequoia's (p.62). Trunk cylindrical but sometimes carrying burrs like great beer-bellies. **Leaves** *Slightly curving*, 10–15 mm, one by one all around the drooping shoots, which their broad bases cover; 4-sided, the inner 2 with grey-green bands. **Cones** 2 cm – abundant on some trees, absent on others.
COMPARE King William Pine (p.58): particularly similar. Taiwania (below); Norfolk Island Pine (p.28).
VARIANTS Chinese Red Cedar (var. *sinensis*; S China) is sometimes considered a separate species (Fortune Cedar, *C. fortunei*) and reached Britain first (1842). Leaves slightly longer, in *soft, dishevelled, slightly weeping*, paler green sprays; scarcer as a younger tree than the Japanese form. Pre-1878 trees are either var. *sinensis* or 'Lobbii', which conveniently represent extremes of habit.
'Lobbii' is a clonal name sometimes used for trees from Thomas Lobb's first consignment of Japanese seed (via Java) in 1853: particularly short leaves in dense *tufts*, bunched on the stiff, narrow crown and with 1 tuft forming the apex.
'Elegans' (1861) is an abundant bush, broad and sprawling but sometimes growing as a single-boled tree, to 20 m. Cloudy dark blue-grey foliage; *purplish* – even shocking maroon – through winter, and making a fine contrast with the peculiarly bright, almost glistening orange bark. Leaves juvenile, held individually and distantly: 2 cm, *soft* and slender, rounded and curling randomly. Habit and long needles suggest a juniper: Temple Juniper (p.54) is closest but is a green, weeping tree. Sawara Cypress 'Squarrosa' (p.42) is an erect tree with much shorter needles in *opposite pairs*.
'Compacta', very occasional, has a *very dense, compact*, round-tipped, fresh green crown; to 15 m.
'Aurescens' has softly yellowish young growth; shapely but very rare; 'Sekkan-sugi' is creamy-yellow in spring.
'Cristata' (1901; very rare) is rather narrow and gaunt. The foliage grows many, curiously attractive, *antler-like masses* of leaves ('fasciations').
Grannies Ringlets, 'Spiralis' (1860), has leaves *twisted* around its shoots. To 20 m, but generally a dwarf. Rare.
'Viminalis' ('Araucarioides', 'Dacrydioides', 'Lycopodioides') is a very rare freak with many long, *unbranching*, spidery shoots. To 20 m; suggesting a giant Brussels Sprout.

Taiwania *Taiwania cryptomerioides*

(Incorporating *T. flousiana*) Taiwan, S China and N Burma; an endangered giant reaching 60 m on Mt Morrison (Taiwan). 1920. Still very rare (milder areas).
APPEARANCE Shape A neat pagoda, the pale grey-green foliage hanging elegantly from upcurving branches; to 20 m so far. **Bark** Pale red-grey; narrow, stringy ridges. **Juvenile leaves** Similar to the typical (adult) foliage of Chinese Red Cedar (above), but fiercely spined, often longer and straight, with a *broad blue-white band* on each side. (Adult scale-leaves – and flowers – have not yet been seen in Europe.)
COMPARE King William Pine (p.58); Norfolk Island Pine (p.28).

'AURESCENS'

gold in spring

'VIMINALIS'

JAPANESE RED CEDAR 'ELEGANS'
fluffy crown often leans

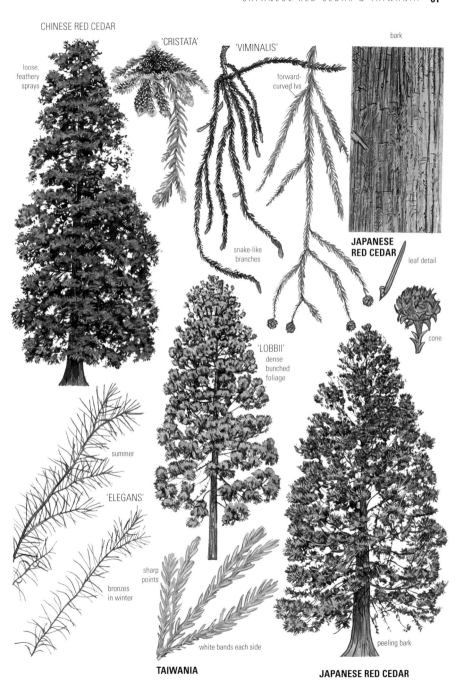

REDWOODS

Coast Redwood *Sequoia sempervirens*

(*Taxodium sempervirens*) The world's tallest tree now that the best Douglas Firs and eucalypts have all been felled for their superior timber. Confined to a fog-bound coastal strip from just into Oregon to S of Monterey, California. 1843. Rather occasional, and rare in dry/exposed areas; in one or two forestry plantations. Unlike most conifers it coppices happily and the rare fallen tree can grow a line of new ones; it very occasionally self-seeds in Britain.

APPEARANCE **Shape** A ragged-sided, dark, misty-leaved column, often broad: sparsely spire-tipped while adding height, then flat-topped; to 45 m only in deep shelter. Turns brown in dry winters, without permanent injury. **Bark** Rufous, deeply spongy; hard purplish surfaces develop in age if few squirrels climb the trunk. **Leaves** Young stems *radially clothed with big scale-leaves*, which have free sharp tips. Side-shoots with yew-like needles in flattened ranks: hard, straight; to 2 cm but *progressively shorter* at shoot tips and bases; 2 white bands beneath. **Cones** 2 cm: unimpressive.

COMPARE Common Yew (p.22). Allies of yew have needles bigger or with no white bands underneath – and no scale leaves up the shoots.

VARIANTS 'Adpressa' ('Albo-spica'; 1867) has small short leaves (cf. the Common Yew 'Adpressa', p.22), white-banded above and *entirely cream-grey on new shoots*. Rather rare: a thin, often gaunt tree, to 22 m.

'Cantab' has much *broader, very blue-grey leaves*. First distributed (1951) as a prostrate dwarf, but readily sending up vertical shoots which grow as vigorously as the type's; to 20 m so far. Rare.

Giant Sequoia *Sequoiadendron giganteum*

(Wellingtonia; Big Tree; Giant/Sierra/Californian Redwood; *Sequoia gigantea*) Another Californian 'living fossil' (from a few groves high in the Sierra Nevada); 'General Sherman' is believed to be the world's largest tree, though a Coast Redwood of similar proportions was 'discovered' in 2000. With their similar names the two trees are much confused, but differ widely. 1853. Already the largest tree in all parts (to 52 m); quite frequent (away from the driest or most polluted lowlands) and very conspicuous, though highly sensitive to salt spray. Never in plantations: useless soft timber, like cardboard.

APPEARANCE **Shape** A dense, dark blue-green spire while adding height, with light branches, *swooping and upcurving*; the oldest trees are now subject to middle-aged spread and have rugged, broken, columnar crowns, except in deep shelter. May shatter in high winds but very seldom blows down. **Bark** Dark red, thickly spongy (proofing the tree against forest fires); harder and darker with age. Bole nearly always *very flared*. **Leaves** Rather inconsequential – sharp radial scales on cord-like shoots, 4 mm wide, with an aniseed scent and no white markings; their tips may curve *outwards*, unlike those of Summit Cedar (p.58) or Patagonian Cypress (p.52). **Cones** Disappointing: 4 cm.

COMPARE Summit Cedar (p.58).

VARIANTS Weeping Giant Sequoia, 'Pendulum' (1863), has an often leaning stem (to 30 m) or long gaunt limbs cloaked in *hanging branches and shoots* – like a green Orang-utan. Rare.

'Aureum' (1856) is *dull yellow* and very rare.

'Pygmaeum' (1891) is dense and *dwarfish*; rare.

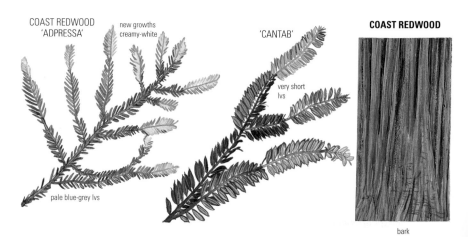

COAST REDWOOD 'ADPRESSA' — new growths creamy-white — pale blue-grey lvs

'CANTAB' — very short lvs

COAST REDWOOD — bark

REDWOODS

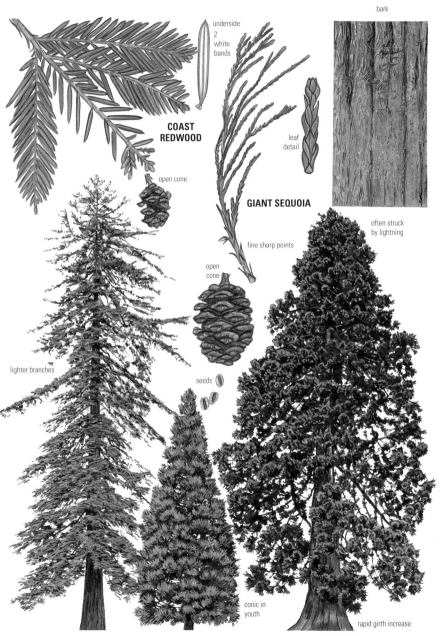

DECIDUOUS CYPRESSES

Swamp Cypress *Taxodium distichum*

(Bald Cypress) Texas to New Jersey. Rather occasional in warm areas. Can cope with waterlogged conditions, throwing up 'knees' to trap silt around its roots (and probably to help the roots to breathe), and it is often assumed to need water. In fact it grows faster in a well-drained soil and thrives on dry sands. Long-lived and healthy, hardly ever blowing down, though trunks and branches easily shatter: one or two English trees are probably 1640 'originals'.
APPEARANCE Shape The European population is now very diverse: sometimes a dense, solid, rounded spire on a *sinuous* bole; sometimes broad with heavy limbs, and often leaning; sometimes open and irregular with lax, fluffy horizontal sprays; sometimes with distant, weeping branches. To 35 m. A *deciduous* conifer, bare and twiggy until late spring, then fresh green, and dark red-brown late in autumn – the tip still green at Christmas. **Bark** Pale orange-grey, shallowly stringy. **Leaves** Main shoots and tip-growths have spirally set leaves with short free apices. On the short side-shoots – which are borne *alternately along the twigs* and drop whole in autumn – the leaves are to 2 cm long, in 2 flat ranks. They are grey-banded beneath, and much softer and fresher green than Coast Redwood's (p.62). **Flowers** Often dioecious. Male catkins, to 20 cm, prominent through winter after hot summers then shedding pollen in mid-spring. **Cones** 3 cm, ripening in 1 year.
COMPARE Dawn Redwood (below). In winter, softer, less craggy bark than larches' (pp.94–8) and less spiky twiggery.
VARIANTS Pond Cypress, var. *imbricatum* (*T. ascendens*; Virginia to Louisiana) is a smaller, less hardy tree, to 22 m. Rather rare and, in summer, quite distinct: the side-shoots *rise vertically* from the branches in a pale green mane, *all closely clothed with 8 mm spirally set leaves*. The shoot-tips *nod* by late summer in the narrowly columnar selection f. *nutans*. Cannot in winter be told from poor, slender examples of the type.

Dawn Redwood
  Metasequoia glyptostroboides

(Water Fir) SW China; a critically endangered giant tree in the wild, discovered only in 1941. Other conifer genera have been introduced at least as recently as this – in fact they continue to be discovered – but the Dawn Redwood is unique in its instant European celebrity, its ease of propagation from cuttings, and the success with which it has grown in all warmer regions: now locally frequent except in the smallest gardens.
APPEARANCE Shape A consistently dense spire so far on a *straight trunk* (forked trees are very rare), rounding off in exposure; to 30 m. In dry open sites the lower trunk often becomes extraordinarily flared and *convoluted*. Earlier in leaf than Swamp Cypress; similar (but earlier) autumn colours. **Bark** *Darker* red than Swamp Cypress's and spongier. **Leaves** Comparable in design to Swamp Cypress's, but the side-shoots (and winter buds – carried beneath the shoots) – and the individual leaves) come in *opposite pairs*. Leaves longer (3 cm) and broader, and a heavier richer green; grey-green underneath. **Flowers** Seldom seen in N Europe.
VARIANTS 'Gold Rush' (2000, from Japan) has *soft, yellow foliage*.

DAWN REDWOOD

conic

long stalk (to cone)

cone – rarely seen

opposite shoots

DECIDUOUS CYPRESSES 65

CHINESE FIR to SILVER FIRS

Chinese Fir   *Cunninghamia lanceolata*

(*C. sinensis*) S and W China to N Vietnam. Rather rare: bigger gardens in warmer areas. One 1804 'original' (grown in a greenhouse pot for 15 years) still thrives in school grounds in Surrey.

APPEARANCE Shape Gauntly conic or multi-stemmed, to 28 m; flat- or dead-topped with age; strong sprouts can grow up from the base. Old, red-brown leaves are retained, so the crown is dense. The cylindrical trunk tapers stage by stage between each branch-whorl. **Bark** Orange-brown, a pretty contrast to the shining, pale green foliage; close stringy ridges. **Leaves** 50 × 4 mm, evenly tapering, clasping the shoot and all around it, but more or less twisted into 2 ranks; broad whitish bands beneath and sometimes 2 narrow white lines above near the leaf-base.

VARIANTS 'Glauca' has metallic blue-bloomed foliage; a particularly striking plant.

OTHER TREES *C. konishii* (Taiwan, 1918) is in a few collections and has *smaller* leaves (30 × 3 mm), standing stiffly all around the shoots, with grey bands on both sides.

Japanese Umbrella Pine   *Sciadopitys verticillata*

Another curiosity with unique foliage; rare and endangered in Japan and usually placed in its own family (Sciadopitaceae). 1853. Very occasional; doing best in high-rainfall areas. Its jizz is strangely like Willow Podocarp's (p.26). (Stone Pine, p.130, is also sometimes called Umbrella Pine.)

APPEARANCE Shape A dense spire, unless the trunk forks low. **Bark** Purple-brown; large, harsh, vertical plates. **Leaves** (Actually pairs of leaves fused together) to 12 cm long, in usually annual whorls. Disparate scale-leaves along the grey-brown twigs appear only as dark knobbles.

Pinaceae is a family of conifers with needles, on twigs that soon become woody and brownish. The needles of silver firs are attached to the smooth shoots by little green sucker-like bases; most will grow in N Europe, but tend to be intolerant of dry air or pollution.

Things to Look for: Silver Firs

- Shape: How narrow-topped is the mature tree?
- Bark: How rugged/flaky? Is it grey, black, orange? Are there rings round the branch-scars?
- Trunk: Is it fluted?
- Shoots: How hairy and what colour? Orange, white, greenish or purple?
- Buds: How long and how resinous? What colour?
- Leaves: Distribution around shoots? How perpendicular to them? Are they curving, notched, rounded, spiny? What white bands/marks (especially above; how bright?)?
- Cones: Do bracts project above each scale?

Key Species

(Leaves tend to be more flat-ranked in shade, and always rise above flowering shoots at the tree-top.)
Grand Fir (p.74): leaves in *flattish ranks* either side of the (non-flowering) shoots. **Noble Fir** (p.88): leaves *sweeping out and up* above the shoot, like a Mohican haircut. **Caucasian Fir** (p.70): straight leaves, *covering the shoot above but with very few below*. **Nikko Fir** (p.86): straight leaves, in several ranks above the shoot but *parted by a 'V'*. **Forrest's Fir** (p.82): forward-angled leaves *all round the shoot (but with more above it)*. **Grecian Fir** (p.76): almost perpendicular leaves *all round the shoot*.

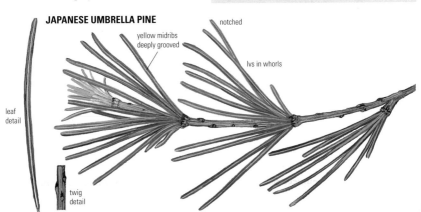

JAPANESE UMBRELLA PINE
- yellow midribs deeply grooved
- notched
- lvs in whorls
- leaf detail
- twig detail

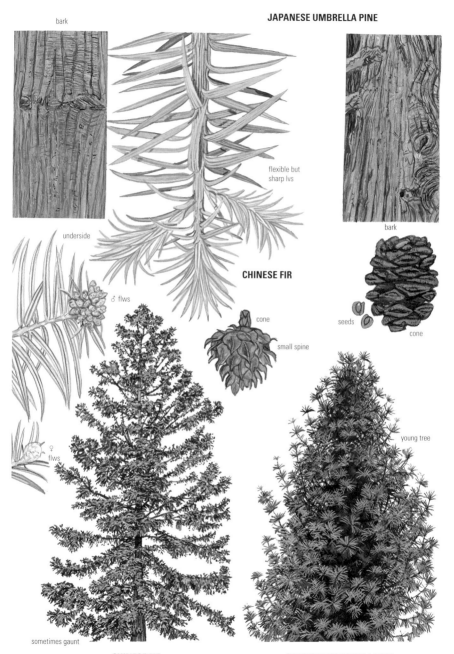

SILVER FIRS (EUROPEAN)

European Silver Fir *Abies alba*

(Common Silver Fir) High ground in France and Corsica; the Pyrenees; the Alps, S and E through the Tatra to the Balkan mountains. Brought to Britain in 1603; becoming rare here except in upland regions where long-lived and still frequent. No longer used in forestry, as in the mild, oceanic climate it has become very vulnerable to defoliation by the aphid *Adelges nordmannianae*. Sometimes naturalizing in humid areas.

APPEARANCE Shape In youth, a regularly whorled spire, the stout leader seldom broken. Old trees sometimes straight, more often heavy-limbed and with *broad*, slanting or fanning tops; dull, *dark* crown, often scraggy and never really luxuriant. A towering tree, which has reached 55 m in Scotland and even more in Germany. **Bark** Shades of grey. Soon cracked into small squarish plates, but sometimes *scalier* than other firs, and suggesting a species of spruce. The bole is usually untidy with dead snags. **Shoots** Usually *dull* grey-buff, with variably dense tiny dark-brown hairs. **Buds** Red-brown, *scarcely resinous*. **Leaves** 2–3 cm; round-tipped or (like the leaves of most silver firs on non-flowering shoots) neatly notched. They are angled forwards at about 30°, in approximately flat ranks on shaded lower branches (cf. Grand Fir, p.74 – but *much shorter*); on flowering shoots they curve and spread above the shoot (like all the leaves of Caucasian Fir, p.70). **Cones** High on old trees: brown, with the bract-scales projecting for 6–7 mm and downcurved, like panting dogs' tongues. Like all silver fir cones, they disintegrate on the tree, so are of limited use in identification. **COMPARE** Grand Fir (p.74): longer, very flat-ranked needles. Caucasian Fir (p.70): leaves always spreading above the shoot and brighter white beneath, conspicuously silvering the branch-ends as you look up; a more luxuriant, glossy tree with less 'scaly' bark). King Boris' Fir (below).

VARIANTS 'Columnaris', derived from French Alpine populations, has *steeply rising branches*; scarcely grown in the UK.

OTHER TREES Sicilian Fir, *A. nebrodensis*, is one of the world's rarest trees: in the 1970s the wild population was down to 21 examples on Mt Scalone, where great forests had stood. Conservation measures are now allowing regeneration to take place, and it is also growing in several UK collections, to 17 m so far. Buds red-brown, with *much white resin*. The leaves (spine-tipped on vigorous growths) tend to curve upwards and spread across the top of all the shoots.

King Boris' Fir *Abies borisii-regis*

S Bulgaria and NE Greece – an intergrade between European Silver and Grecian Firs. 1883. In some collections.

APPEARANCE Shape A rough, vigorous tree. **Bark** Dark grey; close, rugged square plates. **Shoots** With *denser, paler brown hairs* than European Silver Fir. **Leaves** More perpendicular, in several, denser ranks; *narrower*, shinier and seldom if ever notched. There may be a white patch above them near the tip (cf. Cilician Fir and Bornmüller's Fir, p.70). **Cones** Bract-scales spreading level.

COMPARE Apollo Fir (p.76): no hairs on shoots. Pacific Silver Fir (p.74): usually notched leaves and a slender shape.

KING BORIS' FIR

bark
narrow
under
dense pale brown hairs

EUROPEAN SILVER FIR

young tree

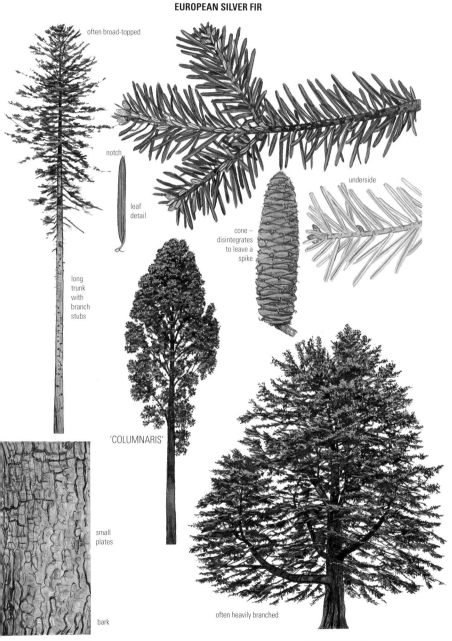

SILVER FIRS (WEST ASIATIC)

Caucasian Fir *Abies nordmanniana*

(Nordmann Fir) NE Turkey and W Caucasus Mountains, where it is Europe's tallest fir (to 70 m). 1848. Occasional and sometimes in small town gardens; more frequent to the N and W. Upmarket Christmas tree, so now in a few plots (the leaves can persist for 25 years so do not fall off in the 12 days of Christmas).
Appearance Shape Nearly always with a *straight trunk to the tip*. Narrow, dense; columnar with age; to 50 m so far. **Bark** Mid/pale grey; shallowly and tidily square-plated with age. **Shoots** Greenish grey-brown, often with tiny dark hairs. **Buds** Pale brown, not resinous. **Leaves** Glossy mid-green; held forwards at 30°; all across the top of most shoots, though with a gap at the vertical on weak shaded ones; few or none below. Clear white bands underneath; usually notched at the tip; light fruity/petrol smell. **Cones** Bract-scales exposed and bent back for 2 cm.
Compare Veitch's Fir (p.72): the least rare of many smaller silver firs with straight leaves held rather forwards and all across the top of the shoot; others (in collections only) are Bornmüller's Fir and Cilician Fir (below), Pacific Silver Fir (p.74), Manchurian Fir (p.84), Pindrow Fir (p.80), Farges' Fir (p.82), Flaky Fir (p.86), and Subalpine Fir (p.78). Douglas Fir (p.120): also with this foliage arrangement, but with slender, long-pointed buds and much softer, slenderer leaves. European Silver Fir (p.68): most leaves often more or less flat-ranked; a rougher, darker, scalier tree.
Variants Trojan Fir, var. *equi-trojani* (*A. equi-trojani*), an isolated population from the Ida Mountains of W Turkey, verging on Bornmüller's Fir and in a few UK collections, has slightly resinous buds; the leaves, just parted above the *clear yellow-brown* shoot (cf. Cilician Fir), are often shiny yellow-green and sometimes have a white patch near the tip.

Bornmüller's Fir *Abies bornmuelleriana*

N Turkey – an intergrade of Caucasian and Grecian Firs. Very rare.
Appearance Shape An often broad, rough tree, to 35 m. **Bark** Differs from Caucasian Fir in being purple-black, less closely cracked. **Shoots** *Shiny red-brown*, never hairy. **Leaves** Longer, more upstanding, often distinctly parted above the shoot, with 2 fine silver lines on the top side towards the base or in a single fuzzy *patch near the tip*.
Compare Cilician Fir (below): paler bark and paler brown twigs (sometimes hairy). Algerian Fir (p.80), Flaky Fir and Korean Fir (p.86): also with white patch above the leaves near the tip, but with shorter, more perpendicularly radiating leaves. Sakhalin Fir (p.72): very slender leaves. Noble Fir (p.88) and Cork Fir (p.78): white band(s) above upcurving leaves; glaucous foliage.

Cilician Fir *Abies cilicica*

SE Turkey to Syria (where it grows with Cedar of Lebanon). 1855. In some collections.
Appearance Differences from Caucasian Fir as follows: **Bark** Smoother, grey, with *particularly conspicuous* black rings around the branch scars (cf. Veitch's Fir, p.72; Maries' Fir, p.74; Noble Fir, p.88). **Shoots** Almost hairless and typically pale *golden-brown* (cf. Trojan Fir, above). **Leaves** More distant; parsley-scented; lightly silver-sprinkled above near their *seldom-notched* tips. **Cones** Bract-scales hidden.

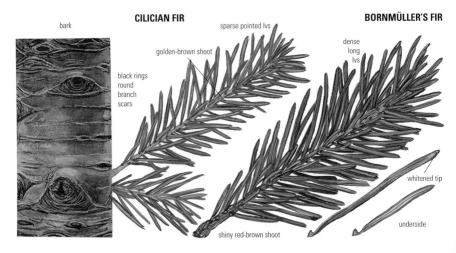

SILVER FIRS (ASIATIC)

Veitch's Fir — *Abies veitchii*

Mountains of central Honshu, Japan. 1879. Rather rare: bigger gardens. Short-lived in the UK, even in humid areas.
Appearance Shape Narrowly conic at its best, to 28 m; the branch-tips sweep up to show their silver-frosted under-leaves. Often, when struggling with dry air, rough and sparse. **Bark** Grey (often pale); finely flaking or with close-packed warts but only square-cracked with age at the base; larger trunks often *fluted and hollowed* under the sometimes dark-ringed branches. **Shoots** Pale grey/buff; very finely hairy. **Buds** *Red-purple, shiny with resin*. **Leaves** A luxuriant and rather glossy Lincoln-green (cf. Forrest's Fir, p.82), with bright white bands beneath and a strong, resinous aroma. They spread all across the top of the shoot (well forwards – at about 45°), with few or none below. End-leaves become abruptly shorter, often making the sprays *broadly square-ended*; each leaf (on non-flowering shoots) also *broadens*, if anything, to its conspicuously notched square tip. **Cones** Small (6 cm) and abundant high up; blue-purple (rarely green) as they ripen; bract-tips *project for 2–3 mm*.
Compare Siberian, East Siberian and Sakhalin Firs (below). Caucasian Fir (p.70): soon square-cracked bole; dull brown buds and cones. Pacific Silver Fir (p.74): more alike.

East Siberian Fir — *Abies nephrolepis*

N China, Pacific Russia and Korea. 1908. In a few collections: a small tree most similar to Veitch's Fir (above).
Appearance Shape Densely and stumpily conic, to 15 m; short level branches. **Bark** Dull pink-grey, closely corky-warted, on a scarcely fluted bole. **Shoots** Fawn, glossy, with tiny hairs. **Leaves** Dull grey-green, narrower and less brightly banded beneath than Veitch's Fir's. Paint-like aroma. Each spray *tapers gradually from halfway up*. **Cones** Purple or (f. *chlorocarpa*) green as they mature; the whisker-tip of each bract projects vertically.

Sakhalin Fir — *Abies sachalinensis*

N Japan, and along the Kuriles to Sakhalin. 1879. Another tree in a few collections which is most similar to Veitch's Fir.
Appearance Shape Typically a narrow spire, to 20 m. **Bark** Grey/blackish brown; smooth but lined with red-brown blisters; often *sprouty* (cf. Min Fir, p.84). **Shoots** Shinier than Veitch's Fir's and shallowly *ribbed*, with fine dark hairs in the grooves. **Buds** Breaking early in the spring in W Europe. **Leaves** Very *dense, soft and slender* (30 × 1 mm; cf. Douglas Fir, p.120), *grass-green for a year*, sometimes with a tiny silver dusting above near their tips; less brightly banded beneath than Veitch's Fir's. Aroma of cedar-wood oil. **Cones** Horizontally projecting bract-tips.
Compare Siberian Fir (below); Manchurian Fir (p.84): less narrow pointed leaves more steeply rising above a smooth shoot. Subalpine Fir (p.78).

Siberian Fir — *Abies sibirica*

Russia, in Taiga forests from the Urals eastwards: one of the world's most abundant trees, growing N to the Arctic Circle, but adapted to prolonged winters and sudden, reliable springs and scarcely surviving in the UK.
Appearance Most like Sakhalin Fir (above), but differences as follows: **Shoots** *Honey-coloured*, often with dense white hairs. **Buds** Yellow-brown; may be stuck together in clusters with white resin. **Leaves** *Yellowish* green with strong, resinous scent. **Cones** Bracts *hidden*.

EAST SIBERIAN FIR — cone, underside
SAKHALIN FIR — underside, buds with white resin
SIBERIAN FIR — underside, ♂ flws
VEITCH'S FIR — bark

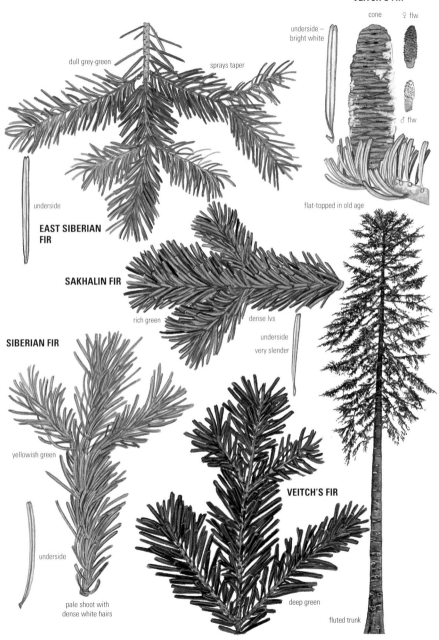

SILVER FIRS

Pacific Silver Fir — *Abies amabilis*

(Beautiful Fir; Red Fir – a name shared with the very different *A. magnifica*, p.88.) S Alaska to N California, near the Pacific. 1830. Rare; a particularly temperamental fir, absent from dry areas.
Appearance Shape A fine, dense deep-green spire when happy, to 40 m. **Bark** Grey: slightly corky, with white horizontal resin-blisters or rounded warts, but few fissures. **Shoots** Slender; grey/orange; *dense short hairs.* **Buds** Like *small grey pearls* (3 mm; cf. Grand Fir, below). **Leaves** To 3 cm; glossy, usually notched, with bright silver bands beneath; spreading all across the top of most shoots (with very few below them) but pressed rather flat and spreading at a wider angle than those of most similar firs, so that the sprays are *wide but shallow*. Aroma of *tangerine rinds* (cf. Grand Fir). **Cones** Purplish as they ripen, with hidden bracts.
Compare Maries' Fir (below). Caucasian Fir (p.70) and Veitch's Fir (p.72): thicker, less hairy shoots and leaves more forward-angled and rising higher above most of their shoots. King Boris' Fir (p.68): hairy shoots, round-ended leaves and a broad crown. Siberian Fir (p.72): pale brown shoots, slender leaves. Smith's Fir (p.82): leaves carried all round many shoots.

Maries' Fir — *Abies mariesii*

Central Japan. 1879. In some collections; to 20 m.
Appearance Differences from Pacific Silver Fir (above) as follows: **Bark** Silver-grey, *freckled* with blackish lenticels and with rings round branch-scars (cf. Cilician Fir, p.70). **Shoots** Brighter orange-pink, *very furry*. **Buds** Similarly pearl-like, but expanding a striking *crimson* colour in late spring. **Leaves** Shorter (2 cm), more forward-angled and less flattened above the shoot, or curving up (making a much *narrower spray*); *gingery* smell.
Compare Smith's Fir (p.82): leaves all round most shoots. King Boris' Fir (p.68).

Grand Fir — *Abies grandis*

(Giant Fir) Vancouver to California. 1832. Rare in dry areas but frequent in the N and W; occasional in plantations; sometimes seeding. The most vigorous silver fir in the UK and the tallest tree in many areas, to 62 m.
Appearance Shape A narrow spire in youth; often rising high above other trees and much broken or lop-sided with age, but with vigorous new tips re-growing close together, even in dry districts, or a single slender wandering top. **Bark** Silvery- or purple-grey, cracking with age into quite close rectangles. **Shoots** Brownish *olive-green* (dull greyish rufous by the second year). **Buds** Like *tiny* (2 mm) grey pearls. **Leaves** Long (25–50 mm) and more strictly *flat-ranked* than any other fir's (but curving above the flowering shoots). Narrow white bands beneath. Delicious tangerine-rind scent. **Cones** Small (8 cm) and high up, on old trees; bracts hidden.
Compare Low's Fir (p.76): grades into Grand Fir (rougher, fissured blackish bark, second year shoots brighter rufous). European Silver Fir (p.68): also has approximately flat-ranked but much shorter leaves; a darker, scalier tree with grey-brown shoots, larger brown buds, and projecting bract-scales. Greenish shoots and ranked leaves recall nutmegs (p.24; needles never notched). Himalayan Fir (p.80): multiple, horizontal ranks of long leaves.

MARIES' FIR

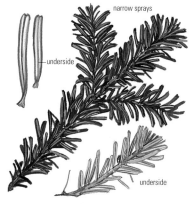

narrow sprays
underside
underside
shoot orange-pink, densely downy

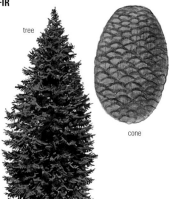

tree
cone

cone

GRAND FIR

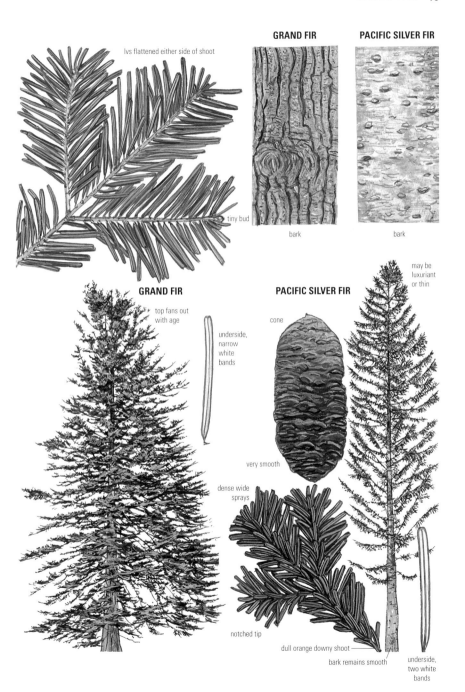

SILVER FIRS

Colorado White Fir — *Abies concolor*

Utah to N Mexico. 1873. Rare: big gardens.
Appearance Shape Open crown, normally conic until old. **Bark** Dark grey, smooth. **Leaves** Thick, grey, all curving vertically above the shoots, slightly recalling Noble Fir (p.88) but much *longer* (5 cm) and *sparser*, and with a lemon-balm aroma. **Cones** Hidden bracts.
Variants Several selections with *very grey foliage* (f. *violacea*) are occasional in gardens everywhere, and highly attractive when young and neatly conic. 'Candicans' has particularly pale silvery leaves; in 'Wattezii' they emerge *cream* then fade to blue-grey.

Low's Fir — *Abies concolor* var. *lowiana*

(*A. lowiana*) Oregon, California: an intergrade between the type and Grand Fir (p.74). 1851. Occasional in larger gardens.
Appearance Shape Gracefully conic when young, especially in northern-sourced trees; dome-topped with age or, in shelter, remaining slender: to 53 m so far. **Bark** Fissured and corky: soon more rugged than other silver firs' and blackish in trees from the N end of the range (almost like Douglas Fir's, p.120). **Leaves** Like Grand Fir's (p.74) in trees from N end of range; leaves of southern trees are greyer, spreading either side of the shoot and curving upwards 45°, but not to the vertical as in the type.
Compare Grand Fir (p.74). Northern trees differ in bark and brighter rufous second-year shoot. Southern trees are more like Colorado White Fir (above) or Noble Fir (p.88), but have longer, less glaucous leaves, not spreading all across the top of the shoot.

Grecian Fir — *Abies cephalonica*

Greek mountains. 1824. Very occasional: coping well in dry areas though troubled by late frosts; rarely seeding.
Appearance Shape Seldom spire-like for long: *massive low branches* and a lumpy, often crooked trunk make a broad, much-broken, rather spiky crown, to 40 m. **Bark** Dull grey; soon closely and deeply cracked into particularly knobbly squares. **Shoots** Stout, hairless, shiny pale brown; often little seen between the leaves' big 'suckers'. **Leaves** *Radiating stiffly all round the shoot* (slightly more above than below); 2–3 cm. Shiny green above with 2 narrow whitish bands beneath, leathery, dully *spined*, and balsam-scented. Their broad angle makes the sprays wider than European Silver Fir's (p.68). **Cones** Abundant high on old trees, 15 cm; bract-scales project and bend back.
Compare Spanish Fir (p.78): the most similar of several silver firs with leaves radiating round the shoot; leaves shorter, rounded, banded/white-dusted above. Vilmorin's Fir (p.78): hybrid of these 2 trees; thicker, more spaced leaves than Grecian Fir, with dull pale green bands beneath. Min Fir (p.84): sparsely radiating leaves; different bark. Algerian Fir (p.80) and Korean Fir (p.86): short, *blunt* leaves all round the shoots. Noble Fir (p.88): variably upcurving leaves.
Variants Apollo Fir, var. *graeca* (var. *apollinis*; not always distinguished), is more widespread in Greece, but scarcer in N Europe. Often strictly columnar; leaves denser, unspined, but more or less parted beneath the shoot: the foliage recalls King Boris' Fir's (p.68), but the stout shoot is *not hairy*.

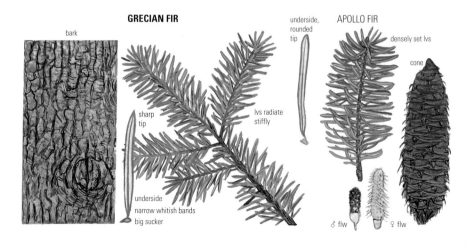

GRECIAN FIR — bark; sharp tip; underside narrow whitish bands; big sucker; lvs radiate stiffly; underside, rounded tip

APOLLO FIR — densely set lvs; cone; ♂ flw; ♀ flw

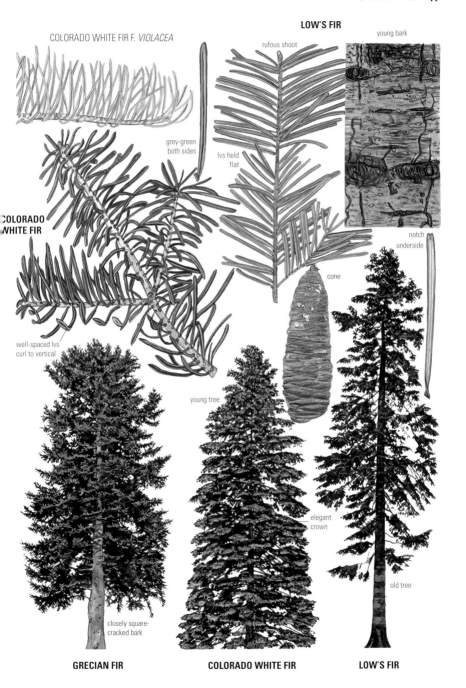

SILVER FIRS

Cork Fir — *Abies lasiocarpa* var. *arizonica*

(*Abies bifolia* var. *arizonica*) Arizona to Colorado. 1903. A desirable *deep blue* tree, confined to big gardens and generally happy only in very cool, humid parts.

APPEARANCE Shape A dense narrow spire, to 20 m; many trees fork, but the several tips remain close and tight. **Bark** Pale and *thickly corky*. **Shoots** Light grey-brown, finely hairy. **Leaves** Mostly held above the shoot and arching closely to the vertical (cf. Noble Fir, p.88): slender, bluntly rounded, greyish above with a central white band and 2 white bands beneath. Balsam/almond scent. **Cones** Hidden bracts.

COMPARE Flaky Fir (p.86): darker grey, more perpendicular leaves; very different bark. Colorado Fir selections (p.76): longer, sparser, uniformly grey leaves.

VARIANTS The semi-dwarf 'Compacta' carries leaves all round its shoots; very rare.

Subalpine Fir, the type of the species, growing as a narrow, straight, deep green spire to minimize breakage by snow in the high mountains of W North America, scarcely survives in most of Britain: bark *grey*, with close rounded warts; leaves greener, with *broken* grey bands above.

OTHER TREES Taiwan Fir, *A. kawakamii* (from the island's high mountains, and in some collections since 1929 as a dense but broadly conic tree), also has a corky but much more finely shredding bark. The slender leaves stand *stiffly in 2 or more ranks* (cf. Himalayan Fir, p.80) on either side of the pale shining brown shoot, whose grooves have dense hairs and which is *almost as thick as the leaves in the top rank are long*. Its cone-bracts do not project.

Spanish Fir — *Abies pinsapo*

(Hedgehog Fir) Confined in the wild to the shady side of a few mountain-tops above Ronda in S Spain, and very vulnerable, but one of the least rare silver firs in large gardens in the UK (since 1839), coping reasonably well with dry conditions and chalky soils.

APPEARANCE Shape Conic, to 30 m, then haphazard with age: dense and gappy, or sometimes spikily open; *greyish* – bright pale blue in grafted selections ('Glauca'). **Bark** Dark grey; developing small, craggy plates. **Shoots** Greeny-brown, the leaf-suckers often almost meeting up. **Leaves** Set densely all around the shoot (slightly parted underneath on young trees or shaded shoots); *perpendicular or leaning slightly backwards*, like an old pipe-cleaner; they are short (15 mm), thick and stiff with blunt tips and broad *grey bands on each side* (but somewhat whiter beneath in 'Glauca'). **Cones** Numerous at the tops of old trees, bright green during summer; bracts hidden.

COMPARE Algerian Fir (p.80): a dark tree with broader leaves, their bands much whiter below than above. Grecian Fir (p.76): longer, sharper leaves, green above and narrowly white-banded beneath. Noble Fir (p.88): sometimes looks similar when foliage is out of reach; retains an often more silvery, less square-cracked bark and has sprays of longer, variably upcurved leaves (often almost solid against the light).

OTHER TREES Vilmorin's Fir (*A.* × *vilmorinii*) is an artificial hybrid with Grecian Fir (Paris, 1867), found in a few collections: a dark, *open, spiky tree*. Slightly longer (25 mm) leaves, much sparser, and sharply pointed; dark green above with pale green bands beneath.

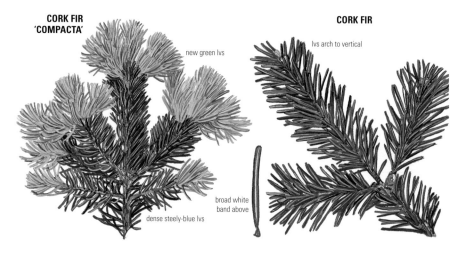

CORK FIR 'COMPACTA' — new green lvs — dense steely-blue lvs — broad white band above

CORK FIR — lvs arch to vertical

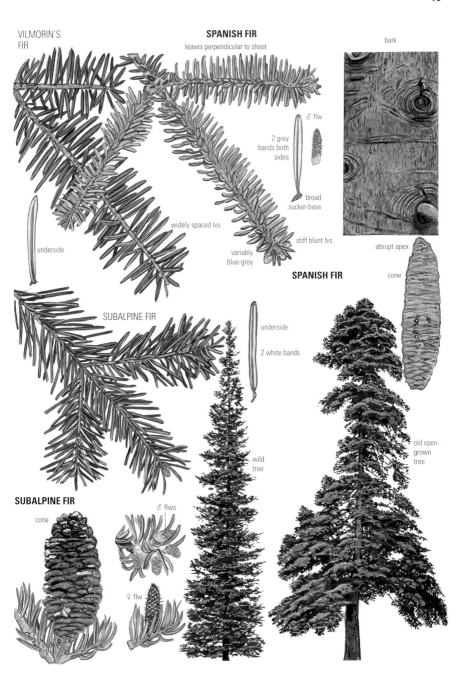

SILVER FIRS

Algerian Fir   *Abies numidica*

Mt Babor, Algeria. 1862. Rather rare, but surviving well in dry parts and on chalk.
APPEARANCE Shape Sturdily conic, then *densely* columnar to 35 m; blackish green. **Bark** Pinkish grey, cracking into *rounded plates*. **Leaves** Short (to 2 cm), thick, leathery; *radiating densely all round* the shiny brown, hairless shoot (though with fewer below). *Roundly square-ended*; dark blue-green with grey bands above or a white dusting near the tip; *brighter* white bands beneath. **Cones** At the tops of older trees; pale green as they mature; bract-scales hidden. **COMPARE** Spanish Fir (p.78): dark grey bark; open/gappy crown; leaves equally grey on both sides. Korean Fir (p.86): small, conic, soon covered in purple cones, has often notched leaves, intensely white beneath. Flaky Fir (p.86): leaves well parted beneath its shoots; papery bark. Grecian Fir (p.76): much longer, sparser, usually sharp leaves.

Himalayan Fir   *Abies spectabilis*

(Webb Fir; *A. webbiana*, *A. brevifolia*) Himalayas, at high altitudes. 1822. Rare: big gardens in wetter areas.
APPEARANCE Shape Typically broad and broken, with heavy, often sprouty branches. To 30 m. **Bark** Soon *craggy* for a fir: flaky pink-grey plates. **Shoots** Very stout, reddish; dark hairs in the *deep cracks*. **Leaves** Very long (to 6 cm; cf. Pindrow Fir, below; Santa Lucia Fir, p.88; Salween Fir, p.84); deeply grooved above and brilliantly white-banded beneath; bluntly pointed, notched or twin-spined. They arch downward like gulls' wings in *several dense ranks* either side of the shoot. **Cones** Bract-scales are almost hidden.
COMPARE Nikko Fir (p.86): smoother bark; much smaller leaves, white-brown shoot. Farges' Fir (p.82): often purple shoots and narrower duller bands beneath its leaves. Faber's Fir (p.82): similar bark but shorter leaves and hairier shoots.

Pindrow Fir   *Abies pindrow*

W Himalayas, at lower altitudes than Himalayan Fir. 1837. Rare: thriving only in the coolest, dampest regions.
APPEARANCE Shape Narrowly conic, rougher with age; to 35 m. **Bark** Dull grey, ridged but not plated. **Shoots** Pale grey-pink, *very stout*; smooth and hairless. **Buds** Large and red, whitened with resin. **Leaves** *Long* and slender (60 × 2 mm), thinly spread above the shoot and arching *forwards and sideways* in ranks on either side. Greenish grey bands beneath; generally with twin-spined tips (cf. Momi Fir, p.84). **Cones** Hidden bracts.
COMPARE Himalayan Fir (above). Farges' Fir (p.82): darker, often *purple* shoots, showing tiny hairs in the *slight grooves*, and straighter leaves parted above the shoot. Low's Fir (p.76): leaves curving upwards, slender shoot and craggy bark. Manchurian Fir (p.84): smaller, slender, more upcurved leaves, not twin-pointed. Other silver firs with leaves spreading above but not below the shoot (see Caucasian Fir, p.70) have leaves not more than 4 cm long.
OTHER TREES Gamble's Fir *A. gamblei* (*A. pindrow* var. *brevifolia*), from higher sites, is in a few collections. Its shorter leaves, rounded or single-spined and with a whitish band above, radiate round the darker, reddish shoots (cf. Forrest's Fir, p.82).

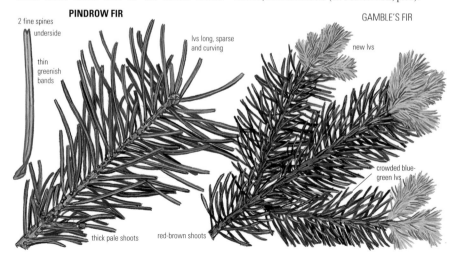

PINDROW FIR — 2 fine spines underside; thin greenish bands; lvs long, sparse and curving; thick pale shoots; red-brown shoots

GAMBLE'S FIR — new lvs; crowded blue-green lvs

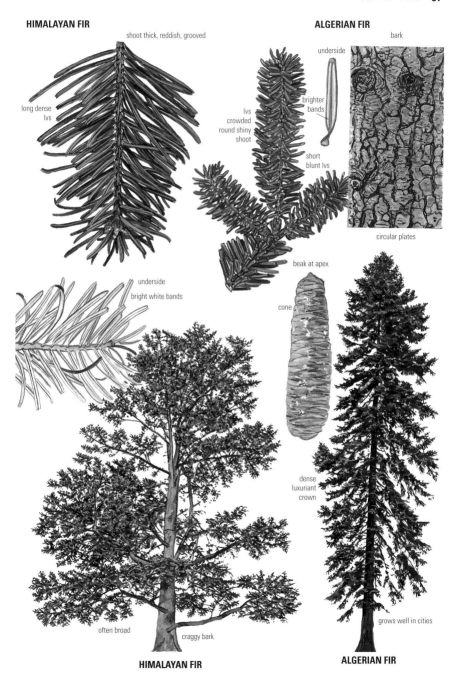

SILVER FIRS (EAST ASIATIC)

Farges' Fir
Abies fargesii

(Incorporating *A. sutchuenensis*) W China. 1901. Rather rare.
Appearance Shape Narrowly conic, to 24 m (can remain a bush in Europe). **Bark** Pinkish grey, flaking finely. **Shoots** *Purple* (rarely more orange), then chocolate-brown; may be finely hairy, but never grooved. **Leaves** Glossy, deep *yellowish* green; fully parted beneath and more or less parted above most shoots; generally dull white bands beneath. **Cones** Projecting bract-spines.
Compare Forrest's Fir (below): bluer leaves all round rufous shoots. Faber's Fir (below): much-fissured bark. Himalayan Fir (p.80).

Forrest's Fir
Abies forrestii

(*A. delavayi* var. *forrestii*) NW Yunnan (China). 1910. The least rare in cultivation of a swarm of closely related and much-confused silver firs from the mountains of SW China; very occasional, in larger gardens.
Appearance Shape Sturdily conic, and remaining so in humid areas; to 28 m so far. The foliage is beautifully luxuriant. **Bark** Grey, smooth; wide, rarely shaggy plates develop with age. **Shoots** (Finely roughened, so matt) *rich orange-red*. **Leaves** Slightly shiny *dark blue-green*, to 40 mm, crowded *all around the shoot* except in shade, though with more above, and *brilliantly* white-banded beneath. Smelling of oranges. **Cones** Frequent even on young trees (but see Korean Fir, p.86) – a regal blue-purple as they ripen; the spines at least of the bract-scales project.
Compare Farges' Fir (above): more purplish shoots and dark yellow-green leaves, well parted beneath them. Gamble's Fir (p.80): lighter green leaves, duller beneath. Veitch's Fir (p.72) and Pacific Silver Fir (p.74): leaves similarly dark above and brilliantly white-banded beneath, but – as in the many similar firs – they are well parted underneath all but the strongest shoots.
Variants Smith's Fir, var. *smithii* (*A. delavayi* var. *smithii*; often grown as *A. georgii/A. delavayi* var. *georgii*; Yunnan, 1923) is a variable tree in many collections, with a browner bark *soon craggily plated* (cf. Himalayan Fir, p.80), *dense short hairs* on its orange-red shoots (cf. Maries' Fir, p.74), and shorter, grey-bloomed leaves.

Faber's Fir
Abies fabri

(*A. delavayi* var. *fabri*) Central W Sichuan (China). 1903. A variable tree in some collections.
Appearance Shape A broad spire. **Bark** Soon craggy (as Smith's Fir, above). **Shoots** Pale shiny brown to purplish; sometimes hairy. **Leaves** Parted below and (narrowly) above the shoot; their margins *sharply downrolled*. **Cones** At least the spines of the bract-scales project.
Compare Himalayan Fir (p.80): larger leaves (with flat edges). Farges' Fir (above): greyer, smoother bark. Other silver firs with similar foliage have more leaves spreading all across the top of most shoots – see Caucasian Fir (p.70).
Variants ssp. *minensis* (*A. minensis*; often grown as *A. faxoniana/A. delavayi* var. *faxoniana*; NW Sichuan, 1911) is in a few collections. Hairs on the *honey-coloured* shoot are confined to grooves, or absent. The bands beneath the *almost flat-edged* leaves are usually duller and there may be a whitish dusting above, near the tip.

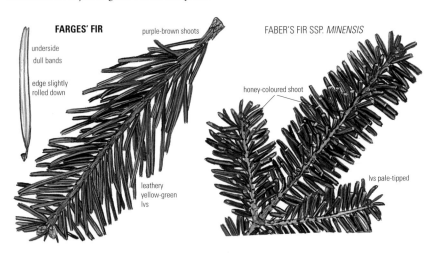

FARGES' FIR — purple-brown shoots; underside dull bands; edge slightly rolled down; leathery yellow-green lvs

FABER'S FIR SSP. *MINENSIS* — honey-coloured shoot; lvs pale-tipped

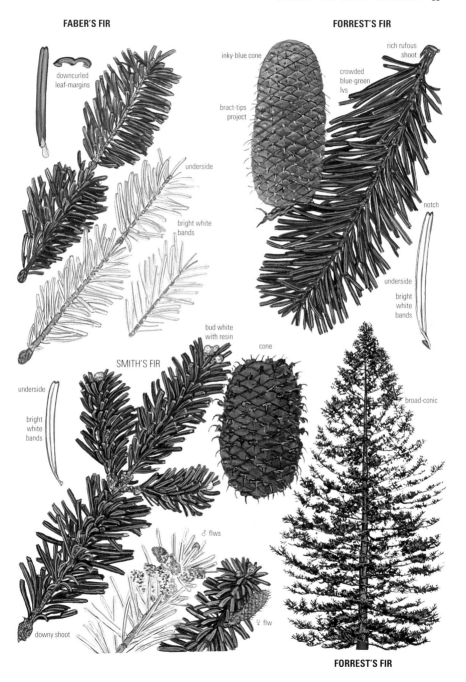

SILVER FIRS (EAST ASIATIC)

Manchurian Fir — *Abies holophylla*

(Needle Fir) Manchuria, Pacific Russia and Korea. 1908. In some collections.
APPEARANCE Shape Openly conic; to 20 m. **Bark** Pale pink- or orange-grey; finely scaling. **Shoots** Stout and hairless, slightly ridged, shiny *fawn-pink*. **Buds** *Big and globular*, pale reddish brown. **Leaves** Bright, glossy green; variably *slender* and relatively long (35 × 1 mm): like Sakhalin Fir's (p.72) but stiffer, quite sharply pointed (*hardly ever notched*; cf. Large-coned Douglas Fir, p.120), and *rising* more steeply and almost vertically over the shoot to show dull narrow *yellowish bands* beneath; almost scentless. **Cones** Yellow-green as they ripen; bracts hidden.
COMPARE Gamble's Fir (p.80): blunt leaves, spreading all round most shoots; purple cones. Flaky Fir (p.86): greyish, shorter leaves; more coarsely peeling bark. Subalpine Fir (p.78): leaves banded brightly beneath and faintly above.

Momi Fir — *Abies firma*

(*A. bifida*) Mountains of S Japan. 1861. Rare: large gardens.
APPEARANCE Shape Broadly and sturdily conic, to 30 m: long, light branches in evenly rising tiers. **Bark** Pink-grey; finely flaking and corkily warted with age, or with some oblong plates near the base. **Shoots** Pale brown, grooved; seldom with odd hairs. **Leaves** Radiating either side of the shoot, with a V-shaped gap above; yellowish; *thick, broad and leathery*, to 30 × 4 mm but longer on young trees. Notched or (on young plants) with twin spines; very faint grey bands beneath and a whitish patch above near the tip. The stalk is *long*, so that the leaves are almost cricket-bat shaped and a distinct fine band of sky shows between them and the shoot. **Cones** Yellow-green as they ripen, the bract-scales projecting 3–4 mm.
COMPARE Nikko Fir (p.86): similar aspect but paler shoot and slenderer leaves with bright white bands beneath. Min Fir (below).
OTHER TREES Salween Fir, *A. chensiensis* ssp. *salouenensis* (E India to SW China, 1907), is a narrower tree in some collections. Leaves as leathery but often richer green and the *longest of any fir* (side ones to 8 cm; upper ranks much shorter); shoots shiny yellow-pink and always *hairless*; cones with hidden bracts.

Min Fir — *Abies recurvata*

Min Valley in W Sichuan, China. 1910. In some collections.
APPEARANCE Shape Narrow, with small level branches; to 22 m. **Bark** Grey-, pink- or orange-brown, with *papery flakes and rolls* (but much finer than those of Flaky Fir, p.86) and some fissures near the base in age; often sprouty (cf. Sakhalin Fir, p.72). **Shoots** *Shiny*, pinkish or orange-grey (often very pale); hairless. **Leaves** Quite broad, distinctly stalked and pale green beneath, like Momi Fir's, but smaller (to 22 mm; diminishing towards the shoot-tip); they may radiate stiffly round weak shoots, though with very few below, while on strong shoots the upper ones *stand at or curl beyond the vertical* (cf. Spanish Fir, p.78). **Cones** Purple at first, with hidden bracts.
COMPARE Momi Fir (above); Nikko Fir (p.86). Apollo/Grecian Fir (p.76) and its hybrid, Vilmorin's Fir (p.78): knobbly grey bark, duller shoots often half-hidden by the dense leaf-suckers, and stiffer/sharper leaves.

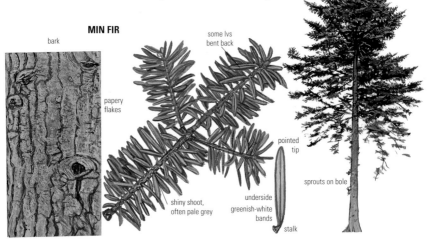

MIN FIR
bark
papery flakes
some lvs bent back
shiny shoot, often pale grey
underside greenish-white bands
pointed tip
stalk
sprouts on bole

SILVER FIRS (EAST ASIATIC) 85

SILVER FIRS (EAST ASIATIC)

Nikko Fir *Abies homolepis*

(*A. brachyphylla*) Mountains of S Japan. 1861. Very occasional everywhere: particularly tolerant of dry and polluted air.
Appearance **Shape** Sturdily and spikily broad-conic until old: light branches shallowly rise close together. To 35 m. **Bark** Tinged salmon-pink and finely flaking (like that of young Norway Spruce); then grey, with packed, corky warts or, rarely, perpendicular cracks; may be sprouty. **Shoots** *Shining* pale brown, even *white*, grooved into *plates* (like a spruce twig, though lacking any 'pegs'). **Leaves** *Well parted above most shoots*, with few or none beneath; spreading *stiffly and straight at a very wide angle*; usually notched at the tip; *bright* white bands underneath. **Flowers** Abundant. **Cones** (Scarcer, but not confined to the tree-top) violet then brown, with hidden bracts.
Compare Momi Fir and Min Fir (p.84): broader leaves scarcely white-banded beneath. Himalayan Fir (p.80): longer, arching leaves; redder shoots; shaggy bark. Korean Fir (below): abundant purple cones; leaves all around most shoots. Faber's Fir and Farges' Fir (p.82): similarly arranged but more forward-angled leaves.

Korean Fir *Abies koreana*

Restricted to S Korea and Cheju Do island. Introduced only in 1913, but now the *most widespread silver fir*, frequent in small gardens for its display of cones.
Appearance **Shape** Plants sourced from Cheju Do are dumpy pagoda-shaped *bushes* with crisp, closely rising branches, while trees from the mainland have reached 15 m in sheltered woodland. **Shoots** Pale pinkish grey, often lightly hairy. **Leaves** Stumpy (12–18 mm), perpendicularly all round most shoots though with few below and some upper ones bent back past the vertical; usually notched. Brilliant white bands underneath *almost coalesce* and there is often a white patch above near the tip. **Cones** Abundant even on young trees; vivid purple-blue cones like Fabergé eggs. These are decorated with golden-brown, outcurving bract-scales and – like all silver firs' – ripen to brown before disintegrating on the tree, so cannot be pulled off for ornaments.
Compare Flaky Fir (below); Nikko Fir (above). Algerian Fir (p.80): a vigorous, dense, dark tree with leaves more densely packed around hairless green-brown shoots.

Flaky Fir *Abies squamata*

Mountains of W Sichuan, China. 1910. In a few collections.
Appearance **Shape** Stiffly conic, to 15 m. **Bark** Soon uniquely shaggy with *big, pinkish-orange papery rolls*, like a lawyer's wig. **Shoots** Purplish, matt (cf. Farges' Fir, p.82). **Leaves** *Grey*-green, spreading and curving almost vertically above most shoots with none below; short (to 25 mm), stiff, and diminishing regularly towards the shoot-tip; broad whitish bands beneath, and a white line or dusting above. **Cones** Violet when ripening; the bract-scales project and bend down.
Compare Hedgehog Fir (p.78) and Algerian Fir (p.80): leaves radiating all round their shoots. Colorado White Fir (p.76): long, sparse leaves the same colour each side. Manchurian Fir (p.84): very slender leaves, bright green above. Min Fir (p.84): papery rolls on the bark are much finer. Subalpine Fir (p.78).

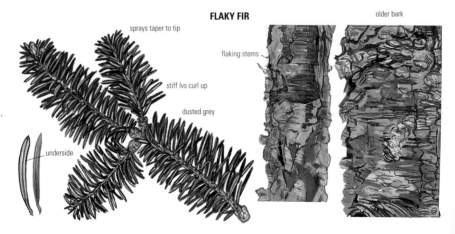

FLAKY FIR — older bark; sprays taper to tip; flaking stems; stiff lvs curl up; dusted grey; underside

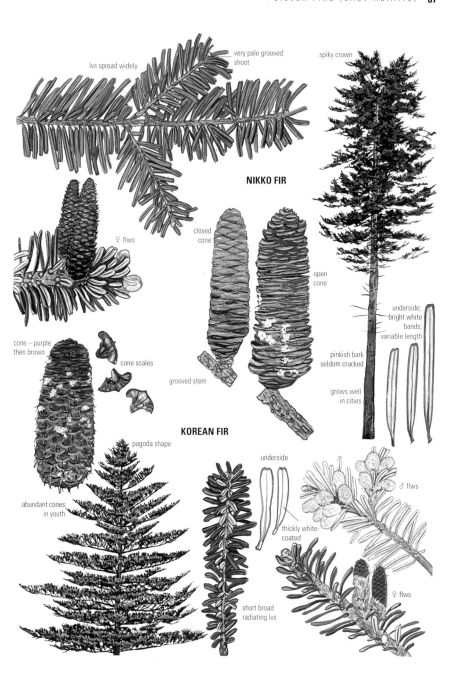

SILVER FIRS (AMERICAN)

Noble Fir   *Abies procera*

(*A. nobilis*) NW USA. 1830. Occasional, in many large and some smaller gardens; very rarely in forestry plantations; sometimes seeding; short-lived and rare in the driest areas.
Appearance Shape Spikily conic, to 50 m in shelter, on a bole which scarcely seems to taper. The heavy cones mean that exposed tops soon break and the tree grows *densely columnar* (sparse in shade or dry conditions). **Bark** *Silvery* or purplish grey; rather distantly cracked; some trees eventually develop rugged ridges. **Shoots** Pale rufous, hairy, but half-hidden by the *very dense leaves*, which are parted beneath the shoot or run parallel with it before usually *sweeping forwards then up*; their rounded (rarely notched) tips often make a close pattern like a hedgehog's back. **Leaves** Always *grey*; grafted selections ('Glauca') are intensely silver. Slender, grooved above and flattened so that they *cannot be rolled* between finger and thumb, with 2 thin grey lines above and 2 whitish bands beneath. Rather oniony aroma. **Cones** Purple-brown with hanging, green bract-scales; ripening golden-brown; *very big, like loaves* (to 25 cm) and carried near the tops even of small trees.
Compare Red Fir (below): corkier bark, more formal spire shape and slenderer, almost *round* leaves. Colorado White Fir (p.76): longer, very *sparse* leaves, the same colour each side. Cork Fir (p.78): a small tree with more flattened, sparser leaves and corky bark. Spanish Fir (p.78): stiffly radiating leaves and square-cracking bark. Manchurian Fir (p.84) and Sicilian Fir (p.68): also with steeply upcurved leaves.

Red Fir   *Abies magnifica*

Oregon and California. 1851. (The very different *A. amabilis*, p.74, is also called 'Red Fir'.) Rare: a beautiful tree, but thriving for long only in deep shelter and high humidity; the distinctively red bark of old wild trees is yet to be seen in Europe. Easily mistaken for Noble Fir.
Appearance Easily mistaken for Noble Fir. **Shape** Dense, narrow, fairy-tale spire, maintained to 45 m in Scotland, on a rather corkily ridged bole sectioned by *rings of black scars* from the *regularly whorled* branches. **Leaves** Longer, slenderer, sparser, laxer than Noble Fir's, and *round enough to be rolled*. **Cones** Bracts normally *hidden*.

Santa Lucia Fir   *Abies bracteata*

(Bristlecone Fir; *A. venusta*) Santa Lucia Mountains, California. 1853. Rare: large gardens.
Appearance The fir that does everything differently, although it may recall Low's Fir (p.76) at first glance. **Shape** A spire, with a long narrow top when adding height, and rather hanging, blackish foliage; to 40 m. **Bark** Purple-black; wrinkled then distantly square-cracking. **Shoots** Smooth, greenish. **Buds** Beech-like: *narrow, pointed and twice as long* as other firs' (2 cm); bright pale brown. **Leaves** Long (50 mm), parted beneath the shoot with only a few spreading above, hard and *spined*; bright white bands beneath. **Cones** Decorated with the bracts' *long spines*, like sea urchins, but rare in Britain – though a seedling tree has grown in Exeter.
Compare Spruces (pp.100–115): sharp buds and sharp leaves but none this large. Californian Nutmeg (p.24); Large-coned Douglas Fir (p.120).

SANTA LUCIA FIR

bark; slender acute beech-like buds; ♀ flw; long apex; spined tip; two white bands; cone

CEDARS

Cedars (4 old-world species) carry their leaves spirally around strong shoots but in rosettes of 10–60 on very short, thick side-shoots (spurs), like the deciduous larches. Pollen is shed late in autumn. ('Cedar' is a name used for many unrelated trees with valuable, aromatic timber – including a broadleaved relative of Mahogany, Chinese Cedar, p.358.) (Family: Pinaceae.)

Things to Look for: Cedars

- Shape (especially branch-tips)?
- Leaves: How long? Are they translucently spined?
- Cones: What is the shape at the tip?

Cedar of Lebanon *Cedrus libani*

The Lebanon (where much reduced by felling – the fine timber is insect-repellent), N to the Taurus Mountains. The essential accessory for a mansion lawn since 1740, when the early British trees were mostly killed by frost. Today's population is hardier. Still quite frequent in lowland areas as an old tree, though currently seldom planted. It has suffered more than most species from the misconception that it will take so long to mature that planting one is futile. In fact, it is very vigorous.

APPEARANCE Shape The great *level plates of foliage* develop with age in open situations. Young trees are rather gauntly conic and the rare woodland specimens have a long straight trunk, to 42 m, but a flat top. To identify a cedar it can be useful to remember 'Lebanon – level; Atlas – ascending; Deodar – drooping', but confusion is possible: Deodar *can* have big level plates, the tip-growths of Cedar of Lebanon *can* droop, and its foliage-plates *can* ascend jaggedly. A dark green tree, or sometimes very grey and as bright as many Blue Atlas Cedars (p.92). **Bark** Black-*brown*, closely ridged and cracked. **Shoots** With very fine down (less than Atlas Cedar's (p.92) and often confined to shallow grooves). **Leaves** About 25 mm, stiff; the *short point green except for its translucent extreme tip*. **Cones** Barrel-shaped, *without a sunken top*, tapering more (above halfway up) than Atlas Cedar's.
COMPARE Other Cedars. Atlas Cedar (p.92): greyer bark, no very broad flat foliage-plates, shorter leaves (to 20 mm in rosettes) with translucent spine-tips, and cones with sunken tops – but some old trees are hard to specify. Deodar (p.92): longer (30–35 mm), softer leaves, and more drooping young growths. Cyprus Cedar (below): much shorter leaves when mature.
VARIANTS 'Aurea' is a very rare, stunted form with soft yellow foliage (cf. Golden Atlas Cedar, p.92). 'Glauca' is a particularly *rich blue* selection; rare (cf. grey examples of the type).

Cyprus Cedar *Cedrus brevifolia*

(*C. libani* var. *brevifolia*) Tripylos Mountain, W Cyprus. 1879. A relatively *small* tree (to 23 m so far), largely confined to collections.
APPEARANCE Shape An approximate spire; bluish or bright green, sometimes rather pendulous. **Bark** *Mid-grey*. **Leaves** On old trees *very short* (7–15 mm) – much smaller than other species', though Atlas Cedars which are clearly sickly and shed most of their foliage in winter can be comparable. Vigorous young trees (with leaves to 25 mm, and no cones) are harder to specify. **Cones** Slender (about 7 × 4 cm) and *lemon-shaped* – with a distinct summit.

CYPRUS CEDAR

CEDAR OF LEBANON

CEDARS

Atlas Cedar *Cedrus atlantica*

(*Cedrus libani* var. *atlantica*) Atlas Mountains, Algeria and Morocco. 1841. Trees of the typical green form are rather occasional.
Appearance Shape Conic in youth; in age still tapering to *narrow*, flat top; to 38 m. The *spiky*-edged, often *rising* plates of foliage are *never as large* or flat as in good Cedars of Lebanon; young shoots rise or droop slightly, though occasional trees have strikingly weeping twigs *and branches* (some may be the clone 'Pendula'; cf. 'Glauca Pendula', below). Blackish to vivid green, or grey (merging into f. *glauca*). **Bark** Usually greyer than Cedar of Lebanon's, cracking often neatly into close plates. **Shoots** With short, dense, blackish hairs. **Leaves** Short (*to 20 mm in rosettes*), with small *translucent* spine-tips. **Cones** Barrel-shaped, with a *dimple at the tip*.
Compare Cedar of Lebanon and Cyprus Cedar (p.90): differences emphasized.
Variants 'Fastigiata' (1890) has *steeply ascending* branches and (dark grey) sprays: a very neat tree, but rare. (The very rare 'Glauca Fastigiata' is bluer but spikier and less narrow.)

Blue Atlas Cedar *Cedrus atlantica* f. *glauca*

Selections of the greyest wild Atlas Cedars (1845 on) are now very frequent in gardens large and small (where they grow with alarming vigour).
Appearance As Atlas Cedar (above) but *bright silver-grey* – much paler than other cedars, and almost pink-tinted when shedding pollen in October.
Compare Blue Colorado Spruce (p.114) and Colorado White Fir (p.76): some selections as bright, but never with needles in rosettes.
Variants 'Aurea' has foliage *emerging pale sunlight-yellow and greying through the year*: an extraordinary 'died-last-week' effect, quite distinct from the golden Cedar of Lebanon. Rare.
'Glauca Pendula' (1900) has branches and silver-grey foliage *hanging in a single curtain* from usually 1 hump-backed main stem. Spectacular, like a lignified spout of water, but rare and difficult to establish.

Deodar *Cedrus deodara*

W Himalayas, where it reaches 80 m. 1831. Very frequent as an old park tree and as a gorgeously soft, silky young plant in small gardens.
Appearance Shape Often a very straight trunk and conic crown into maturity. Infrequently, heavy low limbs and big flat plates develop, like a typical Cedar of Lebanon's, and there may be multiple trunks from the base. Generally *vivid green*; sometimes freshly yellowish (unlike other cedars); sometimes very dark or grey. The leader *droops* (cf. Western Hemlock, p.116) and shoots always spill and hang from the edges of the foliage-plates – but less noticeably on slow-growing old trees. **Bark** Like Cedar of Lebanon's; sometimes scalier and more purple. **Shoots** Quite *densely* hairy. **Leaves** *Long, rather soft*, to 50 mm on extension shoots and 35 mm on spurs, with fine grey lines and a translucent tip. **Cones** Most like Cedar of Lebanon's, but often *absent* even from older trees.
Compare Cedar of Lebanon (p.90): shorter, stiffer leaves and shoots never so weeping.
Variants 'Aurea' has *soft gold* younger foliage; occasional as younger tree.
'Pendula' is dumpy and *very weeping*. Rare.

DEODAR 'AUREA'
DEODAR
BLUE ATLAS CEDAR
♂ flw
single lvs on new growths
ATLAS CEDAR

CEDARS

DEODAR

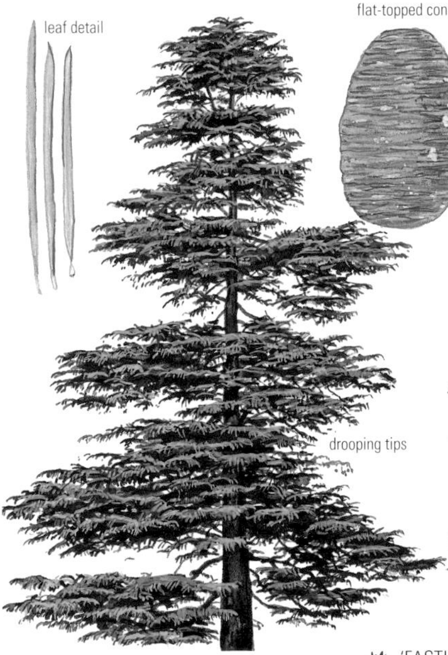

leaf detail

flat-topped cone

drooping tips

ATLAS CEDAR

narrow top

spiky rising branches

leaf detail; tip translucent

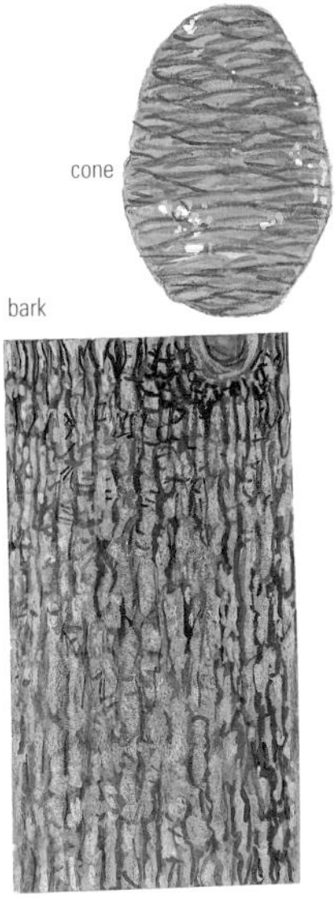

'FASTIGIATA'

upright branching

sunken top

cone

bark

ATLAS CEDAR

BLUE ATLAS CEDAR

leaf detail

LARCHES

Larches (10 species) have deciduous needles, arranged radially around strong shoots at the branch-tips and in rosettes on spur-shoots on older wood (like the evergreen cedars'). Their colours are fresh green in summer, and yellow in autumn. The best means of differentiating the species are the cones, which can generally be picked off the ground all year. (Family: Pinaceae.)

Things to Look for: Larches

- Shoots: Are they hairy? What colour?
- Cones: How big? Are the bracts visible? What colour are they when ripening?
- Cone-scales: How many? Are they straight/curved and are they hairy?

European Larch   *Larix decidua*

(Common Larch; *L. europaea*) The Alps (replaced by Norway Spruce in colder, wetter areas), with varieties in the Tatra and Sudetan mountains and plains of Poland. Long cultivated and locally abundant: older plantations, shelterbelts and parks, away from cities and the driest areas. Rarely naturalizing. The timber is hard and rot-resistant; Tatra and Sudetan forms generally make the best plantation trees.
APPEARANCE Shape Spire-like, on a trunk straight only in the finest, sheltered trees (to 45 m); often broad and characterful in age in dry or exposed sites. The fine shoots *hang* under the branches. *Blond* in winter: more finely and spikily twiggy than Ginkgo (p.20) or Swamp Cypress (p.64). Saplings often grow wildly twisting trunks, which straighten with maturity. **Bark** Pink-brown; wide, often criss-crossing, scaly-topped ridges. **Shoots** *Amber* or pale pinkish, hairless and unbloomed. **Buds** *Low-domed knobbles*. **Leaves** Less than 1 mm wide; vivid green; 2 pale bands beneath. **Female flowers** As bright as rubies in mid-spring among vivid green emerging needles, but easily overlooked. **Cones** Soon brown: oval, 25–40 × 20–25 mm when ripe, the scales *not or scarcely curving*.
COMPARE Japanese and Dunkeld Larches (p.96): redder shoots; broader darker leaves; cones with outcurved scales. Siberian Larch (below) and Dahurian Larch (p.96): normally hairy shoots. Tamarack (p.98): smaller foliage, smoother bark; cones smaller than even Polish Larch's (below), and with fewer scales. Western Larch and Sikkim Larch (p.98): cones with long-protruding bracts. Golden Larch (p.98): a stocky lookalike with much thicker leaves.
VARIANTS 'Pendula' is a broad, depressed-looking tree with *exaggeratedly weeping shoots*; rare. (Japanese Larch, p.96, and *L.* × *pendula*, p.98, have even rarer, spectacularly weeping cultivars.)
Polish Larch, ssp. *polonica* (including Tatra Mountains trees; 1920), is highly distinct: craggier brown bark, often stricter shape, thin ash-blond shoots, and *small cones* (to 20 mm), with slightly incurving scales; Sudetan trees show intermediate features.

Siberian Larch   *Larix sibirica*

(Russian Larch; *L. russica*) Taiga forests of NE Russia to the Tien Shan. 1806. In a few collections.
APPEARANCE Shape In Britain, a poor bush. (Accustomed in the wild to long, hard winters and decisive springs, it leafs out here during mild interludes in mid-winter only to get repeatedly frosted.)
Shoots Pale buff, *hairy at first*. **Leaves** More than 1 mm wide. **Cones** With minutely *hairy scales*, which curve outwards slightly.

EUROPEAN LARCH
summer   autumn

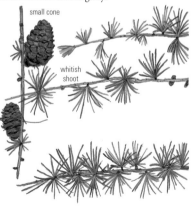

small cone
whitish shoot

POLISH LARCH

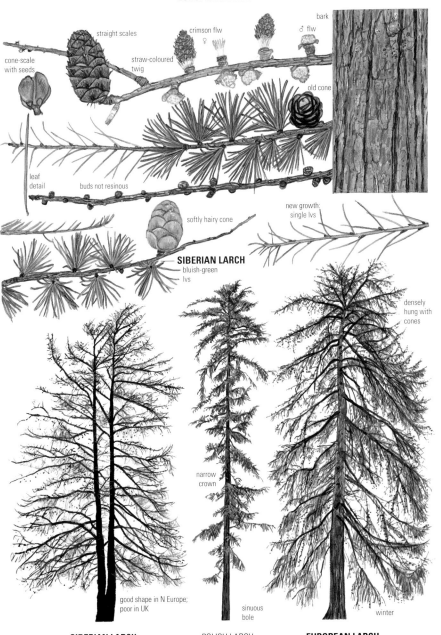

LARCHES

Japanese Larch
Larix kaempferi

(*L. leptolepis*) Mountains of central Honshu, Japan. 1861. Quite rare in gardens but a frequent plantation tree on richer soils: its particularly heavy leaf-fall is able to smother invasive rhododendron growth.
APPEARANCE Shape Often broader-conic than European Larch's when open-grown, to 40 m; in winter, *smoky-red*. **Bark** Reddish or purplish grey-brown; the ridges usually become *shaggier* than European Larch's. **Shoots** *Dark* orange, brown or purplish grey, with a variable *waxy white bloom*. **Buds** Little *resinous* knobbles. **Leaves** A good 1 mm wide; *dark*, heavy green with 2 broad grey bands beneath. **Cones** Almost *spherical*, 30 mm, the scales *curling sharply back* like the petals of a tiny, mummified rose.
COMPARE European Larch (p.94): blond shoots, slenderer bright green leaves and cone-scales not curved back. The 2 trees' hybrid, below, grades into Japanese Larch. The foliage and overall appearance of most larches are very similar and exotic species in gardens are easily overlooked.

Dunkeld Larch
Larix × eurolepis

The hybrid of European and Japanese Larches was first selected at Dunkeld (Perth and Kinross) in 1904, but identified later among trees in a nearby wood which had been planted 7 years earlier. Rather rare as a mature garden tree (to 40 m so far), but for some decades the preferred larch in forestry.
APPEARANCE Shape Usually straight and often slender. **Bark** Reddish brown; most like Japanese Larch's. **Shoots** Pale orange, pale brown or reddish, scarcely bloomed. **Buds** Not resinous. **Leaves** Often longer than either parent's (50 mm); rich green above. **Cones** Oval and *tall*, to 40 × 25 mm, the scales *bending abruptly outwards to some degree but not downcurved at the rim*. Plantation trees (generally back-crosses) often tend towards Japanese Larch, with shorter cones and more curled scales, so that positive identification can be difficult.

Dahurian Larch
Larix gmelinii

(*L. dahurica*) Taiga forests of E Siberia; NE China. 1827. In a few collections.
APPEARANCE Shape Generally stunted in W Europe – wide branches from a bent, picturesque bole, and rich green foliage in *drooping plates* like shawls of green snow. **Shoots** Reddish/yellowish brown, sometimes hairy. **Leaves** Very *slender* (0.6 mm wide), bright or hard green; narrow whitish bands beneath. **Cones** Red/purple in summer as they ripen; smaller and squatter than European Larch's (22 × 18 mm), with slightly outcurved scales; finally a rather glossy brown.
VARIANTS Trees in some collections are var. *japonica*, from S Sakhalin, Pacific Russia and the Kurile Islands (though not Japan), with densely but finely hairy reddish shoots, shorter leaves (2 cm) which have brighter bands beneath, and small cones (2 cm; only 18–25 scales). Others are var. *principis-rupprechtii* (Korea, Manchuria), with cones like European Larch's, and slightly bloomed (hairless) shoots. This is the happiest form in the UK, with sturdy trees to 20 m – slenderer-leaved than European Larches, but otherwise very similar.
COMPARE Siberian Larch (p.94): paler shoot; straight, hairy cone-scales. Tamarack (p.98): also shows poor growth in the UK and has very slender leaves, but its tiny cones have even fewer scales and it can usually be told by its smoother bark and contorted growths.

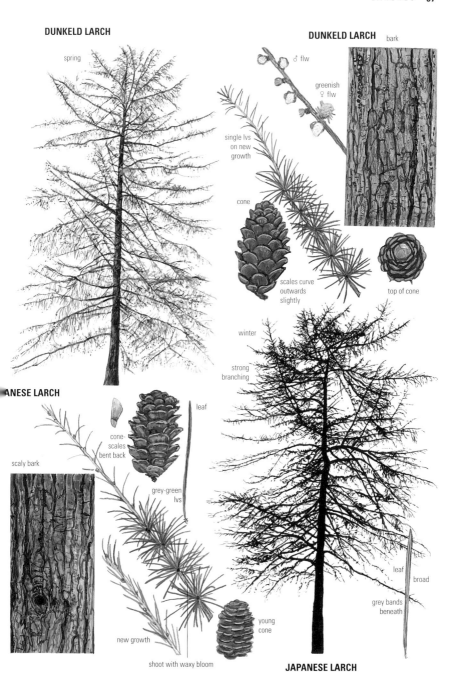

LARCHES; GOLDEN LARCH

Tamarack *Larix laricina*

Alaska to New England; a tough tree, sulking in the soft conditions of W Europe. 1739. Very rare.
APPEARANCE Shape Slender, open, to 20 m; small twisting branches with *curling shoots* – can look dead in winter. **Bark** Pinkish; *finely* scaling. **Shoots** Orange-brown, bloomed pink; slender. **Buds** Resinous. **Leaves** Very narrow (0.8 mm) and dark; grey bands beneath. **Cones** *Tiny* (to 20 mm): *only 15–20 scales* (like an erect Western Hemlock cone (p.116); cf. Dahurian Larch (p.96); red, purple or (rarely) yellow as they ripen.
OTHER TREES *L.* × *pendula* is a putative hybrid with European Larch in a few old gardens: a sturdy tree to 28 m. *Finely* flaking purplish bark; weeping shoots like European Larch's; cones like European Larch's but smaller (25 mm) and *purple* as they ripen. Its clone 'Repens' grows *horizontally* to great lengths – the original at Henham Hall in Suffolk sprawls over a pergola 18 m long.

Western Larch *Larix occidentalis*

NW North America. 1881. Typically for a North American Pacific coast tree, this is the largest larch (to 80 m), but, atypically, its vigour has never been reflected in Europe. Rare.
APPEARANCE Shape Habit much as European Larch. To 33 m, but often stunted. **Bark** Much as European Larch. **Shoots** Pale orange-brown. **Leaves** Bright green, rather triangular in section. **Cones** Highly distinctive: 40 mm tall, the *long straight spine* of each bract projecting (cf. Sikkim Larch, below; these break off the cone on the ground).

Sikkim Larch *Larix griffithiana*

(*L. griffithii*) E Himalayas and Tibet. 1848. In some collections.
APPEARANCE Shape A gaunt tree, to 20 m. **Shoots** *Stout, rufous,* finely hairy, weeping. **Leaves** Large, shiny. **Cones** Distinctively *long, cylindrical* (5–10 cm), with *exposed, arching bracts.*
OTHER TREES *L. potaninii* (W China, 1904; equally rare and gaunt in the UK) has tall cones (to 5 cm) whose protruding bracts mostly *rise vertically* (cf. Western Larch, above); its leaves – on dull tan shoots – are slender and *golden-green.*

Golden Larch *Pseudolarix amabilis*

(*Chrysolarix amabilis*) SE China: an endangered species in its own genus. A relative of silver firs which looks very larch-like. 1853. Rare, in big gardens for its gold autumn colours.
APPEARANCE Shape Sturdily squat-conic, on big *level* limbs; to 22 m, but, needing both warmth and humidity, it remains stunted at many sites in the UK. **Bark** Brown-grey, cracking into long scaly plates. **Buds** Tiny (3 mm), like larches', but slightly *pointed.* **Leaves** All round the extension shoots and in heavy rosettes, like the tentacles of sea-anemones, on the ultimately *massive, long curved spurs,* which are broadest towards their tips; much larger, *thicker* (50 × 3 mm) and darker green or more glaucous than any larch leaves. **Cones** 70 × 45 mm, with thick, *triangular, projecting scales*; they disintegrate on the tree.
COMPARE In winter, Ginkgo (p.20): bigger (5 mm) buds and less scaly bark.

LARCHES; GOLDEN LARCH

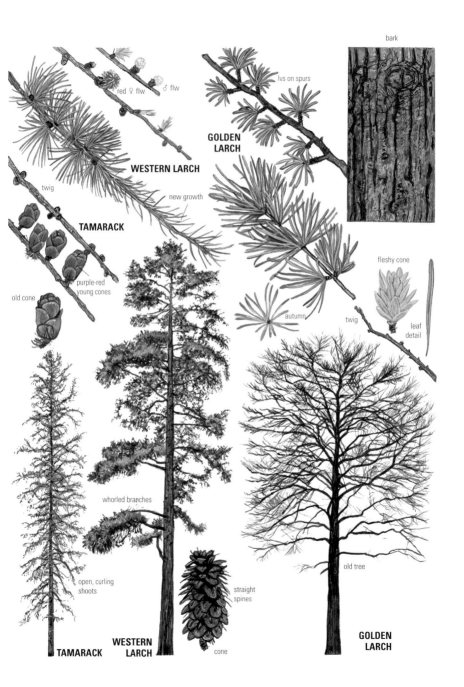

SPRUCES

The Spruces (about 40 species) are an uncomfortable bunch – thin, scaly bark, sharp needles, and woody 'pegs' protruding from the ridged shoots by which the needles are attached. The cones hang, and normally fall when ripe. (Family: Pinaceae.)

Things to Look for: Spruces

- Shape: How broad? Does the tree weep?
- Bark: How shaggy is it?
- Shoots: Are they hairy? What colour are they?
- Leaves: Distribution around shoot? How flattened are they? Are there white bands (on which sides)?

Key Species

Morinda Spruce (below): squarish leaves, all round the shoot and green on each side. **Norway Spruce** (p.102): squarish leaves, nearly all above the shoot and green on each side. **Blue Colorado Spruce** (p.114): squarish very grey leaves. **Alcock's Spruce** (p.108): squarish leaves: green above, white bands below. **Lijiang Spruce** (p.110): flattened-rhombic leaves (bias when rolled between fingers). **Sitka Spruce** (p.106): much-flattened leaves.

Morinda Spruce — *Picea smithiana*

(*P. morinda*) W Himalayas, on drier slopes. 1818. Very occasional: large gardens even in dry areas. **APPEARANCE Shape** Spire-like; ruggedly columnar with age; to 38 m. *Hanging* foliage (less so than Brewer Spruce's, below); scraggy and moping rather than weeping in exposure. **Bark** Purplish grey; harsh round scales, or ridges like Deodar's (p.92). **Shoots** Cream, hairless. **Buds** Purple-brown, to 8 mm. **Leaves** *Longer than other spruces'* (40 mm), *slender and sharp, all round* the hanging shoots; squared, and dull dark green on each face. **Cones** Long (14 cm), taper-tipped; hard smooth-rimmed scales. **COMPARE** Brewer Spruce (below): *flattened* leaves, hairy shoots. Some Norway Spruces (p.102) ('Virgata'; 'Pendula') have leaves all around their strong (*orange*) shoots, but these are seldom 25 mm long. Dragon Spruce (p.112), Colorado Spruce (p.114), Tiger-tail Spruce (p.112), and Black Spruce (p.104): many leaves held below most shoots but much shorter/stiffer needles.
OTHER TREES Schrenk's Spruce, *P. schrenkiana* (Kirgiziya; Tien Shan), is in a few collections, to 20 m. *Denser non-weeping habit*, bright brown buds, and shorter (35 mm), thicker, shorter-pointed leaves, fewer spreading beneath the shoot (cf. Dragon Spruce, p.112, with 20 mm spiny leaves).

Brewer Spruce — *Picea breweriana*

Confined in the wild to a few mountain tops on the Oregon-California border. 1897. Occasional in larger gardens, and admired by many.
APPEARANCE Shape Stumpily conic to 20 m, with massed spiky tops, the many slender branches lugubriously hung with black foliage *like curtains of Spanish Moss*. With age often sparse or malformed. **Bark** Red-black, shedding rather large, smooth circular plates. **Shoots** Pink-brown, finely *hairy*. **Leaves** *Flattened*, 30 mm; dark, dingy green, with 2 narrow white bands beneath; spread all round the shoots. **Cones** 11 cm, taper-based; slender, often curved. **COMPARE** Morinda Spruce (above). No flat-leaved spruce is similar.

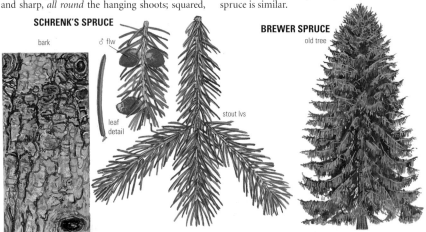

SPRUCES (EUROPEAN)

Norway Spruce *Picea abies*

(Christmas Tree; Spruce Fir) Europe from southern Scandinavia S to cold, wet mountains in the Alps and Balkans, merging in Russia into Siberian Spruce. Long grown in the UK; abundant in wetter areas and still much the commonest ornamental spruce; rarely naturalized. The traditional Christmas tree, regularly planted out in small town gardens though seldom thriving.

Appearance Shape A neat spire, narrow in its dense natural stands; sometimes weeping shoots (Comb Spruce, 'Pendula'); occasionally with low limbs but never broad-topped. To 50 m in Scotland, 68 m in Germany. Heavy foliage except on sickly plants; dark rich green. Odd trees ('Tuberculata') develop hugely swollen branch-attachments (as can Sitka and Lijiang Spruces, pp.106, 110). **Bark** Coppery grey-brown for 50–80 years, very finely shredding; then grey/purple, cracking harshly into small rounded plates. **Shoots** Dark dull *orange*, hairless except in the east of its range. **Leaves** Quite short (15–25 mm), *hard green* with (very) faint white lines on each of the 4 sides (sometimes brighter beneath); stiff and pointed; spreading above the shoot but only below very strong/weeping ones. Rich, hot, sweet smell. **Cones** Long (to 20 cm), slender; scales flimsy, and jaggedly tipped.

Compare Siberian Spruce (below), Koyama's Spruce (p.104) and Red Spruce (p104). Less similar are Tiger-tail Spruce (p.112), Colorado Spruce (p.114) and Dragon Spruce (p.112): thicker, spined leaves, some spreading beneath the shoot. Oriental Spruce (p.108): short, round-tipped leaves. Wilson's Spruce (p.106): slender leaves on white shoots; Alcock's Spruce (p.108): bold white bands on 2 faces of its leaves. Black Spruce (p.104): short bluish leaves on hairy shoots.

Variants Neat dwarfs include some (eg. 'Clanbrassiliana') which slowly mutate into small trees. 'Pyramidata', dense and narrow, has steep branches and closely competing leaders. Rare.
Snake-branch Spruce, 'Virgata', has long leaves all round unbranching, rope-like shoots: quite hideous and very rare.
'Inversa' (even rarer) is a weeping freak whose branches lie almost flat against the trunk.
'Aurea' has much brighter yellow young growths than the type, set off by the dark green interior: rare (cf. Oriental Spruce 'Aurea', p.108).
'Albospica' ('Argenteospica') is extremely rare but spectacular – like a Christmas tree lit up with fairy lights through half the year. *Pure white* young foliage fades slowly to green.

Other trees *P. alpestris* is a relic species from SE Switzerland (not grown in the UK?): paler grey mature bark, downy shoots and shorter, greyer leaves tending to a more radial arrangement.

Siberian Spruce *Picea obovata*

Across N Russia. 1908. In a few collections; often bushy. Intermediates with Norway Spruce (*P. × fennica*) occur across Finland, Norway and N Russia.
Appearance Differences from Norway Spruce (above) as follows: **Shoots** Paler, duller, usually hairy. **Leaves** Side-leaves often angled below shoot, with 1 leaf by the side-buds sticking straight out. **Cones** *Smaller* (5–11 cm), the woodier scales *not toothed*.
Compare Wilson's Spruce (p.106); Red Spruce (p.104).

SIBERIAN SPRUCE — wild tree — dull downy shoot — small cone — ♀ flw — SNAKE-BRANCH SPRUCE

104 SPRUCES

Black Spruce *Picea mariana*

(*P. nigra*) North America – an abundant tree from the Alaskan tundra-edge S to the mountains of Virginia. 1700. Rare.
APPEARANCE Shape *Densely* and often narrowly conic, to 20 m, the lowest branches readily taking root; blue-grey. **Bark** Dark and purplish; quite finely but harshly scaling. **Shoots** Pinkish brown, usually *hairy*. **Leaves** Short (to 15 mm), slender and soft, *crowded around the shoot* though with more above; squared, with whitish blue bands underneath. Lemony aroma. **Cones** *Tiny* (20–35 mm), and abundant even on young trees: purple, maturing a shiny red-brown and sometimes (like only Serbian Spruce's, p.106) *persisting* on the tree for 30 years; scales slightly jagged.
COMPARE White Spruce (p.114): the tips of all its leaves lie *above its shiny shoots*. Red Spruce (below): greener leaves well parted beneath its shoots (intermediate trees occur in the wild). Sikkim Spruce (p.112): another grey tree with needles spreading round its shoots, but *flattened* leaves. Alcock's Spruce (p.108) and Lijiang Spruce (p.110): much brighter whiter bands on the 2 lower faces of their leaves, which are well parted beneath their shoots (Sakhalin Spruce, p.108, is very similar but tends to have brighter *orange* shoots). Dragon Spruce (p.112): wide, gaunt crown, very papery-peeling bark, and stiffer leaves. Oriental Spruce (p.108): leaves even shorter, and rich green on each face.

Red Spruce *Picea rubens*

Nova Scotia and New England; long grown in the UK but confined to collections.

APPEARANCE Shape Narrowly, densely conic, with a long tip when growing well; to 28 m. **Bark** Dark brownish; flaking then closely plated. **Shoots** Variably *hairy*, pale rufous. **Leaves** Short (to 15 mm), slender, square/rounded, parted beneath the shoot and often *upcurving*; glossy, bright green. **Cones** Small (to 5 cm), soon shed; scales not or scarcely toothed.
COMPARE Norway and Siberian Spruces (p.102). Koyama's Spruce (below): less hairy shoots and thicker, stiffer, greyer leaves. Oriental Spruce (p.108): shorter, blunt leaves. Black Spruce (above): greyer leaves spread all round most shoots.

Koyama's Spruce *Picea koyamai*

Mount Yatsuga-dake in central Japan. 1914. A few hundred trees survive in the wild, with more in many European collections.
APPEARANCE Shape Densely conic, to 22 m. **Bark** Purplish grey-brown, flaking vertically. **Shoots** Orange, often with some hairs in the grooves. **Buds** Large and sharp (8 mm), much *whitened with resin*. **Leaves** Rather stiff and often abruptly sharp-tipped, to 18 mm but *diminishing steadily in length towards the shoot's tip* and, more abruptly, towards its base; they are parted beneath the shoot and point *regularly forwards at 45°* above it; square in section, and variably greyish on each face. Strong sweetish smell when crushed, like chrysanthemum leaves. **Cones** To 10 cm, with woody, finely toothed scales.
COMPARE Norway Spruce (p.102): a similar tree with brighter green, softer, shorter leaves and scarcely resinous buds. Dragon Spruce (p.112): stiffer, wide-spreading leaves and distinctively papery-scaling bark.

BLACK SPRUCE

leaf squared
cone persists for years
♂ flws
seed
♀ flws
dense blue-grey lvs
pink-brown hairy shoot
bark

SPRUCES

Wilson's Spruce — *Picea wilsonii*

(*P. watsoniana*) Mountains of N China. 1901. In a few collections.
Appearance Shape Broadly conic and open, with level branches and (frequently) *sprouts* on the bole; to 20 m. **Bark** Grey-pink, cracked into fine, sometimes papery flakes. **Shoots** *Whitish*, deeply grooved, hairless. **Leaves** Square (but easily rolled), *very slender*, shining green, to 16 mm; none spread beneath most shoots. **Cones** To 8 cm, with woody scales.
Compare Siberian Spruce (p.102): darker shoots, sometimes hairy; broader leaves.

Sitka Spruce — *Picea sitchensis*

Alaska to N California, seldom far from the sea. 1831. The commonest forestry conifer in wet areas of Britain; locally frequent in parks and belts but almost absent from dry lowlands. With its massive leader, indifferent to exposure and salt spray, but stunted and yellowish when rainfall drops below 800 mm a year.
Appearance Shape Openly conic and broad, especially in exposure. Sparse, dull blue-grey foliage: even vigorous trees can look half moribund. Huge buttressed bole when happy; to 60 m so far in Scotland. **Bark** Very purple-grey; soon with harsh, scaly plates. **Shoots** White-brown, hairless. **Leaves** Much *flattened*, with 2 bright blue-white bands beneath and narrower lines above; straight and ferociously spiny; to 30 mm. Very few spread beneath the shoot; *side-leaves are held perpendicular to it*; upper ones, *in marked contrast*, lie well forwards and are pressed down above it. **Cones** Short, to 10 cm; thin stiff papery scales, their margins crinkled and toothed.
Compare Hondo Spruce (p.110): the most similar flat-leaved Spruce. Others are Sargent's and Lijiang Spruces (p.110), Sikkim Spruce (p.112), Brewer Spruce (p.100), and Serbian Spruce (below). White Spruce (p.114) has a rather similar bark and foliage colour, but is a small tree with squared, upcurving leaves.

Serbian Spruce — *Picea omorika*

Small, vulnerable population in the upper Drina Valley, Serbia/Bosnia-Herzegovina. Introduced only in 1889 but now widely planted in parks and gardens even in dry areas (where it copes well) for its generally superb shape; rarely in forestry plantations. A short-lived tree.
Appearance Shape Typically a very *narrow solid pagoda* to 30 m, with shoots hanging from long-pendulous then upcurving branches, the lowest of which may layer. (The occasional broad, open specimen may not be recognized.) Rich or bluish green, the rising branch-tips brightly etched with silver from under the leaves – the sprucest of spruces. **Bark** Red-brown, developing big round scales. The foliage of the best trees needs to be parted to see the trunk. **Shoots** Pale brown, *very hairy*. **Leaves** To 22 mm; *broad and flattened* with (older trees) *suddenly bevelled, bluntish tips*; 2 white bands underneath. They spread widely on either side of the shoot and are often pressed down above it. **Cones** Carried even on small trees: about 6 cm, with finely toothed scales and spots of resin.
Compare Lijiang Spruce (p.110): shorter, slenderer, less-flattened leaves. Hondo Spruce (p.110): hairless shoots and slenderer, sharp leaves.

WILSON'S SPRUCE

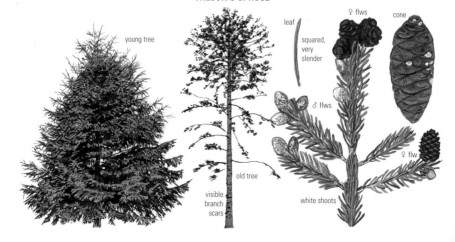

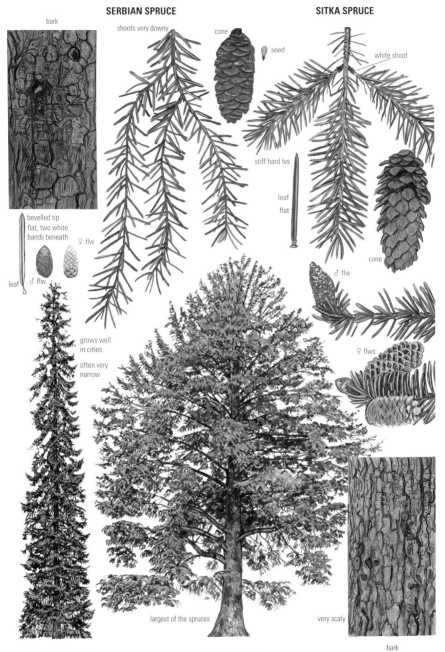

SPRUCES

Oriental Spruce — *Picea orientalis*

(Caucasian Spruce) Caucasus Mountains and NE Anatolia. 1839. Very occasional everywhere, coping relatively well with dry conditions.
APPEARANCE Shape Neatly conic; with age, often densely columnar, the top still bristling with many fine new shoots. A dark, matt green, healthy tree, to 40 m. **Bark** Pinkish brown; very finely shredding, then developing rather regular round plates. **Shoots** Pale brown; densely *hairy* on the ridges. **Leaves** *The shortest of any spruce* (6–12 mm); square in section, *neatly round-ended* on older trees; green on each face. Very few spread below most shoots; those lying above them tend to be pressed down. **Cones** Small (7 cm); rather curved, and often spotted with resin.
COMPARE Norway Spruce (p.102): similar jizz but much longer, sharp leaves. Red Spruce (p.104): longer, pointed leaves and a yellowish crown. Lijiang Spruce (p.110): somewhat flattened leaves, greyer beneath. Serbian Spruce (p.106; also with bevelled needles and doing well in towns): flattened leaves with bold white bands beneath.
VARIANTS 'Aurea' has much brighter gold young growths than the type's, set off by the deep green second-year foliage. Rare (cf. Norway Spruce 'Aurea', p.102).
'Gracilis' grows slowly into a broad, extremely *dense* little tree; rare.
OTHER TREES Maximowicz's Spruce, *P. maximowiczii* (central Japan, 1865; in a few collections) has scarcely hairy, brighter orange shoots and rather longer (10–13 mm) leaves (cf. Norway Spruce; its bark often remains a richer orange-red).

Alcock's Spruce — *Picea alcoquiana*

(*P. bicolor*) Mountains of central Honshu, Japan. 1861. Rather rare.
APPEARANCE Shape Very *sturdily* broad-conic, to 25 m: *long gently rising branches* and gaps between their dense somewhat spiky systems of dark, rather grey-green foliage. **Bark** Brown-grey, with rather squared plates (cf. Hondo Spruce, p.110) – less scaly than most spruces'. **Shoots** Stout, whitish orange and usually hairless. **Leaves** Square in section (rarely flattened-rhombic) but with broad *white bands on the 2 lower sides* and blue-green above; to 18 mm long; parted beneath the shoot and usually curving densely above it. **Cones** To 12 cm, the scale-margins papery and often sharply downcurved (cf. Sargent's Spruce, p.110).
COMPARE Hondo Spruce (p.110): a similar tree but with flat leaves (strong bias when rolled between fingers), a neater shape, and smaller cones. Lijiang Spruce (p.110) and Black Spruce (p.104): dull, dark, hairy shoots. Dragon Spruce (p.112; similar crown): papery-scaling bark, and stiffer leaves almost equally grey on each of their 4 faces. White Spruce (p.114): also with upcurving leaves paler below than above, but with short (12 mm) needles grey on each face.
OTHER TREES Sakhalin Spruce, *P. glehnii* (Sakhalin and N Japan, 1877), is in a few collections (to 24 m). Its similarly bicoloured (but bluer) leaves are shorter (10–15 mm); its shoots *brighter orange* and usually densely hairy; very flaky bark, often a rich purplish brown; cones only about 6 cm. Neat, dense spire shape and small cones recall Black Spruce (p.104), but its shoots are usually brighter and redder.

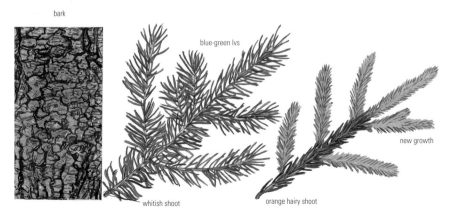

ALCOCK'S SPRUCE — bark; blue-green lvs; whitish shoot

SAKHALIN SPRUCE — new growth; orange hairy shoot

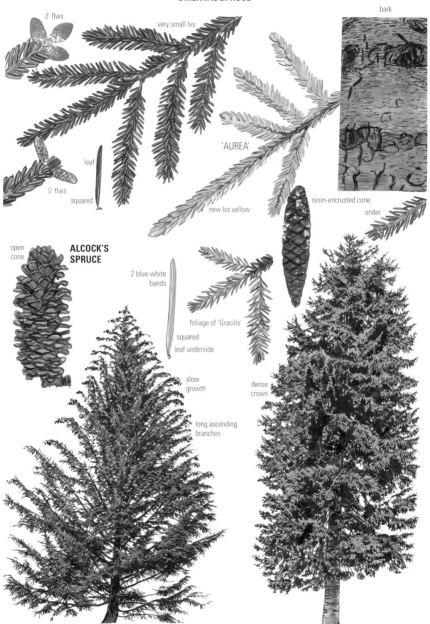

SPRUCES (EAST ASIATIC)

Lijiang Spruce *Picea likiangensis*

SW China, SE Tibet. 1910. Rare, but in some smaller gardens for its dramatic 'flowering' display, from youth.
Appearance Shape Rather openly conic; to 25 m. **Bark** Pale grey; shallowly scaly. **Shoots** Pale grey-brown, usually finely hairy. **Leaves** *Flattened-rhombic* in section (bias when rolled between fingers): bluish green above (bright silver-grey in some Yunnanese trees), with 2 white/grey bands beneath; short (16 mm), and with neatly bevelled tips. Few spread beneath most shoots; those above are pressed *densely forwards and tend to fan outwards*. **Male flowers** Profuse: crimson then yellow with pollen. **Cones** Rich red-purple as they ripen, then to 12 cm, with wavy-edged, firm but papery scales.
Compare Serbian Spruce (p.106): longer, broader leaves. Alcock's Spruce (p.108): squarer, upswept leaves. Black Spruce (p.104): shorter leaves all round its shoots. Sargent's Spruce (below): more alike, but white bands often coalesce under its leaves.
Other trees Purple-coned Spruce, *P. purpurea* (*P. likiangensis* var. *purpurea*; mountains of SW China, 1910), is in some collections: more *orange* bark; *bright green* denser crown often with *upcurled* competing tips; shorter more appressed leaves (6–15 mm), more dully *grey-banded beneath*; and smaller cones (to 5 cm), bluer purple in summer.

Hondo Spruce *Picea jezoensis* ssp. *hondoensis*

Mountains of central Honshu, Japan. 1879. Rather rare: large gardens. (Often labelled as the type, which is widely distributed in E Asia but a sickly tree in a few UK collections: leaves longer, spinier, with duller bands beneath; buds blunter.)
Appearance Shape Broad-conic, to 30 m, on rather level branches. **Bark** Grey, usually rectangularly plated and *less scaly than other spruces'*. **Shoots** Stout, hairless; white/amber. **Leaves** Rich green, *flattened*, with broad white bands beneath; parted below the shoot and *swept above it*. Stiff, sharp; 15 mm. **Cones** 5 cm; the thin, stiff scales have 2 or 3 fine teeth.
Compare Sitka Spruce (p.106): longer, stiffer leaves, the lowest rank *perpendicular* to the shoot. Alcock's Spruce (p.108): squarish leaves. Sargent's Spruce (below): leaves often solidly white beneath and more pressed down above its slender shoots. Other spruces with flattened leaves have duller, darker, hairy shoots.

Sargent's Spruce *Picea brachytyla*

W China, Upper Burma. 1901. Rather rare.
Appearance Shape Openly conic then columnar; rather weeping. A pallid, tousled tree, its silver under-leaves showing in patches. **Bark** Brown-grey; shallow, round scales. **Shoots** Whitish, hairless, *very slender*; scarcely grooved. **Leaves** *Flattened*, narrowly pointed, light green above and with 2 *frequently coalesced brilliant white bands* beneath; to 20 mm. None spread beneath most shoots, but the upper ones are *pressed down* so that some of their tips dip below shoot-level. **Cones** Long-tapered, slightly curved, the woody scale-edges often bent outwards (cf. Alcock's Spruce, p.108).
Compare Sikkim Spruce (p.112): leaves spread all round most shoots. Hondo Spruce (above) has well-separated white bands under rising leaves.

SARGENT'S SPRUCE

SPRUCES (EAST ASIATIC)

Sikkim Spruce *Picea spinulosa*

(East Himalayan Spruce; *P. morindoides*) Sikkim, Bhutan and NE India. c.1878. Rare.
Appearance Shape Very openly columnar to 28 m, with rather weeping foliage. Younger trees (more neatly conic) can be gorgeously silvery. **Bark** Pale grey; scaly, circular plates, or sometimes rough (and even oak-like) ridges. **Shoots** Whitish; less slim than Sargent's Spruce's (p.110). **Buds** Larger (7 mm). **Leaves** *Flattened, slender* and narrowly pointed, to 25 mm, *all around even weak shoots* though with more above them; all point sharply forwards. Grey-green above; *broad white bands beneath* rarely coalesce. **Cones** 7–11 cm; scales thin but woody, and toothed.
Compare Sargent's Spruce (p.110): the same brilliant white undersides to its (broader) leaves which, as in other flat-leaved spruces except for the dramatically weeping, blackish Brewer Spruce, are fully parted beneath the side-shoots. Other very blue spruces (Blue Colorado, Blue Engelmann and White, p.114) have squared leaves and a much denser, spikier habit.

Tiger-tail Spruce *Picea torano*

(*P. polita*) Japan; a scattered tree in the wild. 1861. Rather rare – mostly in collections.
Appearance Shape Untidily and spikily conic with much dead wood; to 28 m but often stunted in dry conditions. Most trees are a distinctively dingy *dark yellow-green* at a distance. **Bark** Brownish grey-purple; large, rough, irregular scales. **Shoots** Stout, whitish brown and hairless, with particularly prominent 'pegs'. **Buds** Shiny chestnut and to 12 mm long. **Leaves** Well spaced all round most shoots at a wide angle (but with more above them), so that the spray appears tiger-striped with sky; but the Japanese name – *tora-no-o* – could just as well allude to the hazards consequent upon grabbing hold of them: they are viciously spined (*painful to grasp lightly*), *thick* (squared), curved and rigid; about 20 mm long. **Cones** Squat (to 12×5 cm), the thinly woody scales finely toothed and wavy.
Compare Colorado Spruce (p.114): a dark green/grey tree with flakier bark and upcurving, less rigid leaves. Dragon Spruce (below): papery bark and greyish leaves less curved and less spiny.

Dragon Spruce *Picea asperata*

W China. 1910. Rare: collections.
Appearance Shape Rather gauntly broad-conic even when growing well, with *gaps* between the dense foliage of one ascending branch and the next; to 20 m. **Bark** *Mauve-grey* (sometimes browner), and usually distinctively *shaggy* with long, flimsily curling scales. **Shoots** Stout and pale; neatly grooved into plates, the 'pegs' usually hairy; **Buds** Big and pointed (10 mm). **Leaves** 10–18 mm, square and rigid, blue-green with fine grey lines on each side; some spread below most shoots but more stand above them – often closely upcurving. **Cones** To 12 cm; thinly woody scales.
Compare Blue Colorado Spruce (p.114): a narrower tree with fine-scaled bark and slightly longer upcurving leaves. Alcock's Spruce (p.108): similarly broad, gappy crown but non-papery bark; its upcurving leaves are only grey-banded on their lower faces. Tiger-tail Spruce (above): very stiff leaves. White Spruce (p.114): softer leaves.

TIGER-TAIL SPRUCE

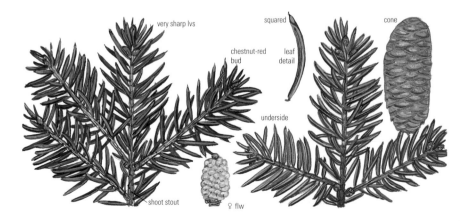

SPRUCES (EAST ASIATIC) 113

SPRUCES (AMERICAN)

Blue Colorado Spruce
Picea pungens f. *glauca*

USA: part of a wild population with small scattered stands along the Rocky Mountains. Abundant in small gardens, dwarfed from selection or from the dry, polluted air (to 28 m in collections). The type (1862; leaves dark green/greyish) is confined to big gardens.
APPEARANCE Shape Columnar-conic; dense and narrow (gaunt when struggling); very spiky. **Bark** Dark purplish: coarse scales or *scaly ridges*. **Shoots** Pale brown, shiny and hairless. **Leaves** *Bright silver-blue* from whitish bands on all 4 sides; spreading all around most shoots but with more above them, and mostly *curving* forwards and up. To 20 mm; stiff and sharp ('pungent'). **Cones** To 12 cm; wavy, papery scales.
COMPARE Blue Engelmann Spruce (below). White Spruce (below): much shorter leaves (13 mm). Black Spruce (p.104): much shorter leaves; hairy shoots. Dragon Spruce (p.112): shorter leaves, a broader crown and peeling bark. Alcock's Spruce (p.108): a darker, broad tree with white bands only beneath its leaves. Sitka Spruce (p.106): white bands under much-flattened leaves; grows much bigger. Blue Atlas Cedar (p.92): the only other common conifer with leaves as intensely silver as some clones; carries most needles in rosettes.
VARIANTS Silver selections include 'Moerheimii' (1912: paler shoots; leaves to 30 mm), 'Hoopsii' (*c*.1958, brightest of all; very slow), and 'Koster' (the descendants of 10 trees selected in 1908).

Blue Engelmann Spruce
Picea engelmannii f. *glauca*

W North America. 1809? Rare. (The type, with greener foliage, is in a few collections.)
APPEARANCE Easily mistaken for Blue Colorado Spruce (above). **Shape** *Neater habit*: dense, slightly weeping; to 30 m. **Bark** *Shallowly* scaly, grey-pink (sometimes orange and finely *flaking*). **Shoots** Pinkish; *hairy* on/near the 'pegs'. **Leaves** *Softer*; more fully parted beneath most shoots (much longer, at 20 mm, than White Spruce's).
OTHER TREES *P*. × *hurstii* is a presumed hybrid with Colorado Spruce in a few collections. Pale reddish hairless shoot; sparser, soft, upcurving leaves. The bark has fine red-purple flakes.

White Spruce
Picea glauca

(*P. alba*) N North America, from the tundra edge S to Wyoming and New York; many shelterbelts in Denmark. Rather rare in the UK: bigger gardens.
APPEARANCE Shape Rather gauntly conic, to 28 m; pale dull grey. **Bark** Purplish grey, growing big round plates. **Shoots** Shiny pinkish white, usually hairless. **Leaves** *Short* (10–16 mm), densely set and *all tending to curve up*; square, with white lines on each surface. **Cones** Small (to 6 cm); smoothly rounded papery scales.
COMPARE Black Spruce (p.104): hairy reddish shoots. Dragon Spruce (p.112) and Blue Colorado Spruce (above): much spinier leaves.
OTHER TREES *P*. × *lutzii* is an intermediate wild hybrid with Sitka Spruce (p.106), in some collections since 1962 (with plantations in Iceland).

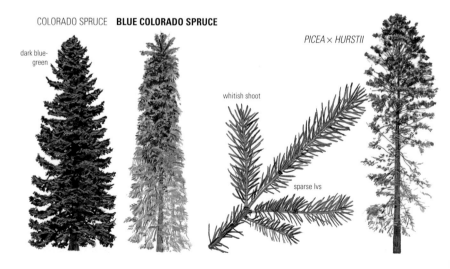

COLORADO SPRUCE — dark blue-green
BLUE COLORADO SPRUCE
whitish shoot
PICEA × *HURSTII*
sparse lvs

HEMLOCKS

Hemlocks (10 species) suggest polished, non-scaly spruces (with smaller cones). Their flat, often silvery leaves have thin stalks that run parallel with the shoot. (Family: Pinaceae.)

Things to Look for: Hemlocks

- Shape?
- Shoots: Are they hairy? What colour?
- Leaves: Notched? Minutely serrated? How broad? Any white/grey bands? Jumbled or ranked? Upside down along the shoot-top?

Western Hemlock   *Tsuga heterophylla*

NW North America, where it makes much the largest species. 1851. Locally abundant away from dry, polluted areas; rarely naturalizing. A frequent forestry conifer: the fine, pale lumber is also ideal for pulpwood. A tree of outstanding beauty when it thrives, though plantations are singularly dark and sinister. **Appearance Shape** Typically straight trunk (rarely forking), fluted with age; narrow pagoda shape, to 50 m in shelter. *Hanging leader* (cf. Deodar, p.92) allows the tree to shoulder its way up through others, while the dense *downswept fans of vivid-green foliage* tolerate deep shade. **Bark** Dark brown, developing cedar-like but more rugged ridges. **Shoots** Coffee-brown; long, fluffy hairs. **Leaves** 6–22 mm, the shortest ones on top (where a few are held randomly upside-down); broadest (2 mm) *midway up*; 2 bright white bands beneath. Aroma of the unrelated poisonous herb Hemlock (sour, parsley-like). **Cones** Drooping from shoot-tips, to 25 mm.
Compare Eastern Hemlock (below). Yunnan Hemlock (p.118): the most similar of the smaller, rare species – paler leaves, broader white bands beneath, and no leaves spreading above most shoots. Chinese Hemlock (p.118): paler leaves *scarcely white* beneath. Carolina Hemlock: slender leaves and a darker, *shining* shoot, hairy only in its grooves. Himalayan Hemlock: longer rigid leaves, spreading above most shoots. The Japanese hemlocks each have stubby, *notched* leaves. Mountain Hemlock: leaves *greyish all round*. Douglas Fir 'Fretsii' (p.120): leaves of more even length carried on green/purple shoots.

Eastern Hemlock   *Tsuga canadensis*

(Canadian Hemlock) E North America: Ontario to Alabama. 1736. Now occasional as an older park tree. **Appearance Shape** Broad, to 30 m, seldom making a good spire; leader droops only slightly; trunk *sinuous or much forked*. **Bark** Greyer and becoming more shaggily fissured than Western Hemlock's. **Leaves** Like Western Hemlock's, but the shoot-hairs are shorter and the leaves (lemon-scented) are often *broadest near the base* then taper evenly, while a row of very short leaves lies *upside-down on top of each shoot*, their silver undersides conspicuous. **Cones** Small: 18 mm.
Compare Western Hemlock (above). The rare species (p.118; see Carolina Hemlock) are also often bushy, but lack the regular line of inverted leaves.
Variants 'Pendula' grows a splendid dense dome of weeping foliage; rather rare. Small forms in collections include 'Fremdii' (blackish, glossy, crowded 9 mm leaves), 'Taxifolia', (long leaves crowded at shoot-tips), and the yellow, compact 'Aurea'. 'Microphylla' has very small leaves (6 mm) and 'Macrophylla' largish, crowded ones.

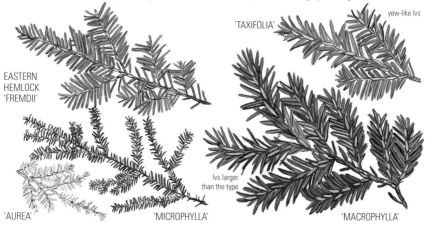

HEMLOCKS; MOUNTAIN HEMLOCK

Carolina Hemlock *Tsuga caroliniana*

Appalachians, USA. 1886. In some collections.
Appearance Shape Often bushy; to 13 m. **Shoots** Rufous; long-hairy in grooves. **Leaves** Often sparse and held at random, typically *wide* angles; *slender* and *parallel-sided to a round tip*; brightly white-banded beneath. Lemon-scented. Unlike the common American species', they are *not minutely serrated* – no drag against a fingernail.
Compare Western Hemlock (p.116): broader leaves. Yunnan and Chinese Hemlocks (below).

Chinese Hemlock *Tsuga chinensis*

Central and W China. 1900. In some collections.
Appearance Shape To 20 m; often very densely bushy. **Leaves** Resemble Western Hemlock's (p.116), but yellower glossy green above with only *pale green bands beneath*; margins minutely serrated near the (very slightly notched) tip, or not at all. Angled well forwards at 30–40°.
Compare Yunnan Hemlock: white under-leaves.

Southern Japanese Hemlock
Tsuga sieboldii

1861. Rare: collections.
Appearance Shape Broad; often bushy; to 22 m. **Bark** Grey; square-cracked with age. **Shoots** *Shining buff, hairless*. **Leaves** Broad, to 20 mm; with *well-notched tips* and broad *dull* white bands beneath; no minute serrations.
Other trees The rarer Northern Japanese Hemlock, *T. diversifolia* (higher altitudes; 1861), has a paler, more orange bark, markedly *orange* shoots (very finely *hairy* at first), and more neatly parallel leaves which are *whiter-banded* beneath (cf. Korean Fir, p.86).

Himalayan Hemlock *Tsuga dumosa*

Himalayas to N Vietnam. 1838. Rare: big gardens. Needs humidity and shelter to thrive for long.
Appearance Shape To 28 m but often very low, on sinuous/multiple stems; rather weeping. **Bark** Larch-like: coarse, flaky ridges. **Shoots** Whitish brown, with *scattered* hairs. **Leaves** *Rigid*, long and dark, to 30 mm; held in ranks above the shoots; brilliant broad white bands beneath.
Other trees Yunnan Hemlock, *T. yunnanensis* (W China, 1908; sometimes conflated) is in a few collections: its leaves (in rather flattened ranks) are more like Western Hemlock's (p.116) but longer, paler and more brightly white-banded beneath; their tips may be minutely notched.

Mountain Hemlock
Hesperopeuce mertensiana

(*Tsuga mertensiana*) NW North America, in high mountains. 1854. The jumbled needles recall the true hemlocks (or cedar's extension-shoots), while the cones are more spruce-like. Rather rare: vigorous (to 36 m) only in Highland Scotland, but grown elsewhere for its svelte habit and colour.
Appearance Shape Slender-conic, especially when not thriving, or sometimes bending/forking; *separate drooping plates of foliage*. **Bark** Ruggedly scaly; chocolate-brown – contrasting with dark *grey* leaves. **Shoots** Pale brown, shining, with some long hairs. **Leaves** *Irregularly held on either side of shoot*, to 20 mm; narrow, thick and *grey all round*. **Cones** To 8 cm; flimsy, downy scales.
Variants Jeffrey's Hemlock, var. *jeffreyi* (*Tsuga × jeffreyi*; in some collections, to 18 m), has smaller, sparser, flatter leaves, *yellowish green each side*.

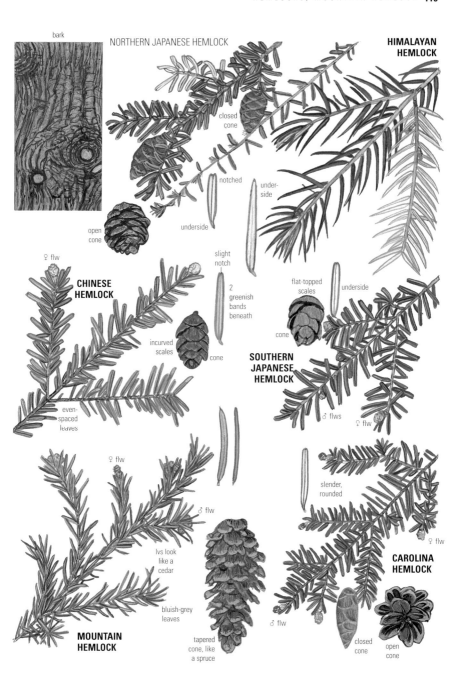

DOUGLAS FIRS

Douglas Firs (5 species) have foliage most like that of the Silver Firs' (but the 'suckers' are tiny and only evident when a leaf is pulled off). Their buds are long-conic. (Family: Pinaceae.)

Douglas Fir *Pseudotsuga menziesii*

(*P. taxifolia*, *P. douglasii*) W North America. 1827. The coastal type was once the world's tallest conifer, with trees of 120 m – all now felled for their timber. Abundant away from dry, polluted parts in parks and plantations; rarely naturalizing; Britain's tallest trees now 62 m in Welsh and Scottish mountains.
APPEARANCE Shape Spire-like in shelter, on a flag-staff bole. The wispy leader is easily blown out, so old trees often become broad, with heavy, drooping masses of foliage; a few are very weeping. Blackish green; occasionally steel-grey. **Bark** Grey and smooth in youth, then with wide, orange fissures; finally *black- or grey-purple*, and *massively craggy* (cf. Low's Fir, p.76). **Shoots** Slender, grey-brown, finely hairy. **Buds** Very *slender*, pale brown: almost beech-like. **Leaves** *Soft, flexible and slender* (unlike any silver fir's), to 3 cm, with a *narrowly rounded* tip and narrow, white-green bands beneath; they spread all round most shoots, though with few below. Strong, hot, sweet, fruity aroma fills the air. **Cones** Dropping when ripe; 6 cm. *3-pronged snakes'-tongue bracts*, 15 mm long, point *towards the tip*, but soon break off on the ground.
VARIANTS Blue Douglas Fir, var. *glauca* (E Rocky Mountains, 1876) is healthier than the type (and as frequent as it) in dry lowland areas: a slender, lightly branched, blackish/grey tree, to 30 m only. Bark dark *fawn-grey; scalier* and only shallowly fissured; leaves thicker, less aromatic; often greyish above, with greyer bands beneath; cones small, the bracts *bent to the horizontal* – the most reliable distinction, odd type trees being very blue.
'Fretsii' is a semi-dwarf with very *short, broad leaves* (10×3 mm), brightly banded beneath (suggesting a hemlock's, but of even length); 'Brevifolia' has *short* but slenderer leaves, some *curving backwards*. Both are in a few collections.
'Stairii', very rare, has *pale yellow* younger foliage.
OTHER TREES Large-coned Douglas Fir, *P. macrocarpa* (SW California, NW Mexico; 1910) is in a few collections. Buds redder; leaves longer (30–50 mm), incurved, *long-pointed* (cf. Morinda Spruce, p.100) and less aromatic. The large (to 18 cm) cones are not seen in Europe.
Japanese Douglas Fir, *P. japonica* (a rare tree from SE Japan; 1910) has made a gaunt plant to 15 m in a few collections. Its grey bark soon grows *close, scaly ridges*. Shoots *hairless*; the leaves all round them, with *notched tips* and little fragrance.

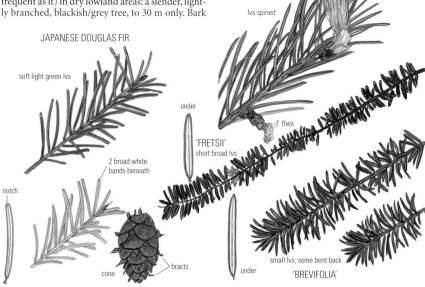

LARGE-CONED DOUGLAS FIR
♀ flw
lvs spined
♂ flws
'FRETSII'
short broad lvs
JAPANESE DOUGLAS FIR
soft light green lvs
under
2 broad white bands beneath
notch
cone
bracts
under
small lvs, some bent back
'BREVIFOLIA'

PINES

Of the pines grown in Britain, all except Singleleaf Nut Pine (p.144) have adult leaves split into 2–8 needles; the number of needles per bundle offers the easiest way to subdivide these trees. The needles are bound by a papery sheath, and juvenile leaves (saplings; sprouts) are undivided. Cones are usually woody, but more spruce-like in the 'soft pines'. (Family: Pinaceae.)

Things to Look for: Pines

- Shape? Bark?
- Buds (if 2-needled): Are the scale-tips free? Do they curl outwards?
- Shoots (if 5-needled): Colour? Bloom? Hairs?
- Leaves: In 2s, 3s, 5s, or combinations? How long? Twisted (if in 2s)? Drooping, or serrated, or white-banded (if in 5s)?
- Cones: How big? Spiny? With scales (if 5-needled) curving inwards or outwards?

Key Species

5-NEEDLE PINES: '**Soft pines**' have silky leaves, white-lined on their inner surfaces. Leaves of **Bhutan Pine** (p.138) are long enough (12–20 cm) to droop; **Weymouth Pine's** (p.138) are shorter and held stiffly; **Limber Pine** (p.146) has faint white lines on *each surface* of its unserrated leaves. **Foxtail Pines** (p.136): short (2–4 cm) stiff leaves. **Endlicher Pine** (p.132): long, stiff leaves.
3-NEEDLE PINES: Ponderosa Pine (p.146): long, stiff leaves. **Monterey Pine** (p.144): bright, slenderer/shorter leaves. **Lacebark Pine** (p.148): short, stiff leaves. **Jelecote Pine** (p.148): slender hanging leaves.
2-NEEDLE PINES: Scots Pine (below): short often grey leaves. **Lodgepole Pine** (p.126): short deep-green leaves. **Corsican Pine** (p.124): long leaves; buds with appressed scale-tips. **Maritime Pine** (p.132): long leaves; buds with bent-back scale-tips.

Scots Pine *Pinus sylvestris*

Europe (Spain to the Caucasus; Lapland to Siberia), in southern mountains and northern heaths. Scattered natural forests remain in the Scottish Highlands (var. *scotica*); abundantly planted throughout Britain for some 300 years. Common forestry tree; 'weed' on lowland heaths, spreading from old way-marking roundels; yellowish or moribund on chalk. The contrast of soft grey foliage and orange-pink bark makes it one of the prettiest pines.
APPEARANCE Shape Spire-like in plantations and in the Baltic var. *rigensis* (preferred for ships' masts; once much planted in Brittany); open-grown trees (especially var. *scotica*) soon picturesquely rounded; to 40 m. **Bark** Red-grey scales at first. The *papery orange-pink bark* intensifies with age in the top half of the tree, while the lower trunk grows big papery-surfaced mauve plates, or sometimes rugged purple ridges. **Shoots** Clear green-brown; hairless. **Buds** With some papery-white scales just free at their tips. **Leaves** In 2s, short (5–7 cm), *thicker and often more twisted* than other 2-needle pines' except Lodgepole Pine (p.126); pale *blue-grey* in var. *scotica*; blackish and very short in var. *engadinensis* (the Alps) and var. *lapponica* (N Scandinavia). **Cones** Slim; 5–8 cm.
COMPARE Japanese Red Pine (p.142): longer, brighter green, straight leaves. Red Pine (p.128) and Chinese Red Pine (p.142): sometimes with red upper bark, but leaves longer (10–15 cm), straight. Many pines have leaves as short but deeper green, and a dull grey bark (see Lodgepole Pine, p.126). Maritime Pine (p.132) and Stone Pine (p.130): often confused, but long leaves and bark are distinct.
VARIANTS All retain orange-pink flaking bark down to a graft-point:
'Aurea' has pale gold leaves in winter; fading a curious pale sickly colour as the foliage emerges grey. Rare.
'Fastigiata' is very slender, with erect branching – in silhouette like a miniature Lombardy Poplar. Rare.
'Watereri' (rather rare) has *very dense short needles* in a picturesque bonsai-like dome (eventually to 15 m), on twisted limbs.

SCOTS PINE 'AUREA'
yellow through winter

'FASTIGIATA'
vertical branches

BLACK PINES

Corsican Pine   *Pinus nigra* ssp. *laricio*

(*P. n.* var. *maritima*/var. *corsicana*) Corsica; Calabria; Sicily. 1759. Abundant: forestry plantations, shelterbelts, parks.
Appearance Shape Typically *a straight, clean stem*, often unforked to great heights (45 m); *light* level branches carry quite delicate plates of soft, grey-green foliage, never very dense. **Bark** *Grey*-mauve; big, finely scaly plates and wide fissures. **Shoots** Stout, pale yellow-brown. **Buds** Lacking free scale-tips. **Leaves** In 2s, 12–18 cm, rather twisted; slender, greyish. **Cones** Conic, to 8 cm; dull grey-brown.
Compare Other subspecies. Austrian Pine (below) is distinguishable by its jizz, but the intermediate features of the 2 rare forms in cultivation may perplex. Red Pine (p.128): the most similar 2-needle pine with long leaves and no free bud-scales. Chinese Red Pine (p.142): low, domed crown; Bishop Pine (p.128): very rugged bark. Maritime Pine (p.132): reddish, closely plated bark on a sinuous trunk; stiffer leaves; free bud-scales. Ponderosa Pine (p.146): longer leaves *in 3s*; stiffer habit.

Austrian Pine   *Pinus nigra* ssp. *nigra*

(Black Pine; *P. n.* var. *austriaca*) S Austria to central Italy and the Balkans. 1835. Very frequent: old parks and shelterbelts. A rough dark pine, but excelling in coastal exposure and on chalk.
Appearance Shape Trunk often long but *seldom quite vertical*; snaggy and generally *heavily limbed*; open-grown trees often broad and much forked from the base, pointed or tabular; to 43 m. *Dense, tufty foliage, hard, blackish green.* **Bark** *Darker* than Corsican Pine's; the plates often lift shaggily. **Buds** With papery scales, loose but gummed with resin. **Leaves** Shorter (8–14 cm) and thicker than Corsican Pine's. **Cones** As Corsican Pine.
Compare Other subspecies. Japanese Black Pine (p.142): very gaunt crown; leaves only to 11 cm; free bud-scales. Bosnian Pine (p.134): smoother grey bark; leaves to 9 cm; indigo young cones.

Crimean Pine   *Pinus nigra* ssp. *pallasiana*

The Crimea and (var. *caramanica*) Turkey, Cyprus, Greece and Macedonia. 1790. Very occasional in old parks and big gardens.
Appearance Shape A straight trunk often divides at 2–10 m into *close, vertical 'organ-pipe' stems* (as Corsican Pine will only after early damage). Often narrowly flat-topped, to 42 m. Foliage *slightly denser and darker* than Corsican Pine's, in rather level plates: *fuzzy effect*. **Bark** Like Corsican Pine's; often with more *yellows* in it. **Leaves** 12–16 cm, slightly grey. **Cones** Large (to 10 cm); sharply ridged, pale scales make them *knobbly*.

Pyrenean Pine   *Pinus nigra* ssp. *salzmannii*

(*P. n.* var. *cebennensis*; 'salzmannii' is sometimes used to cover all the western forms, including Corsican Pine). S France to central Spain. 1834. In some collections.
Appearance Shape Typically low, rough and broad, to 25 m; quite dense, blackish foliage on *descending branches*. **Bark** Purplish, the plates often shaggy. **Shoots** *Bright orange* (yellow/brown in the other ssp.). **Leaves** *Slender*, 150–170 × 1.1 mm, *not prickle-tipped*. **Cones** Purplish brown; rather *flat scales*.

PYRENEAN PINE — not sharp; 2s; slender; branches descend

CORSICAN PINE — light branches; open cone; bark; usually a good bole

BLACK PINES

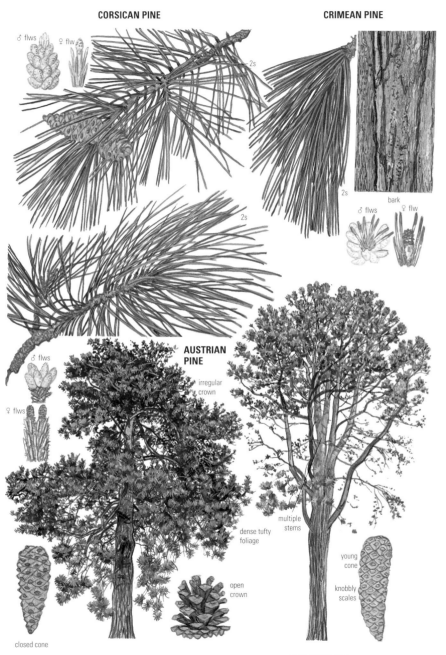

LODGEPOLE PINES

Lodgepole Pine
Pinus contorta ssp. *latifolia*

Rocky Mountains, from W Alaska to Colorado, where it colonizes in the wake of forest fires. 1854. Abundant in upland plantations; rare in gardens.
APPEARANCE **Shape** A neat, narrow, open spire to 28 m with small, rising branches, though the trunk often forks. **Bark** Red-brown, *finely scaly*; sometimes very shallowly square-cracked. **Shoots** Green-brown, hairless and glossy. **Buds** With smoothly appressed scales, encased in resin. **Leaves** In pairs, 6–10 cm, *broad* and twisted; bright dark green; the needle-sheath is about 5 mm long. **Cones** Small and slender (50 × 25 mm); a minute prickle wears off each scale.
COMPARE Other subspecies. Pines with short bright green or blackish-green leaves in 2s include Japanese Red and Japanese Black Pines (p.142), Jack Pine and Scrub Pine (p.128), Aleppo Pine (p.130), Mountain Pine and Bosnian Pine (p.134), and Alpine and Scandinavian forms of Scots Pine (p.122). Mountain Pine is most similar: greyer bark; rigid, straighter, blacker leaves retained for up to 10 years with sheaths 8 mm long; and cones with 'stretched' lower scales. Bosnian Pine has smoother grey bark and indigo immature cones. Japanese Black Pine develops a unique, gaunt habit and has golden-brown shoots. Aleppo Pine is an open, wispy tree soon with plated bark, while the upper barks of Japanese Red and Scots Pines are orange-pink. Jack and Scrub Pines have very short (3–5 cm) needles. Austrian Pine (p.124) has longer (8–14 cm) leaves.
OTHER TREES Another American pine (New York to Texas) with short (5–9 cm) leaves is Shortleaf Pine, *P. echinata* – in a few collections since 1739. Shoots *bloomed blue-white* then orange and *peeling* in their second year; needles usually paired but *sometimes in 3s* (cf. Yunnan Pine, p.142); its bark scales more coarsely and the wide branches carry *many sprouts* (cf. only Northern Pitch Pine, p.144).

Shore Pine
Pinus contorta ssp. *contorta*

Near the Pacific from S Alaska to N California. 1855. Rather occasional everywhere in gardens and belts; in many upland and a few southern plantations.
APPEARANCE **Shape** Dense, often bushy but sometimes to 32 m, and rounded with age; vivid, blackish green; shoots sometimes twisted. **Bark** Black-brown, soon knobbly with very *small square plates*. **Buds** Often *twisted* as they expand in spring. **Leaves** Shorter (4–6 cm), denser and pressed closer to the shoots than Lodgepole Pine's.
COMPARE Jack Pine (p.128): often forward-pointing cones. Scrub Pine (p.128): yellowish, more open crown. Mountain Pine (p.134): longer, blacker leaves and greyish, flakier bark.
OTHER TREES ssp. *bolanderi* (Mendocino plains, California), is a vigorous form in a few collections: slenderer leaves (*none* with a resin-canal – never a white spot appearing when they are snapped, as there will be for some leaves of the other forms).

Sierra Lodgepole Pine
Pinus contorta ssp. *murrayana*

(Murray Pine) Washington to N Mexico, W of the Rockies. 1853. A form grown in a few UK plantations (sometimes by accident: it tends to be very slow), and in the odd collection.
APPEARANCE **Shape** Densely conic. **Bark** Pinkish brown, very shallowly scaly. **Leaves** Short (5–8 cm), *broad and rigid*, dark yellowish green, and dropping in their second year. **Cones** Normally *dropping when mature*.

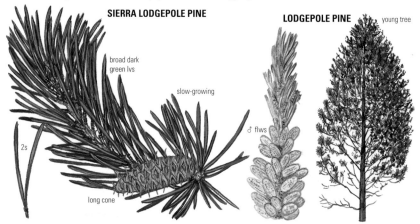

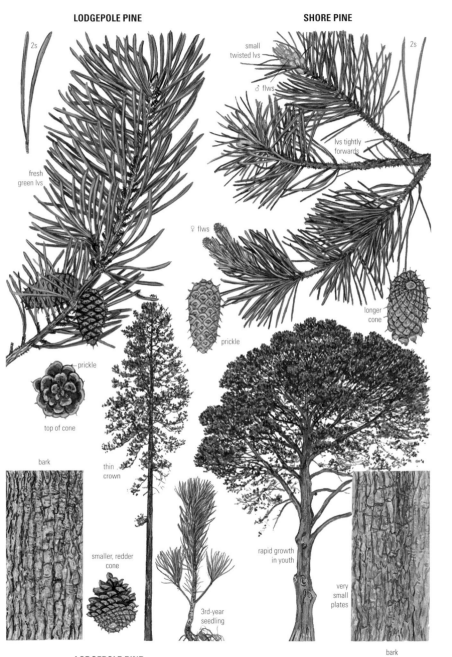

PINES (AMERICAN)

Bishop Pine *Pinus muricata*

(*P. remorata*) Confined to 7 Californian clifftops and islands. 1846. Occasional in large gardens, and in a few trial plantations – can grow 2.5 m in a year. Probably even tougher than Monterey Pine (p.144) in coastal exposure, though less tried.
APPEARANCE Shape Trees from the 2 northernmost stands (Big Lagoon and Noyo River, Mendocino; var. *borealis*) are bluish and densely spire-shaped to 30 m, then rounded and leaning; southern trees (var. *muricata*) quickly grow low, picturesque, dark grey-green domes on twisting branches from often very short boles, with *fuzzier foliage* than Monterey Pine's. **Bark** Blackish and often very ruggedly fissured – like Monterey Pine's. **Buds** Reddish, with much white resin; scales appressed. **Leaves** *In 2s*, 8–15 cm; *stiff* and curved. **Cones** Broad and persistent, like Monterey Pine's, but with a small *prickle* on each scale: they open only after forest fires, and can be engulfed *en masse* by the expanding branches.
COMPARE Monterey Pine: leaves in 3s (except for Mexican island forms). Japanese Black Pine (p.142): gaunt, not domed, crown. Maritime Pine (p.132): *recurved bud-scales* and similar if longer leaves. The usual bark is unlike any other 2-needle pine's. Coulter Pine (p.148): a little like a young var. *borealis*, but with needles in 3s.

Red Pine *Pinus resinosa*

(Norway Pine) NE North America, from Newfoundland and Manitoba to West Virginia. 1756. In some collections.
APPEARANCE Shape A thin, straggly spire, to 20 m; ultimately domed. **Bark** Purplish, with *fine orange-pink scales* (less confined to the upper trunk than in Scots Pine; cf. Korean Pine, p.134). **Shoots** Stout and *orange*. **Buds** Large, chestnut; scales appressed. **Leaves** Straight, dark and yellowish; in pairs, 10–15 cm; *snapping cleanly when bent in a hoop*. Strong lemony scent.
COMPARE Corsican Pine (p.124): a vigorous, sturdy tree with more flexible leaves and duller bark. Ponderosa Pine (p.146): needles in 3s. Chinese Red Pine (p.142): soon broad-crowned.

Jack Pine *Pinus banksiana*

(*P. divaricata*) Canada and NE USA, where it covers vast tracts. A rare plant in Europe.
APPEARANCE Shape Irregularly rounded, to 16 m, or bushy: a real scallywag of a tree, with a sparse, lumpy-spiky crown. **Bark** Grey-brown; rather shallow vertical fissures. **Shoots** Thin, green-purple/brown. **Leaves** Very short (3–4 cm), quite broad and twisted; rather yellowish. **Cones** Persistent, minutely prickled, with a slender curving tip; pointing *forwards* up the shoot (cf. only Calabrian Pine, p.130) or standing perpendicular to it.
COMPARE Shore Pine (p.126): square-cracked bark, dense crown and cones always pointing somewhat backwards. Mountain Pine (p.134): longer (6 cm) black-green leaves and backward-pointing cones.
OTHER TREES Scrub Pine *P. virginiana* (New York to Alabama, 1739) is rarer: *very broad* and often bushy, to 11 m, with dull, often *yellow-green* foliage. Shoots have a *startling violet bloom* (duller purple by third year); leaves very broad and often even shorter (35 mm); the cones, with a longer bristle on each scale, point backwards somewhat.

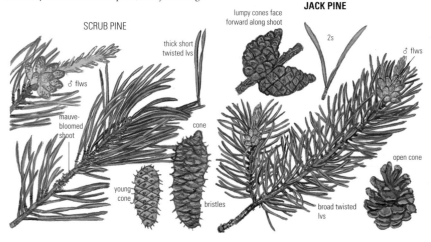

SCRUB PINE — thick short twisted lvs; ♂ flws; mauve-bloomed shoot; cone; young cone; bristles

JACK PINE — lumpy cones face forward along shoot; 2s; ♂ flws; open cone; broad twisted lvs

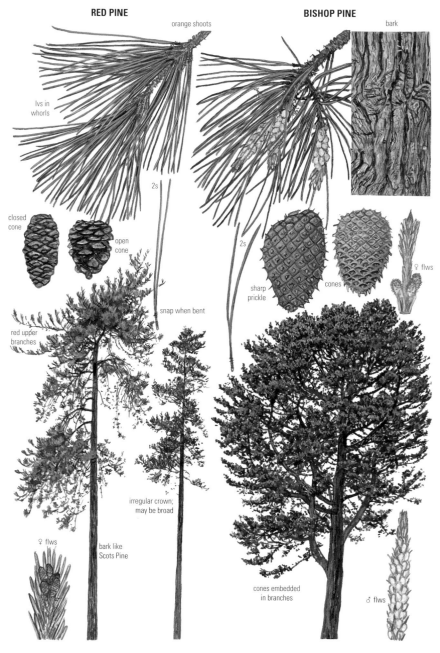

PINES (EUROPEAN)

Stone Pine *Pinus pinea*

(Umbrella Pine – cf. the unrelated Japanese Umbrella Pine, p.66). Mediterranean Europe; Black Sea coasts of Turkey; long grown in warmer parts of Britain. Very occasional, but in many small gardens, often raised from cones collected on holiday. Saplings, like those of Aleppo Pine, retain single, juvenile leaves for 4–6 years (cf. Singleleaf Nut Pine, p.144).
Appearance Shape *Many equal leaders* compete by 5 m; rather open, dull and twiggy in shade but in full sun growing a dense, wide *parasol of foliage* (to 20 m) on bare, much-forking limbs which radiate straightly from the often short trunk. **Bark** Deeply fissured from the first and then, at its best, developing *big, crisp, flat orange-purple plates*; in N Europe often darker, greyer and remaining closely fissured. **Shoots** Orange-green. **Buds** *Free, curling scale-tips*, matted together with silvery whiskers (cf. Maritime Pine, p.132). **Leaves** In rather distant pairs, to 16 cm; stiff, straight and dark grey-green, with an oniony scent. **Cones** 10 cm, *very broad, like a fist*, the large seeds a delicacy.
Compare Maritime Pine (p.132): taller growth on an often longer, sinuous trunk, longer paler leaves, and slender cones. Aleppo and Calabrian Pines (below): scalier bark; wispily open or spire-shaped crown; slenderer, brighter green leaves. Scots Pine (p.122): sometimes confused, but a very different tree. Old thriving Stone Pines are unmistakable; the thinner, sparser crowns of struggling examples in N Europe may go unrecognized, but the many joint leaders are characteristic.

Aleppo Pine *Pinus halepensis*

W Mediterranean Europe – with outlying populations E to Aleppo (Syria); very drought-tolerant. Long grown in the UK but very rare; rarely in small gardens from cones collected on holiday.
Appearance Shape An open, rough spire in youth, often forking; *spiky, wispy new growths* with short needles held closely to the shoots; then broad, to 20 m, but keeping a loose, jagged outline. **Bark** Vertically fissured and scaly; greyish orange-brown. **Shoots** Slender, remaining smooth; grey with a slight bloom. **Buds** Non-resinous; scale-tips free but often straight. **Leaves** In 2s (rarely some 3s), straight, slender, 6–11 cm; *bright pale green*; grassy aroma. **Cones** *Small*, slender (7–12 cm); never pointing forwards.
Compare Calabrian Pine (below); Stone Pine (above). Japanese Red Pine (p.142): denser crown and orange, papery bark. Jack Pine (p.128): more stunted habit and much shorter, twisted leaves.

Calabrian Pine *Pinus brutia*

(*P. halepensis* var. *brutia*) Greece E to the Lebanon; Turkey; the Crimea; long naturalized further W (including Calabria). 1836. In a few collections.
Appearance Shape With slightly darker, denser, but fluffier foliage than Aleppo Pine. **Bark** As Aleppo Pine. **Shoots** Stout, green-brown, *pimply after 2 years* from the leaf-scars. **Leaves** Longer than Aleppo Pine (15 cm). **Cones** Pointing *forwards* (cf. only Jack Pine, p.128), or standing perpendicular to the shoot.
Compare Maritime Pine (p.132): crisper bark, stout leaves, very rounded crown. Chinese Red Pine (p.142): stiffer, domed crown.

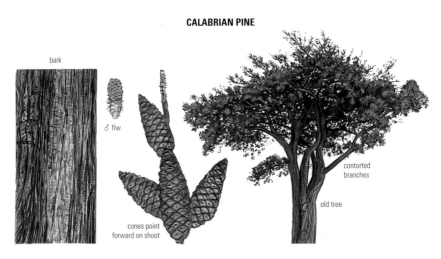

CALABRIAN PINE

bark

♂ flw

cones point forward on shoot

contorted branches

old tree

PINES (EUROPEAN) **131**

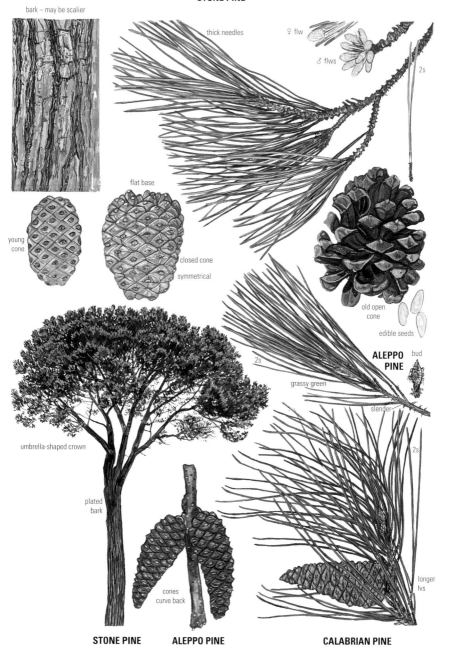

PINES

Maritime Pine — *Pinus pinaster*

(Cluster Pine; *P. maritima*) Mediterranean coasts from Portugal to Greece, and Morocco. The source of turpentine and rosin. Long grown in Britain; occasional but locally frequent in SW England (plantations behind some dunes); rarely naturalized.
APPEARANCE Shape Open, whorled and often leaning; with age, rounded on a typically long sinuous bole, to 30 m – like Scots Pine (p.122) but much more open; greyish. **Bark** Soon purplish and deeply fissured. Old trees may grow beautiful lizard-skin patterns of small *flat, orange-purple plates* and crisp, black fissures. **Shoots** Stout, pale green-brown. **Buds** Upper scale-tips bent outwards and matted with silvery whiskers (cf. Stone Pine, p.130). **Leaves** In sparse pairs, *long* (12–25 cm), *stout and stiff; pale grey-green*. **Cones** Persistent; *slender*; 10 cm, *shining brown*: often collected as ornaments.
COMPARE Stone Pine (p.130): bark scalier or with longer plates, crown with numerous leaders from youth, cones broad. Bishop Pine (p.128): darker, more rugged trunk, prickled cones, appressed bud-scales. Bent-back bud-scales also help distinguish from Corsican Pine (p.124), the commonest 2-needle pine with long, greyish leaves. Scots Pine (p.122) is sometimes confused. Other rare pines with longer needles and free bud-scales are Calabrian Pine (p.130), with scalier bark and forward-pointing cones; and Japanese Red, Japanese Black and Chinese Red Pines (p.142).

Endlicher Pine — *Pinus rudis*

(*P. montezumae* var. *rudis*) Central and N Mexico. Rare, but in large gardens in milder parts for its sometimes spectacular foliage; usually grown as *P. montezumae*.
APPEARANCE Shape A broad dome from youth, densely surfaced with often *huge, grey foliage tufts*; to 25 m. **Bark** Blackish, ruggedly plated. **Shoots** Stout; brown with some purplish bloom (cf. Jeffrey Pine, p.146). **Buds** About 15 mm. **Leaves** In 5s, usually 10–16 cm in the wild but *to 30 cm* on older trees in the UK; slender but spreading stiffly, with white lines *on each face*. Lemon-scented. **Cones** To 12 cm; a tiny prickle drops off each scale.
COMPARE Coulter Pine (p.148): leaves as long and blue when thriving, but in 3s. Apache Pine (p.146). Whitebark Pine (p.136): leaves also in 5s and white-lined on each side, but much shorter.
OTHER TREES The following also belong to this Mexican complex and are in some collections.
The true Montezuma Pine *P. montezumae* (S to Guatemala) has greener shoots and a more open crown of *hanging, bright-green leaves* (to 25 cm), in 5s (some 4s/6s), with a slight *onion smell*.
P. pseudostrobus (1839; S to Guatemala) has slenderer hanging leaves (cf. Jelecote Pine, p.148), 18–30 cm long and in 5s *on strongly grey-bloomed shoots*; its greyer bark is smooth at first.
Durango Pine, *P. durangensis* (1962) can be the most spectacular of all: *soft grey* drooping needles *to 40 cm long* (cf. Apache Pine, p.146), in 5s on strongly white-bloomed shoots.
Hartweg's Pine (*P. hartwegii*; S to Guatemala and El Salvador, 1839) is a gaunt, narrower, dark yellowish tree, to 23 m; leaves in *3s, 4s or 5s on the same shoot* (cf. Apache Pine, p.146) and only 9–15 cm, slender but stiff; *purple* cones ripen blackish.

MARITIME PINE

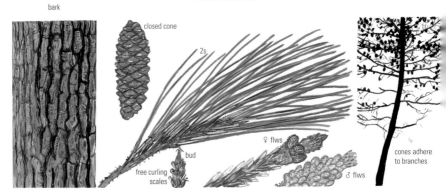

ENDLICHER PINE

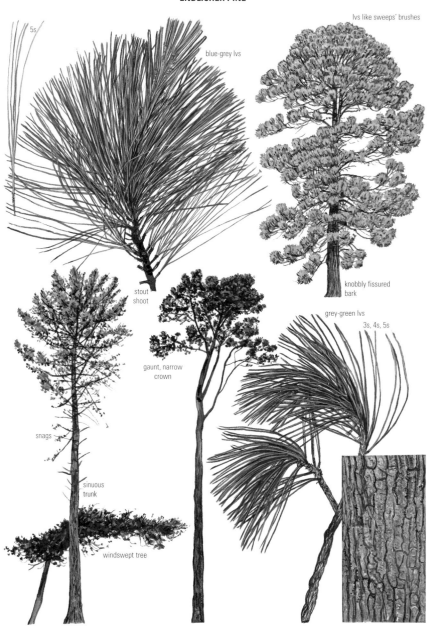

MARITIME PINE HARTWEG'S PINE

PINES

Arolla Pine — *Pinus cembra*

(Swiss Stone Pine) The Alps and Carpathian Mountains. 1746. Rather rare: bigger gardens.
Appearance Shape In youth, a dense spire/column on short, level branches; broad or leaning at maturity (to 30 m). **Bark** Dark grey-brown; soon with craggy, harshly flaking plates. **Shoots** *Brown with very dense hairs*. **Leaves** In 5s; densely held. *Short* (8 cm) and almost straight, their inner faces blue-white. **Cones** Squat, to 8 cm, the big triangular scales still closed when they fall. The nuts are eaten by terrestrial mammals (including humans), or pillaged by Nutcrackers.
Compare Korean Pine (below). Foxtail Pines (p.136): the only 5-needle pines with needles even shorter, denser and darker, but finely white-lined on *each side*. Macedonian Pine (p.138; also densely conic in youth): silkier leaves on hairless shoots. Sugar Pine (p.140) and Western White Pine (p.146): shoots with quite dense but *shorter* reddish hairs; longer, lax needles.

Korean Pine — *Pinus koraiensis*

NE Asia; Japan. 1861. In some collections.
Appearance Shape A spire, to 20 m: light, level, pale plates of fuzzy foliage from a straight, slender trunk (cf. Chinese White Pine, p.142). **Bark** *Pink and grey papery flakes* (duller than the upper bark of Scots Pine, p.122; cf. Red Pine, p.128). **Leaves** most like Arolla Pine's; the slightly longer, laxer spreading needles are less sharply pointed and if cut show *3 resin-drops, not 2*. **Cones** Similar to but larger than Arolla Pine's (12 cm).

Bosnian Pine — *Pinus heldreichii*

(Incorporating *P. leucodermis*) Balkans; Calabria. 1890. Big gardens, but now much planted.
Appearance Shape Often a *neat, vividly Lincoln-green* spire, to 25 m. **Bark** *Ash-grey*; slowly cracking in *neat, shallow squares*. **Shoots** *Grey-bloomed* then *pale* grey-brown. **Buds** Large (15–25 mm); chestnut, abruptly *long-pointed*, the papery scale-tips appressed. **Leaves** In 2s, 6–9 cm; stiff and densely forward-angled. **Cones** *Indigo*, finally rufous.
Compare Mountain Pine (below). Lodgepole Pine (p.126): yellower leaves, flaky bark, short-pointed buds and unbloomed shoots. Austrian Pine (p.124): longer leaves, rugged bark and dark rough shoots. Japanese Black Pine (p.142): gauntly spiky.

Mountain Pine — *Pinus mugo*

(*P. montana*; *P. mughus*) High mountains from Spain to the Balkans. 1779. Dwarf selections occasional in small gardens; in collections to 20 m, especially as the western ssp. *uncinata* (*P. uncinata*).
Appearance Shape Often multi-stemmed; open, straggling; ssp. *uncinata* can be sturdily conic. **Bark** Grey (pinkish/blackish), *finely scaling/square-cracked*. **Shoots** Shiny orange-brown. **Buds** Short-pointed; the scale-tips usually resin-caked. **Leaves** 4–6 cm, stiff, straight; dark grey-green; sheaths 8 mm. **Cones** Those of ssp. *uncinata* have generally hooked (but not spined) scales, the lowest ones *curiously 'stretched'*; those of the type have scales with a central blunt boss.
Compare Bosnian Pine (above; differences emphasized). Lodgepole Pine (p.126). Jack Pine (p.128): yellower, shorter leaves; cones pointing forwards.

MOUNTAIN PINE

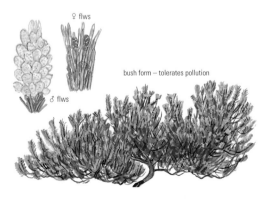

♀ flws
♂ flws
bush form – tolerates pollution

2s
hard lvs in whorls
scales drawn downward
cone (ssp. *uncinata*)

BRISTLECONE TO KNOBCONE PINES

Rocky Mountain Bristlecone Pine
Pinus aristata

W Colorado, N Arizona, and N New Mexico. 1863. In many big gardens: a talking point, though little else. Exceedingly long-lived in the wild on mountains which are too cold and dry for many pests or wood-rotting fungi to survive, though not proving so in the UK. It is thriving on the N coast of Iceland.
APPEARANCE Shape Thin, spiky, to 12 m; very slow. **Shoots** Rufous-hairy, little-branched. **Leaves** In 5s, *very short* (2–4 cm), incurved, and *densely covering* the shoots for up to 30 years; white bands on their inner surfaces; margins not serrated. Each needle's single resin duct is peculiarly near the surface, and when these break (usually by their second year) the foliage becomes *spotted with white resin, like dandruff*. **Cones** With 5 mm level *spines*.
COMPARE Arolla Pine (p.134).

Bristlecone Pine
Pinus longaeva

White Mountains of E California, S Nevada and central Utah. One of the world's longest-lived trees, to 4900 years (perhaps exceeded by Huon Pines, p.26, and possibly by the Yew at Fortingall in Tayside): what turned out to be the oldest known one was chainsawed by a pioneering researcher so that its annual rings could be counted. Making slow progress (since 1972) in a few big gardens, but likely to prove of sentimental value only.
APPEARANCE Differences from Rocky Mountain Bristlecone Pine (above) as follows: **Leaves** *Largely unspotted* (cf. Foxtail Pine). **Cones** With more slender spines *which soon break off*.

Foxtail Pine
Pinus balfouriana

California, with 2 widely separated populations in the northern Coastal Ranges and the central Sierra Nevada. 1852. An unprepossessing plant in a few collections.

APPEARANCE Differences from Rocky Mountain Bristlecone Pine (above) as follows: **Shape** Growth stronger; to 20 m. **Leaves** *Unspotted*, sharper, with a sweeter (marmalade) smell. **Shoots** Less downy. **Cones** Longer (9–13 cm); scales carry only *small, stout prickles*.

Whitebark Pine
Pinus albicaulis

Mountains of W North America. 1852; reintroduced 1900. In a few collections.
APPEARANCE Shape Bushy; slow. **Bark** White-grey on old wild trees. **Shoots** Yellow-brown, *hairless*. **Leaves** In 5s, 3–6 cm, very dense, shiny dark green with fine white lines *on each surface*. Unlike the commoner 5-needle pines' leaves (but like Foxtail and Bristlecone Pines', above, and Limber Pine's, p.146), these are *not serrated*: there is no drag when they are pulled between finger and thumb. **Cones** 4–8 cm, with soft scales still closed when they drop (cf. Arolla Pine, p.134).

Knobcone Pine
Pinus attenuata

(*P. tuberculata*) SW Oregon; California; NW Mexico. 1847. In a few collections.
APPEARANCE Shape An *open, long-branched tree*, to 24 m, usually decked with persistent whorls of long cones. **Bark** Pinkish, flaky and shallowly fissured. **Leaves** *In 3s*, slender, 10–18 cm, but quite stiff; *pale green*. **Cones** *Slender*, to 18 cm: sharply pointed scales like the back of an armoured dinosaur.
COMPARE Northern Pitch Pine (p.144): shorter leaves, denser crown, small cones. Monterey Pine (p.144): also with persistent cones and leaves in 3s, but tall dense crown and very rugged bark.

BRISTLECONE PINE

ancient wild tree

young tree

ROCKY MOUNTAIN BRISTLECONE PINE

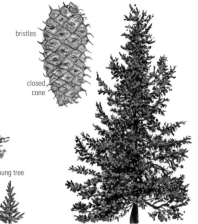

bristles

closed cone

138 SOFT PINES

Bhutan Pine *Pinus wallichiana*

(Blue Pine; *P. excelsa*, *P. griffithii*) Himalayas. 1823. Quite frequent (even in small gardens). Rarely naturalizing.
APPEARANCE Shape *Openly* conic, then broad and broken/leaning, to 32 m. **Bark** Purple- or orange-grey; *scaly ridges*. **Shoots** Grey-green, hairless; some *mauve-grey bloom*. **Leaves** Silky, grey-green. In 5s, with white-lined inner surfaces; very slender, 10–20 cm – *long enough to droop so most tips lie below shoot-level*. **Cones** Long (10–30 cm); a few small basal scales usually bent back, the rest *straight*.
COMPARE Weymouth Pine (below). More similar are Mexican White Pine (p.140): unbloomed, minutely hairy shoots; and the hybrid Holford's Pine (p.140). Chinese White Pine (p.142): more open, with squat cones. Montezuma Pine (p.132): leaves the same colour on each face. Jelecote Pine (p.148): 3-needled; also with silky, hanging leaves.

Weymouth Pine *Pinus strobus*

E North America: once the region's tallest tree (80 m); but much felled for its fine timber and reduced by Blister-rust (see p.140). Introduced by Captain George Weymouth in 1705: occasional (in some small gardens); rarely naturalizing.
APPEARANCE Shape A spire, soon broad/broken; to 42 m. Crisp *plates of fuzzy foliage*. **Bark** Smooth dark grey, then with rugged but shallow *black* ridges, or redder (cf. Scots Pine's lower bark, p.122). **Shoots** Slender, green; a transient *tuft* of down behind each leaf-bundle. **Leaves** In 5s, with white-lined inner surfaces; 8–12 cm. Slender but short enough to stand *almost straight around the shoot* (forward-angled). **Cones** Seldom to 20 cm, *slender*; a few *small basal scales often curve outwards*.
COMPARE Macedonian Pine (below). The other short-leaved 'soft pines' – Sugar (p.140), Western White (p.146), Arolla and Korean (p.134) – have shoots with dense reddish hairs; Sugar Pine is closest. Japanese White Pine (p.142): *twisted* leaves. Chinese White Pine (p.142): broad cones and sparse needles. Limber and Western White Pines (p.146): leaves the same colour each side.
VARIANTS 'Contorta' (dense, curved leaves on curiously *wiggling* twigs), 'Fastigiata' (*steep* limbs make a tight spire), and 'Pendula' (*weeping* small branches), among others; all are rather rare.

Macedonian Pine *Pinus peuce*

SW Balkan mountains. 1864. Very occasional: larger gardens and a few high-altitude plantations; rarely seeding. Generally healthier than most 5-needle pines, even in exposed/polluted sites.
APPEARANCE Shape *Dense* and often columnar on a good bole; rich, dark, slightly *spiky* foliage (cf. Limber Pine, p.146), not tabular. To 41 m. **Bark** Blackish; more quickly cracking than Weymouth Pine's *in squares/circles;* finally with rough ridges. **Shoots** Fresh green (rarely grey-bloomed, like Bhutan Pine's, above), *hairless*. **Leaves** As Weymouth Pine. **Cones** To 15 cm; *all the scales slightly incurving*.

BHUTAN PINE

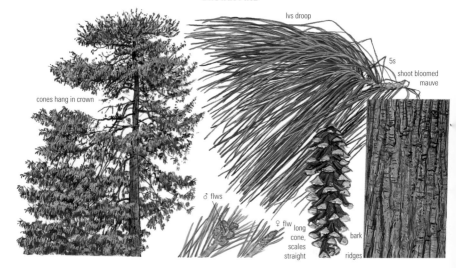

SOFT PINES

Mexican White Pine   *Pinus ayacahuite*

Mountains of S Mexico to Honduras. 1840. Big gardens. The hardiest and least rare of legion silky-leaved Mexican pines.
APPEARANCE Shape Sturdily broad-conic, to 28 m; sinuous, level lower limbs. **Bark** Purple- or grey-brown, cracking generally into *squares* (cf. Macedonian Pine, p.138, and Chinese White Pine, p.142). **Shoots** Greenish, not bloomed; very *fine*, buff hairs. **Leaves** In 5s, the inner sides brightly white-lined; quite long (12–18 cm), drooping and beautifully silky. **Cones** Magnificent but treacherously resinous; 20–45 cm, tapering to a long point. A few basal scales always *strongly bent outwards*. In var. *veitchii* (central Mexico; in some collections) *all the scales* are elongated and curve outwards.
COMPARE Holford's Pine (below). Bhutan Pine (p.138): hairless shoots and few if any cone-scales bending outwards. Chinese White Pine (p.142): sparser crown, hairless shoots and stumpy cones. Other widely grown 5-needle pines have shorter, stiffer, generally denser leaves (see Bhutan Pine); Western White Pine (p.146) is the most similar.

Holford's Pine   *Pinus × holfordiana*

Mexican White Pine (in its var. *veitchii*) and Bhutan Pine (p.138) crossed at Westonbirt Arboretum, Gloucestershire, in 1904, but the seedlings' hybridity was only recognized when they began coning in 1932. Repeat-crosses are now in many big gardens and a few plantations; differentiation from the parents is not always possible.

APPEARANCE Shape Vigorously conic, or broad; strong branches can rise from the base. Fairly dense, but with a long, open apex; to 36 m. **Bark** With often *orange*-purple square plates or scaly ridges (cf. Jelecote Pine, p.148). **Shoots** As Mexican White Pine – unbloomed; usually minutely hairy. **Leaves** Long, bluish, drooping: like Bhutan Pine's. **Cones** *Big and quite broad*, to 30 cm; only the tiny scales at the base bend outwards.

Sugar Pine   *Pinus lambertiana*

W USA: the world's largest pine. The sugary sap oozes freely from old wild trees. Unfortunately, it is particularly susceptible to Blister-rust (*Cronartium ribicola*, a fungus of European origin whose generations alternate between soft pines and currant bushes), but a few healthy trees remain in UK collections. 1827.
APPEARANCE Shape Most like the Weymouth Pine (p.138), but with a tidier, blacker crown when growing well. **Bark** Tends to crack early into *small*, blackish squares (cedar-like). **Shoots** Green, *covered in fine short red-brown hairs*. Strongly lemon-scented when bruised. **Leaves** About 10 cm long, pressed closely to the young shoots then spreading stiffly. **Cones** Huge (to 45 cm); rarely seen in N Europe.
COMPARE Western White Pine (p.146): a more open tree with browner shoots, sharper buds and many cones from an early age. Mexican White Pine (above): longer, spreading leaves on *minutely* hairy shoots. Arolla Pine (p.134): shorter, stiffer leaves (8 cm) on shoots with denser, longer hairs.

MEXICAN WHITE PINE

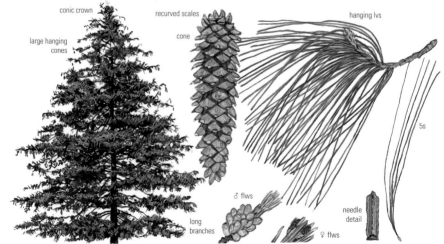

PINES (EAST ASIATIC)

Chinese White Pine — *Pinus armandii*

(Armand's Pine; David's Pine) China, N Burma, Taiwan. 1897. In some collections. An elegant, 5-needled soft pine most like Bhutan Pine (p.138).
APPEARANCE Shape *Very openly conic*, to 25 m, on level branches. **Bark** Grey; distantly square-cracked. **Shoots** Unbloomed, usually *hairless*. **Leaves** 12–14 cm; straight and grass-green in northern trees, drooping and blue in those from SW China. **Cones** *Barrel-shaped,* to 20 cm, with thick, incurved scales (cf. Arolla Pine, p.134).

Japanese White Pine — *Pinus parviflora*

Japan. 1861. Very occasional: larger gardens.
APPEARANCE Shape Trees selected in Japanese gardens, often very spreading, have *picturesque, level plates* of dense, *grey* foliage. Wild-sourced plants are columnar and spikier, to 20 m. **Bark** Pale grey-mauve; large, curling scales. **Shoots** White-brown, usually very finely hairy. **Leaves** In 5s, *short* (4–8 cm) and *very twisted* (cf. Scots Pine, p.122). Inner sides with blue-white lines; outer surface also glaucous. **Cones** Small (6 cm), barrel-shaped.
COMPARE Arolla Pine (p.134); Whitebark Pine (p.136).

Chinese Red Pine — *Pinus tabuliformis*

(*P. sinensis*) N China. 1862. In a few collections.
APPEARANCE Shape *Often broad*; level branches and dense, spiky foliage. **Bark** Like Corsican Pine's (p.124), but with finer, dull orange-red flakes on the branches. **Buds** Rich brown, with some free scale-tips. **Leaves** In 2s, shiny grey-green, 9–15 cm.
COMPARE Japanese Red Pine (below); Calabrian Pine (p.130); Red Pine (p.128).

OTHER TREES Yunnan Pine, *P. yunnanensis* (W China, 1909), sometimes treated as a variety, is very different in cultivation but equally rare. Leaves usually *in 3s*, to 20 cm and *elegantly drooping*: luxuriant trees can recall Jelecote Pine (p.148), but have thicker, heavier green foliage; their dark reddish scaly bark is duller.

Japanese Red Pine — *Pinus densiflora*

Japan; Korea; NE China and Pacific Russia. 1861. Rather rare.
APPEARANCE Shape Spikily domed, to 20 m; easily taken for Scots Pine (p.122). **Bark** *Bright orange-pink upper bark* distinguishes this from Chinese Red Pine (above), Aleppo and Stone Pines (p.130) and Japanese Black Pine (below). **Leaves** *Slender*, straight, often sparser than Scots Pine's; *shining green* (8–12 cm). **Cones** Small (5 cm) and more persistent than Scots Pine's.
COMPARE Red Pine (p.128): leaves to 15 cm and a stiff, thin crown.

Japanese Black Pine — *Pinus thunbergii*

(Kuro-Matsu) Japan, Korea; a tough tree from coastal habitats. 1861. Rare.
APPEARANCE Shape Singularly *gaunt* from an early age; distant *wandering branches*; blackish. To 25 m. **Bark** Purple-grey; rugged interlacing fissures. **Buds** *White* with silky, free-tipped scales (cf. Maritime Pine, p.132). **Leaves** In 2s, rigid and *spined*; 7–12 cm. **Flowers** Prolific. **Cones** Small (5 cm), with few, large, rather spongy scales, and carried in *clusters of up to 50* on some trees.
COMPARE Austrian Pine (p.124); Bosnian and Mountain Pines (p.134); Shore Pine (p.126); Japanese Red Pine (above). Habit and spiny leaves distinguish.

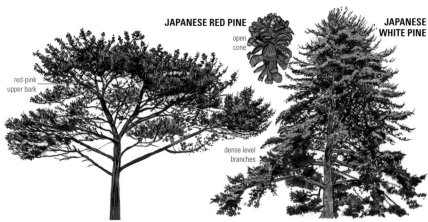

JAPANESE RED PINE — open cone; red-pink upper bark

JAPANESE WHITE PINE — dense level branches

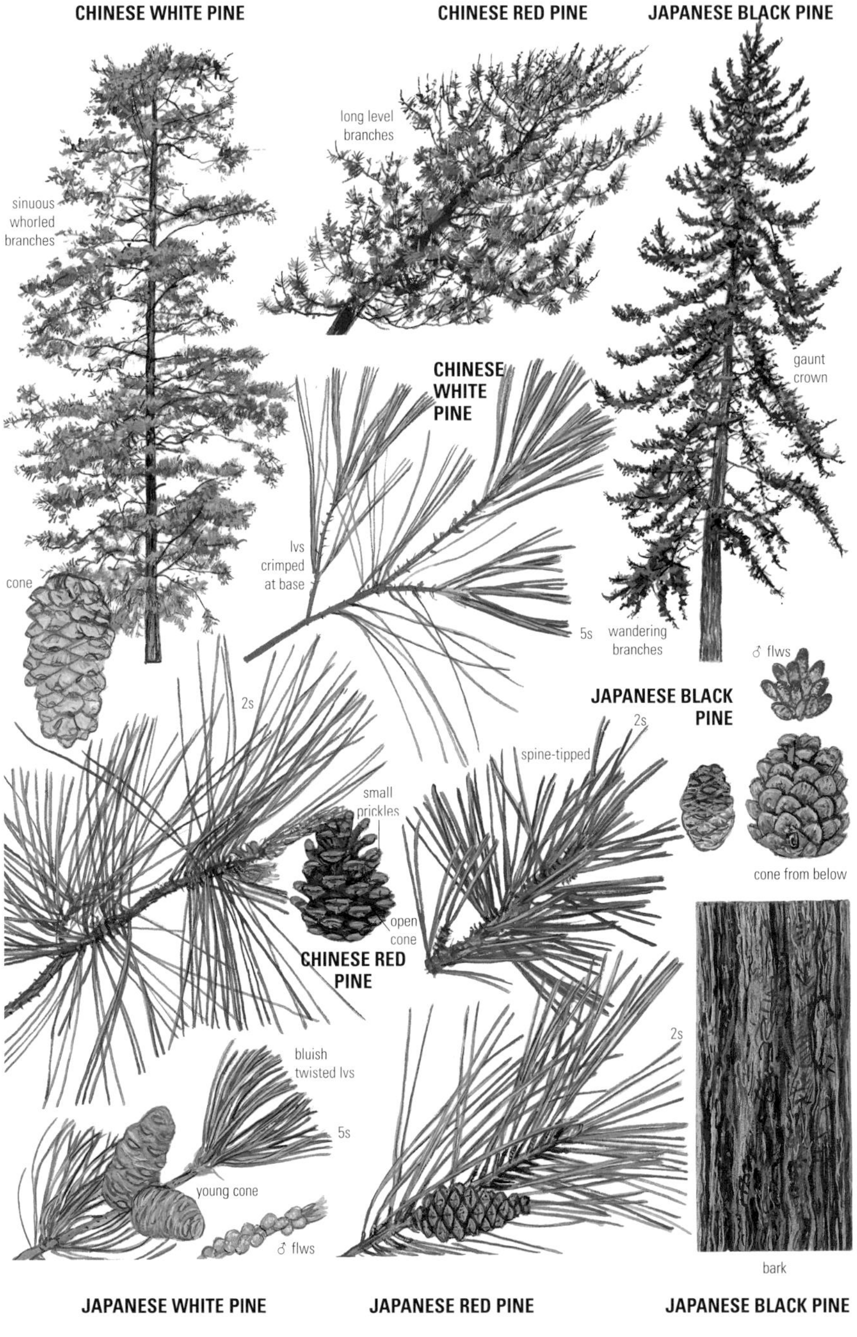

PINES (EAST ASIATIC)

CHINESE WHITE PINE

CHINESE RED PINE

JAPANESE BLACK PINE

sinuous whorled branches

long level branches

gaunt crown

CHINESE WHITE PINE

cone

lvs crimped at base

5s

wandering branches

♂ flws

JAPANESE BLACK PINE

2s

cone from below

2s

spine-tipped

small prickles

open cone

CHINESE RED PINE

bluish twisted lvs

5s

young cone

♂ flws

2s

bark

JAPANESE WHITE PINE

JAPANESE RED PINE

JAPANESE BLACK PINE

PINES (AMERICAN)

Monterey Pine *Pinus radiata*

The type is confined to three cliffs around Monterey, California: like Monterey Cypress (p.48) it seems to have gone the wrong way when migrating north in its preferred climate zone after the last glaciation, only to find its route cut off by the ocean and by arid coastal zones. 1833. Frequent in milder areas and growing much faster in the UK than in the wild; abundant near some coasts, withstanding salt spray very well; rarely seeding. Occasionally in forestry plantations. (It is New Zealand's principal forestry tree, having reached 60 m in 41 years.)
APPEARANCE Shape Spikily conic at first, then *densely domed* and sometimes very broad, on heavy, twisting limbs; to 45 m. The bole may be long and straight but is more often short and twisted. The bright green needles appear very dark but brilliant *en masse*. **Bark** Grey; purple-black with age and *very ruggedly fissured*. **Leaves** In 3s (but in 2s in var. *binata* from Guadalupe Island and var. *cedrosensis* from Cedros Island, Mexico, which have scarcely been grown in the UK); 10–16 cm long; very slender and nodding. **Cones** Fist-sized, with a minute prickle wearing off each scale; persisting abundantly in whorls and opening only in the wake of forest fires.
COMPARE Bishop Pine (p.128): a *2-needle* pine often with very similar bark and crown but stiffer, bluer leaves and strongly prickled cones. Other 3-needle pines with similarly slender/short leaves are Knobcone Pine (p.136), Northern Pitch Pine (below), Yunnan Pine (p.142), with thicker leaves, and Jelecote Pine (p.148), with hanging, grass-green foliage. Lace-bark Pine (p.148): 8 cm leaves and unique bark.
OTHER TREES Loblolly Pine, *P. taeda* (New Jersey to Texas, 1713) is a similarly domed tree in a few collections. Bark less rugged in age; shoots glossier and paler brown; cones strongly spined. The slender but greyer leaves are much sparser.

Northern Pitch Pine *Pinus rigida*

E North America – Ontario to Georgia. *c.*1743. Rare: collections.
APPEARANCE Shape Irregular, often broad; curving branches crusted with cones; *numerous sprouts* often grow on the bole, as in very few other conifers: it regrows if coppiced. **Bark** Dark brown: craggy ridges never as deep as Monterey Pine's. **Leaves** In 3s, 7–14 cm; rather stiff; dull grey-green. **Cones** 3–8 cm, with prickled scales.
COMPARE Knobcone Pine (p.136). Lace-bark Pine (p.148): shorter leaves, and different bark.
OTHER TREES Gregg Pine, *P. greggii* (Mexico, *c.*1905) is a rather similar tree in some collections. Brighter green leaves lie closer against whitish shoots, the bark is paler, and the cones have only tiny spines. The crown is stiffer, on steep branches without the characteristic sprouts.

Singleleaf Nut Pine *Pinus monophylla*

SW USA. A principal source of American pine-nuts (pinyons). 1848. In a few collections.
APPEARANCE Shape A dense, bushy but attractively glaucous tree. **Bark** Ruggedly square-cracked, blackish-purple. **Leaves** 5 cm, carried *singly* (a few split into pairs by their second year). Rounded (as the semicircular leaves of 2-needle pines and the triangular segment-like needles of 3- to 8-needle pines are if fitted back together). White lines all round.
COMPARE Saplings of Stone, Aleppo and Calabrian Pines (p.130).

SINGLELEAF NUT PINE — needles held singly; closed cone; open cone

NORTHERN PITCH PINE — bark; sprouts on bole

PINES (AMERICAN)

Western White Pine — *Pinus monticola*

Rocky Mountains. 1831. In some big gardens, but very susceptible to Blister-rust (see p.140).
APPEARANCE Shape A soft pine most like Weymouth Pine (p.138); a softly grey-green *spire*, to 40 m. **Bark** Smoother than Weymouth Pine's; grey. **Shoots** First year: brownish green with *fine dense rusty hairs* (cf. Sugar Pine, p.140). **Leaves** In 5s; 11 cm, slender and slightly drooping; faint white line on outer as well as inner side. **Cones** *Abundant*; slender, to 30 cm; lower scales much bent back.
COMPARE Limber Pine (below); Mexican White Pine (p.140).

Limber Pine — *Pinus flexilis*

(*P. reflexa*) W North America. 1851. In some collections.
APPEARANCE Shape Broad, dark and dense, to 20 m; *particularly spiky* with projecting shoots (newer leaves closely appressed). **Shoots** Green, usually with fine brown hairs; exceptionally 'limber' and *can be tied in knots*. **Leaves** Different from the otherwise similar 5-needle pines (see Weymouth Pine, p.138): dark, with very fine white lines equally *on each surface*. 6–12 cm; *unserrated* (no drag when drawn between fingernails). **Cones** *Short (to 12 cm)*.
COMPARE Whitebark Pine (p.136): shorter, denser, greyer leaves. Western White Pine (above): also with a (faint) line on the third side of each *serrated* leaf.

Ponderosa Pine — *Pinus ponderosa*

(Western Yellow Pine) W Rocky Mountains. 1827. Occasional in parks and gardens everywhere.
APPEARANCE Shape A *sparse spire*, often to great heights (40 m); sometimes becoming irregular or rounded but still quite slender. **Bark** At first resembles Corsican Pine's (p.124). Older trees more craggily fissured, blackish or warm red. No distinctive smell. **Shoots** Shiny orange/green. **Buds** Red-brown, resinous; may expand to 5 cm before spring. **Leaves** In 3s, long (12–22 cm) and stiff, *like chimney brushes* but sometimes more forward-angled. **Cones** Only *to 15 cm*, a tiny backward-pointing prickle on each scale.
COMPARE Jeffrey Pine (below); Coulter Pine (p.148). The three (without cones) are confusing – foliage is soon out of reach. The bark is often smoother in Jeffrey Pine (with a distinctive smell), and craggier in Coulter Pine. Hartweg Pine (p.132) has a rougher shape and some leaves in 4s/5s.
OTHER TREES Apache Pine, *P. engelmannii* (Royal Pine), is one of several exciting recent introductions from Mexico (1962). Needles very stout and grey but long enough at *40 cm* to droop; in 3s, 4s and 5s in the same thick, scaly-rough, red-brown shoot (cf. Endlicher Pine, p.132).

Jeffrey Pine — *Pinus jeffreyi*

S Oregon to N Mexico. 1853. Very occasional in larger gardens.
APPEARANCE Differences from Ponderosa Pine (above) as follows: **Bark** Always blackish; never very craggy. The resin contains heptane: in summer the shoots and crumbled bark *smell of wine-gums*, often the most useful distinguishing feature. **Shoots** Blue-bloomed. **Buds** Non-resinous. **Leaves** Foliage often *very sparse*, bluer. **Cones** *Larger* (to 25 cm), with *backward*-pointing prickles.
COMPARE Coulter Pine (p.148): huge cones with forward hooks.

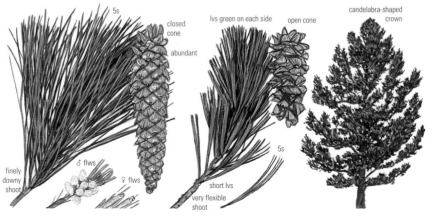

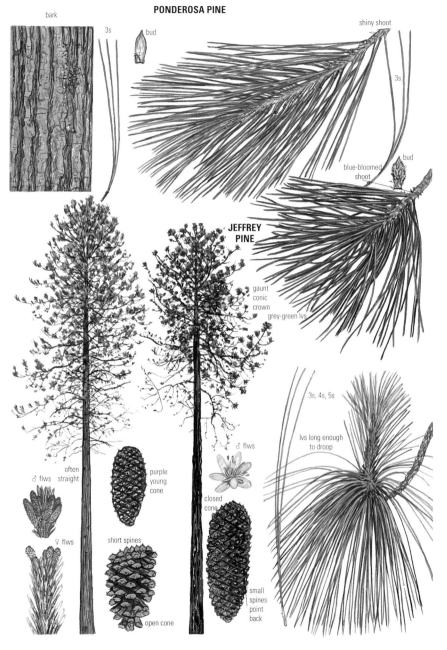

¹⁴⁸ PINES

Coulter Pine *Pinus coulteri*

(Big-cone Pine) S California into NW Mexico. 1832. Very occasional: larger gardens in warmer areas.
APPEARANCE Shape Broad-conic on a good but sometimes forking bole, with chimney-brushes of bluish-green foliage; rather open even when healthy and sometimes very sparse. A tremendously vigorous, short-lived tree, to 30 m. **Bark** Blackish, soon *deeply* craggy; almost like Monterey Pine's (p.144), and lacking the lemony smell of Jeffrey Pine's (p.146). **Shoots** Stout, often mauve-bloomed. **Buds** Large and orange (like Ponderosa and Jeffrey Pines', they may expand to 5 cm before winter sets in). **Leaves** In 3s, long and stiff, to 30 cm. **Cones** Pineapple-sized (to 35 cm), with *big forward hooks* on each scale. They drop when ripe, making the tree distinctly unsuitable for public parks.
COMPARE Jeffrey Pine (p.146): a similar but generally slenderer tree, its smaller cones with little backward-pointing prickles. Ponderosa Pine (p.146): less rugged, often reddish bark, a stricter shape, and small cones. Apache Pine (p.146): rough, unbloomed shoots; leaves often in 5s. Endlicher Pine (p.132): wide-domed crown; needles in 5s. Endlicher Pine's shorter-leaved ally Hartweg Pine is more alike.
OTHER TREES Digger Pine, *P. sabiniana* (California, 1832), is a *singularly sparse* conic tree in a few collections, whose smaller cones (15–20 cm) have *backward* hooks. Its lemony aroma is like Jeffrey Pine's (p.146). The edible pine-nuts were once a staple of the Digger Indians.

Lacebark Pine *Pinus bungeana*

N China (much grown in temple gardens). 1846. In a few southern gardens for its unique bark, which gradually develops thin scales of *white, ginger and blue*. Young plants are 'ugly ducklings', with pale, dull grey bark; old Chinese trees' bark is almost silver.
APPEARANCE Shape Generally bushy, but with dense spire-tips; to 14 m. **Buds** Standing *5 mm clear of the end leaves*; outward-curved scales. **Leaves** In 3s, *sparse*, to 8 cm only; bright dark green with faint white lines on each face.
COMPARE Mountain Pine (p.134): a 2-needle pine whose bark remains dull grey.

Jelecote Pine *Pinus patula*

(Mexican Weeping Pine) Cloud forests of E Mexico. 1837. Perhaps the most beautiful hardy pine, now occasional as a young garden tree. It can succumb to hard frosts.
APPEARANCE Shape Often multi-stemmed, on level sinuous branches; to 20 m. **Bark** *Orange*, papery-scaling, then duller and widely fissured. **Leaves** In 3s, to 25 cm, very slender and *hanging over the white-bloomed shoots like a pony's mane; pale, shimmering grass-green* to bluish. The tips often brown in dry cold spring winds. **Cones** Persistent, long-conic, woody and bright brown; to 10 cm.
COMPARE Yunnan Pine (p.142): darker crown. Foliage suggests a 5-needle 'soft pine', but none has the lovely contrast of grass-green leaves and orange bark. Montezuma Pine (p.132): long bright green hanging leaves, but needles in 4s–6s.

LACEBARK PINE

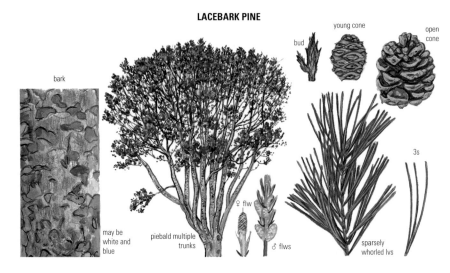

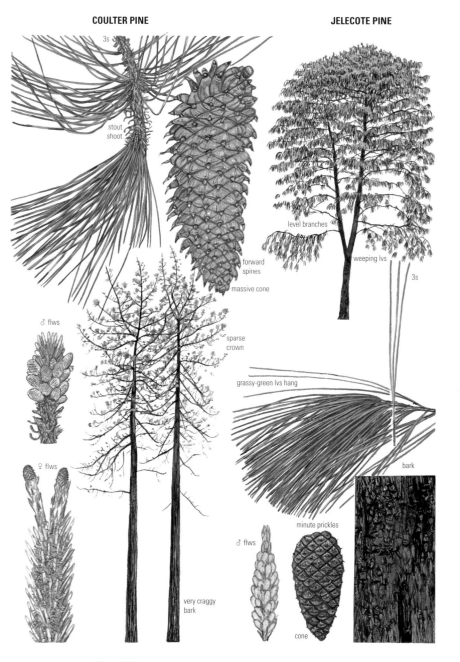

POPLARS

Salicaceae is a family of nearly always dioecious trees and shrubs. Silky hairs on the tiny seeds carry them on the wind; they germinate if they land in wet mud, and all species like a rich, wet soil. Most tree forms suggest a big branch stuck in the ground. The 35 poplars have very sharp buds with many scales, and long, wind-pollinated catkins.

Things to Look for: Black Poplars

- Bark: How rugged?
- Shoots: Are they rising or drooping?
- Leaf-stalk: Glands at top? Hairy?
- Catkins (what gender)?

Key Species: Poplars

White Poplar (below): lobed leaves, white-hairy under. **Aspen** (p.152): almost round leaves, green under. **Aspen** saplings: small triangular leaves, downy under. **Balsam Spire Poplar** (p.160): leaves smoothly white under. **Wild Black Poplar** (p.152): triangular leaves, green under. **Chinese Necklace Poplar** (p.162): big, very downy leaves.

White Poplar   *Populus alba*

(Abele; *P. nivea*) W and central Eurasia (including Britain?); Tunisia. Frequent. The whitest tree in the landscape.
APPEARANCE Shape Never straight; twiggy and suckering. *Light*, wayward branches; to 28 m only. **Bark** Pale grey in youth, pitted with lines of diamonds (cf. Aspen, p.152; Sallows, p.168). Old trees have creamy-white limbs above black, rugged boles. **Shoots** *White with close wool, which lasts through winter.* **Buds** Dumpy; white-woolly. **Leaves** Unfold furry; hairs rub off the shiny dark grey-green upper side but *plaster the under-leaf all season*. With maple-like lobes on strong shoots; rather rounded on weaker ones (cf. Grey Poplar).

VARIANTS Bolle's Poplar, 'Pyramidalis', is an 1872 selection from wild trees in Turkmenistan; locally frequent. *Lombardy Poplar-shaped*, then a narrow funnel of *erect shoots*. More vigorous, with brighter cream-white upper bark and shoots hairless by autumn; bigger, more lobed leaves, soon glossy above, create a darker crown. Female (green spring catkins). 'Raket' ('Rocket'; by 1956) is a rare spire-shaped improvement; *shoots spreading*.
'Richardii' (1918) has *sunny yellow* leaves (white beneath) – like the type's in autumn. Rare; to 17 m.

Grey Poplar   *Populus canescens*

(*P. × canescens*) Central Europe – a stable hybrid of White Poplar and Aspen. Frequent; naturalized.
APPEARANCE Shape *Massive, clean, high branches; to 40 m*; dense *dark grey-green foliage, jagged/weeping*; top branches typically sweep over like a Catherine-wheel. **Bark** Like White Poplar's; base soon very rugged. **Shoots** *Red-grey*; white wool *tends to rub off* by winter. **Leaves** Normally rounded, with big, wave-shaped teeth; more maple-shaped on strong growths. They emerge woolly cream-grey; by summer the older ones are *almost hairless*, the newest still patchily grey-woolly beneath. **Flowers** Most trees are male: purple-grey 4 cm catkins in early spring.
COMPARE White Poplar (above). Can be taken for a giant Aspen (p.152): check for white-hairy young leaves.
VARIANTS Picart's Poplar, 'Macrophylla', has strikingly *large* leaves, to 15 cm wide. Very rare.

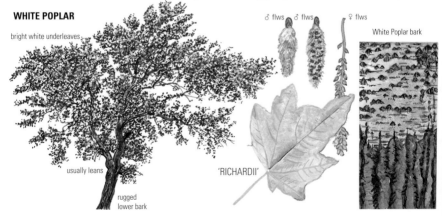

WHITE POPLAR
bright white underleaves
usually leans
rugged lower bark
♂ flws   ♂ flws   ♀ flws
White Poplar bark
'RICHARDII'

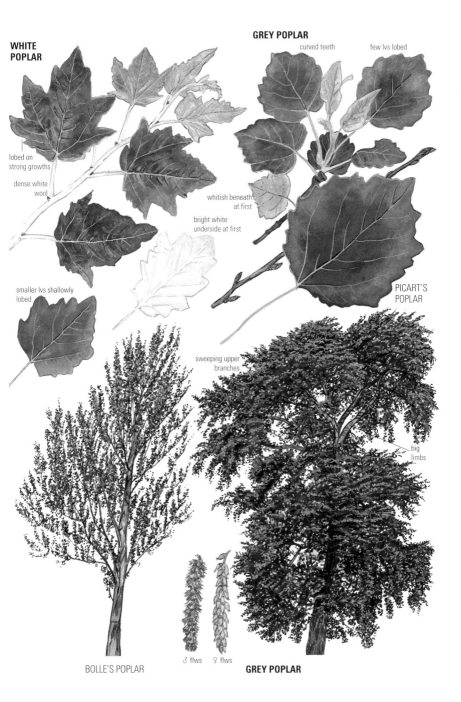

POPLARS

Aspen
Populus tremula

W Eurasia (including Britain and Ireland); Algeria. Abundant: suckering stands in scrub, coppices and shingle banks; seldom planted.
Appearance Shape *Slender* (exceptionally to 30 m); long bole and *small*, rather perpendicular branches; spiky, gaunt and knobbly with catkin-buds in winter. Soft green; very short-lived. **Bark** Cream, pitted with bands of small black diamonds (cf. White Poplar, p.150; sallows, p.168); then grey and rugged at the base. **Shoots** Shiny brown on adult tree. **Buds** Long-pointed, *painfully sharp* (like Bird Cherry's, p.342, but soon growing *short side-shoots at right-angles*). **Leaves** *Round*, with big wave-shaped teeth (cf. Grey Poplar, p.150), fluttering (even more than other poplars') in the slightest breeze, so trees are often heard but not seen. Leaves open late and coppery, with grey down, but are *soon hairless*. **Suckers** With triangular green, softly furry leaves on velvety stems: foliage adult by 2 m.
Variants Weeping Aspen, 'Pendula', has rather stiffly hanging branches. 'Erecta' has *very steep* branches from a straight bole. Both are rare.

Berlin Poplar
Populus × *berolinensis*

Hybrid of Lombardy Poplar and *P. laurifolia* (a narrow-leaved Asiatic balsam poplar), c.1800. Now very occasional and hardly planted in the UK (but still widespread in central Europe). The original cross was female; males are now commoner.
Appearance Shape Steep-limbed and narrowly rounded, to 30 m, on an often long, sprouty bole; quite densely leafy. **Bark** Grey-brown; quite rugged, criss-crossing ridges. **Leaves** *Diamond-shaped* (or round-based), to 12 cm; pale grey-green beneath and finely hairy at first.

Compare Simon's Poplar (p.162); Noble Poplar (p.158); 'Androscoggin' (p.162).

Wild Black Poplar
Populus nigra ssp. *betulifolia*

(Downy Black Poplar) NW Europe, including England and Wales. Old trees (all planted?) locally abundant in flood-plains, old parks and some cities but there are only about 6000 in Britain; younger plantings rather occasional – recreation ground shelterbelts, gardens.
Appearance Shape The normally short bole nearly always leans. Soon a huge broad tree to 38 m, of immense presence: heavy, outward-arching limbs with *many small burrs* and *massed, rising shoots*. In summer, richer green and *much leafier* than its hybrids (p.156). Odd trees (accidental crosses with Lombardy Poplars?) have steep, paler grey limbs and more open crowns. **Bark** Greyish brown (some old trees nearly black): short deep fissures swirl round burrs and snags. **Shoots** *Amber*, knobbly, with long orange-grey buds. **Leaves** *Small* (7 cm): conspicuously long-tipped, with *no glands at the base*. The young (green) shoots, leaves and stalks have *tiny fine hairs* (cf. the hybrids 'Robusta', p.156, and 'Florence Biondi', p.158), shed by autumn; type trees (S/E Europe to central Asia; Tunisia) are hairless. Sweet balsam scent in spring (but milder than in balsam poplars, p.160). **Flowers** Males with red catkins in mid-spring. Old female trees outnumbered by 100:1 ('Manchester Poplar' clones planted in polluted N English cities were all male); green catkins with seed-drop *in May*.
Compare Its cultivars (p.154). Railway Poplar (p.156) is much confused.

BERLIN POPLAR

long tapered base

ASPEN

big curved teeth

soon hairless

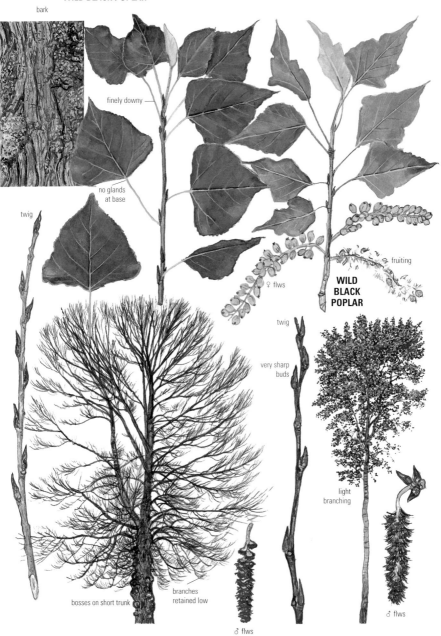

BLACK POPLAR CULTIVARS

Lombardy Poplar *Populus nigra* 'Italica'

A sport probably of a central Asian race of *P. nigra* (see p.152). 1758. Abundant in warmer areas; sometimes naturalizing by suckers.
APPEARANCE Shape Branches remain more tightly vertical than any poplar's except 'Serotina de Selys' (p.158); to 38 m. **Bark** Greyish *brown*, moderately craggy and distantly burred with age. **Shoots** Rather *amber; hairless from the first*. **Leaves** *Small, long-pointed*, hairless, in rather dense masses. **Flowers** A *male tree*: red catkins in mid-spring and no 'fluff'.
VARIANTS 'Elegans' is a particularly tight clone – little more than a trunk with twigs. Sickly examples of common clones can look almost as gaunt.
Golden Lombardy Poplar, 'Lombardy Gold', is a stumpy sport (to 12 m so far), found in 1974 as a branch of a mature tree in Surrey; very rare.
Plantier's Poplar, 'Plantierensis', is a hybrid (Metz, 1884) of the original Lombardy Poplar with the W European Wild Black Poplar (p.152), whose *finely hairy young growths* it inherits: the hairs are best seen on the leaf-stalks and are gone by autumn. Probably frequent in some areas and always worth looking for. Tends to be slightly broader and leafier in maritime climates than typical Lombardy Poplars; the limbs (often sprouty) carry the Wild Black Poplar's *frequent* small burrs. *Stronger balsam smell* in summer (though weaker than Balsam Poplars, p.160). Male and female clones exist.

Female Lombardy Poplar
  Populus nigra 'Foemina'

('Gigantea') Another hybrid of the original Lombardy Poplar.
APPEARANCE Shape *Narrowly funnel-shaped*, to 32 m, with heavier limbs diverging to the tip. **Bark** Paler grey-brown than original Lombardy Poplar's. **Leaves** Hairless from the start; slightly larger and *sparser*. **Flowers** *Green* catkins shedding drifts of cotton wool in May.
COMPARE Bolle's Poplar (p.150): similar jizz; also has a pale bark.

Populus nigra 'Vereecken'

A male clone, selected in Holland in the mid 20th century. Rare.
APPEARANCE Shape Straight trunk and *light branches rising at 60°* to make an open, shapely, narrow crown, to 30 m; young growths not downy.
COMPARE Prince Eugene's Poplar and 'Robusta' (p.156).

Afghan Poplar *Populus nigra* 'Afghanica'

('Thevestina') A female clone of Asiatic origin, much planted across S Europe and growing poorly in a few northern collections.
APPEARANCE Shape Like Lombardy Poplar. **Bark** *Bright white* (cf. Bolle's Poplar) on the branches, dark and fissured at the base. **Leaves** More rounded than western forms'.

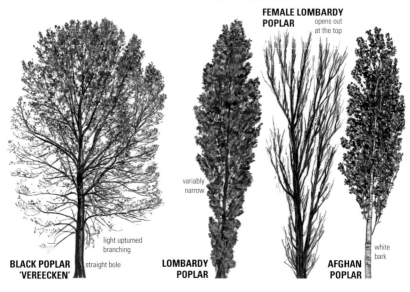

BLACK POPLAR 'VEREECKEN' — straight bole, light upturned branching

LOMBARDY POPLAR — variably narrow

FEMALE LOMBARDY POPLAR — opens out at the top

AFGHAN POPLAR — white bark

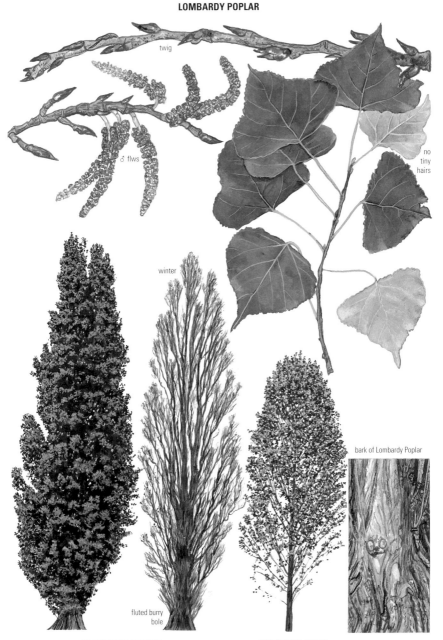

HYBRID BLACK POPLARS

Hybrid Black Poplars *Populus × canadensis*
(*P. × euramericana*) The cross between the European Black Poplar (p.152) and forms of the North American Eastern Cottonwood (*P. deltoides*: now confined to a few collections in Europe) has occurred or been made many times since 1750. With their 'hybrid vigour' the offspring grow furiously, and have been used abundantly in shelterbelts and plantations on rich soil: tending to tower above other trees, they currently dominate most lowland landscapes. The overall appearance (even of known back-crosses with Wild Black Poplars) is much closer to that of the American parent: all develop a *greyer* bark, relatively *regularly* though deeply fissured, and an *airy* crown of constantly susurrating, metallic-green leaves, mid-*green* underneath, *between which the sky often remains visible*. Shoot-tip leaves are very triangular, older ones more wedge-shaped at the base; they are biggest (like other poplars') at the crown-tip, where they drop last in autumn. Young leaves are *fringed with fine hairs*, and at the top of the stalk there are often *1–3 small knobbly glands*. Yellowish-*grey* winter shoots carry 1 cm, sharp, narrow buds. Gender is the safest way of distinguishing the several common clones. Females have thin yellow-green catkins in spring which *around mid-summer* shed snowstorms of cotton-wool seeds (the Wild Black Poplar usually sheds its seeds in May); males carry fat red catkins in early spring.

'Robusta' (1895) is now much the most planted clone in plantations and belts; it is also the most distinctive, and so attractive that it earns a place in many gardens.
APPEARANCE Shape Straight, rather twiggy trunk, seldom forking; in youth (and to 40 m) roughly *spire-shaped* with regularly whorled, erect branches like a silver fir's. The distance between the whorls indicates one summer's growth and can be 2.2 m. Old open-grown trees (few as yet) develop relatively *light*, clean, steep limbs, sometimes from low down, and may lean. **Bark** More *shallowly*, shortly and crookedly fissured than in most clones; paler and a browner grey. **Shoots** Very finely hairy when young. **Leaves** Flush ahead of the other clones, in *midspring*, a *pale coppery-red*; rather *large* (10 cm), soon rich green, and relatively *dense*, on stalks minutely hairy at first (cf. 'Florence Biondi' and Wild Black Poplar, p.152). **Flowers** A *male* clone.
COMPARE Prince Eugene's Poplar (below). 'Balsam Spire' (p.160): also spire-shaped and planted everywhere, has a smoother dark grey bark and larger leaves *white* beneath.
VARIANTS A similar male tree in a few plantations in S England is 'Heidemij', with a slightly darker grey bark. Its leaf-stalks are particularly long and flattened, making the foliage shiver in any breeze.

'Railway Poplar'. 'Marilandica' (May Poplar) arose around 1800; it has since been submerged in various scarcely distinguishable backcrosses with 'Serotina', known collectively as 'Regenerata' or 'Railway Poplar'. All are *female clones*, with variably dense snowstorms from June to August (some trees look otherwise very like the *male* Black Italian Poplar, p.158). Unlike the other hybrid poplars they cope well with exposure, and are locally abundant as mature trees in coastal areas.
APPEARANCE Shape Trunk seldom quite straight; often low-forking; typically *more densely domed* than the other clones; to 40 m. Twiggily and roughly conic when grown together (but infrequent in lines and scarce in forestry). Heavy, *outcurving* limbs; many fine shoots *descend*. Sometimes grassy-green summer foliage. **Bark** Grey, often pale or greenish, with deep, generally rather crooked fissures, sometimes some burrs, and often many sprouts. **Leaves** Sparse but *untidily clustered*; they flush often very late, hairless and brownish yellow.
COMPARE Black Italian Poplar and 'Florence Biondi' (p.158); Wild Black Poplar (p.152); Noble Poplar (p.158).

Prince Eugene's Poplar, 'Eugenei' (Metz, 1832), probably has Lombardy Poplar as its father. Sometimes in plantations; otherwise rare.
APPEARANCE Shape *Long, usually straight trunk* and light but *spreading* limbs make a *columnar crown* (to 42 m). **Bark** Dark grey. **Leaves** *Small*, generally tapered at the base, appearing particularly sparse. Pale brown, rather late leaf-flush. **Flowers** A *male* clone.

'Gelrica' is in a few plantations.
APPEARANCE Shape Broader than Prince Eugene's Poplar (above), but similar in its small, very sparse leaves. Very clean trunk and limbs, the smaller branches *silvery-grey* (as they can be in 'Robusta'). To 30 m. **Leaves** A higher proportion *heart-shaped* at the base, on red-flushed stalks; they flush rather late, a pale coppery colour. **Flowers** Usually male; one variant is female.

HYBRID BLACK & NOBLE POPLARS

'Florence Biondi' ('OP226') is a very vigorous, straight-stemmed Hybrid Black Poplar selected by 1950, and now occasional in plantations and belts.
Appearance Shape Cleaner and more graceful than any Railway Poplar clone (p.156). Dark, very sparse foliage; some sprouts make the slightly jagged crown denser along the steeply ascending limbs. **Leaves** Stalks at first minutely *downy* (cf. only 'Robusta', p.156). **Flowers** A *female* tree.

'I-78' ('Casale 78') is another recent *female* clone in a few younger plantations.
Appearance Shape As straight as 'Florence Biondi', vigorous and airy, but lacks the sprouts on the limbs. **Leaves** Stalks not downy.

Black Italian Poplar, 'Serotina' (France, 1750), features in many tree guides as the common Hybrid Black Poplar, but these clones are short-lived and as this one was little planted in the 20th century it is now distinctly occasional in most parts.
Appearance Shape A long trunk, seldom straight, typically carries huge, *clean, incurving* limbs; shoots rather thick and ascending. Has reached 45 m. **Bark** Dark/mid grey, soon with long, rather *regular*, very deep fissures. **Leaves** Unfold *very late, a pale coppery brown*; in summer sparse but evenly spread, a dark *sea-green*. **Flowers** A *male* clone.
Compare Railway Poplar and Prince Eugene's Poplar (p.156).

Golden Poplar, 'Serotina Aurea' (Ghent, 1871), can revert piecemeal to 'Serotina'. (Poplars, growing readily from cuttings, are seldom grafted, so most Golden Poplars cannot be identified from November to June.)
Appearance Shape Domed, twiggy habit, as Railway Poplar (p.156). A locally occasional and eye-catching ornamental: the biggest golden tree (to 30 m). **Leaves** Flush very late and pale brown, then *acid yellow*, finally soft green. (Green clones on chalk can be patchily yellow with chlorosis.)

'Serotina de Selys' ('Serotina Erecta'; by 1818) is a rare male clone.
Appearance Shape At first glance an unusually sharply pointed, sparsely foliaged, glistening Lombardy Poplar (to 37 m). **Bark** *Grey*, as in 'Serotina' (rather blackish and closely fissured near the base). **Leaves** Larger and shinier than Lombardy Poplar's, sea-green, often with 1–3 basal glands; unfolding pale coppery-brown and *very late*, as in 'Serotina'.

Noble Poplar *Populus* × *generosa*

Eastern Cottonwood (see p.156) was crossed with Western Balsam Poplar (p.160) at Kew Gardens in 1912. Descendants (of both sexes) are now occasional as older trees (to 40 m).
Appearance Shape At first glance a poor Railway Poplar (p.156), snaggy and often leaning and cankered. **Bark** Pale grey, *less deeply fissured*. **Leaves** Larger, somewhat denser (to 20 cm), *pale whitish green underneath* between a network of green veins; always with 2–3 glands at top of leaf-stalk.
Variants 'Beaupré' and 'Boelare' are recent Belgian selections with big leaves and *straight, smooth, pale trunks*: hugely vigorous but confined as yet to collections and trial plots.
Compare 'Rochester' and 'Oxford' (p.162).

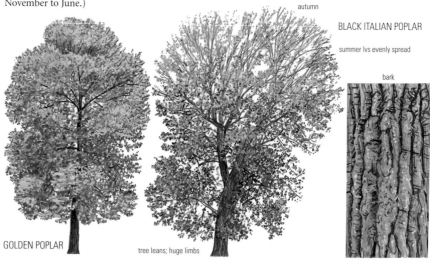

autumn

BLACK ITALIAN POPLAR

summer lvs evenly spread

bark

GOLDEN POPLAR

tree leans; huge limbs

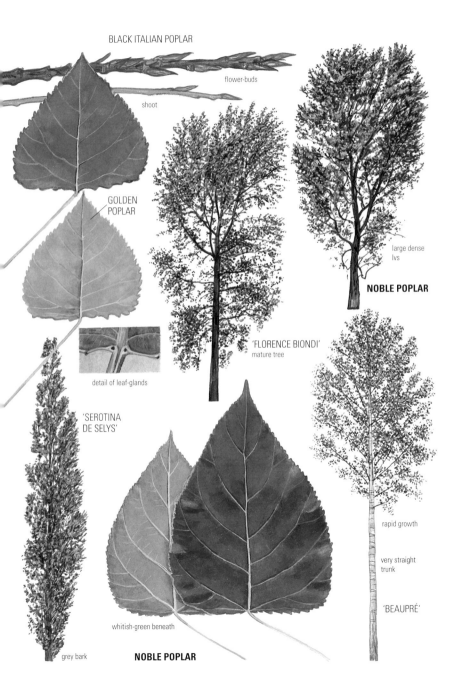

BALSAM POPLARS

Western Balsam Poplar *Populus trichocarpa*

W North America, where it is one of the world's tallest broadleaves (to 70 m). 1892. Now (as var. *hastata*) rather occasional in the UK: susceptible to bacterial canker (*Aplanobacter populi*), short-lived, and largely replaced by its hybrids (below). The sweet scent of the buds' resin, copious in all balsam poplars, *fills the air* through spring.
APPEARANCE **Shape** Narrow, to 42 m, with steep limbs, soon much broken, on an often densely sprouty bole which usually leans with age; glossily luxuriant foliage *densely clothes* the stems. **Bark** Silvery, or pale brown; shallowly and closely ridged. **Shoots** Red-brown, hairless; slightly *angled* at first. **Buds** Long, sharp and sticky. **Leaves** Variably triangular or oval/heart-shaped, 10–25 cm long; *oily-white underneath between a fine network of green veins* but soon almost hairless; flushing acid-green *early* in spring and briefly yellow in early autumn. **Fruits** Opening into 3 segments (female trees).
COMPARE Noble Poplar (p.158); Doronoki (p.162).
VARIANTS 'Fritzi Pauley', a selection from Mt Baker, Washington, is now sometimes used in forestry – especially in cool wet areas – for its canker-resistance, extreme vigour and *straight (but sprouty) trunk* (cf. 'Balsam Spire'), which carries light, rather *level* branches. The bark is *brown*, and soon *rather scaly*; a male clone.
OTHER TREES Eastern Balsam Poplar, *P. balsamifera* (*P. tacamahaca*; E North America, 1689), now a rarer tree, differs in *prolific suckering* (sometimes naturalized) and in smoothly rounded young shoots and almost untoothed leaves, under which a few hairs *persist*. The clearest distinction (but present only in female plants) is that the seed-pods on the catkins open in 2 segments, not 3.

Balsam Spire Poplar *Populus* 'Balsam Spire'

('TT32') The cream of a series of artificial hybrids of Western and Eastern Balsam Poplars, now very abundant: plantations, shelterbelts, parks. Generally canker-resistant.
APPEARANCE **Shape** At first a dense *spire*; the *straight* trunk seldom forks. Light, *steeply rising* branches make a *spiky fan* at the top of old trees (to 35 m). **Bark** *Silver-black, smooth* for many years, then finely fissured. **Leaves** Shortish, rounded, dark above (but to 30 cm on the strong top-growths). **Flowers** A *female clone*; seed-drop in high summer.
COMPARE 'Robusta' (p.156): a less dense spire, flushing later and coppery-red, with very little balsam scent; 'Fritzi Pauley' (above); 'Androscoggin' (p.162).

Variegated Poplar *Populus × jackii* 'Aurora'

(*P. candicans* 'Aurora') Jack's Poplar (or Balm of Gilead) is most probably a natural hybrid of Eastern Balsam Poplar with Eastern Cottonwood (see p.156). This cultivar (*c.* 1920) is now frequent.
APPEARANCE **Shape** A roughly conic, snaggy, fragile, much-cankered female tree, to 20 m. **Bark** *Pale grey*, shallowly ridged. **Leaves** *Less white beneath* than the balsam poplars', on finely *downy* stalks. Unexcitingly dark green in spring (albeit balsam-scented), but the big later growth-tip leaves are *splashed with, or entirely, cream*. (Green, reverted trees are lower and more densely leafy than Noble Poplar, p.158.)

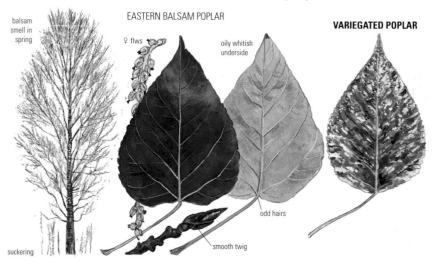

EASTERN BALSAM POPLAR — ♀ flws — oily whitish underside — odd hairs — smooth twig

VARIEGATED POPLAR

balsam smell in spring — suckering

POPLARS (ASIATIC)

Simon's Poplar *Populus simonii*

(*P. przewalskii*) N China. 1862. Rare, but in the odd park or shelterbelt for its graceful *weeping habit* and balsam smell.
Appearance Shape A sprouty, usually cankered, often slanting trunk and steep main limbs carry fine, hanging shoots. To 20 m. **Shoots** Angular, hairless. **Leaves** Narrowly *diamond-shaped* (often *broadest above halfway*); green-white beneath and hairless; very variable in size – to 12 cm but with many *only 5 cm*, on a stalk just *1–2 cm long*. Flushing early.
Compare Berlin Poplar (p.152): a leafier tree of stronger, stiffer growth, its leaves usually broadest below halfway up.
Variants 'Fastigiata' is a roughly, untidily *columnar* clone (whose finer branches still weep).
Other trees *P.* × *acuminata* (a hybrid balsam poplar from W North America; in a few collections) shares the graceful, weeping habit and diamond-shaped leaves, which are *more uniform in size and shining green beneath*.

Doronoki *Populus maximowiczii*

(Japanese Balsam Poplar) NE Asia, Japan. *c.*1918. Rare: big gardens.
Appearance Shape Domed; often low. **Bark** Very grey; rather flat ridges. **Leaves** Carried densely and *level*; they flush very early. Rounded (slightly diamond-shaped on stronger growths): *short, sudden points usually bend down and sideways* (cf. Goat Willow, p.168); whitish green beneath; *finely hairy along the sunken veins on both sides* at least at first.
Flowers Spectacularly profuse catkins.
Other trees Hybrids from Doronoki selected in Maine in 1934 inherited the distinctive leaf-shape (though more triangular on strong growths) and are still seen as much-cankered trees in some big gardens. They include the hugely vigorous, *straight* and more open *P.* 'Androscoggin' (male parent Western Balsam Poplar, p.160), the dense and irregularly upright female *P.* 'Oxford' (male parent Berlin Poplar, p.152), and the slightly more open, tall-growing female *P.* 'Rochester' (male parent Plantier's Poplar, p.154).

Chinese Necklace Poplar *Populus lasiocarpa*

Central and W China. 1900. Rather rare: large gardens in warmer parts.
Appearance Shape Gaunt: a few level branches on a usually straight trunk; to 25 m. **Bark** Soon *shaggy*; grey, scaling ridges. **Shoots** Thick, with *fawn velvet* for several years. **Leaves** Like blotting paper; *huge, heart-shaped* (to 35 cm), *woolly beneath*; normally with pink/red hairy veins and on stalks almost like rhubarb-sticks. **Flowers** The commonest clone, unlike all other poplars, bears catkins (to 20 cm long) with both male and female flowers, and is self-fertile.
Compare Other trees with giant heart-shaped leaves: American Lime (p.402); Idesia (p.408).
Variants Var. *thibetica* (*P. szechuanica*; *P. violascens*) is a less spectacular tree in a few collections: twigs soon hairless; hairs under the leaves soon confined to the main veins.

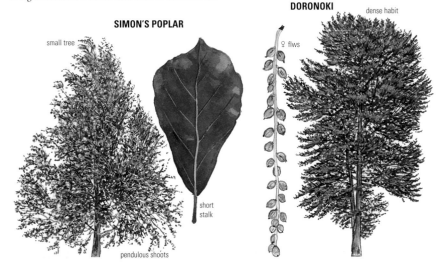

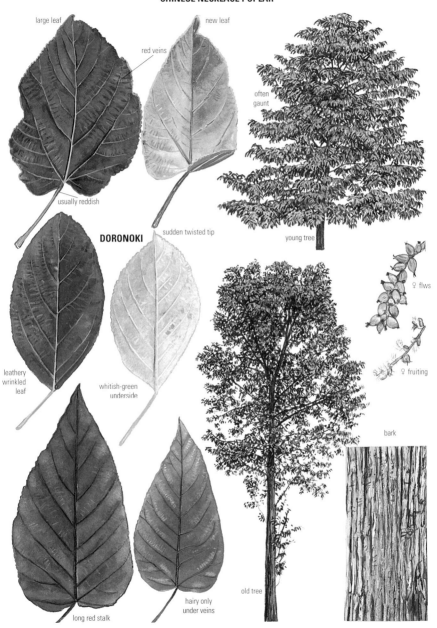

WILLOWS

Willows (400 species, from trees to prostrate sub-shrubs) grow in all continents except Australia and Antarctica. Like poplars, they tend to suggest a big branch stuck in the ground. Their buds have one smooth, flattened scale; the shoots curve evenly and taper to an aborted tip. Short female catkins ('pussies') attract pollinating insects. (Family: Salicaceae.)

Things to Look for: Willows

- Shape/size?
- Shoots/young bark: Colour?
- Leaves: Downy beneath? How broad? How glossy?

Key Species

White Willow (below) and **Crack Willow** (p.166): tall trees with narrow, boat-shaped leaves. **Golden Weeping Willow** (p.168): similar; weeping habit. **Sallows** (p.168): oval leaves, finely grey-woolly beneath. **Bay Willow** (p.166): broadly boat-shaped, hairless leaves. **Common Osier** (p.170): bushy; very narrow leaves.

White Willow   *Salix alba*

Europe including Britain and Ireland; W Asia, N Africa. Locally abundant but generally confined to the banks of rivers, larger ponds and marshland ditches; the wild form is seldom planted.
APPEARANCE Shape A short leaning bole carries big, *ascending* limbs; the fine shoots *droop*. A large tree, to 30 m, flushing yellow; dark, misty grey in summer, and dull grey-brown in winter; often pollarded. **Bark** Dark *grey*; rugged, criss-crossing ridges. **Shoots** Very slender, grey; softly hairy for a year. **Buds** Slender, flattened, silky. **Leaves** Silky-hairy above at first, and remaining (wild forms) *silver-downy beneath*, *with odd hairs above*; about 8 cm.
COMPARE Crack Willow (p.166); Willow-leaved Pear (p.320); Oleaster (p.412).
VARIANTS Silver Willow, var. *sericea* (var. *argentea*), is a rather occasional garden tree, whose leaves *stay silky-hairy above*. Slower (to 25 m), with *non-drooping shoots* – like a puff of pale smoke.
Cricket-bat Willow, var. *caerulea*, is frequent, often in lines and plantations in wet ground – growing with extraordinary vigour to 30 m but also providing the best timber for bats; sometimes naturalized. An almost *straight* trunk creates a rough *spire* but can fork low into steep limbs; shoots dark *purplish red* (though finely grey-downy); leaves *almost hairless by late summer* but vivid blue-grey beneath. *Female*: green spring catkins quickly shed fluffy seeds.
Coral-bark Willows (including 'Britzensis' and 'Chermesina') are conspicuous in parks and gardens, to 28 m: shoots *brilliant orange* (amber towards spring; cf. Basford Willow, p.166), the tree in winter sunlight is like a giant flame. Browner bark than the type's; wispily spire-shaped when young, the twigs always *rising and incurving*. Leaves are *soon almost hairless* but blue-grey beneath; crown a distinctive pale *greyish yellow* in summer. *Male trees*, gold in early spring with short catkins. (The female 'Cardinal' is much rarer.)
Golden Willow, var. *vitellina*, is a name sometimes used for Coral-bark Willows but more precisely refers to clones (now rare) with *clear yellow* shoots.

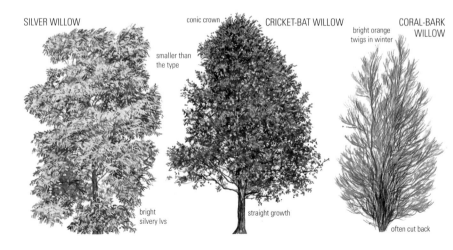

SILVER WILLOW — smaller than the type — bright silvery lvs

CRICKET-BAT WILLOW — conic crown — straight growth

CORAL-BARK WILLOW — bright orange twigs in winter — often cut back

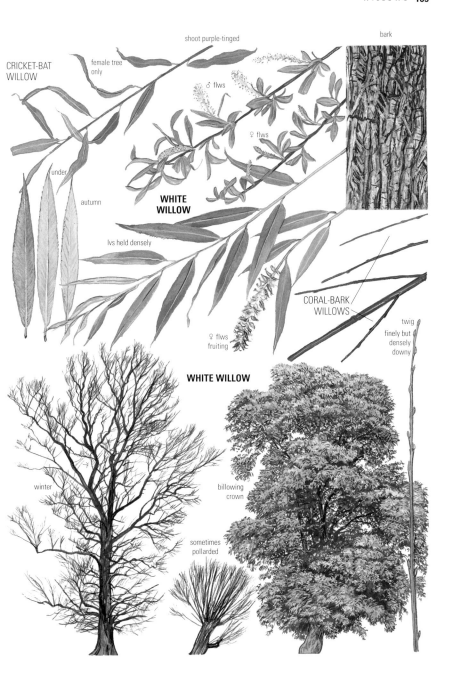

WILLOWS

Crack Willow *Salix fragilis*

Europe E to Romania, including Britain and Ireland. Abundant in wetter ground, though seldom planted. Most tall-growing wild trees may be hybrids (*S.* × *rubens*) with White Willow, or may descend from the selection var. *russelliana* ('Duke of Bedford's Willow').
Appearance Shape Broad and short-boled except in woodland, with *wide-spreading* branches and an open interior; twigs not weeping. Glossy-green foliage; dull orange in winter (cf. Coral-bark Willow, p.164). Leaning and much shattered with age; often pollarded. **Bark** Dark *brown*; very rugged criss-cross ridges. **Shoots** Yellow-brown, shiny, soon *hairless*; buds narrow, smooth. Side-twigs snap off readily (whence the common name), and lodging point-up in river-mud will root and grow a new tree (the seeds are often sterile). **Leaves** *Long* (to 15 cm), shiny; slightly silky at first then *hairless* though blue-green underneath.
Compare White Willow (p.164); Bay Willow (below); Almond Willow (p.170).
Variants Basford Willow, *S.* × *rubens* 'Basfordiana' (*c.*1863), is Crack Willow's (very occasional) answer to Coral-bark Willow (p.164). *Brilliant orange* shoots; very *spreading*, twiggy habit; a male tree. The female 'Sanguinea' is much rarer and has smaller leaves (8 cm). The often bushy var. *decipiens* is a (rare?) wild form whose red young shoots become duller grey by winter.

Bay Willow *Salix pentandra*

N Eurasia (including N England, Scotland, N Ireland and N Wales); in high mountains further S. Frequent by rivers and in wet woodland; in S England a very occasional park tree (rarely naturalized).
Appearance Shape A *dense* dome, to 20 m, or bushy; blackish, almost luminously glossy foliage. **Bark** *Dark grey*; scaly criss-cross ridges. **Shoots** Glossy green-brown, hairless. **Buds** Glossy brown. **Leaves** *Hairless*, thick, shiny – *rather like Bay's* (p.276), but deciduous; finely toothed; blue-grey beneath. **Flowers** Catkins on *leafy shoots in late spring* (opening with/before the leaves in most willows).
Compare Crack Willow (above). The two trees' hybrid, *S.* × *meyeriana*, is intermediate and very rare (as is *S.* × *ehrhartiana* – the hybrid of White Willow, with *odd hairs on both leaf-surfaces*). Almond Willow (p.170): less glossy leaves.
Other trees *S. lucida* (E North America) is much rarer: *long fine leaf-tips*; flowering shoots downy.

Dragon's Claw Willow
Salix babylonica 'Crispa'

('Annularis') China. The true *S. babylonica*, the Chinese 'Willow Pattern' weeping willow, is confined to a few collections in the UK and needs more summer warmth to grow well.
Appearance Shoots Tending to arch gracefully rather than hang straight. **Leaves** Slenderer, *brighter pale green* than other 'weeping willows' (p.170–72), soon hairless. Dragon's Claw Willow is a rare (and scarcely weeping) Chinese selection: its leaves *curve in circles*.

Corkscrew Willow
Salix babylonica var. *pekinensis* 'Tortuosa'

(*S. matsudana* 'Tortuosa') N China/Japan. 1925. Abundant, but very short-lived.
Appearance Shape An electrocuted-looking shock of *corkscrew* branches, then broad-domed and slightly weeping; the fine shoots still curl crazily (cf. Corkscrew Hazel, p.198). Pale, soft green until December. **Bark** Pale brown; relatively shallow criss-cross ridges. **Leaves** To 8 cm, variously buckled but not curving in rings. Golden Weeping Willow (p.168) can rarely grow buckled leaves.

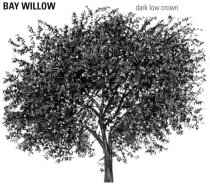

BAY WILLOW — dark low crown

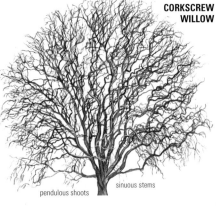

CORKSCREW WILLOW — sinuous stems, pendulous shoots

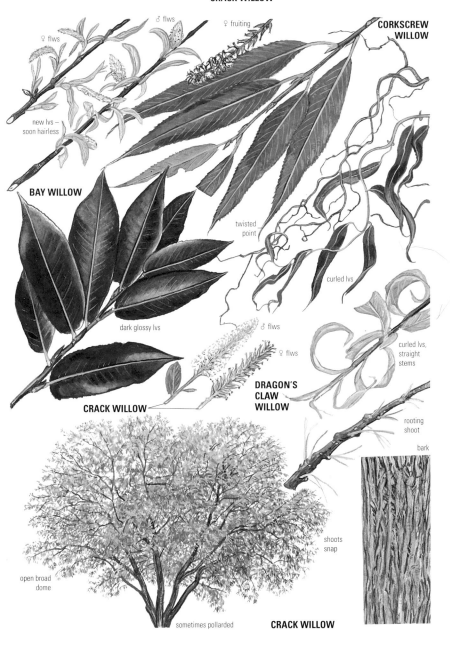

GOAT TO GOLDEN WEEPING WILLOWS

Goat Willow — *Salix caprea*

(Great Sallow; Pussy Willow) NW Eurasia, including Britain and Ireland. Very abundant except on the lightest soils.
Appearance Shape Domed, to 22 m; weak arching branches on usually a single trunk. **Bark** Grey: at first banded with small diamond-shaped pits; soon with rather shallow, criss-cross ridges. **Shoots** Grey (red/yellow in sun), thicker than most Willows', *soon hairless*. **Buds** Rather rounded (*downy* in var. *sphacelata* (*S. coaetanea*), from N Europe (Highland Scotland included) and the Alps, which has narrower, almost *untoothed* leaves). **Leaves** *Not more than twice as long as broad*, the *abrupt tip bent sideways*; dark, *wrinkly*; a very fine grey-green felt beneath; few or no teeth. **Flowers** *Precede the leaves*. Male trees' 'pussies' gold; females' silver, quickly shedding fluffy seeds. **Compare** Grey Sallow (below). A range of rare hybrids (*S.* × *reichardtii*) appear to link the two. **Variants** Weeping Sallows, 'Kilmarnock' (male) and 'Weeping Sally' (female), are abundant but very short-lived: weak stems *arch down* from the graft-point (cf. Weeping Osier, p.170).

Grey Sallow — *Salix cinerea* ssp. *oleifolia*

(Rusty Sallow; *S. atrocinerea*) W Europe, including Britain and Ireland. As abundant as Goat Willow, except on dry sites (both are known as 'Common Sallow'); differences as follows.
Appearance Shape *Bushy*, seldom long-trunked; to 15 m. **Bark** Grows darker, shallower ridges. **Shoots** Minutely *hairy for a year*; 2-year twigs *ridged* under their bark. **Buds** Minutely *hairy for a year*. **Leaves** Usually much smaller; *2–3 times as long as broad* (cf. Smith's Willow, p.170), and broadest above halfway up; a fine felt beneath plus odd *rusty* hairs under the veins. Semicircular stipules are common (as in Almond Willow (p.170) and Violet Willow). **Flowers** Starting later than Goat Willow's, the 'pussies' slightly smaller.
Other trees The type is local and bushy: downier shoots; greyer leaves lacking rusty hairs.
S. pedicellata (Mediterranean Europe) has larger leaves (*10–12 vein-pairs*).

Violet Willow — *Salix daphnoides*

Central Europe. 1829. A very occasional garden tree; rarely naturalizing.
Appearance Shape A *low*, rough, dark grey-green dome (to 18 m), short-boled. **Bark** Grey; relatively shallow criss-cross ridges. **Shoots** Green/brown, downy then *bloomed purple-grey* (spectacular if coppiced; cf. Purple Osier, p.170). **Buds** Shiny dark red. **Leaves** *Small* (to 12 cm), narrowly boat-shaped (cf. Cricket-bat Willow, p.164): dark and shiny; blue-grey beneath; soon hairless, but on woolly stalks.

Golden Weeping Willow — *Salix* × *sepulcralis* 'Chrysocoma'

(*Salix alba* 'Tristis'; *S. alba* var. *vitellina pendula*) Berlin. By 1888. Abundant in warm areas, deriving its habit from Chinese Weeping Willow (p.166) and its vigour and twig-colour from Golden Willow (p.164).
Appearance Shape A broad head of twisting limbs (to 24 m) clothed in *long, straight hanging shoots* – 6 m long in ideal conditions. In leaf March–December, *pale yellow-grey-green*. **Bark** Grey-brown; deep criss-cross ridges. **Shoots** Green, then (in sun) greyish *gold* for many years. **Leaves** Unfolding silky-grey; soon hairless above; hairless beneath in 3 months (but blue-grey). **Flowers** A male clone (infertile; it grows odd female catkins). **Compare** Dragon's Claw Willow (p.166); Salamon's and Thurlow Weeping Willows (p.170).

GOAT WILLOW — spring ♂ catkins

VIOLET WILLOW

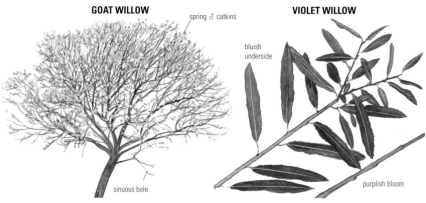

bluish underside

sinuous bole

purplish bloom

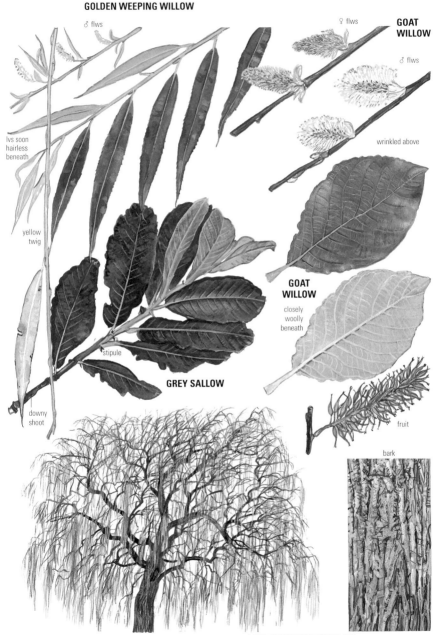

WEEPING TO SMITH'S WILLOWS

Salamon's Weeping Willow
Salix × sepulcralis 'Salamonii'

France. By 1864. Occasional as an older park tree, now replaced by Golden Weeping Willow (p.168).
APPEARANCE Shape Weeping, but with much shorter 'streamers'; *greyer* in summer with downy new growths. **Bark** *Darker grey*; more rugged. **Shoots** Green then *dull greyish brown* (cf. Thurlow Weeping Willow, below, and Chinese Weeping Willow, p.166). **Leaves** Slower to become hairless.

Thurlow Weeping Willow *Salix × pendulina*

(*S. × elegantissima*) Weeping willows with big, glossy leaves are presumably hybrids of Chinese Weeping and Crack Willows. Rather rare.
APPEARANCE Shape Weeping, though much less dramatically and tidily than Golden Weeping Willow (p.168). **Bark** Brown-grey; very *coarse* criss-cross ridges. **Shoots** *Dull grey*-brown (cf. Salamon's Weeping Willow, above). **Leaves** *Large, rather glossy*, to 15 cm, quickly hairless.
VARIANTS Wisconsin Weeping Willow, 'Blanda' (*S. × blanda*), is a low, less weeping form with broader, *very glossy* leaves; very rare.

Common Osier *Salix viminalis*

Mid Europe, including S England. Locally abundant (marshes); much planted/hybridized for basketry.
APPEARANCE Shape Generally bushy, to 10 m: vigorous *wand-like stems* from a short, gnarled, scaly trunk. **Shoots** Soon hairless, yellowish. **Buds** Silky-white, rounded; closely set *like strings of pearls*. **Leaves** *Very narrow* (to 20×1 cm); dark and wrinkled above, silky-white under; *untoothed* (cf. Oleaster, p.412).
OTHER TREES Hoary Willow, *S. elaeagnos* (central Europe, Asia Minor; in some gardens), has *minutely toothed* leaves, woolly- not silky-hairy beneath.

Almond Willow *Salix triandra*

(French Willow; *S. amygdalina*) Eurasia, including England, E Wales, S Scotland and S Ireland. Once much grown/hybridized for basketry; now rather occasional in marshes/flood-plains.
APPEARANCE Shape Often bushy (rarely to 20 m). **Bark** Grey-brown, *peeling and flaking orange* on small branches; later ruggedly ridged. **Shoots** Shiny brown; soon hairless. **Leaves** Like Crack Willow's (p.166) but smaller (5–10 cm) and *hairless*; green or bluish beneath. **Male flowers** *3 stamens*.

Purple Osier *Salix purpurea*

Eurasia, including Britain and Ireland; N Africa. Rather occasional in wet places, including upland bogs. Once much planted/hybridized for basketry.
APPEARANCE Shape Bushy. **Bark** *Shining grey*, with lenticel-bands; not fissured. Particularly bitter with salicylic acid (the active principle in aspirins). **Shoots** Soon hairless; yellow, or shining *red-purple* in sun (cf. Violet Willow's *bloomed* twigs, p.168). **Leaves** Soon hairless; like Crack Willow's (p.166) but smaller (to 12 cm), broadest towards the often blunt tip, and toothed *only* (except in ssp. *lambertiana*) near the apex; *often in opposite pairs*.
VARIANTS Weeping Osier, 'Pendula', is an occasional garden tree whose weak purplish stems arch down from the graft-point (cf. weeping sallows, p.168).

Smith's Willow *Salix × sericans*

(*S. × smithiana*) The largest and most distinct of several sallow/osier hybrids (*S. caprea × viminalis*): a wild tree (?), and in odd gardens for its display of pussies (normally a female clone).
APPEARANCE Shape Surprisingly sturdy; to 12 m. **Bark** Dark grey; big criss-cross ridges. **Shoots** Soon hairless. **Leaves** Broadly *boat-shaped*, to 15 cm; wrinkled; *green* beneath though *very thinly* felted.

ALMOND WILLOW **THURLOW WEEPING WILLOW**

winter

peeling bark

big glossy lvs

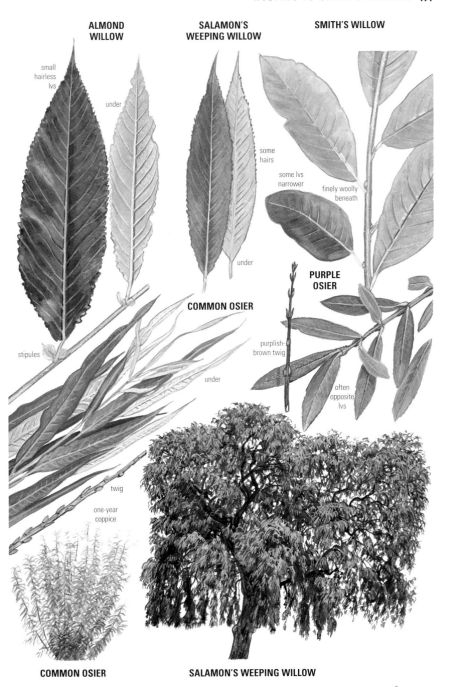

WINGNUTS

Wingnut trees (about 8 species) have big, compound, alternate leaves; male (and usually female) flowers in catkins. Like walnuts (but not hickories), wingnuts have a chambered pith – an obliquely cut one-year-old twig shows a series of close divisions. (Family: Juglandaceae.)

Things to Look for: Wingnuts

- Bud: Does it have scales?
- Leaves: How many leaflets? Is the central leaf-stalk grooved? Is it winged?

Caucasian Wingnut *Pterocarya fraxinifolia*

Caucasus; N Iran. 1782. Rather occasional, but more recently in many town parks for its almost vandal-proof vigour.
Appearance Shape Often many-stemmed, or with a cup of heavy limbs from a short lumpy bole. Coarse and gaunt, but magnificent when to 35 m; a vivid, luxuriant green. Trees in unmown areas hide themselves in huge *thickets of suckers*. **Bark** Grey-brown: very *coarse* ridges criss-crossing over one another. **Shoots** Thick, almost hairless. **Buds** Often some distance above last year's leaf-scar. *No scales* – mere miniature leaves, *on short stalks* and clad in *rufous hairs*. **Leaves** Distinguished from those of other trees with large, long, alternate, compound foliage (walnuts, Pecan, sumachs, rowans, Tree of Heaven, Chinese Cedar) by the *oblong*, stalkless, overlapping, *floppy* leaflets (up to 25), shiny above and with some pale, long, star-like hairs under the midrib. The central stalk is sometimes finely grooved above but *rounded*. Autumn colours clear yellow. **Female flowers** Catkins, to 50 cm long and conspicuous through summer; strung with green nuts which have two 1 cm *angled* wings.

Hybrid Wingut *Pterocarya × rehderiana*

(*P. fraxinifolia* × *stenoptera*; USA 1879, and several times since.) Very occasional; one of the most vigorous of hardy trees, to 30 m.
Appearance Shape Short single trunk (sometimes a bush) and often very broad crown; suckering. **Bark** Grey; relatively *neat*, shallow, criss-crossing ridges. **Leaves** Fewer more *rounded* and distant leaflets than Caucasian Wingnut's. The central stalk has 2 interrupted *vertical flanges, with a deep groove between*.
Compare Pecan (p.174); Butter-nut (p.180). The flanged leaf-stalk distinguishes.

Chinese Wingnut *Pterocarya stenoptera*

China; N Vietnam. 1860. Rare, but in a few town parks in warm areas as well as bigger gardens.
Appearance Shape Sometimes bushy; sometimes a short-boled, broad, shapely tree, *not suckering*. Most closely resembles Hybrid Wingnut (above). **Shoots** With long, brown hairs when young. **Leaves** Central leaf-stalk has *spreading, toothed wings*.

Japanese Wingnut *Pterocarya rhoifolia*

One of Japan's largest trees. 1888. A handsome plant in one or two collections.
Appearance Very like Caucasian Wingnut, with round, unwinged central leaf-stalks, but differences as follows: **Buds** (Until midwinter and unlike other wingnuts') with 2 or 3 *dark brown scales*. **Leaves** Not more than 21 leaflets, more distant, with *long slender points*; sometimes hairy beneath but sometimes only with tufts under the vein-joints. **Fruit** The wings of the nuts are *horizontal*, not angled.

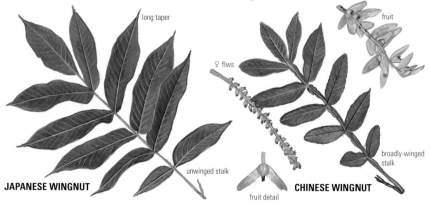

JAPANESE WINGNUT — long taper, ♀ flws, unwinged stalk
CHINESE WINGNUT — fruit, broadly-winged stalk, fruit detail

HICKORIES

Hickories (25 species) hail mostly from Eastern North America. Trees have slender limbs, large end buds and often huge leaflets. Their timber is strong but flexible – the best for tool-handles. Their name derives from a Native American word for 'nut paste'. (Family: Juglandaceae.)

Things to Look for: Hickories

- Bark: How 'shaggy' is it?
- End bud: What is its size, shape and colour?
- Leaves: How many leaflets? Hairy/scented? End leaflet stalked? Main stalk downy?

Pecan — *Carya illinoinensis*

Iowa south to N Mexico. *c*.1760. The tallest hickory (to 52 m in the wild), with the best nuts. Very rare: growing well only in the warmest areas.
Appearance Shape Broader-domed than most hickories, on a slender trunk and branches. **Bark** Grey-brown; close, scarcely shaggy ridges. **Shoots** Furry, then shiny brown. **Buds** End bud (relatively) small, chocolate brown; downy at its tip. **Leaves** Typically with *11–15* smallish leaflets, long-pointed, the lower ones at least *aerofoil-shaped* (curving backwards); finely downy beneath. **Fruit** The commercial pecan, seldom ripening in the UK.
Compare Black Walnut (p.178); Varnish Tree (p.360); Chinese Cedar (p.358); Hybrid Wingnut (p.172). None has curved leaflets.

Mockernut — *Carya tomentosa*

(Big-bud Hickory; *C. alba* K. Koch) Ontario to Texas. 1766. Rare: big gardens in warm parts.
Appearance Shape Domed with age, but on a long bole; to 27 m. **Bark** Purple-grey with (in youth) shallow, rounded, criss-cross ridges; eventually slightly 'shaggy' (cf. Pignut, p.176). **Shoots** Dull brown, with short, stiff hairs. **Buds** End bud very *big* (2 cm; twice the shoot's width), velvety. **Leaves** With 7 (9) huge, thick, drooping leaflets, shiny dark green above, the end one on a *slim stalk* to 4 cm. Central stalk has *hard dense hairs*; hairs under the leaf and above midrib. Usually a *sweet paint/grass smell*. **Fruit** Very thick-shelled inedible nuts.
Compare Big Shellbark Hickory (below): end leaflet scarcely stalked. Bitternut (p.176): beaked yellow buds; smaller leaves usually with 9 leaflets. Shagbark Hickory (p.176): 5 leaflets. Red Hickory (p.176): much less downy.

Big Shellbark Hickory — *Carya laciniosa*

(Kingnut; *C. sulcata*) New York to Oklahoma. 1804. Rare: large gardens in warm parts.
Appearance Shape Slender and irregular, on a long bole; to 27 m. **Bark** Usually soon shaggy and *scaly*; narrow ridges lift at top and bottom, like skis. **Shoots** Yellow/pink; downy at first. **Buds** End bud *huge* – to 25 mm; downy green/brown scales. **Leaves** Often the largest of any hickory, with 7 (5 or 9) thick, hard leaflets to 35 cm, the end one on a stalk only to 1 cm long; almost hairless above and finely downy beneath. Central stalk *finely and softly downy*. **Fruit** Nuts almost as good as pecans.
Compare Shagbark Hickory (p.176): 5 leaflets. Mockernut (above).

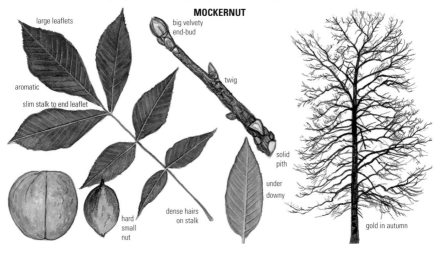

MOCKERNUT

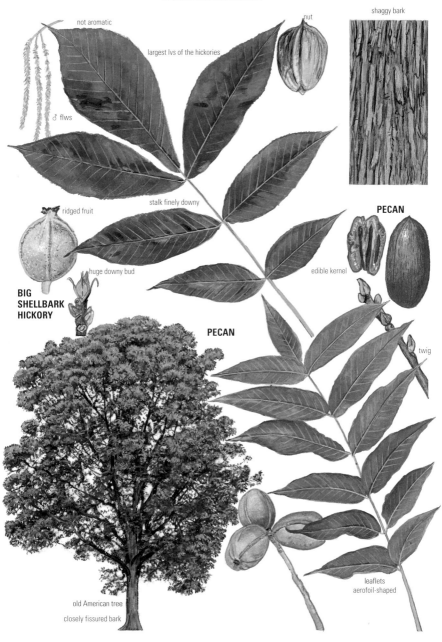

HICKORIES

Bitternut *Carya cordiformis*

(*C. amara*) Quebec to E Texas. 1689. Rather rare: large gardens.
APPEARANCE Shape Long-trunked to dense, often long-conic tips; leafier in the UK than other hickories; to 30 m. **Bark** *Close*, criss-cross ridges become slightly 'shaggy' with age. **Shoots** Slender, olive, soon *hairless*. **Buds** End bud hairy, often *rich yellow, with long, curved, flattened beaks*. **Leaves** Typically with 9 *relatively small ash-like leaflets* (rarely 5 or 7), the end one *scarcely stalked;* they are downy at least under the veins. Central stalk finely and persistently hairy. **Fruit** Small, bitter nuts.
COMPARE Big Shellbark Hickory (p.174): shaggy bark, brown buds and finely hairy shoot; usually 7 leaflets (other hickories seldom have 9 leaflets). White Ash (p.440): similar jizz, but *leaves opposite*.

Shagbark Hickory *Carya ovata*

(*C. alba* Nutt.) Quebec to NE Mexico. 1629. Rather rare: large gardens in warmer parts.
APPEARANCE Shape Long, light, arching branches from a long trunk; rather gaunt and easily storm-broken; to 27 m. **Bark** Narrow criss-crossing plates usually curl from the trunk at top and bottom, like a flotilla of gondolas – probably to deter squirrels from reaching the nuts; rarely remaining smooth. **Shoots** Thick, variably velvety; often with a ring of red-brown down by each bud. **Buds** End bud *fat*, conic, 15 mm, green-brown/dull gold, sometimes beautifully silky. **Leaves** With 5 (very rarely 7) leaflets, often large – end leaflet to 30 cm (on a *stout*, 1 cm stalk) – and sometimes almost hairless, though the margin retains rufous tufts between the teeth; often thick, with an oily finish. The stout central stalk usually becomes hairless *except for dense hairs on the bulbous base*. **Fruit** Nuts as good as pecans, though seldom grown commercially.
COMPARE Big Shellbark Hickory and Mockernut (p.174): typically 7 leaflets. Pignut (below) has small buds and almost hairless foliage.

Pignut *Carya glabra*

(Smoothbark Hickory; *C. porcina*) Ontario to E Texas. *c.*1750. Rare: large gardens in warm areas.
APPEARANCE Shape Light rising branches from a long trunk. To 26 m; dark and glossy in summer. **Bark** A beautiful, almost metallic grey-purple, wrinkling into generally smooth, shallow criss-cross ridges; or shaggy with age. **Shoots** *Hairless* from the first. **Buds** End bud *small* (7 mm), yellowish green. **Leaves** With 5 (7 or 3) leaflets, the end one scarcely stalked; often smaller than most hickories'. Central stalk *hairless*; leaflets *hairless* except under the main veins. **Fruit** Sweet nuts.
Compare Mockernut (p.174): very hairy shoots and central leaf-stalks. Sweet Chestnut (p.212): similar jizz; similar bark in youth; leaves match Pignut's largest (atypical) leaflets.
OTHER TREES Red Hickory, *C. ovalis* (*C. glabra* var. *odorata*), the commonest hickory in the wild and perhaps a hybrid with *C. ovata*, is rarer here. Bark shaggier with age; young shoots *finely hairy*, stouter, reddish; leaflets (typically 7) finely hairy beneath at first. A much less downy tree than Mockernut (p.174).

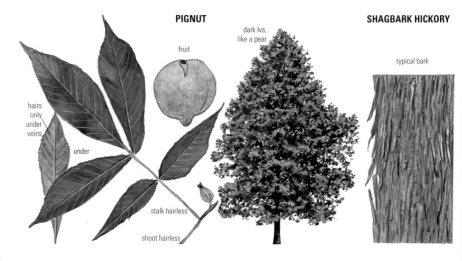

PIGNUT — fruit; hairs only under veins; under; stalk hairless; shoot hairless
dark lvs, like a pear
SHAGBARK HICKORY — typical bark

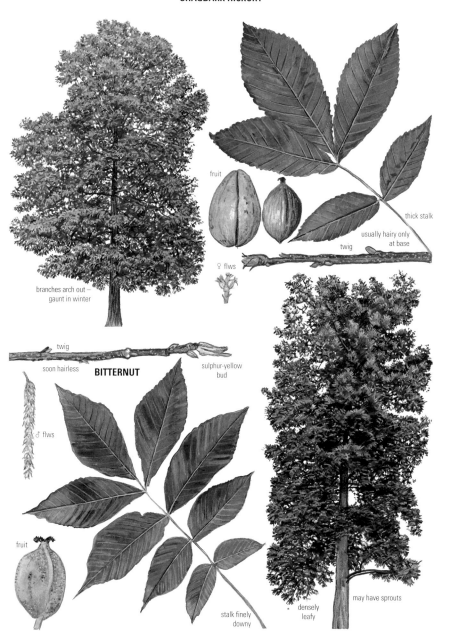

WALNUTS

Walnuts (15 species) have big compound leaves and broadly triangular buds. Fungi associated with the roots can poison the soil, which limits nearby competition. (Family: Juglandaceae.)

Things to Look for: Walnuts

- Bark
- Shoots: Are they hairy or sticky?
- Leaves: Are the leaflets serrated?
- Fruit: Shape of nuts? How smooth? Are they single or in clusters, or on tails?

Common Walnut *Juglans regia*

SE Europe through the Himalayas and N Burma to SW China. Long cultivated in the UK and rarely naturalized on limestone in S England; more frequent in countryside gardens than in towns. It likes a dry climate but some of the best of the few big trees are in Scotland. Sometimes long-lived. The timber is the most valuable grown here: trees are dug out rather than felled, as the best wood is at the base.

Appearance Shape Wide and gaunt, on heavy, *twisting*, oak-like branches; to 30 m. Leaves flush late and coppery. **Bark** Shallow, rounded *silvery-grey* ridges; darker grey and rougher with age. **Shoots** Stout, curving, almost *hairless*. **Buds** Squat, purple-brown. **Leaves** With 5–13 (*usually 7*) leaflets: the end one *very large* (to 20 cm), the basal pair *much smaller*. Leaflets oval, *untoothed* (odd peg-teeth on lower pairs and saplings); shiny and *leathery*, hairless except for tufts under vein-joints. Strong smell of shoe-polish. **Fruit** Nuts ripen only in long, hot summers. Many fruiting clones are grown; in orchards, trees traditionally 'distressed' by thrashing the foliage to stimulate fruiting.
Compare Yellow-wood (p.352); Tree of Heaven (p.358): similar in winter; darker smoother bark.

Variants Cut-leaved Walnut, 'Laciniata', is rare but attractive: a fresh-green, feathery little tree.

Black Walnut *Juglans nigra*

E and central USA; another valuable timber species. Long grown in the UK; rather occasional as a magnificent specimen in old parks but now a little more widely planted in warmer areas.
Appearance Shape A grand dome, to 35 m, on less twisting branches than Common Walnut. By early summer a very leafy, rich vivid green. **Bark** Grey or blackish; deep criss-crossing ridges. **Shoots** Brown, hairy; pale grey velvety buds. **Leaves** With 10–23 slender leaflets, but often no end one (cf. Chinese Cedar, p.358); leaflets finely toothed and downy beneath, on minutely hairy central stalks. Little aroma. **Fruit** Nuts abundant and strong-flavoured (when, in warm climates, they ripen properly), but in a very thick husk – special nutcrackers are sold in North America to open them. They are carried singly or in pairs except in the clone 'Alburyensis' (original tree at Albury Park, Surrey, with scions in some collections), which, like Butter-nut (p.180), carries clusters of up to 5.
Compare Chinese Varnish Tree (p.360): untoothed leaves (and irritant sap: any potential foliage should be handled with care). Chinese Cedar (p.358): untoothed leaves with garlic smell. Pecan (p.174): paler bark, aerofoil-shaped leaflets. Some other walnuts (p.180): sticky red hairs, broad leaflets, and clustered fruits. A very different tree from Common Walnut (above), though hybrids (*J. × intermedia*) are known.

COMMON WALNUT

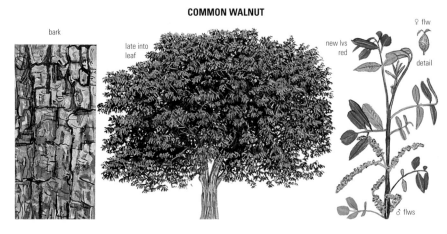

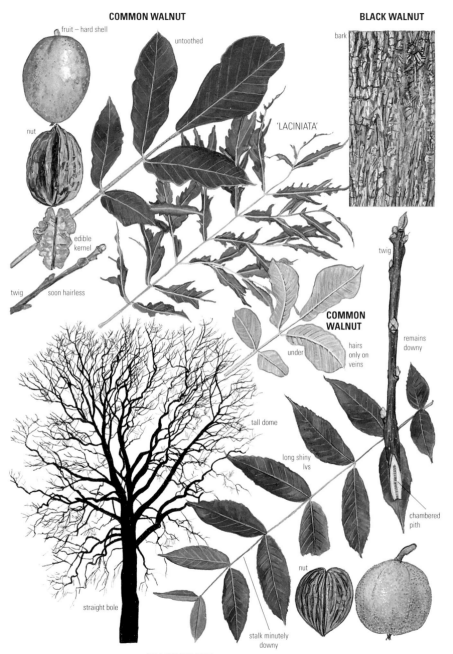

WALNUTS

Butter-nut *Juglans cinerea*

E North America – Minnesota to Georgia. 1633. Rare. Leaves and bark were once widely used in North America as a laxative.

Appearance Shape A slender tree to 24 m, its foliage most like Black Walnut's (p.178). **Bark** *Like Common Walnut's* (p.178) though sometimes rougher and darker. **Shoots** With a long, red, rather sticky felt which wears off after a year, and *a persistent hairy band* between each leaf-scar and *pink-white bud*. **Leaves** Very big (to 70 cm); 11–17 bright green, broad, flimsy leaflets, hairy above at first and grey underneath with starry down; the central stalk is *covered in sticky red hairs. Strong, sweet, paint-like smell*. **Fruit** Nuts in a *cluster (3–5;* cf. Black Walnut 'Alburyensis', p.178), the rinds also covered in sticky hairs; husk tapering to a short point. Sweet and buttery, but seldom ripening in the UK.

Compare Japanese Walnut (below): even stickier leaf-stalks.

Japanese Walnut *Juglans ailanthifolia*

(*J. sieboldiana*) Japan, Sakhalin. 1860. Rare, but growing well in Scotland.

Appearance Shape Broad but sometimes gaunt; to 20 m; suckering. **Bark** Grey, with shallow ridges: slightly rougher than Common Walnut's (p.178). **Shoots** With fine white streaks, and short dense sticky whitish hairs for up to 3 years. The big, pale leaf-scars are trefoil-shaped. **Leaves** Huge (to 1 m); 9–21 big, rather oblong leaflets, bright and shiny but finely downy above and very downy beneath; they taper *abruptly to a short point*. The central leaf-stalk is covered more densely than Butter-nut's in dark red, sticky down. **Fruit** Nuts *clustered on a long* (15 cm) *hanging tail*, in rinds with sticky hairs; husk pointed, with a rather *smooth surface but a prominent ridge where the halves meet*. In late spring the flower-head is erect and the female flowers (*12–20*) are quietly spectacular – paired, 1 cm crimson plumes. In the Heart-nut (var. *cordiformis*, cultivated in Japan for its fruit, and not found wild), the husk is *flimsy* and heart-shaped at the base.

Compare Manchurian Walnut (below): differences emphasised. Downy Tree of Heaven (p.358): leaflets with 1–6 big glandular teeth at the base.

Manchurian Walnut *Juglans mandshurica*

(NE mainland Asia, 1859. Very rare in the UK but less so in continental Europe.

Appearance Shape As Japanese Walnut (above): often very low and wide. **Bark** *Pinkish-grey*; shallowly ridged. **Leaves** As big as Japanese Walnut's; the matt yellowish-green leaflets have *longer, finer points* and *very shallow toothing*. **Fruit** Larger but shorter, and often rounded; in *short hanging clusters that develop from heads of 5–10 flowers*; the husk is *deeply pitted and longitudinally grooved*, but *not prominently ridged* where the halves meet.

Other Trees *J. cathayensis*, from central China, is equally frequent in UK collections, but is very difficult to distinguish and sometimes merged.

JAPANESE WALNUT

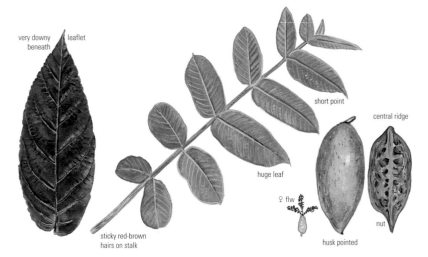

WALNUTS

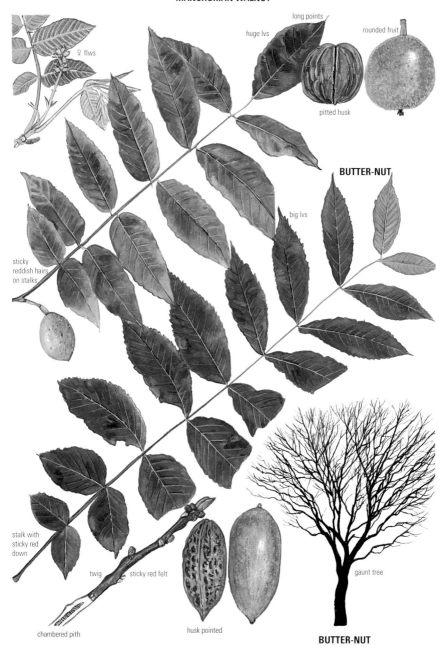

BIRCHES

Birches (60 species; hybridizing freely) have light airy crowns and typically bright peeling bark. The trees display their male catkins through the winter months. Their shoots are slender, with big, conic, often sticky buds. (Family: Betulaceae.)

Things to Look for: Birches

- Bark: What colours? How rugged?
- Shoots: Are they downy?
- Leaves: How dark/glossy? Are they doubly (regularly) toothed? Is the leaf-stalk downy? How many leaf-veins (how close?, how parallel?)?
- Fruiting catkin-scales – Downy? Lumpy?

Silver Birch *Betula pendula*

(Warty Birch; *B. verrucosa*; *B. alba* in part) Europe including Britain and Ireland (this and Downy Birch are the only wild trees in Iceland); NW Asia. Transiently dominant on sandy soils. One of the prettiest, airiest birches, planted everywhere.
APPEARANCE Shape Twigs soon *weep*, like fountains on the best trees. To 30 m; short-lived except in Highland Scotland. **Bark** Orange-red on young stems; soon white, but growing *rough black arrows/diamonds* until the whole butt is *dark and rugged*. **Shoots** Hairless (except for strong sprouts); purple-brown; *scrubby with little white warts* especially in sun. **Leaves** *Hairless, on hairless stalks*; very triangular: *double teeth up the straight sides*.
COMPARE Downy Birch (below). (The hybrid, *B. × aurata*, is widespread and intermediate.) Weeping habit distinguishes from the exotic species, but accidental intermediate planted hybrids occur.
VARIANTS Young's Weeping Birch, 'Youngii', is abundant: *a mop of long, hanging shoots*. Old trees make some upward progress and develop extraordinary, fluted trunks with big black and white zones, like a Right Whale's head. 'Tristis' is more dignified, but occasional: as weeping as the best wild trees; smoothly white-barked above the graft. 'Purpurea' is occasional and strangely ineffective, its maroon foliage muddy and sparse.
'Golden Cloud' has soft yellowish foliage; very rare (birches on chalk are often yellow from chlorosis). 'Fastigiata' is rather rare: *vertical wriggling branches*, as if electrocuted, with some drooping shoots, make a *narrow balloon-shape* (cf. Downy Birch). 'Obelisk' is a newer improvement.
Swedish Birch, 'Laciniata' (Cut-leaved Birch; 'Dalecarlica'), is occasional: an airy, shapely grey crown of cut leaves on a smoothly rounded, very white trunk. 'Birkalensis' (less cut-leaved; cf. Mongolian Lime, p.404) is in some collections.
OTHER TREES *B. obscura* (E Europe; in some big gardens in the UK) usually has a duller grey bark and dark, rounder leaves, more tapered at the base.

Downy Birch *Betula pubescens*

(White Birch, Brown Birch; *B. alba* in part.) Europe (including Britain and Ireland); W Asia. Abundant everywhere on poor or damp non-chalky soils – more widespread than Silver Birch, but less planted.
APPEARANCE Shape Twiggy, to 28 m; the fine branches scarcely weep. **Bark** Purple-red when young, taking longer than Silver Birch's to whiten. Old trunks have *bands of grey*, but little sharp, vertical patterning. **Shoots** *Softly hairy* (hairless *but with sticky brown warts* in ssp. *carpatica* from upland regions, including Highland Scotland); **Leaves** *Rounded*-triangular, *single-toothed*, on *downy* stalks.
COMPARE Silver Birch (above). Hybridizes freely with exotic species (which usually have bigger leaves – but see Japanese White Birch, p.188).
OTHER TREES *B. celtiberica* (N Iberia; in a few collections) has leaves like Downy Birch's, but on shoots which have Silver Birch's *white* warts.

SILVER BIRCH 'TRISTIS'

'PURPUREA'

SWEDISH BIRCH

long hanging shoots

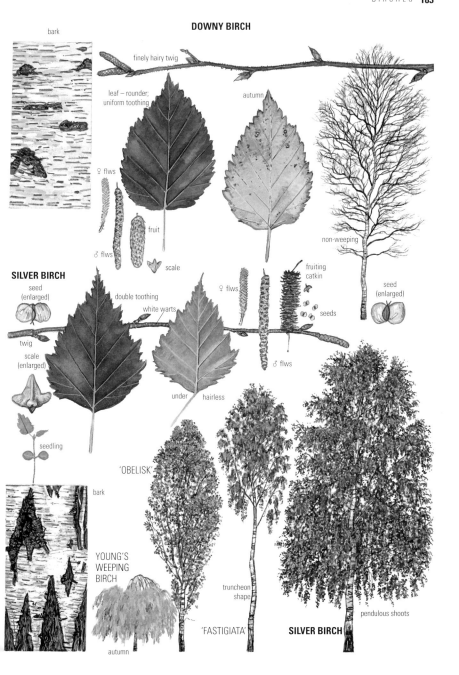

BIRCHES (AMERICAN)

Paper-bark Birch   *Betula papyrifera*

(Canoe Birch) Across North America, S to New York. 1750. Locally frequent: gardens, streets.
APPEARANCE **Shape** Stiffly rising, spreading branches, to 20 m. Lacks much of Silver Birch's grace (p.182): thicker shoots and *sparse*, big, dark leaves. **Bark** Slightly whiter than the European native birches' (but shiny brown in some varieties), peeling horizontally: fine dark lenticel-bands, but rough zones only if the 'paper' is torn off. **Shoots** Warty; some long hairs at first. **Leaves** Large (to 10 cm in some forms), the *disproportionately few main veins* (5–10 pairs) imperfectly parallel. *Dull dark green*; scattered black dots (glands) beneath. *Quite long hairs on the leaf-stalk*, under the vein-joints, and scattered above the leaf.
COMPARE Himalayan Birch forms (p.186): *glossier* leaves with closer veins; hairs extend under the veins. Oriental White Birch forms (p.188; more alike): more triangular leaves on *hairless* stalks. Erman's Birch (p.186): fresher green leaves and *no hairs* on its mature shoots or (usually) its leaf-stalks.
OTHER TREES Grey Birch, *B. populifolia* (E North America), is in a few big gardens: smaller, hairless, doubly toothed leaves with *conspicuously long, jaggedly toothed tail-like points*. Blue Birch, *B. coerulea-grandis* (E North America, 1905; very rare), differs in its short-pointed, singly toothed leaves. (Stiffer habit than Silver Birch, p.182; larger (8 cm) leaves; scarcely patterned white bark.)
B. × koehnii is a rare *rather weeping* tree with more triangular leaves than Paper-bark Birch's, presumably a hybrid with Silver Birch: similar trees can arise in Paper-bark Birch plantings.

River Birch   *Betula nigra*

(Black Birch) E USA. 1736. Occasional, but increasingly planted in parks and streets.
APPEARANCE **Shape** Broad-limbed with drooping shoots; to 16 m. **Bark** Cream at first, soon with great *scrolled scales*; rufous or *almost black* in age. **Shoots** Often downy. **Leaves** *Long, with double, scalloped teeth*, hairy at least under the veins and on the stalks.
OTHER TREES *B. davurica* (N China, Korea; 1822) is a small tree in some big gardens with similar bark and less markedly double-toothed leaves.

Yellow Birch   *Betula alleghaniensis*

(*B. lutea*) Manitoba to Georgia. 1767. Very occasional in bigger gardens for its autumnal gold.
APPEARANCE **Shape** Sometimes a giant bush, to 20 m; ovoid, on strong rising branches. **Bark** Papery-smooth, horizontally banded; *yellowish grey-brown* then pale dull creamy-grey. **Shoots** Hairy at first; disinfectant (oil of wintergreen) aroma when bruised. **Leaves** Hornbeam-like: finely and *irregularly* double-toothed; long hairs at least on the stalks and under the veins. **Flowers** Female catkins have hairy scales.
COMPARE Transcaucasian Birch (p.188). Green Alder (p.192), hornbeams (p.194) and hop-hornbeams (p.196): lacking the papery bark.
OTHER TREES Cherry Birch (Sweet Birch), *B. lenta* (Ontario to Alabama, 1759), is rare and has darker brown-purplish bark (ultimately dull dark grey), and regular leaf-toothing, the *bigger tooth at the end of each vein* with a *whisker-tip*. Leaf-hairs are more quickly confined to beneath the veins; the female catkin-scales are *hairless* (cf. Japanese Cherry Birch, p.188).

PAPER-BARK BIRCH   **YELLOW BIRCH**

variably white bark — twig — rough warts — downy stalk — fruiting ♀ flws — distant veins — autumn — young bark

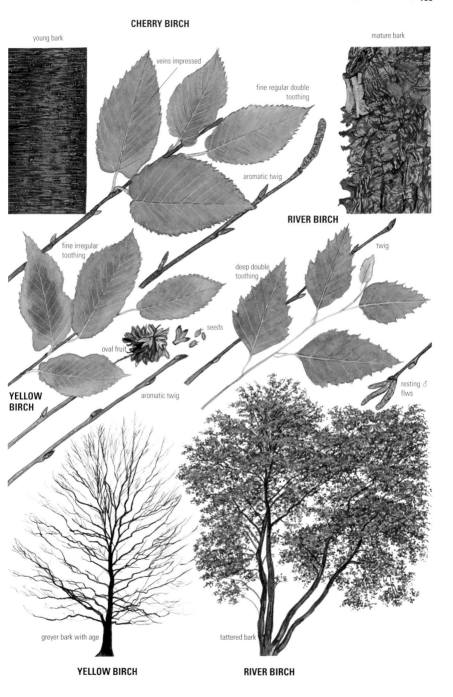

BIRCHES (ASIATIC)

Monarch Birch — *Betula maximowicziana*

(Maximowicz's Birch) Japan, the Kuriles. c.1890. Very occasional in larger gardens.
APPEARANCE Shape Strong, rising branches; to 24 m. **Bark** Papery, with horizontal lenticels; red-brown then a rather dull grey-white. **Shoots** Hairless; disinfectant (oil of wintergreen) scent when bruised. **Leaves** *Big* (to 14 cm), *heart-shaped*, soon hairless except for tufts under the vein-joints; dark and rather shiny.
COMPARE Paper-bark Birch (p.184): leaves never so big/heart-shaped. The brown-barked, strongly growing sapling is un-birchlike and recalls Dove Tree (p.412). (Lime leaves, p.400, are *lop-sided*.)

Himalayan Birch — *Betula utilis*

Himalayas to W China. 1849. Planted for its attractive bark: named selections are increasingly frequent in smaller gardens, parks and streets.
APPEARANCE Shape Somewhat stiffly rising branches to 22 m; quite sparse leaves. **Shoots** Relatively stout; *very hairy* at first. **Bark** Amazing palettes of brilliant, glistening colours, even on 3 cm stems: dead white in the W Himalayan var. *jacquemontii* (*B. jacquemontii*); golden; pale mauve; salmon-pink; crimson or (var. *prattii*) purple. Horizontally marked with small grey/amber lenticels; often with papery, peeling rolls and, rarely, harder curling scales; the white pigment (betulin) may rub off on the hand. **Leaves** Dark, rather glossy, on hairy stalks; 5–9 cm; hairs scattered above them, and under the veins which are in 7–14 pairs (most in eastern trees; 7–8 pairs in var. *jacquemontii*). **Flowers** Male catkins in winter have embossed scales.

COMPARE Chinese Red-barked Birch (p.188): *hairless* young shoots *or* very narrow leaves. Erman's Birch (below): shoots soon hairless. Paper-bark Birch (p.184): *dull* leaves with distant veins, scarcely hairy beneath. Oriental White Birch forms (p.188): more triangular leaves, the stalks *at least not hairy*.
VARIANTS Most named selections ('Grayswood Ghost'; 'Silver Shadow'; 'Jermyns' – neatly steep-branching, with catkins to 17 cm) are intensely white-barked forms of var. *jacquemontii*, often grafted (at the base) on Silver Birch.

Erman's Birch — *Betula ermanii*

NE Asia, Japan. 1890. Occasional: larger gardens.
APPEARANCE Shape A vigorous, potentially long-lived birch, to 22 m; *dense, often yellowish foliage*. **Bark** Often brilliant *golden*-white, horizontally peeling and *shredding*; sometimes pinkish; rarely shining orange. **Shoots** Warty, *soon hairless*. **Leaves** Triangular/heart-shaped and particularly *neat*; soon almost hairless, with 7–11 often *closely parallel, rather impressed vein-pairs* (14–15 in var. *japonica*); stalks usually hairless. **Flowers** Male catkins in winter fat and *smoothly scaled*.
COMPARE Himalayan Birch (above): hairy shoots. Paper-bark Birch (p.184): hairy shoots and relatively distant, imperfectly parallel leaf-veins.
OTHER TREES *B. costata* (NE Asia) has slender triangular leaves, very *long-pointed* and *never heart-shaped at the base*, with 10–14 vein-pairs. In some collections (though most labelled trees seem to be 'Grayswood', a white-barked clone of *B. ermanii* var. *japonica*). Many older planted 'Erman's Birches' are clearly hybrids with Silver and Downy Birches.

MONARCH BIRCH

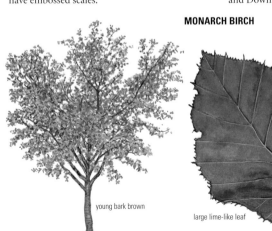

young bark brown

large lime-like leaf

mature bark

BIRCHES (ASIATIC)

HIMALAYAN BIRCH

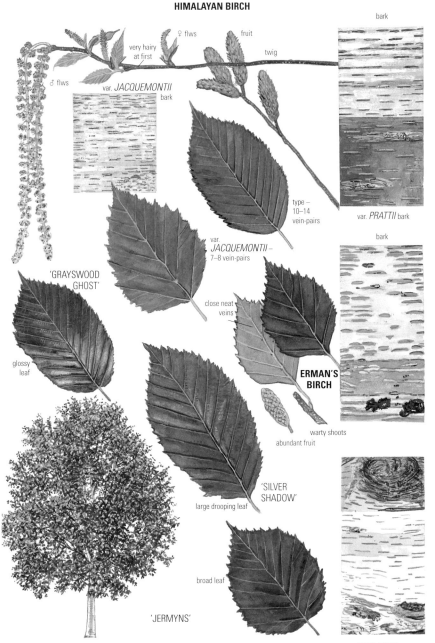

BIRCHES (ASIATIC)

Chinese Red-barked Birch
Betula albo-sinensis

NW China. Effectively a NE continuation of Himalayan Birch (p.186), with the same rainbow bark. 1901. Very occasional.
APPEARANCE Shape Usually slender. **Bark** Glistening pinks, mauves and oranges. Northern trees (var. *septentrionalis*) have darker, bloomed barks – smoky rich pinks and crimsons – in big, exfoliating sheets. Others (hybrids?) can have dull white barks. **Shoots** *Hairless* in the type. **Leaves** *Slender* (particularly in var. *septentrionalis*), long-pointed, with 10–14 vein-pairs.
COMPARE Himayalan Birch (p.186): hairy shoots *and* broader leaves; broader, more knobbly-scaled winter male catkins.
VARIANTS 'Hergest' is a rare but lovely selection with pale mauve, silky-smooth bark.

Japanese Cherry Birch
Betula grossa

Japan. 1896. In a few collections. An Asiatic answer to Cherry Birch (p.184).
APPEARANCE Shape Often broad. **Bark** Dull, purplish-brown, papery. **Shoots** Aroma of disinfectant (oil of wintergreen) when bruised. **Leaves** Neatly double-toothed, hornbeam-like; with more persistent silky hairs than Cherry Birch's, all over the upper sides of the smaller (5–10 cm), thicker leaves. **Flowers** Female catkin-scales *fringed with hairs* (cf. those of Yellow Birch, p.184).

Japanese White Birch
Betula mandshurica var. *japonica*

(Manchurian Birch; *B. platyphylla* var. *japonica*; *B. resinifera*) Mountains of Japan. *c*.1887.
APPEARANCE Shape Rather stiffly upright branching and shoots. **Bark** Very white; some horizontal grey shading. **Shoots** Very warty (cf. Silver Birch, p.182; Paper-bark Birch, p.184); hairy at first. **Leaves** To 7 cm, dark and matt; slightly hairy along the veins on both sides; the *stalk is hairless*.
COMPARE Sichuan Birch (below); Downy Birch (p.182). Paper-bark Birch (p.184) and Himalayan Birch (p.186): more oval leaves on hairy stalks. Erman's Birch (p.186): usually with more closely parallel veins.

Sichuan Birch
Betula szechuanica

(*B. platyphylla* var. *szechuanica*; *B. mandshurica* var. *szechuanica*) SW China to Tibet. 1908. Rare.
APPEARANCE Shape A rather ungainly tree with sparse, dark leaves. **Bark** *Dead-white* (unlike Japanese White Birch), the white betulin *coming off like old whitewash on the hands* (cf. some Himalayan Birches, p.186). **Leaves** Rather bluish, *white*-green beneath, *hairless*, and long-pointed (though less strikingly so than Grey Birch's, p.184).

Transcaucasian Birch
Betula medwediewii

Caucasus Mountains. 1897. A handsome, sturdy little tree, often the last of all into leaf; rare.
APPEARANCE Shape *Globular*. Dense, stiffly rising shoots on upcurving branches from a typically very short bole; to 9 m. **Bark** Papery, *silvery-brown* and flaking (like some hazels', p.198). **Shoots** Slightly hairy at first; disinfectant (oil of wintergreen) scent when bruised. **Buds** *Glossy green, to 12 mm*. **Leaves** Large (to 13 cm), dark green, with a few hairs along the 8–11 pairs of sunken veins; on *short* (12 mm) hairy stalks; bright gold in autumn.
COMPARE Yellow Birch (p.184).

TRANSCAUCASIAN BIRCH — autumn — short stalk

fruit — ♂ flws

CHINESE RED-BARKED BIRCH

ALDERS

Alders (30 species) carry their buds (except in the Green Alder group) on small slim stalks; male catkins exposed colourfully through winter; female catkins ripening into little woody 'cones'. Their roots develop nodules, enhancing the fertility of poor soils. (Family: Betulaceae.)

Things to Look for: Alders

- Shoot: Is it hairy?
- Leaves: Pointed? Does the rim curl under? Are they downy/glossy beneath?

Key Species

Common Alder (below): leaves often with indented tips. **Italian Alder** (p.192): heart-shaped leaves. **Grey Alder** (p.192): well-pointed oval leaves. **Green Alder** (p.192): well-pointed, finely toothed, oval leaves.

Common Alder — *Alnus glutinosa*

(Black Alder) Europe including Britain and Ireland; W Asia; N Africa. Abundant but little planted except in reclamation schemes on spoil heaps or landfill sites; often dominant in bogs, 'carrs' and on river-banks. By 2002, many trees had been killed by a new water-borne *Phytophthora* root pathogen.
APPEARANCE Shape An approximate spire when young; old trees (to 28 m) can be broad, with twisting oak-like branches, but in woodlands retain long straight boles. Often coppiced (the timber – white when cut, but oxidizing orange-red in minutes – made ideal charcoal for gunpowder). **Bark** Brown; pale horizontal lenticels, then closely and deeply square-plated, the verticals predominating. **Shoots** Hairless. **Buds** All stalked. Lumpily *club-shaped*, exquisitely *mauve-bloomed* (sometimes dull and greyer). **Leaves** Dark, leathery, and *racquet-shaped*; the end never pointed and *often indented* (cf. Alder Buckthorn, p.398). On vigorous growths, they can have very shallow, rounded lobes. **Flowers** Male catkins densely wine-red through winter.
VARIANTS 'Imperialis' (occasional) is a gaunt, 'Japanese-looking' tree with soft green, very *feathery foliage* – with the jizz of Swamp Cypress (p.64). 'Laciniata' (rarer) grows like the type; regular triangular lobes run halfway to the midrib (cf. Wild Service, p.296). The equally rare and similar if rather more elegantly lobed Grey Alder 'Laciniata' (p.192) has *downy shoots* and different bark. 'Pyramidalis' is a very rare steep-branched spire. 'Aurea' is rare. Chlorotic trees of the type may also have patchily yellow foliage; Golden Grey Alder (p.192) is commoner.
OTHER TREES Downy Alder, *A. hirsuta* (Japan, Manchuria; 1879), in some collections and to 20 m, shares racquet-shaped, somewhat lobed leaves, but these are doubly toothed, larger, and variably *rufous-hairy beneath*. Its catkins open in mid-winter.

Red Alder — *Alnus rubra*

(Oregon Alder; *A. oregona*) S Alaska to California. Occasional; large gardens and a few shelterbelts.
APPEARANCE Shape A very vigorous, broad, leafy spire, on light, rising branches. To 24 m; short-lived in Britain. **Bark** Grey; rather smooth. **Shoots** Waxy, angular; long hairs *soon shed*. **Leaves** Large (to 15 cm), deeply double-toothed or shallowly lobed; dark rich green above and grey beneath (though downy only under the impressed veins); the edge is *minutely but sharply rolled down* so that the under-surface is rimmed with dark green. The stalks are shorter than in most alders.
COMPARE Grey Alder (p.192): hairier shoots and smaller, less lobed leaves (edges not rolled down).

RED ALDER — leaf margin decurved — stalked bud — seeds — cone — grey beneath — COMMON ALDER 'AUREA'

COMMON ALDER

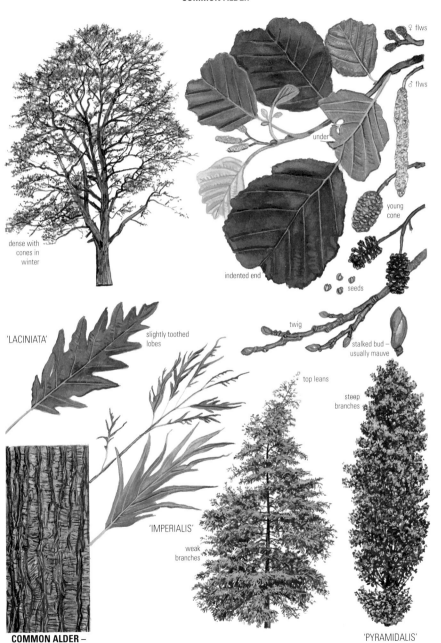

ALDERS

Grey Alder — *Alnus incana*

Europe except Britain; the Caucasus. 1780. Locally abundant in belts, reclamation schemes, etc; rarely naturalizing.
Appearance Shape Vigorous trees broad-conic; often leaning/stunted and *suckering*. To 24 m: short-lived. **Bark** *Grey*; distantly cracked at maturity. **Shoots** *Grey-hairy* when younger. **Leaves** To 10 cm; often broad but always pointed; deeply toothed/shallowly lobed; dull above, grey and more or less downy beneath.
Compare Red Alder (p.190). The hybrid with Common Alder, *A.* × *hybrida* (*A.* × *pubescens*), crops up occasionally and is intermediate.
Variants 'Aurea', rare and slender (to 12 m), has yellow foliage; its yellow shoots turn orange in winter. 'Ramulis Coccineis' (less scarce) is similar in summer; in winter a heart-warming sight – catkins *salmon-pink*, shoots *fiery-red*.
'Pendula', rare and low, weeps rather gauntly.
'Laciniata', rare, has sharply and deeply lobed leaves (cf. Common Alder 'Laciniata', p.190).

Italian Alder — *Alnus cordata*

S Italy; Corsica; NW Albania. 1820. Locally abundant in parks, streets and shelterbelts. A plant of vigour and polish – even on dry chalk – which still blends well with wild trees.
Appearance Shape A spire at first; eventually leaning but usually still narrow; to 28 m. **Bark** Pale brown-grey; vertically plated with age. **Shoots** Bright brown, grey-bloomed. **Leaves** *Dark, glossy, heart-shaped*; like Common Pear's (p.316) but bigger (4–12 cm) – pointed in Italian trees and more rounded in Corsican ones; hairless except for orange tufts under the vein-joints. They flutter on their long stalks. **Flowers** Male catkins to 10 cm, fawn-yellow and showy in spring. **Cones** Big (to 3 cm).
Other trees Caucasian Alder, *A. subcordata* (1838), remains rare. Craggier bark; downy shoots; leaves more *oblong*, the base rounded/very slightly heart-shaped; *hairy* at least under the veins.
Spaeth's Alder, *A.* × *spaethii*, a hybrid (Berlin, 1908) of Caucasian with Japanese Alder (*A. japonica*), is rare but very handsome. Leaves *boat-shaped*, to 15×6 cm, dark, glossy; a few hairs beneath; small, *distant teeth* and *distant, curving veins*. Japanese Alder itself (in some collections) has smaller, *hairless*, long-pointed leaves, a gaunt habit, and bark cracking into large squares.
Oriental Alder, *A. orientalis* (Cyprus; the Middle East; 1924), is very rare. Bark grey-brown, soon harshly square-cracked; shoots hairless; oblong leaves smaller than Caucasian Alder's, rather dull above but usually *shiny beneath* and hairless except for tufts under the vein-joints.

Green Alder — *Alnus viridis*

(*Betula viridis*) Mountains of central and SE Europe. 1820. In a few collections.
Appearance Shape Normally a suckering bush. **Bark** Brown. **Shoots** Hairless. **Buds** *Pointed, scarcely stalked*; shiny purplish. **Leaves** *Sharply double-toothed* (cf. hop-hornbeams, p.196); green beneath, with tufts in the vein-joints.
Other trees *A. firma* (Japan, 1894; very rare) has even more hop-hornbeam-like leaves (finely double-toothed; *up to 24 pairs* of impressed veins), and a cracked brown bark. To 15 m; often bushy.

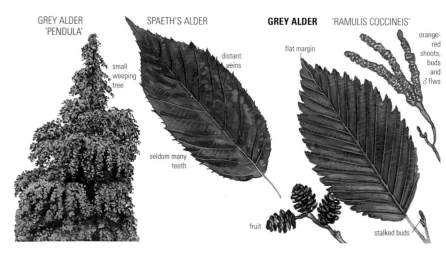

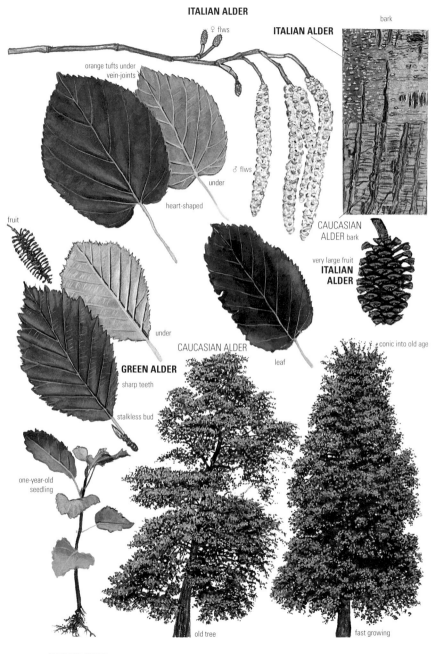

HORNBEAMS

Hornbeams (up to 70 rather similar species) hail mostly from East Asia. Their bark is usually smooth and grey; their leaves have many parallel veins. Catkins are hidden until spring; short strings of small nuts are backed by long, green, variously lobed bracts. (Family: Betulaceae.)

Things to Look for: Hornbeams and Hop-hornbeams

- Buds: How long and how sharp? Appressed?
- Leaves: How many veins? Are they hairy beneath?
- Fruit: Nut-bracts – are they lobed (with what toothing? wrapped around the nut?)?

Common Hornbeam *Carpinus betulus*

Europe including SE England; Asia Minor. In its native range often the dominant tree on heavy clay, having been selected and coppiced for charcoal production; locally abundant on richer soils in hedges and scrub and much planted throughout Britain and Ireland. Hornbeam means 'hard tree', and anybody who has tried sawing one will know why: the timber was used for chopping blocks, cog-wheels – and as an exceptionally calorific firewood.
APPEARANCE Shape Branches seldom heavy or spreading, though old trees in the open (to 30 m) can be very broad and shapely; delicate traceries of fine twigs. **Bark** Grey, smooth, with vertical, wriggling dull silver or orange snake-marks (cf. *horizontal* markings of young beech); odd wide fissures and shallow, smooth criss-crossing ridges develop with age. Trunks soon grow *lumpy, muscular flutings*. **Shoots** Slender, with long hairs at first. **Buds** Slender and long (8 mm), the tips *slightly incurved*; never spreading widely from the shoot like beech-buds. **Leaves** 7–12 cm; double-toothed, and corrugated by the 10–13 pairs of close, *impressed* veins; there are some long hairs above them, and on their stalks. Pale yellow in autumn. **Flowers** Male catkins expand in spring as yellow curtains. **Fruit** Nut-bracts 3 cm long, randomly toothed; *1 short basal lobe on each side*.
COMPARE Common Beech (p.204): different foliage, but more alike in winter. Trees with similar leaves include Green Alder (p.192), Yellow and Cherry Birches (p.184), Japanese Cherry Birch (p.188), Hornbeam-leaved Maple (p.384), the hop-hornbeams (p.196), and Alder-leaved Whitebeam (p.294); only the other true hornbeams, Alder-leaved Whitebeam and Hornbeam-leaved Maple retain a smooth grey bark, and none of these are large trees in N Europe. Its 3-lobed bracts *or* its long buds distinguish Common Hornbeam from most other hornbeam species (p.196).
VARIANTS 'Fastigiata' ('Pyramidalis'; 1885) is an abundant street tree, to 24 m, with a neat, dense, stylish *ace-of-spades shape* (but ultimately a broad balloon, and bushy if not trained to a 2 m stem in the nursery). Many light branches spring *from a single point* and curl in at their tips – the branches of woodland-grown wild trees can approximate to this habit, but will not all leave the trunk together. 'Columnaris' is rare, making a *densely rounded shape*, with big, broad, *very oblong* leaves.
'Incisa' ('Quercifolia') makes a slender, irregular tree, its leaves variably *deeply lobed*. Rather rare; sometimes reverting.
Other deservedly very rare cultivars include a 'Pendula' which does not weep, a 'Purpurea' with (after a few days) ordinary green leaves, and a 'Variegata' with (given a year or two) no traces of variegation.

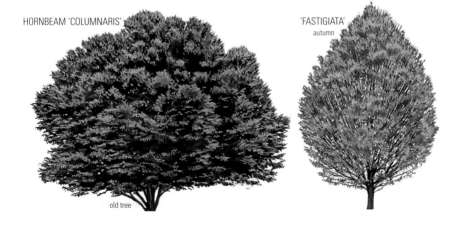

HORNBEAM 'COLUMNARIS'
old tree

'FASTIGIATA'
autumn

HORNBEAMS & HOP-HORNBEAMS

Oriental Hornbeam *Carpinus orientalis*

Sicily, and NE Italy E to Iran. 1739. In some collections, to 20 m.
APPEARANCE Differences (subtle!) from Common Hornbeam are as follows: **Shoots** Very slender. **Buds** Smaller (4 mm), sharp. **Leaves** Smaller (2–6 cm). **Fruit** Nut-bracts *unlobed* (but coarsely toothed).
OTHER TREES *C. turczaninowii* (N China, Japan; *c.*1914) is in some big gardens, to 12 m: an exceptionally pretty tree with tiny (3–4 cm) leaves. Its nut-bracts have a small basal lobe *on one side only*; the other side is jaggedly toothed.

American Hornbeam *Carpinus caroliniana*

(Blue Beech) E USA; NE Mexico. 1812. In a few collections.
APPEARANCE To 14 m. Autumn colour can be scarlet. Easily mistaken for Common Hornbeam; differences are as follows: **Buds** *Small* (3 mm), *blunt*. **Fruit** Larger (3-lobed) nut-bracts, the main lobe often sharply toothed only down one side.
OTHER TREES Aka-shide, *C. laxiflora* (Japan, Korea; 1914), is a dainty, spreading tree in some collections, its little (4–7 cm) leaves with *elongated slender points*; the *small* nut-bracts (to 2 cm) normally have a *very small* basal lobe on each side and jagged teeth up one side of main lobe.
C. viminea (*C. laxiflora* var. *macrostachya*; China, 1900) is equally rare: larger leaves (to 10 cm) and longer bracts (to 25 mm) than those of Aka-shide.

Kuma-shide *Carpinus japonica*

(Japanese Hornbeam) Japan. 1879. In some collections.
APPEARANCE Shape A bushy tree to 8 m. **Bark** Sometimes *scaly brown*. **Leaves** Dark, glossy, slender, with *20–24 pairs of strikingly parallel sunken veins*; every third tooth is whisker-tipped. **Fruit** The nut-bracts – in big, hop-like bunches – have only *one small rounded basal side-lobe*, but *enfold the nut*.
COMPARE Hop-hornbeams (below); Green Alder (p.192).
OTHER TREES Sawa Hornbeam, *C. cordata* (Japan and E Asia; 1879), differs in the very *heart-shaped base* of its broader leaves, and in its *big* (15 mm) buds.

European Hop-hornbeam *Ostrya carpinifolia*

SE France E to the Caucasus. 1724? Very occasionally seen in larger gardens.
APPEARANCE Often passed over as a Common Hornbeam (p.194); but conversely hornbeams in full fruit are sometimes assumed to be hop-hornbeams. To 24 m. **Bark** Brown-grey, cracking into *square, ultimately shaggy plates*. **Buds** Fat, *not pressed against the shoot* like Common Hornbeam's. **Leaves** Very like Common Hornbeam's but with a few more main veins (12–15 pairs); these *may branch*. **Flowers** Male catkins *exposed through winter* before they open (as in birches and alders). **Fruit** In *hop-like clusters of bladders* studding the crown almost like flowers through summer, *white*/greenish then red-brown; nuts (*sealed in each bladder*) 6×3 mm.
COMPARE Kuma-shide (above); Green Alder (p.192).
OTHER TREES Ironwood (American Hop-hornbeam), *O. virginiana* (E USA, 1692), is in some big gardens: its leaves often have fewer veins and more long hairs, each with a tiny *swollen tip* (gland), and its nuts are larger (to 8 mm).
Japanese Hop-hornbeam, *O. japonica* (1888), is the rarest of the three in the UK; its leaves have fewer, distant veins (9–12 pairs), and are more *closely hairy underneath*.

JAPANESE HOP-HORNBEAM

fruit bladders

downy beneath

9–12 vein-pairs

HAZELS

Hazels (about 15 species) expose their male catkins through winter; female flowers sit on the twigs like crimson sea-anemones and lead to large nuts in leafy cups. (Family: Betulaceae.)

Common Hazel — *Corylus avellana*

Europe including Britain and Ireland; Turkey; N Africa. Very abundant, except on poor or waterlogged soils, as a woodland understorey and in hedges. Once much planted for coppicing; the wand-like 3- to 6-year-old growths – so pliable that a strong hand can knot them – were indispensable for wattle hurdle-making and (in the south) to bind laid hedges. Hazel's fertility is now compromised in much of Britain by introduced Grey Squirrels, which habitually strip the nuts before they ripen.
Appearance Shape Usually multi-stemmed; old plants to 15 m. **Bark** Often burnished bronze when young but harsh to the touch and finely peeling; old stems pale brown, with some shallow, flat ridges. **Shoots** Pale green-brown, with long, rather harsh hairs. **Buds** Green, *fat, oval*. **Leaves** Soft, hairy and floppy; nearly *round* (to 12 cm), with a *sudden sharp point*; on short (1 cm), long-hairy stalks. **Flowers** The yellow male catkins expand and open in late winter. **Fruit** Squirrels permitting, the nuts ripen early in autumn. They are sheathed in shucks *about their own length*.
Compare Filbert (below). Wych Elm (p.242): sometimes confused as bushy regrowth, but its thicker, rougher leaves have asymmetrical bases.
Variants Weeping Hazel, 'Pendula', is rare: an umbrella of shoots from twisting limbs on a single, trained stem or grafted on a trunk of Filbert.
Corkscrew Hazel, 'Contorta', is occasional and much used in flower-arranging: its leaves and shoots curl madly (cf. Corkscrew Willow, p.166).
Golden Hazel, 'Aurea', is rather rare but vigorous (to 11 m): yellowish young foliage fades to dull pale green through summer.
Cut-leaved Hazel, 'Heterophylla' ('Laciniata'), is very rare and has jaggedly lobed leaves; not striking at a distance.

Filbert — *Corylus maxima*

Balkans; Turkey. 1759. Occasional in old gardens.
Appearance Shape A more vigorous plant than Common Hazel. **Bark** *Greyish and distantly cracked*. **Leaves** Often more distinctly lobed. **Fruit** The longer, narrower nuts are hidden in 'Christmas stockings' which are nearly *twice their length* and jaggedly toothed. Cob-nuts (cottage gardens; a few commercial plantations in W Kent) are mostly hybrids with Common Hazel.
Variants Purple Filbert, 'Purpurea', is frequent and has purple leaves and catkins. (This is sometimes misnamed *Corylus avellana* 'Purpurea', a plant which does exist but is very rare and has brownish-pink *young* leaves and purplish catkins.)

Turkish Hazel — *Corylus colurna*

SE Europe and Asia Minor. Grown in a few gardens since 1582 but now a locally frequent street tree.
Appearance Shape A broad but symmetrical *spire*. *To 26 m*: a straight trunk, which is rarely forked, carries light, level branches with dark, luxuriant, hanging foliage; broadening or leaning only in age. **Bark** Pale brown, with *close scaly ridges* from the first – rather like Field Maple's (p.368). **Leaves** Recognizably a hazel's, but more lobed and shinier. **Fruit** Like nuts of Common Hazel but a little larger; in frilly, bristly cups.
Compare Scarlet Thorn (p.288); Wild Service (p.296); Yunnan Crab (p.308).

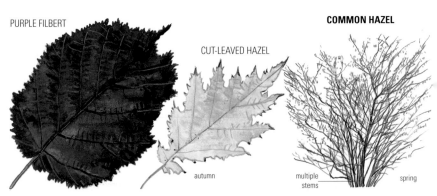

PURPLE FILBERT

CUT-LEAVED HAZEL — autumn

COMMON HAZEL — multiple stems — spring

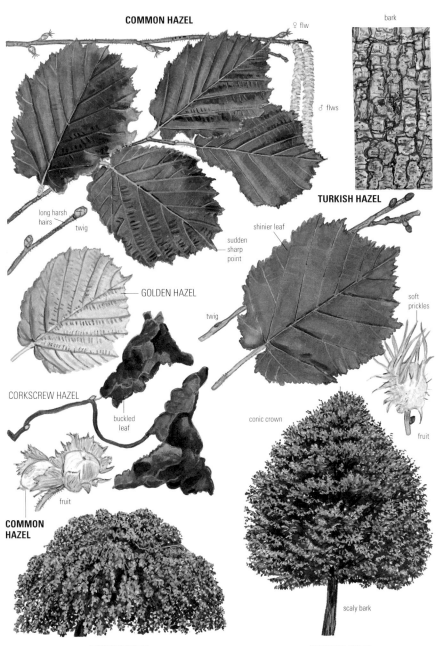

SOUTHERN BEECHES

The southern beeches are a genus of at least 40 southern-hemisphere trees, with beech-like nuts and diverse leaves. (Family: Fagaceae.)

Things to Look for: Southern Beeches

- Bark. What is it like?
- Leaves: Deciduous? How many main veins? Do they have teeth (Regular? How big?)?

Antarctic Beech *Nothofagus antarctica*

(Nirre) S Andes – to Cape Horn. 1830. Occasional.
APPEARANCE Shape *Sparse, gaunt, irregular*, like a lapsed bonsai, to 20 m; crisped peninsulas of fine twiggery; often bushy and usually leaning. **Bark** Dark brown, with lenticel-bands, then rough, grey plates. **Leaves** Tiny (2–4 cm), crinkled; *3–5 pairs* of veins and large, blunt teeth *each with 4 small teeth*; dark, shiny, and looking evergreen, they flush early with a powerful scent of cinnamon and cheap soap.
COMPARE Red Beech (p.202): *simple teeth.*
OTHER TREES Lenga, *N. pumilio* (same habitats; c.1960), is very rare. Vigorous but often multi-stemmed, with *smooth grey bark*; its leaves, to 4 cm, have *5/6 vein-pairs*, the regular blunt main teeth *each with 2 small teeth*.

Rauli *Nothofagus nervosa*

(*N. procera*) S central Andes. 1910. Occasional, doing best in mild, wet areas and freely naturalizing. The fine timber is beech-like.
APPEARANCE Shape Usually strictly conic when growing fast; later with rather level branches on a *strong often straight trunk*, and dense foliage. To 36 m so far. **Bark** Grey; soon with *long, flat, vertical plates*. **Shoots** Green, *long-hairy for a year*; sturdier than Roblé Beech's. **Buds** Birch-like: big (1 cm). **Leaves** To 9 cm; *15–18 pairs* of boldly impressed veins and *blunt* teeth/little lobes. They flush bronze, and in autumn turn pale gold with some reds.
COMPARE Roblé Beech (below). Blunt toothing distinguishes from Hornbeam, etc. (see p.194).

Roblé Beech *Nothofagus obliqua*

(Coyan; Hualle) S Andes. 1902. Locally frequent (some plantations): a hugely vigorous (but short-lived) tree which seems to 'belong' in N European landscapes. Readily naturalizing.
APPEARANCE Shape Wispily conic, then openly *irregular*, to 30 m; seldom broad; fine, downward-fanning branches. **Bark** Brown/silvery, soon with square/rounded *harshly curling plates*. **Shoots** As slender as any tree's; fine white hairs. **Buds** Appressed, 4 mm – trees can look dead until these expand, branch by branch, in late winter. **Leaves** Small (4–8 cm); *6–11 pairs* of sunken veins; *sharply, irregularly toothed*/slightly lobed. Crimson and yellow in autumn.
COMPARE Rauli (above). The putative hybrid (*N. × alpina*; in odd collections) is intermediate. Irregular toothing distinguishes from all oaks/zelkovas.
OTHER TREES Hualo, *N. glauca*, is a lovely, tender Chilean tree in a few collections: bright *orange papery bark*; leaves (often heart-shaped at the base) *pale blue-grey beneath*.

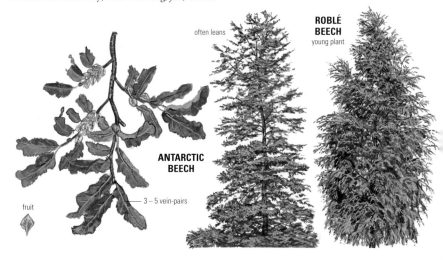

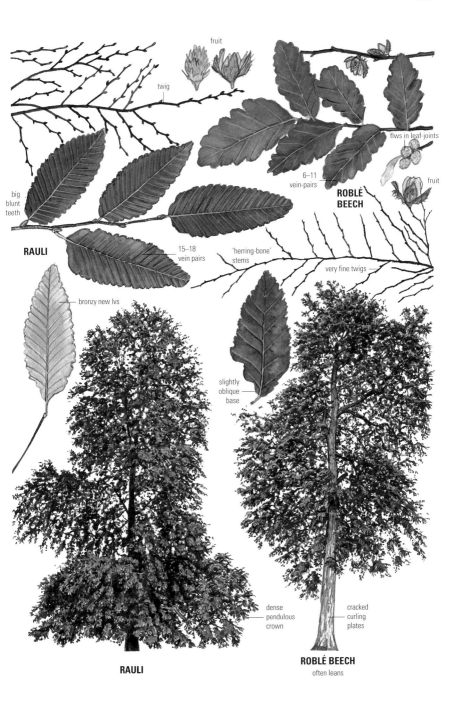

SOUTHERN BEECHES (EVERGREEN)

Coigüe — *Nothofagus dombeyi*

(Dombey's Beech) S central Andes. 1916. Very occasional: larger gardens away from cold parts. A tree of beauty and vigour.
Appearance Shape Neatly ovoid, or wispy on a sinuous bole, or a giant bush; to 36 m. **Bark** Blackish at first; horizontally wrinkled. Harsh red/grey plates develop, set off by the brilliant, dark, evergreen foliage. **Leaves** Small (2–4 cm), thick, flat, broadest *below* halfway; small, jagged, *irregular* teeth. Shiny above; underneath, smooth, pale, *matt green*, and *speckled black* (use a lens). **Male flowers** Usually *in 3s*. **Fruit** Finely *downy* nuts.
Other trees Guindo, *N. betuloides* (Oval-leaved Southern Beech; Coigüe de Magellanes), growing S to Cape Horn, was one of the first southern beeches to be tried in Europe (*c.*1830), but is now confined to a few big gardens; to 20 m. Denser, oval leaves (broadest *halfway up*) have more or less *regular blunt teeth* and usually *white* speckles underneath. The crimson male flowers hang *singly*, like tiny fuchsias; nuts *hairless*.

Red Beech — *Nothofagus fusca*

New Zealand. *c.*1910. Rare: large gardens away from cold parts.
Appearance Shape Slender-domed to 24 m; often wispily open. **Bark** Smooth grey; scaly ridges in age. **Leaves** Quite matt and flimsy – looking deciduous but dying off yellow and red one by one *throughout the year*. To 5 cm; *3–6 pairs of big, slightly hooked, simple teeth*.
Compare Antartic Beech (p.200): double round-toothing.

Silver Beech — *Nothofagus menziesii*

(Menzies' Beech) New Zealand. *c.*1850. Rare: large gardens in mild areas.
Appearance Shape Broad and bushy, to 25 m. **Bark** Remaining smoothly burnished, with fine lenticel-bands; silvery (as in the wild) or more often purplish (like Wild Cherry, p.322). **Leaves** *Tiny* (1 cm), glossy, *round*, deeply and bluntly *double-toothed*; 2 curious hairy pits lie under the base of the midrib.
Compare Black Beech (below): untoothed leaves. No other large hardy trees have broad leaves as tiny.
Other trees Myrtle Beech, *N. cunninghamii* (Tasmania; much rarer), has vertical, *scaly bark-ridges* and slightly triangular, singly toothed leaves, fringed with tiny hairs and lacking pits beneath.

Black Beech — *Nothofagus solanderi*

New Zealand. By 1917. Very rare: large gardens in milder parts.
Appearance Shape Spiky, the tiny leaves in flat layered sprays, as if bonsaied. Often a giant bush; to 25 m. **Bark** Smooth, blackish; vertically fissured with age. **Leaves** *Tiny* (1 cm) and *untoothed* (recalling the – opposite – leaves of *Lonicera nitida*); *taper-based, round-tipped*. Glossy above but pale underneath, with microscopic wool.
Variants Mountain Beech, var. *cliffortioides*, of earlier introduction from higher sites and slightly less rare in the UK, has *round-based* leaves, the sides often rolled down and the *pointed tip rearing up*.
Compare Small-leaved Azara (p.408). Silver Beech (above): double-toothed leaves.

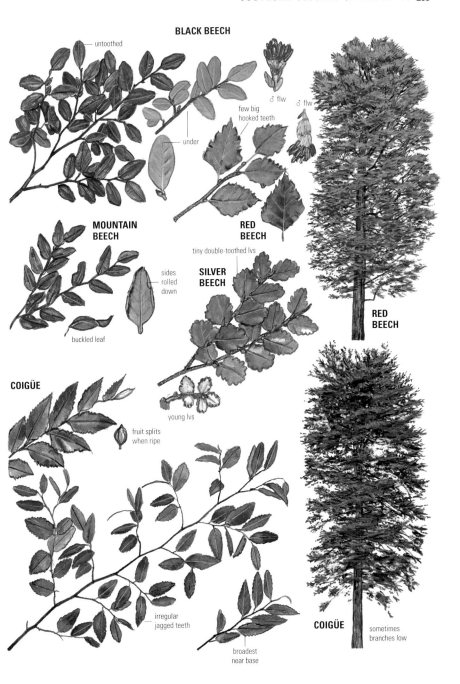

BEECHES

Beeches (all northern-hemisphere trees) have smooth grey bark, long buds and edible nuts in four-parted husks. (Family: Fagaceae.)

Common Beech — *Fagus sylvatica*

S and W Europe, including (probably) S and Middle England; planted abundantly throughout Britain and Ireland in plantations, parks and shelterbelts, and well naturalized; often dominant on mineral soils, but disliking wet ground. Sadly this is one tree whose younger bark frustrated male Grey Squirrels now habitually rip off in their breeding season, preventing tall straight growth. **APPEARANCE Shape** In woodland often with a long, slightly sinuous trunk, but readily growing a huge dome on strong, smooth limbs; one of the UK's tallest broadleaves, to 40 m. The singularly dense foliage shades out competition and shields the thinly clad under-bark from the sun's heat. Many old trees are pollards, but beech is shorter-lived than most big trees, blowing down easily and quickly decaying. Seldom successfully coppiced. Twigs within 3 m of the base retain their dead leaves in winter (as do hornbeams and some oaks), so make a perfect hedge. **Bark** Silver-grey, with slight *horizontal* etchings; some trees develop shallow or rugged criss-crossing ridges. **Shoots** Slender, grey (silky at first) and zig-zag. **Buds** *Torpedo-shaped, 2 cm, copper/grey, spreading at 60°* (some may be a camouflaged snail, *Cochlodina laminata*). **Leaves** To 10 cm, with odd, tiny, distant teeth; hair-fringed, and silky all over as they unfold; 5–9 vein-pairs. **Fruit** Nuts in prickly husks, on 2 cm stalks. **COMPARE** Other beeches (p.210): 7–15 vein-pairs. Hornbeam (p.194) is often confused.

VARIANTS Many and popular. Copper Beech or Purple Beech, f. *purpurea*, is abundant and crops up frequently in the wild and in plantations, in colours ranging from a quiet pinkish brown (f. *cuprea*) to royal purple, depending on the proportion of purple xanthocyanins in the leaves. Ungrafted trees of dubious origin become dull and dark after the first flush of wine-red, with inner leaves dark green; named clones include the bright 'Rivers' Purple' (Sawbridgeworth; by 1870). In winter, a grafted beech more than 25 m tall is very likely to be Copper; a *single stem often persists*, with *numerous*, light, *ascending* branches from swollen bases and a crown *narrowing to a small flat top*. The nut-husks are also purplish. Compare 'Purpurea Tricolor' (p.208).

Dawyck Beech, 'Dawyck', was discovered around 1860 in a wood next to Dawyck Gardens near Peebles by the head gardener, and since the 1930s has become a frequent municipal and garden tree. The branches rise vertically but twist and turn, like a Lombardy Poplar reflected in choppy water; mature trees (to 28 m) remain very narrow only in shelter. In winter the steel-grey cast and long buds distinguish it from Cypress Oak (p.216) and other big fastigiate trees.

Golden Beech, 'Zlatia' (1892, its name deriving from the Serbo-Croat for 'gold'), is undeservedly rare: the fresh yellow of the young leaves fades by mid-summer but is more even than the yellowing often seen on drought-stressed wild beeches.

best forms are grafted

COPPER BEECH

greens in summer

GOLDEN BEECH

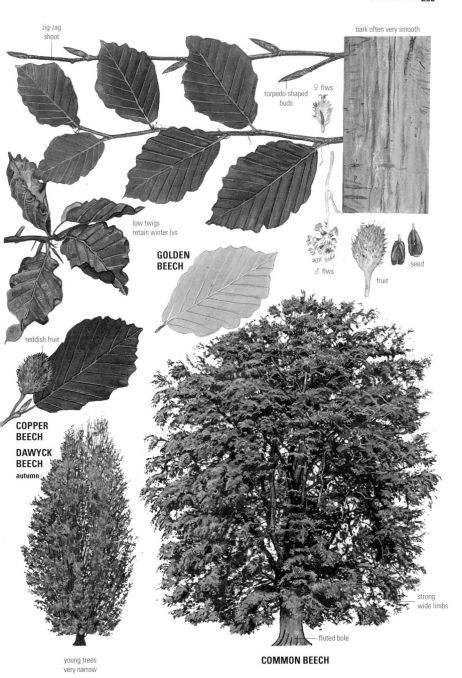

BEECH CULTIVARS

Fern-leaved Beech, 'Aspleniifolia' ('Heterophylla'), is only locally frequent but is a tree of great distinctiveness and beauty, to 28 m, generating interest and sometimes bewilderment. The depth of the lobing varies from clone to clone (and compare Oak-leaved Beech, p.208). In the commonest and most feathery form (seldom grafted), the shoot-tip leaves are narrower or even linear, and the crown is distinctively *pale, matt and fluffy* even when seen at a distance; it colours early in autumn. This tree is a 'chimaera', with inner tissues of typical Beech enveloped by cells of the sport, so that sprouts with normal leaves will often grow from the trunk and branches, especially after an injury; unlike ordinary reversions, these seldom or never take over the whole crown. In winter, the tree is typically broad with a skirt of fine branches almost sweeping the ground, and has very *dense*, fine, horizontal or slightly rising shoot-systems; the distinctive leaves are also very slow to rot. Compare Common Alder 'Imperialis' (p.190); Cut-leaved Hornbeam (p.194); Cut-leaved Zelkova (p.252). No oak has leaves as small and as deeply lobed, or the characteristic 2 cm beech buds.

Large-leaved Beech, f. *latifolia*, has strikingly big, rather clumsy leaves (to 17 cm, but with the usual number of vein-pairs – seldom as many as 9; the equally large leaves of Oriental Beech, p.210, can have up to 14). 'Prince George of Crete' is a grafted clone found in some big gardens.

'Grandidentata', in a few collections, has convex leaves with large scalloped teeth rather than lobes. Compare American Beech (p.210), which has several more vein-pairs.

Cockscomb Beech, 'Cristata' (1836), is rare and has a slender, extraordinarily gaunt habit (but to 28 m) and tufts of densely bunched, crumpled leaves, some round but most deeply and asymmetrically toothed like a cockerel's crest. A very slow-growing purple-leaved sport which arose in a Cumbrian churchyard, 'Brathay Purple', is now also in commerce.

Small-leaved Beech, 'Rotundifolia' (1870), is also rare. The leaves, with *usually 4 vein-pairs*, are nearly round and only 2–4 cm long. The crown is usually broad and shapely. (The upper branches of sickly wild beeches can have leaves almost as small, but will have more – 5–9 – pairs of main veins.) 'Cockleshell' (1960) has even smaller leaves (consistently about 2 cm long). Compare Red Beech (p.202).

OTHER TREES include 'Miltonensis', very locally occasional as an old tree. It too has small (6 cm), rounded, untoothed leaves. Its bark is distinctive: flat, criss-cross silver ridges, making the graft particularly conspicuous. The dense outer branches weep (cf. Weeping Beech, p.208), but these trees are probably of German origin and not the same as the weeping beech originally discovered at Milton Park, Northamptonshire, in 1837.

COCKSCOMB BEECH

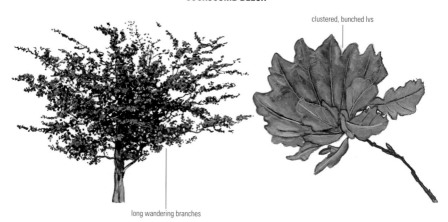

clustered, bunched lvs

long wandering branches

BEECH CULTIVARS

'Albomarginata' is very rare and has leaves with an irregular whitish margin; 'Albo-variegata' has white-splashed leaves. Variegated beeches are large-growing but not showy, and they tend to revert; 'Luteovariegata' has yellow markings, mostly around the leaf-margin.

'Purpurea Tricolor' ('Roseomarginata'; 1879) has purple leaves finely margined with pink and some white; slower than Copper Beech but attractively different only at close range. It should not be confused with the exceedingly rare, dainty 'Tricolor' (fine white and pink margins to *green* leaves).

Weeping Beech, f. *pendula*, is the grandest of pendulous trees, to 25 m; locally frequent. Densely leafy shoots cascade from parabolically curving limbs; these layer given the chance, and the original tree brought from France to the Knaphill Nursery in Surrey in 1826 has grown into a whole cathedral of columns with walls of foliage 60 m apart. A few clones (producing many weeping seedlings) are more symmetrical, with shorter hanging shoots from widely arching branches. 'Miltonensis' (p.206) is also pendulous.

Similar clones include 'Bornyensis', in some collections as a little igloo, the foliage hanging from twisted branches on a high graft (cf. Contorted Beech, below). Weeping Copper Beech, 'Purpurea Pendula', is a rather muddy purple and very occasional variant on 'Bornyensis', to 5 m. 'Purple Fountain' (1975; as dark and now more planted), is shaped like 'Pendula': narrowly rising and cascading limbs.

Golden Weeping Beech, 'Aurea Pendula', has the fountain-like habit of 'Pendula' and the soft yellow younger leaves of 'Zlatia'. 1900; still very rare.

Contorted Beech, f. *tortuosa*, crops up in many places through France, Germany, Denmark and Sweden and has crazily *twisting* branches; sometimes tall but generally a low umbrella. Grafts ('Remillyensis', 'Pagnyensis') are in a few collections.

Oak-leaved Beech, f. *laciniata* ('Quercifolia'), is a very rare and less spectacular variation on Fern-leaved Beech (p.206) with rather shallow, regular, triangular lobes; it lacks Fern-leaved Beech's unique texture. Compare 'Grandidentata' (p.206) and American Beech (p.210).

'Rohanii' (probably a hybrid of purple and oak-leaved clones) is slow and very occasional, with dull, brownish purple leaves; the lobes are often raggedly toothed. The crown (to 17 m) is dense, jagged and rather narrow, with many rigidly rising, tiny branches.

'Rohan Gold' (1970) differs in its yellow then green leaves; very rare to date.

'Dawyck Purple' (Holland, 1969) has the shape of Dawyck Beech (p.206) but good purple foliage: occasional so far as a young tree.

'Dawyck Gold' (a sister seedling), with yellow younger leaves, is rare as yet.

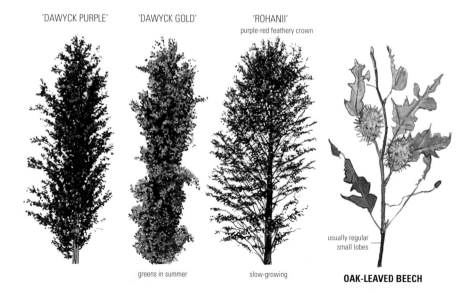

'DAWYCK PURPLE'

'DAWYCK GOLD'
greens in summer

'ROHANII'
purple-red feathery crown
slow-growing

usually regular small lobes

OAK-LEAVED BEECH

BEECHES

Oriental Beech — *Fagus orientalis*

E Balkans to Iran. 1910. Rather rare. A particularly vigorous tree, to 30 m so far, easily mistaken for Common Beech (p.204).
APPEARANCE Shape Rather narrow, with upswept branching. Dark *matt crown* of big clumsy leaves – almost like Huntingdon Elm's (p.246) at a distance; the short trunks of open-grown trees are often deeply fluted, like a giant hornbeam's. **Buds** Slender and spreading, but less elongated than those of Common Beech (1 cm). **Leaves** *Longer*, darker and less glossy than Common Beech's; the broadest point comes above midway and they have up to 14 and *frequently 10 pairs of veins* (cf. 5–9 for Common Beech); the margins are hair-fringed (as in Common Beech) but normally lack any tiny teeth. **Fruit** Nuts are held on thin *stalks to 4 cm long* and the husks' bristles (towards the stalk at least) are slightly *spoon-shaped*, and leaf-green when young.
OTHER TREES *F. moesiaca* (many leaf-veins; husks most like Common Beech's) and *F. taurica* (few leaf-veins; husks most like Oriental Beech's) are controversial species from the Balkans intermediate with *F. sylvatica* (which this far south is restricted in the wild to high hills); not grown in the UK?

Japanese Beech — *Fagus crenata*

(*F. sieboldii*; 1892) In a few collections.
APPEARANCE Shape A tree (to 25 m) most like Oriental Beech. **Leaves** With 7–11 vein-pairs; tending to be broadest *below* the middle, and more *broadly tapered (even rounded)* at the base than other beeches'. **Fruit** The stalk of the nut – to 4 cm long in Oriental Beech – is short and thick (15 mm). The base of the nut-husk has the same *flattened green whiskers*.

Chinese Beech — *Fagus engleriana*

Central China. 1911. A plant of great beauty in some large gardens.
APPEARANCE Shape Smaller than other beeches (to 17 m; sometimes branchy from the base). **Shoots** Dark brown and hairless (silky at first in Common Beech, p.204). **Buds** Particularly long and slender. **Leaves** Longer and *slenderer* than Common Beech's (12×6 cm; 10–14 vein-pairs) and a *paler sea-green* (slightly silvery beneath but hairless except for tufts in the vein-joints); they tend to hang elegantly from the light, horizontal branches. The margin, deckled but scarcely toothed, is *hairless*. The graceful habit most recalls Keaki (p.252), which has strongly toothed leaves.

American Beech — *Fagus grandifolia*

E North America. 1766. Needs maximum summer heat to grow well: like the American Plane (Buttonwood), it was long considered virtually uncultivable in Britain but shapely young trees now thrive in a few southern collections.
APPEARANCE Shape Often stunted. Unlike other beeches it *suckers freely*. **Leaves** To 12×7 cm with *regular teeth* but otherwise resembling Oriental Beech's (11–15 pairs of veins).
COMPARE Common Beech 'Grandidentata' (p.206).
OTHER TREES *F. lucida* (Hubei, China; 1905), in a few collections, also has regular teeth to its leaves which are *slender* (8×4 cm) and *glossy green on both sides*.

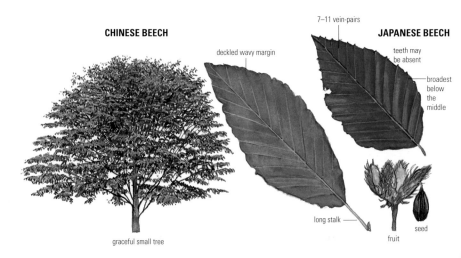

CHINESE BEECH — graceful small tree

JAPANESE BEECH — 7–11 vein-pairs; deckled wavy margin; teeth may be absent; broadest below the middle; long stalk; fruit; seed

SWEET & GOLDEN CHESTNUTS

Sweet chestnuts (10–12 species) have big nuts in spiny sheaths. Chestnut blight (Endothia parasitica)*, a fungal infection accidentally spread from East Asia (and is still spreading northwards across Europe), has made the American Chestnut* (Castanea dentata) *almost extinct in its native range. (Family: Fagaceae.)*

Sweet Chestnut *Castanea sativa*

(Spanish Chestnut) S Europe, N Africa; Asia Minor; brought to Britain by the Romans and completely naturalized in warmer parts (though not elsewhere in N Europe). Abundant except on wet or very chalky soils, living to a huge age; dominant in many sandy woods in SE England where it was planted and coppiced for its timber: the vigorous, straight stems begin to grow a very durable heartwood after 2–5 years.

APPEARANCE Shape Often slender for a big tree, at least near the top, with a long bole in woodland, to 36 m; dense, glossy foliage. **Bark** In young trees dull silvery-purple, horizontally banded then vertically cracked. By 60 years brown, with a vertically stretched network of crisp, criss-crossing ridges which eventually *spiral* (usually clockwise). **Shoots** Grey, the tip aborted (cf. limes); knobbly – on stronger growths – from *long prominent buttresses* below each blunt, few-scaled, hairless bud. **Leaves** Bigger than any other wild tree's, with *spine-teeth 1 cm apart*; scurfy underneath, at least at first. **Flowers** Crowns turn fawn in high summer with stiffly spraying male catkins which smell strongly of frying mushrooms; good crops of chestnuts follow only in warm parts. (Fruiting chestnuts – 'Gros Merle', 'Paragon', 'Marron de Lyon' – are grafted trees with usually only 1 nut per husk; very scarce in the UK.)
COMPARE Japanese Chestnut Oak (p.230).

VARIANTS Variegated Chestnut, 'Albomarginata' (1864), is a rare but bright, delicate tree, apt to revert. The creamy leaf-margin is yellow as the leaves unfold. 'Variegata' ('Aureomarginata'), whose leaf-margins remain gold, is much rarer to date.
Cut-leaved Chestnut, f. *heterophylla*, is a group of rare, often feeble clones whose strongly toothed leaves can (especially at branch-tips) be *reduced to strips*. 'Laciniata' (1838) is distinct, if liable to revert: all its teeth are *drawn out into long threads*.
OTHER TREES American Chestnut, *C. dentata*, is confined in the UK to a few collections. Its leaves are slightly narrower and more tapered at the base than Sweet Chestnut's, and *completely hairless*, but its buds are finely downy.

Golden Chestnut *Chrysolepis chrysophylla*

(Giant Chinkapin; *Castanopsis chrysophylla*) Oregon and California. 1844. In a few big gardens.
APPEARANCE Shape To 20 m, but often bushy. **Leaves** Small (3–13 cm), shiny; coated beneath with minute *golden wool*. Odd specimens have racquet-shaped leaves, with indented tips.
COMPARE *Castanopsis cuspidata* (p.230): dull yellow sheen under the leaves.
OTHER TREES The bright gold under-leaves are shared with an even rarer (and bushy) evergreen, the Golden Oak of Cyprus (*Quercus alnifolia*), whose small (5 cm), rounded, convex leaves are *toothed* near their tips.

GOLDEN CHESTNUT — evergreen — ♂ flws — fruit — gold scales beneath

CUT-LEAVED CHESTNUT 'LACINIATA' — long filaments

VARIEGATED CHESTNUT — inner lvs completely white

OAKS

Oaks (500 species) all carry 'acorns' in 'egg-cups'. Their buds cluster at shoot-tips (any can dominate next season to create the twisting, wide limbs of many species). (Family: Fagaceae.)

Things to Look for: Oaks

- Shoots: Hairy?
- Buds: Whiskered?
- Leaves: Evergreen? Shape (especially at base)? Any bristle-tipped teeth? Hairs/wool underneath? Leaf-stalk – how long is it?

Key Species

Sessile Oak (below): rounded, untoothed lobes. **English Oak** (p.216): similar leaves with auricles. **Turkey Oak** (p.218): jaggedly lobed leaves (fine felt underneath). **Chestnut-leaved Oak** (p.226): regular triangular teeth/lobes. **Red Oak** (p.234): lobes each with several whiskered teeth. **Willow Oak** (p.238): deciduous, unlobed leaves. **Holm Oak** (p.222): evergreen leaves, untoothed at maturity. **Lucombe Oak** (p.220): evergreen, lobed leaves.

Sessile Oak *Quercus petraea*

(Durmast Oak) Mid Europe, including Britain and Ireland. Much more locally dominant than the familiar English Oak (p.216), avoiding heavy/alkaline soils, and absent from many lowlands. Seldom planted in forestry or until recently for ornament, for all its poise and stature.

APPEARANCE Shape Cleaner and less twiggy than English Oak; larger, glossier leaves *evenly spread*. Often taller (to 42 m), but equally gigantic with age. **Bark** As English Oak; can be more shallowly scaly. **Shoots** As English Oak. **Buds** Tend to have more scales. **Leaves** With regular, rather shallow lobes; main veins hairy beneath at first and generally only running to the lobe-tips. Base *broadly tapered* (only sometimes and faintly showing the backward-pointing auricles of English Oak), on a *12–20 mm stalk*. **Fruit** Acorns sit on the twigs, *with short stalks or none* ('sessile').

COMPARE English Oak (p.216). Downy Oak (p.218): leaves much hairier beneath. Mirbeck's Oak (p.228): longer leaves with closer, smaller lobes. Caucasian Oak (p.228): persistently downy shoots/leaf-stalks. Turner's Oak (p.220): forward-angled shallow lobes. 'White oaks' (p.232): narrowly tapered leaf-bases.

VARIANTS Medlar-leaved Oak, 'Mespilifolia', is rather rare and can puzzle. Dark, dense, often irregular crown of long, narrow undulant leaves, seldom lobed (most like Star Magnolia's, p.268). Rugged bark (though more closely square-cracked than most type trees') and blackish foliage distinguish it from Willow Oak (p.238), etc, with unlobed leaves. Moscow Oak, 'Muscaviensis', is a much scarcer variant: its high-summer 'lammas-growth' leaves *are* lobed.

'Laciniata' (very rare) has narrower, less extremely lobed leaves than Cut-leaved English Oak (p.216). 'Columna' is much rarer and less vigorous than Cypress Oak (p.216); leaf-shape distinguishes it.

OTHER TREES *Q. dalechampii* (4–7 lobe-pairs) and *Q. polycarpa* (7–10 lobe-pairs), from SE Europe, have *downy* acorn-cup scales. *Q. iberica* (Transcaucasus) has leaves with *big tufts* under its vein-joints. *Q. mas* (Pyrenees; N Spain) has *downy acorn-stalks*. In the UK, all these trees are confined to a few collections.

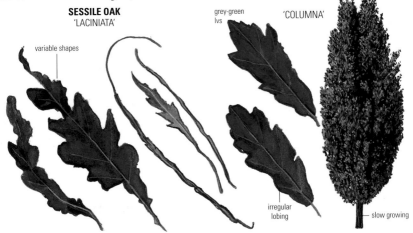

SESSILE OAK
'LACINIATA'
variable shapes
grey-green lvs
'COLUMNA'
irregular lobing
slow growing

ENGLISH OAK

English Oak *Quercus robur*

(Pedunculate Oak; Common Oak; *Q. pedunculata*) Europe, including Britain and Ireland, to the Caucasus. Abundant except on marshy, chalky or very light soils; the dominant big tree across much of Britain, supporting a greater variety of leaf-eating insects than any other. Planted and/or pollarded (along with Sessile Oak) in old deer parks. Often the most valuable of forestry trees, with selected 'standards' mixed through most coppiced woods; everywhere in parks and gardens. Lifespan 1000+ years. In contrast to Sessile Oak a 'weed' species in scrub etc, though currently reluctant to reproduce within woodland: seedlings in shade succumb to the mildew (*Microsphaera alphitoides*) which can turn mature crowns grey late in hot summers.

APPEARANCE Shape Heavy, spreading, *twisting* branches make a broad crown, to 38 m; foliage in *blobby bunches*. **Bark** Grey; short, deep, knobbly ridges. **Shoots** Silvery; orange-brown oval buds cluster at their tips. **Leaves** With *irregular, deep lobes* (pointed only on sprouts); 2 *tiny lobes at the base* (auricles) flank the *short* (4–10 mm) stalk. Small veins extend towards the *deep, narrow sinuses* between each lobe. **Flowers** Curtains of yellow male catkins as the leaves unfold orange. **Fruit** Acorns (skipped in many years in N Europe to limit populations of acorn-predators) often paired, on a 5–12 cm stalk ('peduncle').

COMPARE Sessile Oak (p.214). The hybrid (*Q.* × *rosacea*) is probably occasional (in continental Europe, Sessile Oaks tend to flower a fortnight after English Oaks, making crosses rare). Turkey Oak (p.218): even as a seedling, has long-whiskered buds. Among widely grown oaks, only English Oak and Daimyo Oak (p.224: huge leaves, hairy shoots) have pronounced 'auricles', but exotic species raised from seed may turn out to be hybrids with English Oak, with auricles but usually bigger leaves, less deeply lobed.

VARIANTS Cypress Oak, f. *fastigiata*, is at its best often taken in winter for a Lombardy Poplar, but the dense shoots are thicker and the branches twist; solid and blackish in leaf – like an Italian Cypress. Many are broader and open, but with *erect writhing shoots from steep branches*. Occasional; to 30 m.
Cut-leaved Oak, 'Filicifolia', is rare: a feathery, ghostly-grey tree, often gaunt, to 17 m. Hairs under its young leaves suggest it is a clone of *Q.* × *rosacea*.
Purple Oak, 'Atropurpurea', is rare and stunted. Trees with leaves flushing red-purple then green (f. *purpurascens*) grow larger and may be seen in the wild. They redden again during the species' flush of 'lammas-growth' in high summer (which repairs the ravages of spring leaf-eating caterpillars).
Golden Oak, 'Concordia' (1843), has yellow foliage, fading through summer; very rare. (Yellow chlorotic foliage is often seen on oaks, but all the growths will never be so evenly coloured.)

OTHER TREES *Q. brutia* (S Italy) and *Q. pedunculiflora* (Balkans to the Caucasus) have downy under-leaves. *Q. thomasii* (S Italy) has more deeply lobed leaves. *Q. hartwissiana* (Bulgaria to the Caucasus) has leaves more like Sessile Oak's (but long acorn-stalks). (All are confined to a few collections.)

very feathery

uniformly greenish yellow by late summer

seldom grows well

CUT-LEAVED OAK   **GOLDEN OAK**

OAKS (EUROPEAN)

Turkey Oak — *Quercus cerris*

SE France E to Turkey. 1735. Abundant in parks and belts; often aggressively colonizing on sandier soils. Host to alternate generations of the Knopper Gall wasp (*Andricus quercuscalicis*), whose caterpillars now turn the acorns of English Oaks within flying-distance into oozing, lumpy galls: eradication of Turkey Oak from the New Forest is consequently planned. Vigorous and very tough, thriving in coastal exposure, but worthless as a timber tree.

Appearance Shape Sometimes with a long, straight bole and pointed tip; to 40 m. Straighter, slenderer branches than wild oaks' are swollen where they leave the trunk. *Blackish* and slightly feathery in summer; the clustered, mop-head buds tipping the straight shoots are conspicuous in winter. **Bark** Palish mauve-grey: *deep, wedge-shaped fissures*, often orange at the base. **Shoots** Slender; densely grey-hairy. **Buds** *All* with *big twisting whiskers*. **Leaves** Thick; rough but quite shiny above and minutely grey-felted beneath. Slender, but variable: often with simple, pointed/rounded lobes but sometimes elaborately cut (strong growths/f. *laciniata*). **Fruit** Acorns stalkless, the cup-scales carrying mop-heads of more *big whiskers*.

Compare Lucombe Oak (p.220): can grade towards its parent but usually has unwhiskered side-buds, regular triangular lobes to *evergreen* leaves, and shorter, twisting branches. Pyrenean Oak (p.228): unwhiskered side-buds and leaves softly hairy both below *and above*. Chestnut-leaved Oak (p.226): the largest of a group whose leaves always have small, regular triangular lobes; projecting veins tip each with a minute bristle (scarcely seen in Turkey Oak).

Variants 'Argenteovariegata' ('Variegata') is one of the brightest of white-variegated trees; slow, but reverting readily. Rare.

Other trees Valonia Oak, *Q. ithaburensis* ssp. *macrolepis* (*Q. macrolepis*; *Q. aegilops* ssp. *macrolepis*; Albania, Greece, and Turkey – plus SE Italy?; in a few collections since 1731), is equally whiskery: a smaller tree, with twisted, drooping, sprouty branches, a ruggedly square-cracked bark, smaller leaves *with bristle-tipped lobes*, and *huge* (4 cm) acorns – in Italy once grown commercially for their tannin content.

Downy Oak — *Quercus pubescens*

(Green Oak; *Q. lanuginosa*) S Europe, W Asia; often dominant, especially on limestone. Long grown in the UK but now rare.

Appearance Shape Wide, twisting branches like English Oak's (p.216); to 22 m. **Bark** Almost like Holm Oak's (p.222): black-grey and quite closely square-cracked. **Shoots, Buds** Softly grey-downy. **Leaves** Long stalks (like Sessile Oak's (p.214) but hairy), and the irregular lobes and deep narrow sinuses typical of English Oak: they are downy above at first and normally remain densely woolly beneath. (Some in cultivation may be hybrids – their leaves are almost hairless beneath by autumn.) **Fruit** Acorns nearly stalkless; cup-scales grey-downy.

Compare Pyrenean Oak (p.228): paler, more rugged bark. Caucasian Oak (p.228).

Other trees *Q. brachyphylla* (the Aegean) has smaller leaves. *Q. congesta* (Sicily, Sardinia, S France) has larger leaves and more whisker-like acorn cup-scales. *Q. virgiliana* (Corsica E to Turkey) has very big leaves (to 16 cm) and sometimes *edible* acorns. These are scarcely grown in the UK.

closely-cracked bark

fruit

grey downy cup

DOWNY OAK

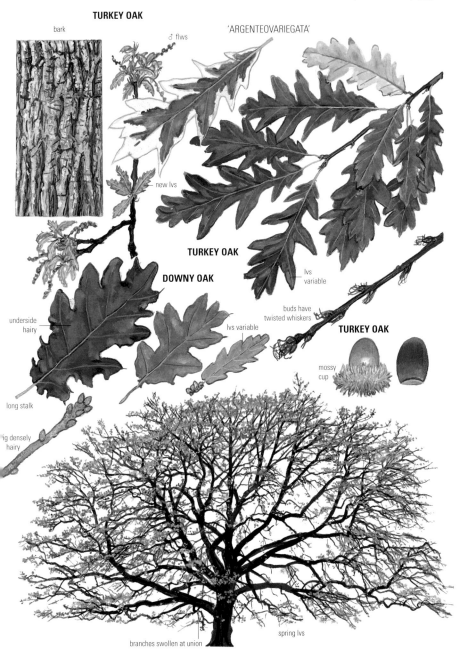

HYBRID EVERGREEN OAKS

Lucombe Oak
Quercus × *hispanica* 'Lucombeana'

(*Q.* × *crenata* 'Lucombeana'; *Q. lucombeana*) Turkey and Cork Oaks hybridize freely in S France/Italy, but the locally frequent trees in Britain and Ireland mostly derive from a cross raised at his Exeter nursery by William Lucombe in 1762. This is fertile, and today's population incorporates backcrosses with both parents; seedlings can also be found. The nomenclature is confused, but several variants can be distinguished. The original clone (provisionally 'William Lucombe') is the most frequent, especially in E Devon.

APPEARANCE Shape Tall (to 35 m): *heavy, twisting branches* are much swollen as they leave the trunk, which is *never long*; interior clean and *open*, surfaced with shining dark leaves which look deciduous but *hang on thinly until spring*. **Bark** *Purple*-grey, with deep triangular ridges; not corky. **Shoots** Hairy-grey; the end bud has big whiskers, the side-buds *never do*. **Leaves** Finely grey-felted beneath; the rather regular triangular lobes have *minute whisker-tips*.
COMPARE Turkey Oak (p.218; differences emphasized). Chinese Cork Oak (p.230). Turner's Oak 'Spencer Turner' (below). Chestnut-leaved Oak (p.226): leaves have *more (9–14) pairs* of shallower, regular lobes.
VARIANTS 'Crispa' (1792) is almost as frequent; it differs from Cork Oak (p.224) in stronger growth and more substantially lobed leaves. A broad, low and dense tree, with hanging outer shoots and many wandering twisting branches; fully evergreen. Bark cream/grey; soon *deeply corky*. Leaves small (7 cm); oval, with blunter lobes than 'William Lucombe'; a small proportion reduced to straps.

'Fulhamensis' (distributed by Whitley and Osborne of Fulham by 1783 and perhaps of independent origin; now certainly rare) resembles 'William Lucombe' but has a corkier bark, weeping outer shoots, and neatly oval leaves with usually 6 pairs of *precise, triangular lobes*.
Some trees have many deeply and irregularly lobed leaves, often with a narrow violin-like 'waist' and sometimes thread-like. 'Diversifolia' (very rare) is a lightly crowned, feathery tree, its branches rising then arching. Another clone resembles 'William Lucombe' but is a smaller, glossier, very dense, non-weeping tree, with a dark, slightly corky bark; its side-buds may be whiskered.
'Cana Major' (1849) is a fully deciduous clone.

Turner's Oak
Quercus × *turneri* 'Pseudoturneri'

English Oak hybridized with Holm Oak at Spencer Turner's nursery in Leyton (now in London) some time before 1776. The original cross, 'Spencer Turner', is now almost extinct; 'Pseudoturneri' (a back-cross?) remains occasional: stylish, but easily overlooked.

APPEARANCE Shape Dense (to 25 m), with twisting branches from *low on the bole; sparse fresh green leaves through winter* (cf. Lucombe Oak). **Bark** Dark grey; knobbly/square-cracked ridges. **Shoots** Densely hairy. **Leaves** Look 'stretched': *long-tapered base*; neat, shallow, *forward-pointing rounded lobes*. Dark and rather shiny above; downy under the main veins.
COMPARE Mirbeck's Oak (p.228) and Basket Oak (p.232): larger leaves with more lobes (8–12 pairs).
VARIANTS 'Spencer Turner' has a taller, open crown, a coarser bark, and shorter leaves with rounded bases, and can confuse if encountered.

LUCOMBE OAK
usually evergreen

bark variant

HOLM OAK TO MACEDONIAN OAK

Holm Oak
Quercus ilex

(Mediterranean Oak) S Europe; introduced early and abundant away from the coldest parts in shrubberies and belts, especially near the sea (where it copes outstandingly with salt spray); quite well naturalized in milder areas.
APPEARANCE Shape Often bushy; closely rising, rather *straight* branches from a sinuous bole; to 30 m. Very dense; the outer shoots may weep. **Bark** Blackish; becoming closely and shallowly square-cracked. **Shoots** Slender, with fawn wool. **Buds** Tiny; the end one has curling whiskers. **Leaves** Evergreen: blackish, a *fawn-grey felt* coating the often concave underside; spinily lobed on sprouts and saplings, then untoothed. Very variable in breadth. **Flowers** Fireworks of golden male catkins in early summer as last year's leaves drop. **Fruit** Acorns small (15–20 mm), with felted cup-scales. The acorns of var. *ballota* (*Q. rotundifolia*), whose leaves are smaller and more rounded, are *edible*: once much grown in S Spain and N Africa.
COMPARE Californian Live Oak (below); Cork Oak (p.224); Golden Oak of Cyprus (p.212). Phillyrea (p.442): similar bark and crown but leaves, *in opposite pairs*, hairless. Most evergreen oaks (see Japanese Evergreen Oak, p.230) have longer, elegant, glossy leaves.

Californian Live Oak
Quercus agrifolia

(Encina) California; NW Mexico. 1843. Rare.
APPEARANCE Shape Resembles Holm Oak but foliage fresher green; to 17 m. **Bark** Dark grey, rather *distantly* fissured. **Leaves** Broader, persistently spiny and often very convex. The underside is *smooth*; mid-green with *tufts of hairs only in the vein-joints*.
OTHER TREES Kermes Oak (Holly Oak; Grain Tree), *Q. coccifera* (W Mediterranean regions), has long been grown in N European collections; to 10 m. Once economically important as the food-plant of the Kermes Insect from which Scarlet Grain dye was extracted. Differs from Californian Live Oak in its rougher bark (like Holm Oak's, but paler) and smaller (2–4 cm) *quite hairless*, very glossy leaves. In the E Mediterranean its place is taken by the slightly larger-leaved Sindian or Palestine Oak, *Q. calliprinos*. Ubame Oak, *Q. phillyreoides* (Japan, China; 1861), is a cheerful little plant in some big gardens, differing from Californian Live Oak in its brown, coarsely ridged bark and *spineless* oval leaves (usually with tiny *blunt* teeth), deep green above and minutely scurfy under their lower midrib.

Macedonian Oak
Quercus trojana

(Trojan Oak; *Q. macedonica*) SE Italy, S Balkans, and W Turkey. *c.*1890. In some big gardens.
APPEARANCE Shape *Dense and spiky*, to 20 m; green until Christmas and retaining many dead leaves through winter (as with many oaks on twigs within 3 m of the base). **Bark** Dark grey; ruggedly square-cracked. **Shoots** Grey-felted. **Buds** Without whiskers except for a few on the end buds. **Leaves** 4–9 cm, dark shiny grey-green and soon quite hairless, with small but regular, usually incurving lobes tipped by 1 mm bristles; on *very short stalks* (2–8 mm).
COMPARE Lebanon Oak (p.224): larger, longer-stalked leaves; more open crown. Lucombe Oak (p.220): leaves always finely felted beneath.

CALIFORNIAN LIVE OAK

MACEDONIAN OAK

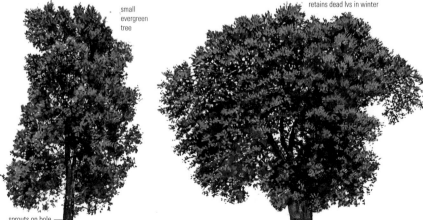

small evergreen tree

retains dead lvs in winter

sprouts on bole

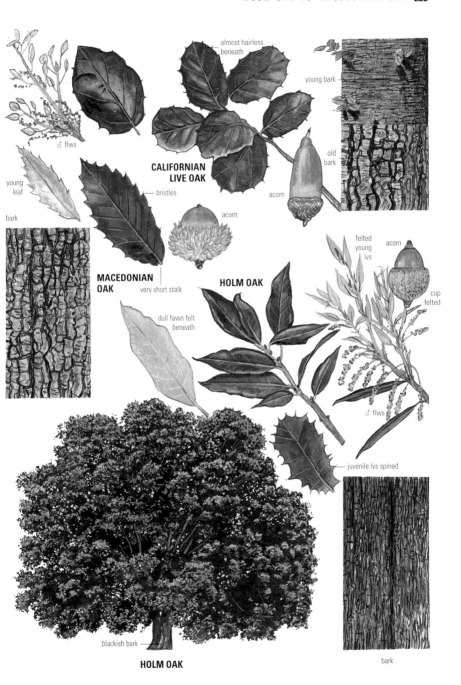

OAKS

Cork Oak — *Quercus suber*

Mediterranean Europe from Portugal to Croatia; Morocco. Long grown in the UK: very occasional, though hardy and much admired. In Mediterranean plantations the cork is stripped off every 7–10 years without damaging the underlying cambium, though the advent of plastic corks has threatened this ancient, wildlife-rich landscape.
APPEARANCE Shape A low, dark, matt dome on heavy, *twisting* branches; to 23 m. **Bark** Cream, orange or grey, soon *deeply convoluted* like a Brancusi sculpture. Annual rings can be counted in the broken edges. **Shoots** Woolly. **Leaves** Evergreen, with small, often spiny lobes; blackish above; *pale grey* beneath with a dense felt.
COMPARE Holm Oak (p.222): very different bark; leaves untoothed at maturity. Lucombe Oak (p.220): sometimes as corky; has larger, consistently triangular leaf-lobes. Chinese Cork Oak (p.230): well-lobed deciduous leaves.

Armenian Oak — *Quercus pontica*

(Pontine Oak) Caucasus. 1885. Some collections.
APPEARANCE Shape A bushy plant. **Shoots** Stout. **Bark** Bronze. **Leaves** Splendidly big (15×9 cm) and bright; the *15–17 parallel sunken yellow veins* running – half of them branch – to curved spiny teeth; greyish beneath, and hairy when young.
OTHER TREES Some trees in collections with inferior foliage are *Q.* × *hickelii*, the hybrid with English Oak: slightly fewer vein-pairs run to little lobes which may have intermediate teeth, and there is a hint of auricles at the base.

Lebanon Oak — *Quercus libani*

Syria; Asia Minor; Kurdistan. 1855. Rare.
APPEARANCE Shape A slender, neat tree to 20 m, suggesting a diminutive Turkey Oak (p.218). **Bark** Similar to Turkey Oak. **Buds** *Only the end buds* have the long whiskers. **Leaves** *Slender, quite small* (10–12 cm), *glossy*, with 10–12 pairs of regular *bristle-tipped* triangular lobes; stalks 6–10 mm.
COMPARE Chestnut-leaved Oak (p.226): a spreading tree with larger leaves, their lobes scarcely bristle-tipped. Japanese Chestnut Oak (p.230): longer, very glossy leaves and gaunt habit. Macedonian Oak (p.222): dense, broader leaves with 2–8 mm stalks.

Daimyo Oak — *Quercus dentata*

(*Q. daimio*) Japan; Korea; NE China. 1830. Rare.
APPEARANCE Shape Often gaunt because of annual frost damage, with tufts of shoots from a few twisting limbs; some trees make respectable domes, to 18 m. The *huge, rich green leaves* are often retained, dead, through winter. **Bark** Dark grey, ruggedly ridged. **Shoots** Stout, *densely hairy*. **Leaves** *25–40 cm* and as thick as parchment; rather downy, variably deeply lobed and with a hint of auricles beside the *hairy* 10–15 mm stalks.
COMPARE Caucasian Oak (p.228): narrower lobes to more normal-sized leaves, auricular only if there has been hybridization with English Oak. Burr Oak (p.232). English/Sessile Oak hybrids: can have 'hybrid vigour' and extra-big leaves, but the shoots and leaf-stalks will not be hairy.

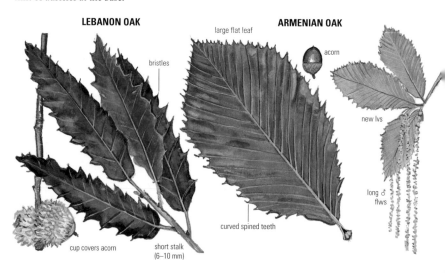

LEBANON OAK — bristles; cup covers acorn; short stalk (6–10 mm)

large flat leaf; curved spined teeth

ARMENIAN OAK — acorn; new lvs; long ♂ flws

OAKS

Hungarian Oak — *Quercus frainetto*

(*Q. conferta*) The Balkans, Romania, Hungary and S Italy; the specific name is a corruption of the Italian word for the tree, 'farnetto'. 1837. Occasional, but beginning to be more widely grown. In many ways the grandest of all oaks.

APPEARANCE Shape Typically a broad and splendid glossy-leaved wheel of *straight*, rather stiff, radiating branches, to 38 m, creating a slightly jagged outline; woodland-grown trees can have long boles and are slenderer. **Bark** Pale/purplish grey, cracking into rather small, neat square plates. Many older trees (the clone now marketed as 'Hungarian Crown') are grafts on English Oak. **Shoots** Softly hairy, at least at first. **Buds** Large (1 cm), grey-brown, with numerous loose, hairy scales (cf. Mirbeck's Oak, p.228), but no whiskers. **Leaves** Big (to 25 cm), *elaborately cut into many rather square, narrow lobes*; downy beneath and with a few harsh hairs above; in autumn gold and biscuit-brown. (Some trees are hybrids with English Oak: smaller leaves with fewer lobes and slight auricles. Acorns collected in cultivation also yield hybrids with Turkey Oak.)

COMPARE Caucasian Oak (p.228): whiskered buds, shaggy bark and less squarely lobed leaves. Mirbeck's Oak (p.228): most similar in winter, but has darker bark. Sessile Oak (p.214): the bark is sometimes finely cracked, but trees are more rugged and twisting (with *smoothly tiled* bud-scales).

Chestnut-leaved Oak — *Quercus castaneifolia*

Caucasus, Iran. Rare but magnificent – the 1846 original at Kew is now one of the biggest trees in Britain.

APPEARANCE Shape Domed (to 32 m) on steep, slightly twisting branches; some are slender/irregular. Dead leaves may persist through winter. **Bark** Purple-grey; rugged, twisting, triangular ridges. **Shoots** Hairy at first. **Buds** Nearly *all with big whiskers* (as Turkey Oak, p.218). **Leaves** 12–20 cm, slender; the veins run to 9–14 pairs of small regular triangular lobes with *minute* bristle-tips; dark above, greyish and usually finely downy beneath. Stalk 2–4 cm, finely hairy. (Some garden trees with paler bark and deeper, irregular lobing are probably Turkey Oak hybrids.) **Fruit** Acorns with 'shaggy' cups, as Turkey Oak's.

COMPARE Turkey Oak (p.218): fewer, less regular lobes. Lucombe Oak (p.220): *only about 6 pairs of* triangular lobes but sometimes a similar habit. Lebanon Oak (p.224): similar, but with smaller, short-stalked leaves, each lobe strongly bristle-tipped. Japanese Chestnut Oak (p.230): glossier leaves with long bristles. Chinkapin Oak (p.232): gland-tipped bristles. The comparison of the leaves to Sweet Chestnut's (p.212) is rather fanciful.

VARIANTS 'Greenspire' (1948) is very occasional: the branches rise at about 60°, making (to date) a narrowly columnar/funnel-shaped tree.

HUNGARIAN OAK

branches radiate — neatly cracked grey bark

CHESTNUT-LEAVED OAK

bark

OAKS

Caucasian Oak   *Quercus macranthera*

Caucasus to Iran. 1873. Another handsome, vigorous, undeservedly rare species.
Appearance Shape An often untidy, spiky dome, to 30 m. **Bark** Grey-brown; *coarse, scaly plates*. **Shoots** With *persistent dense hairs*. **Buds** *Large* (to 15 mm) and dark chestnut, with white hairs and a few *long whiskers* (cf. the much *smaller* buds of Turkey Oak, p.218). **Leaves** Big (to 22 cm) and particularly thick, with irregular *forward-angled* lobes (cf. Turner's Oak, p.220), often deep and narrow, or sometimes like Mirbeck's Oak's; softly downy beneath and on hairy stalks. (Hybrids occur with English Oak, and have smaller, auricular leaves.) **Compare** Mirbeck's Oak (below); Hungarian Oak (p.226). Pyrenean Oak (below): deeper lobing to flimsier leaves. Downy Oak, p.218.

Mirbeck's Oak   *Quercus canariensis*

(*Q. mirbeckii*) S Spain, Portugal and N Africa (but not the Canaries). 1844. A magnificent yet rare oak.
Appearance Shape Typically a narrow dome, to 30 m, on a sturdy bole (though some trees are shapeless); *retaining some or many green leaves* until spring. **Bark** Dark purplish grey; rather close square plates, crisp or rugged. **Shoots** Ridged; brown wool soon rubs off. **Buds** To 1 cm, numerous scales fringed with white hairs (cf. Hungarian Oak, p.226). **Leaves** Rather glossy and often convex, to 18 cm, with *8–14 pairs* of regular quite *small* lobes diminishing neatly towards the tip; they are covered at first in loose reddish wool, traces of which may persist under the midrib. (Many older plantings are hybrids with English Oak: smaller, auricular leaves, a broader crown of twisting branches, and paler grey, vertically ridged bark.) **Compare** Sessile Oak (p.214): paler, coarser bark; fewer lobes. Caucasian Oak (above): shaggier bark. Basket Oak (p.232); Turner's Oak (p.220).
Other trees Portuguese Oak, *Q. faginea* (*Q. lusitanica*; Iberia), has much shallower, less regular lobing/toothing and leaves often *grey-felted* beneath, on woolly shoots; in a few collections.

Pyrenean Oak   *Quercus pyrenaica*

(*Q. toza*) W France; Iberia; Morocco. 1822. Rare.
Appearance Shape Slender, irregular. The grafted clones most often seen (f. *pendula*) have strongly weeping outer shoots. The last oak into leaf; grey all over in early summer as if mildewed. **Bark** Pale grey; close but quite craggy. **Shoots** Densely grey-downy. **Buds** With long whiskers *soon shed*. **Leaves** 8–20 cm, flimsy; becoming stiffly glossy above but *covered on both sides* with soft, light-catching down (denser beneath); lobing rounded but deep and irregular; stalk very downy. **Flowers** Male catkins make showy golden cascades in early summer.
Compare Turkey Oak (p.218) and Caucasian Oak (above): buds persistently whiskered. Downy Oak (p.218): darker, more closely cracked bark. White Oak (p.232): foliage soon hairless.

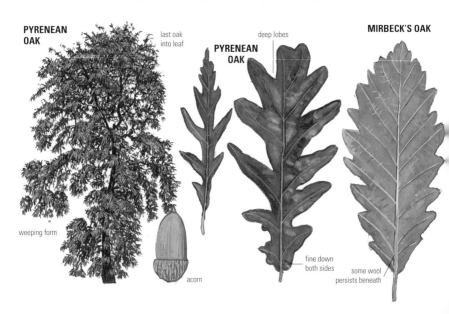

PYRENEAN OAK — weeping form — last oak into leaf — acorn
PYRENEAN OAK — deep lobes — fine down both sides
MIRBECK'S OAK — some wool persists beneath

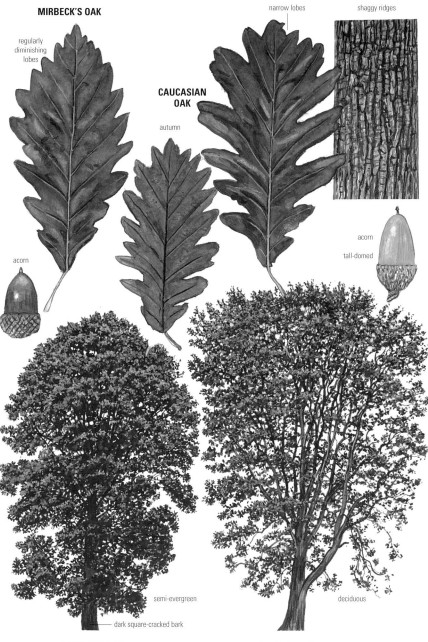

OAKS (EAST ASIATIC)

Japanese Chestnut Oak — *Quercus acutissima*

NW India to Japan. 1862. Rare: big gardens.
APPEARANCE Shape Irregular; *gaunt and open*; to 23 m. **Bark** Dark grey; *very deeply* and ruggedly ridged. **Shoots** *Soon hairless*. **Leaves** *Long* (20 cm) and slender; *very glossy*; about 15 pairs of parallel veins end in strong, *2–6 mm bristles*; pale *green beneath* with fine hairs only on the veins. The slender, hairless stalk is 7–20 mm long.
COMPARE Chinese Cork Oak (below). Chestnut-leaved Oak (p.226): massive crown; lobe-bristles barely 1 mm long. The several other similar oaks have smaller, less glossy leaves.

Chinese Cork Oak — *Quercus variabilis*

N China, Korea and Japan. 1861. Very rare.
APPEARANCE Shape Often rather gaunt, on strong level branches; to 22 m. **Bark** Pale fawn-grey and soon *deeply corky* (cf. Cork Oak, p.224). **Shoots** Slightly hairy. **Leaves** To 20 cm; parallel veins end in *2–5 mm bristles*; palely *grey-felted beneath*.
COMPARE Lucombe Oak (p.220): smaller, more or less evergreen leaves, whose lobes scarcely bristle-tipped. Armenian Oak (p.224) and Japanese Chestnut Oak: almost hairless leaves; different bark.

Japanese Evergreen Oak — *Quercus acuta*

(Akagashi; *Q. laevigata*) Japan. 1878. Rare: big gardens in milder areas.
APPEARANCE Shape Usually bushy; densely domed, to 14 m. Foliage and habit suggest anything but an oak. **Bark** Smooth; dark grey. **Leaves** Unfolding brown-woolly then becoming hairless; thick; dark and glossy above; *dull* yellowish beneath. They narrow abruptly to a long, *finely rounded tip*. **Flowers** On a stiff 5 cm catkin.

COMPARE Tarajo (p.366); Fortune's Rhododendron (p.428); Chinese Tree Privet (p.442).
OTHER TREES *Castanopsis cuspidata* (Japan and China, 1830) is a representative of a big genus of E Asian trees (intermediate between oaks and sweet chestnuts). It has grown rather reluctantly to 13 m in a few mild gardens. It looks very like Japanese Evergreen Oak, but has a *broader* (3 mm), rounded point to its long-tipped leaf (which may show odd teeth towards the tip); the yellowish-green underleaf is slightly *metallic* with microscopic smooth scales (cf. Golden Chestnut, p.212).
Lithocarpus edulis is a Japanese member of another big evergreen genus scarcely seen here. Also like Japanese Evergreen Oak, it has an often *silvery* sheen under its leaves which are bigger (to 15 cm) and less abruptly pointed, yellowish but glossy above and (like the shoots) hairless from the start. The pale grey bark cracks into shallow plates; the crown may be taller and more open (to 15 m); its edible acorns, in 3s on stiff 5–10 cm catkins, seldom develop in the UK.

Bamboo Oak — *Quercus glauca*

(Thonp Oak) Himalayas E to Laos and N to Japan. 1854. Another un-oaklike rarity in mild areas.
APPEARANCE Shape Slender, or densely but gracefully bushy-domed; to 15 m. **Bark** Smooth, dark grey; odd narrow fissures develop. **Leaves** Glossy; glaucous beneath, sometimes silky-hairy; with *distant teeth in the upper half*. They hang elegantly, flushing maroon, and are particularly delicate, *slender* (100×25 mm), fresh green and bamboo-like in the form – most often grown in the UK – which is sometimes distinguished as *Q. myrsinifolia*.
Compare Wheel Tree (p.274): long leaf-stalks. Peruvian Nutmeg (p.276): opposite leaves.

JAPANESE EVERGREEN OAK

JAPANESE CHESTNUT OAK

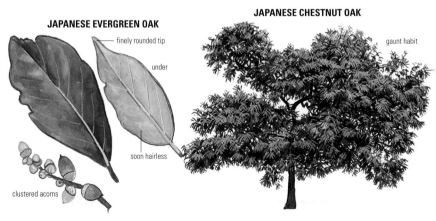

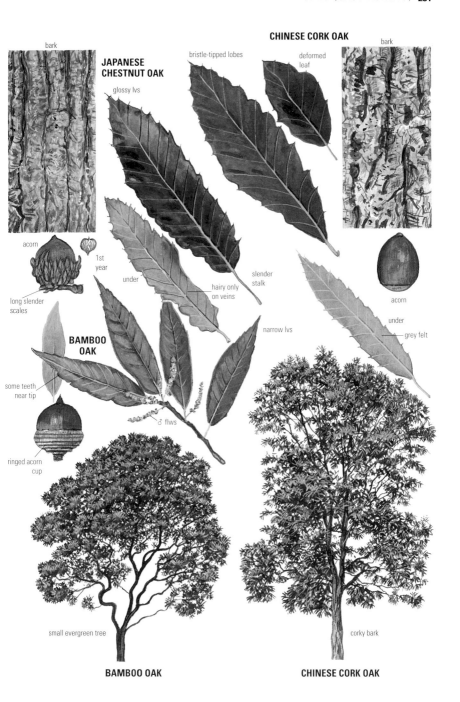

WHITE OAKS

White Oak *Quercus alba*

Ontario to Florida, where it is often the most important oak in the landscape. 1724. In a few big gardens; the New World 'white oaks' (related to English Oak) seldom thrive in the UK.
Appearance Shape Domed, on twisting branches. **Bark** Dark grey; coarse *shaggy* ridges. **Leaves** Most like English Oak's (p.216) but usually larger, with a little down underneath at first, and *narrowly tapered* to their *10–25 mm* stalks. They often turn rich purple in autumn.
Compare Pyrenean Oak (p.228); Turner's Oak (p.220).

Basket Oak *Quercus prinus*

(Chestnut Oak) E USA. *c.*1688. In a few collections.
Appearance Shape Domed, to 19 m; often gaunt. **Bark** Blackish brown; close but rugged ridges. **Shoots, Buds** Hairless. **Leaves** Bright green and variably finely downy beneath; *narrowly tapered* at the base and *long-pointed*; big, *rounded*, forward-pointing teeth/small lobes each tip a prominent parallel vein; stalk yellow and *slender*, 4 cm. **Fruit** Large acorns, the long cup with downy but closely appressed scales.
Compare Mirbeck's Oak (p.228): oval leaves *broadly* tapering at their bases, crisper bark. Swamp White Oak (above).
Other trees Chinkapin Oak, *Q. muehlenbergii* (E USA, often on limestone; 1822), is in a few collections and has narrower leaves, variably downy beneath; the lobes are each tipped by a *tiny knobbly gland* and may be sharper/incurved.

Also with *gland-tipped teeth* to leaves of similar shape is *Q. serrata* (*Q. glandulifera*; China, Korea, Japan, 1877) – a singularly handsome tree in a few big gardens (to 20 m), its glossy, hanging leaves are grey-hairy beneath (cf. Chinese Cork Oak, p.230). The bark develops *wide*, rugged, grey-brown ridges.

Swamp White Oak *Quercus bicolor*

E North America. 1800. Very rare.
Appearance Shape In the UK, usually a good, slender dome, to 25 m. **Bark** Grey; shaggy, criss-cross ridges. **Leaves** Flimsy, to 18 cm; shiny above, grey beneath with a *minute felt*. Always broadest near the tip: irregular shallow lobes above a long-tapered *unlobed base*.
Other trees Burr Oak (Mossy-cup Oak), *Q. macrocarpa* (E North America, 1811; equally rare), has a narrower leaf (to 26 cm), variously more deeply/irregularly lobed from *near the base*. *Thread-like scales* form a mossy *fringe* around the acorn-cup.

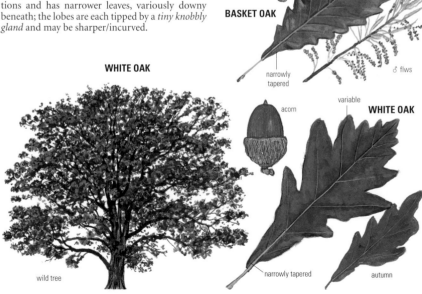

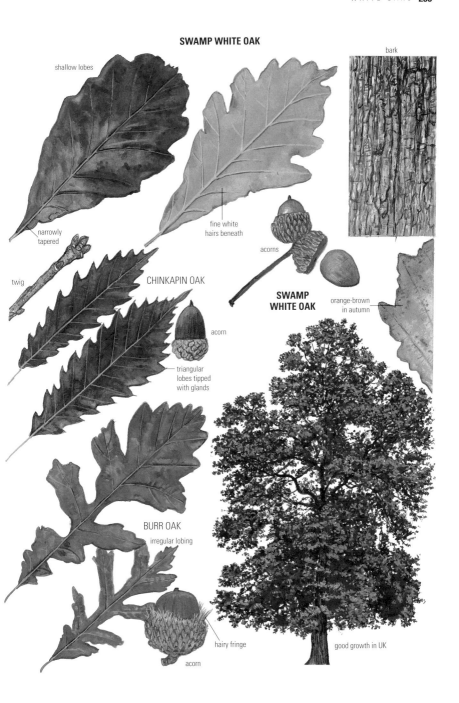

RED OAKS

Red Oak — *Quercus rubra*

(*Q. borealis*; *Q. maxima*) E North America. 1724. Abundant in warmer parts: parks, gardens, forestry rides (and in a few plantations). Growing as large in Britain as in its native habitat (like several other of the 'red oaks' but unlike the 'white oaks' on p.232); rarely naturalizing.

APPEARANCE Shape Soon broad, on *strong, wide but clean and straight branches*; to 32 m. The American 'red' and 'willow oaks' (collectively 'black oaks') lack the rugged, crooked appearance of many oak species; their timber is soft and they are relatively short-lived. **Bark** Silver-grey and smooth at first. The British population is now diverse and some old trees retain smooth bark, with hemispherical warts; others develop shallow or even scaly grey ridges between orange fissures. **Shoots** Slender, grey, quickly hairless. **Buds** Chestnut, the scales with slightly hairy tips. **Leaves** Big (often 20 cm long), the variably *shallow* lobes each with *2 or more whiskered teeth*; soon *hairless* except for minute buff tufts under the vein-joints. Seldom glossy above and always a *matt pale green beneath*. They unfold late and are pale yellow for a week; autumn colours can be orange-brown or deep red, but are a disappointing warm brown in many seasons in the UK. **Fruit** Acorns, like those of other 'black oaks', are small (2 cm) and take 2 years to ripen.

COMPARE Other 'red oaks': Scarlet and Pin Oaks (p.236); Spanish and Shumard Oak (p.238); Black Oak (below). These are slenderer trees with usually smaller, more deeply lobed leaves *glossy beneath* (but dull *and downy* in Spanish Oak).
VARIANTS 'Aurea' has leaves which *remain* brilliant gold through early summer. Very rare.

Black Oak — *Quercus velutina*

(Quercitron Oak) Ontario to Texas and Florida. 1800. Rather rare.

APPEARANCE Shape Most like Scarlet Oak's (p.236): seldom broad; to 30 m. **Bark** Grey; usually darker and more closely square-cracked than that of other 'red oaks'. The yellow/orange under-bark shows in the cracks and was the source of the dye quercitron. **Shoots** Remaining *woolly for some months*. **Buds** Conspicuously *large (6–10 mm) and fawn-furry*. **Leaves** Glossy and blackish; *thick*, and leathery, on thick yellow stalks; about 15 cm long, with variably deep whisker-toothed lobes; they emerge covered in scurfy wool of which patches remain underneath through summer, while tufts persist under the vein-joints. Deformed (forking/scarcely lobed) leaves are frequent on some trees. **Fruit** Acorn-cups have hairy scales.
COMPARE Scarlet Oak (p.236); Shumard Oak (p.238). Spanish Oak (p.238): the only other 'red oak' with such downy buds, leaves and shoots.
VARIANTS Champion Oak, 'Rubrifolia', is a rare grafted selection with *huge leaves* (to 40 cm).
OTHER TREES Californian Black Oak, *Q. kelloggii*, is confined to collections. It has less scurfy leaves, and buds which (like Red and Scarlet Oak's) are hairy *only at their tips*.

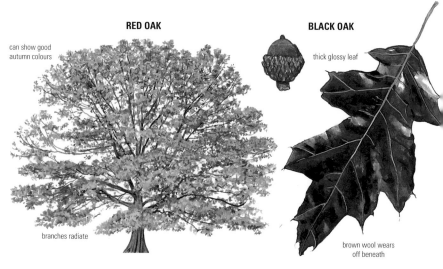

RED OAK
can show good autumn colours
branches radiate

BLACK OAK
thick glossy leaf
brown wool wears off beneath

RED OAKS

Scarlet Oak *Quercus coccinea*

SE and central USA. 1691. Rather occasional; rare in colder areas.
APPEARANCE Shape Irregular and quite slender, generally on a long, *sinuous* trunk with rather *few, long but slender wandering limbs*; untidy with ascending small branches and twigs. To 30 m. Sprinkled in high summer with *very yellow* late growths. **Bark** Silver-grey; sometimes remaining smooth except for hemispherical warts; more often is shallowly rugged and purplish, with orange fissures. **Shoots** Slender, soon hairless. **Buds** 5 mm, red-brown, the scales with hairy tips. **Leaves** Typically 13 cm long and *rising on each side of the stalk; deep, rounded bays* between perpendicular whisker-toothed lobes; glossy both sides, with *small* buff tufts under main vein-joints. Autumn colours smouldering reds at least near the branch-tips: the most reliably colouring oak in the UK.
COMPARE Pin Oak (below): often most easily distinguished by crown shape. Its leaves (and those of Shumard Oak, p.238) have *big* tufts under the vein-joints; 'Splendens' (below) confuses the picture but should always have a visible graft. Red Oak (p.234): bigger leaves more or less *matt* underneath. Black Oak (p.234): blackish, leathery leaves, often scurfy beneath.
VARIANTS 'Splendens' was selected for vigour and autumn colour at the Knap Hill Nursery, Surrey, by 1890 and has been much planted at least in SE England. A grafted tree with bigger leaves (to 18 cm) which have *bigger tufts* under the vein-joints (cf. Pin Oak and Shumard Oak).

Pin Oak *Quercus palustris*

Ontario to N Carolina and Kansas, in wet sites. 1800. Frequent in warm areas of the UK, especially as a young tree, but almost absent from Scotland.
APPEARANCE Shape Usually very distinctive: trunk rather straight; crown broad-conic then (in the open) densely domed, with many very fine dead branches retained in the interior and fine live ones forming a *distinct descending skirt from 5 m*. Old trees become more open and irregular after breakages, but often retain traces of the skirt, as do crowded woodland trees. To 28 m; the small leaves make the tree look taller. **Bark** Silver-grey; darker and more ruggedly ridged only in age. **Shoots** Slender, soon hairless. **Buds** Small (3 mm), *dull* brown, more or less *hairless*. **Leaves** Typically only 11 cm long, with deep, narrow, perpendicular lobes; fresh green, and glossy at least underneath; there are always *big* (2–4 mm wide) *buff drifts of hair* under the main vein-joints. Autumn colour a rich, uniform scarlet-brown in good years. **Fruit** Acorns in very *shallow* cups.
COMPARE Scarlet Oak (above): *small* tufts under the vein-joints, except in 'Splendens', and a more open, irregular crown. Shumard Oak (p.238): Red Oak (p.234): bigger leaves matt underneath.
OTHER TREES Northern Pin Oak, *Q. ellipsoidalis* (Ontario to Missouri, 1902), is in some gardens as a young, autumn-colouring tree; the cups of the *almost stalkless* acorns are *deeper* and less saucer-like than Pin Oak's, enclosing at least a third of the fruit.

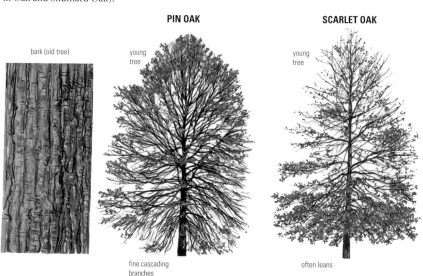

bark (old tree)

PIN OAK

young tree

fine cascading branches

SCARLET OAK

young tree

often leans

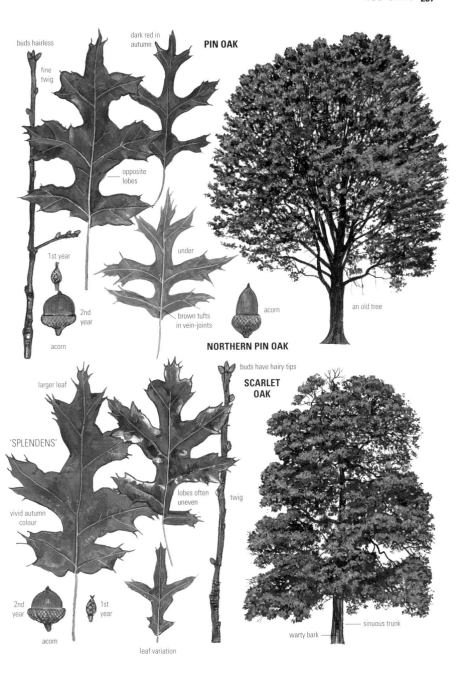

RED & WILLOW OAKS

Shumard Oak — *Quercus shumardii*

SE USA. 1897. A 'red oak' in a few collections.
Appearance Similar to Scarlet Oak (p.236) but differences as follows: **Buds** *Pale/greyish-yellow*. **Leaves** With big tufts under the vein-joints; the lobes tend to be more triangular.
Compare Pin Oak (p.236): different crown, more perpendicular lobing. Spanish Oak (below).

Spanish Oak — *Quercus falcata*

(Southern Red Oak) E and S USA. 1763. In a few big gardens.
Appearance Shape An open, often irregular dome; to 24 m. **Bark** Dark grey; soon with knobbly ridges. **Shoots** With scurfy wool when younger. **Leaves** May be 3-lobed; the lobes, more *triangular and curved* than those of other 'red oaks', often *lack* whisker-tipped teeth. Dark and glossy above once loose wool has been shed, but dull grey and *remaining downy beneath* (cf. Black Oak, p.234).
Compare Ludwig's Oak (below); Black-jack Oak (p.240).
Other trees Pagoda Oak, *Q. pagoda* (*Q. falcata* var. *pagodifolia*; similar habitats), has more regular, perpendicular lobes; in some collections.

Ludwig's Oak — *Quercus × ludoviciana*

SE USA (natural hybrid of *Q. pagoda* with Willow Oak). 1880. One of the loveliest and most distinctive oaks, but still very rare.
Appearance Shape Slender, dense; usually on a long straight trunk. To 23 m so far. **Bark** Dark grey; remaining rather smooth. **Leaves** Unfold deep coppery-brown, the growing tips retaining a *purple tinge* until late summer: glossy dark green with late, orange-red autumn colours, or semi-evergreen (like neither parent, but cf. Water Oak, p.240). Typically there is *1 particularly big triangular lobe each side*, a little more than halfway up; most of the teeth are whisker-tipped. Some scurfy down beneath, with longer hairs under the main veins.
Compare Spanish Oak (above); Lea's Oak (p.240).

Willow Oak — *Quercus phellos*

E USA. 1723. Rather rare: the most often seen of a group of American oaks which have unlobed leaves but are close to the 'red oaks' (the leaves emerge late and yellow; the acorns take 2 years to mature); trees from both groups interbreed freely in the wild.
Appearance Shape Domed or gaunt, to 26 m; twiggy, on few big, steep limbs. **Bark** Pale grey; developing shallow, lumpy ridges. **Shoots** Slender. **Buds** Tiny (2–4 mm). **Leaves** Untoothed; very narrow, to 10×2 cm; dull yellowish green above and soon hairless beneath. They flush gold, and high-summer lammas growth turns vigorous trees yellow again. **Fruit** Acorns minute – 1 cm.
Compare Shingle Oak and Laurel Oak (p.240); Almond-leaved Pear (p.318). Medlar-leaved Oak (p.214): jizz very different.
Other trees Bartram's Oak, *Q. × heterophylla* (*c.*1822), is the natural hybrid with Red Oak. Very rare; an open-crowned tree to 21 m, differing from Willow Oak in its more closely fissured bark and larger, darker, reddish-flushing leaves which (especially at the shoot-tips) have *odd shallow, spined triangular lobes/teeth* (cf. Lea's Oak, p.240, whose harder leaves are downy beneath).
Q. × schochiana (*c.*1894) is the natural hybrid with Pin Oak; also very rare. Small-domed (to 15 m); its broader, brighter leaves have random *rounded* lobes (cf. Laurel Oak and Water Oak, p.240).

RED & WILLOW OAKS

LUDWIG'S OAK

WILLOW OAK

variable

bark

new lvs copper-brown

1st year — 2nd year acorns

smooth bark

LUDWIG'S OAK

1st-year acorn

WILLOW OAK

WILLOW OAKS & ELMS

Shingle Oak *Quercus imbricaria*

A 'willow oak' from the SE USA. 1786. Rare: only in big gardens.
APPEARANCE Shape Irregularly domed and rather sprouty; to 27 m; dark green in summer, with yellow lammas growths. **Bark** Pale grey; smooth then with knobbly ridges. **Leaves** Untoothed; oval (very rarely 3-lobed at the tip); flushing late and yellow then glossy green above, but remaining finely *grey-woolly beneath*.
COMPARE Willow Oak (p.238): narrower hairless leaves. Tupelo (p.412); Willow-leaved Magnolia (p.268).

Water Oak *Quercus nigra*

E USA, in wet places. 1723. Rare: big gardens.
APPEARANCE Shape Sturdily domed; to 18 m. **Bark** Purplish grey; developing knobbly ridges. **Leaves** Dark, glossy and hairless (except for little tufts under the vein-joints); broadest near the lobed tip (as illustrated) or narrow (like Willow Oak, p.238) but *nearly always* with odd, rounded lobes (cf. Q. × schochiana, p.238). They remain green at least until Christmas and can hang on thinly until spring (cf. Turner's Oak, p.220).
COMPARE Black-jack Oak (below): much broader, thicker leaves.
OTHER TREES Laurel Oak, *Q. laurifolia* (1786), in a few big gardens, is perhaps a natural hybrid of Water and Willow Oaks. Its leaves too are almost hairless: willow-like or broadening towards a rounded tip, and rarely with odd rounded lobes. Most like Q. × schochiana, (p.238) whose leaves are more frequently and rather more narrowly lobed.

Lea's Oak *Quercus × leana*

SE USA. A natural hybrid of Shingle and Black Oaks, in a few collections.
APPEARANCE Shape A tall, open dome, to 23 m. **Bark** Dark grey, developing small knobbly plates. **Leaves** Glossy and rather leathery, with some *scurfy down beneath*; they usually have *1 or 2 big, simply triangular lobes* each side.
COMPARE Bartram's Oak (p.238): hairless leaves with usually fewer and smaller lobes. Ludwig's and Spanish Oaks (p.238): more elaborate lobing.

Black-jack Oak *Quercus marilandica*

E USA. c.1739. In a few collections.
APPEARANCE Shape Low, on sturdy, rather erect branches; to 20 m. **Bark** Soon ruggedly square-cracked. **Leaves** Glossy and (like Water Oak's) widest near the tips, but much *harder and thicker* and often *very broad*; golden with fine down underneath. The thick stalk is *only 1–2 cm long*.
COMPARE Spanish Oak (p.238): leaves sometimes 3-lobed but flimsier, on slender, long stalks.

*There are up to 60 species of elm. Their seeds are surrounded by a round, aerodynamic wing. All the forms included here, except for 'Ulmus × diversifolia' (p.242) and Siberian Elm (p.250), have asymmetrical leaves, meeting the stalk consistently higher on one side than the other. Some forms have leaves scrubby above, like sandpaper, from very short stiff hairs (cf. Paper Mulberry, p.256); in the majority only juvenile leaves (on sprouts and low branches of older trees) are rough. Winter shoots are generally dark grey with darker, purple-brown buds. Since 1966, a new virulent strain of Dutch Elm Disease (DED) has destroyed most old elms in many parts of Europe. DED is due to a fungus (*Ophiostoma novo-ulmi*) that is transported from tree to tree on the mouthparts of bark-boring beetles (*Scolytus* species); in an attempt to isolate the infection, the tree shuts down its sap-conducting vessels, and the crown above the blockage is starved of sap and dies within days. The root system usually remains alive and most elms are able to sucker vigorously, but the new plants become vulnerable to infection after about 10 years when their trunks are thick enough for beetle attack. (Family: Ulmaceae.)*

Things to Look for: Elms

- Leaves (adult): How rough? How many secondary teeth? How downy beneath? How asymmetrical at the base? Stalk – how long?
- Fruit: Downy-winged? What shape? Stalk – how long?

Key Species

Wych Elm (p.242): big, always scrubby leaves.
Huntingdon Elm (p.246): big, glossy adult leaves.
Smooth-leaved Elm (p.246): narrow, glossy adult leaves. **English Elm** (p.244): nearly round, variably scrubby leaves.

WILD ELMS

Wych Elm — *Ulmus glabra*

(Scots Elm; *U. montana*) Europe; W Asia; the one elm indisputably native to Britain and Ireland. Abundant in upland areas; more local to S but much planted. Old trees now rare except in N/W Scotland: seldom suckering, so much reduced by DED.
APPEARANCE Shape Young plants broad, on sinuous stems. Old trees domed and billowing; thicker shoots than other wild elms (*never corky-winged*), and dull, black-green foliage. A giant tree (to 40 m) – even in coastal exposure. **Bark** *Smooth and grey for 20 years*; then grey-brown shaggy ridges. Often sprouty. **Shoots** Dark grey, with hard hairs. **Buds** Black-purple, hairy; *broad* and squat. **Leaves** The largest of any native tree's (to 18 cm; 14–20 vein-pairs), hard, rather *oblong* and often (though *never* in the N British/Scandinavian ssp. *montana*) with horn-like lobes at the 'shoulders': matt and *permanently scrubby* above; some stiff, thin, white down underneath. *Almost stalkless* (2 mm on the 'short' side). **Fruit** Usually abundant: the wing downy at its notched tip, the seed central.
COMPARE European White Elm (p.248); Belgian Elm (p.246).
VARIANTS 'Camperdown' (1850) sometimes resists DED and is now a very occasional garden tree. A green igloo: branches writhe from a high graft, the hanging shoots dense with big leaves almost hairless beneath.
'Pendula' ('Horizontalis'; 1816) is also now very occasional. To 18 m, grafted; the finer branches in wide, *gently descending sprays*: stiff yet elegant like the fingers of a Javanese dancer.
Golden Wych Elm, 'Lutescens', is now rare. Leaves emerge soft spring-green and *intensify to brilliant yellow* through summer. (Much larger-leaved and more elegantly domed than 'Louis van Houtte', p.244, and other golden elms.)
Exeter Elm, 'Exoniensis' (by 1826), is now rare: a quirky balloon-shape (to 20 m) of twisting, *erect* branches, carrying dense bunches of smaller, rounded, twisted leaves (cf. 'Dampieri', p.246).
OTHER TREES Field Elm (*U. minor*; Europe; W Asia; N Africa) is the other species found wild in Britain and Ireland. It has (much) smaller/ slenderer leaves than Wych Elm's, distinctly stalked and typically smooth except on sprouts; craggier bark, and hairless fruits, the seed near the end notch: the common elm of lowland England (invading some woods but scarcely a woodland tree), perhaps introduced as a fodder crop by early farming tribes. It spreads mainly by suckering so many local clones exist, several (e.g. 'Boxworth Elm' and 'Dengie Elm') showing considerable immunity to DED; only the most widespread and stable are treated on pp.244–6. Trees turn gold at the end of autumn and (except for English Elm) flush *very late*.
Hybrid Elms (*U.* × *hollandica*) occur when fertile Field Elms grow near Wych Elms; suckering clones dominate in parts of East Anglia, combining the parents' features variously (i.e. large, smooth adult leaves or smaller, scrubby ones, usually on distinct stalks), and tending to have some DED-resistance. Named types include '*U.* × *diversifolia*', (mainly Hertfordshire, Cambridgeshire and Suffolk) with *small*, very slightly rough leaves, downy beneath, and *all symmetrical at the base* on at least 1 side-shoot in 10, a rather square-cracked bark and an open, spreading crown; and '*U.* × *elegantissima*' from the E Midlands (a variable grouping to which the popular tiny-leaved bush 'Jacqueline Hillier' is believed to belong). See pp.246–8 for some ornamental clones.

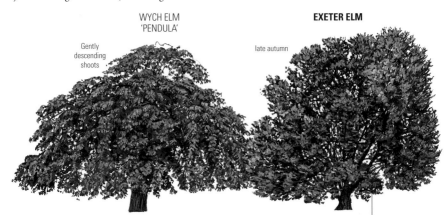

WYCH ELM 'PENDULA'
Gently descending shoots

EXETER ELM
late autumn
erect branches

FIELD ELMS

English Elm *Ulmus minor* var. *vulgaris*

(*U. procera*; *U. campestris*). The common elm across the richer farmed soils of middle England (and the only Field Elm – see p.242 – with often *permanently scrubby leaves*); once dominant in many landscapes, spreading by suckers but hardly ever by seed: the genetically identical trees turned out to be the most susceptible elms of all to virulent DED and have now been entirely reduced in the south (outside disease control zones) to locally abundant regrowth; surviving as a rare planted tree further north and west.

APPEARANCE Shape Suckers spikily conic: regular fish-bone patterns of fine, stiff shoots; twigs in sun develop *corky wings* much more often than other Field Elms (cf. Dutch Elm, p.248; Field Maple, p.368; Sweet Gum, p.278). Old trees (once to 45 m) tall-domed like thunderheads, on an often straight bole with many sprouts and *a few big limbs*; blackish with a smoky multitude of dense leaves. **Bark** Soon with close, knobbly, pale brown ridges, then grey, *coarsely square-cracked*. **Shoots** Fine, downy. **Buds** *Tiny*, grey/purple. **Leaves** *Often stay scrubby above*; finely downy beneath; *rather round* (cf. Coritanian Elm, p.246), 5–10 cm, and puckered/*crumpled*; on short (5 mm) stalks. The vein-hairs can sting, like the related Nettle's.

VARIANTS 'Louis Van Houtte' (1880) is a rather spikily erect golden-leaved sport (cf. 'Dampieri Aurea', p.246); now almost extinct in gardens.

Cornish Elm *Ulmus minor* var. *cornubiensis*

(*U. stricta*) A group of Field Elms locally abundant in Cornwall, W Devon and SW Ireland; nearly extinct as planted trees elsewhere. Typically narrow-domed on a straight bole; the branches rise quite steeply with (in exposure) sky visible between dense systems of *vivid green* foliage. The bark grows very scaly grey-brown ridges, which may *curl free at each end*. The shoots are finely hairy only at first; the leaves small (about 6 cm) and sometimes *cupped*; leathery-smooth above (but scrubby on suckers and low shoots) and downy only under the midrib. On good 1 cm stalks downy above. The teeth (with 0–2 secondary teeth) are rather *blunt*.

OTHER TREES Goodyer's Elm, var. *angustifolia*, from near the coast in Hampshire (and in Brittany) differs in its rounded, darker crown and longer-stalked leaves which have *2–3 secondary teeth*; no mature trees are now known.

Wheatley Elm, 'Sarniensis' (Jersey Elm; Guernsey Elm), from the Channel Islands, differs from Cornish Elm in its blackish-green *spear-shape* (to 37 m) on light, steep branches (cf. 'Lobel', p.248), the tip narrowly rounded even on forking trees; its leaves have *1–3 secondary teeth*. An abundant street tree since 1836; survivors are now rare.

Lock Elm, var. *lockii* (Plot Elm; var. *plotii*), from the N Midlands, is a slender tree unlike the Cornish Elm in its light, more spreading branches, *open, tilted tip* and weeping shoots: many *side ones*, instead of producing consistent rounded patterns of 3–6 leaves, *continue to grow* in plumes (cf. asiatic elms, p.250). Its leaves are dull above and *minutely roughened*, and *white-hairy beneath* at least in patches near the midrib. Rarely planted outside the natural range and now very scarce.

winter

conic into old age

WHEATLEY ELM

scrubby lvs

'LOUIS VAN HOUTTE'

gold through summer

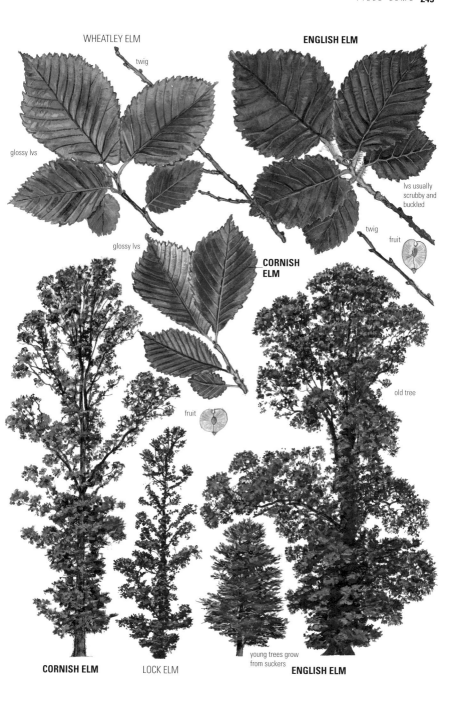

FIELD & ORNAMENTAL HYBRID ELMS

Smooth-leaved Elm
Ulmus minor var. *minor*

(*U. carpinifolia*) This name covers the common elms in parts of Suffolk, Essex, Hertfordshire, Kent and E Sussex (and across much of Europe, W Asia and N Africa; planted elsewhere); often spreading from seed, these are more varied than other Field Elms, and may resist DED.
APPEARANCE Shape Domed, on usually sinuous boles; dark, glinting foliage. Some forms weep dramatically (f. *pendula* – still widespread in Hertfordshire); some are stiff and straight. **Bark** Grey-brown; scaly sometimes criss-cross ridges develop slowly. **Shoots** Soon hairless, occasionally corky-winged (cf. English Elm, p.244). **Buds** Slender, downy. **Adult leaves** Glossy, flat and leathery-smooth; 6–15 cm (depending on the clone); variably *narrow* and *narrowly tapering to the base on the short side*, on a good 1 cm stalk. Compare the new hybrid elms (p.248).
VARIANTS 'Viminalis' is a rare slender wild variant which has *jagged, hooked 1 cm teeth*.
Coritanian Elm (rarely given specific status – *U. coritana* – but often subsumed in *U. minor* var. *minor*/var. *vulgaris*) has scattered populations in E England (little planted elsewhere). Raggedly spreading; leaves rounded (like English Elm's – but *always smooth* except on sprouts and low branches; cf. Dutch Elm, p.248) and often slightly *heart-shaped at the base*, on 1 cm slender curved stalks; up to 4 blunt secondary teeth.
In the E Mediterranean, ssp. *canescens* (*U. canescens*) has *very downy shoots*; its leaves (grey-downy when young) have *simple teeth* and up to 18 vein-pairs.

Huntingdon Elm
Ulmus × *hollandica* 'Vegeta'

(Chichester Elm) Raised in a Huntingdon nursery c.1760. For long the most planted ornamental Hybrid Elm (see p.242): it has some DED-resistance so remains occasional in town parks (and some hedges/belts).
APPEARANCE Shape Tall-domed on often *straight, clean main limbs*. Most trees have dark, quite sparse foliage. **Bark** Grey; *regular* criss-cross ridges. **Leaves** Glossy above (but rough on low sprouts and the *frequent suckers*) and hairy *only in tufts under the vein-joints*; large (to 15 cm), on 15 mm stalks; the margin more often than in other elms (except 'Plantijn', p.248) curves in to meet the first vein *on the 'short' side*. **Fruit** With seed near wing's centre.
COMPARE European White Elm and 'Commelin' (p.248). Some wild hybrids (see p.242) are similar.
VARIANTS Belgian Elm, 'Belgica' (c.1694), little planted in the UK, is closer to Wych Elm: leaves rough above, downy beneath; *long, long-pointed*; but distinctly stalked, and on shoots hairless by autumn. 'Dampieri Aurea' ('Wredei'), rare as a mature tree, is still being quite widely planted, as it shows good DED-resistance. Straggling, erect limbs make a broad column/funnel, to 16 m. Leaves small, rather round; jaggedly toothed and very crumpled, but soon smooth, *shiny* (cf. 'Louis van Houtte', p.244) and almost hairless: *brilliant rich gold* especially in later summer. Branches can revert to the otherwise almost extinct 'Dampieri' (similar, but with dull dark green leaves).
OTHER TREES Dickson's Golden Elm, *U. minor* 'Dicksonii', is a very rare, slow, golden sport of Wheatley Elm (p.244) with *flat*, shiny leaves.

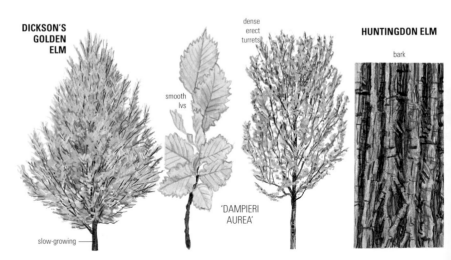

DICKSON'S GOLDEN ELM

slow-growing

smooth lvs

dense erect turrets

'DAMPIERI AUREA'

HUNTINGDON ELM

bark

FIELD & ORNAMENTAL HYBRID ELMS 247

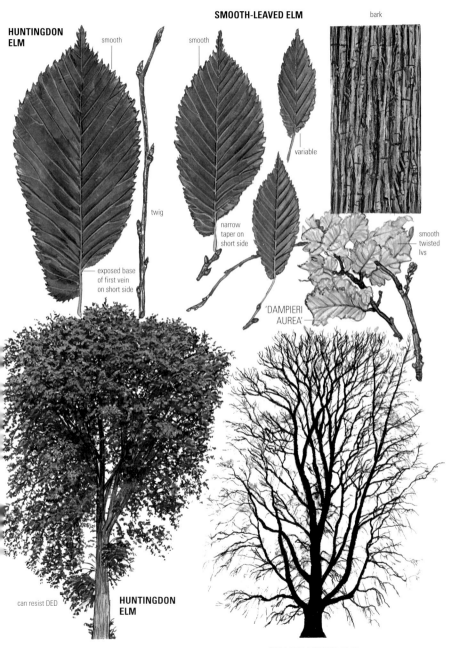

ORNAMENTAL HYBRID & WHITE ELMS

Dutch Elm — *Ulmus* × *hollandica* 'Hollandica'

('Major') A much planted Hybrid Elm (see p.242) since 1680; very occasional suckers remain.
Appearance Shape A straggling, dark dome (once to 43 m); strong shoots corky-winged. **Bark** Brown; cracking into *smaller* shallow plates than English Elm's. **Leaves** Often buckled like English Elm's but much longer (to 15 cm); adult ones more or less smooth (cf. Coritanian Elm, p.246); downy only under the midrib. **Fruit** Seed *touches the fruit's notched margin*.
Other trees *U.* 'Commelin' (Holland, 1940; Huntingdon × Smooth-leaved Elm) is a rare but often DED-resistant street tree. Steep branches with *dark grey*, closely ridged bark make a narrow, rather untidy and open crown, to 22 m (cf. 'Plantijn', below). Shoots matt brown; leaves slightly smaller and more oblong than Huntingdon Elm's; scarcely showing the curvature of the short side onto the first vein.
U. × *hollandica* 'Groeneveld' (Holland, 1963) is one of a group of newer, largely DED-resistant hybrid elms (with 'Dodoens', 'Lobel', 'Plantijn', below). It is locally occasional: *columnar* on *sinuous* erect branches, then widening; bark smooth at first (cf. 'Dodoens'). Leaves about 8 cm; glossy above; *finely downy* beneath; fruits very freely.
U. 'Dodoens' (Exeter Elm × Himalayan Elm, *U. wallichiana*; Holland 1973) is rare. Rather open, on steep branches; bark smooth and silvery at first (cf. 'Groeneveld'). Leaves about 10 cm; blackish, glossy and *deeply toothed*; hairless except for tufts under the vein-joints (cf. Dutch Elm).
U. 'Lobel' ('Dodoens' × *U.* × *hollandica* 'Bea Schwarz'; Holland 1973) is locally frequent. *Narrowly oblong/funnel-shaped* in youth (cf. 'Plantijn'); steep, straight main limbs; *stiff* shoots spread at narrow angles; bark quickly ridged then square-cracking. Leaves about 8 cm; glossy, blackish, almost smooth above, with a thick dark margin; *only 2 mm of basal asymmetry* (cf. asiatic elms, p.250).
U. 'Plantijn' (Holland 1973; mostly Smooth-leaved Elm) is rare. Funnel-shaped (laxer than 'Lobel'; cf. 'Commelin'). Leaves about 9 cm; *slightly rough above*; drifts of white hair under the vein-joints; slightly pie-crust margin has big teeth with up to 4 *secondary teeth* and *curves in to meet the first vein on the 'short' side* (like Huntington Elm, p.246).

European White Elm — *Ulmus laevis*

(Fluttering Elm; *U. effusa*) E France to the Caucasus. Very rare in the UK as an old wild tree (possibly native).
Appearance Shape A billowing dome: often with fine sprouts and small burrs up the branches. **Bark** As Wych Elm's (p.242); may scale more finely. **Shoots** Downy at first. **Buds** *Orange*-purple, *long-pointed*. **Leaves** Long (to 13 cm; up to 19 vein-pairs), broadest below halfway; very asymmetrical at the base, the margin (with big, *very hooked* double teeth) sometimes curving in to meet the first vein *on the 'long' side*; rich green and quite shiny but slightly rough above, and often downy beneath; stalk 15 mm. **Flowers** *Long-stalked*; seeds *fluttering on their stalks; hair-fringed, with 2 converging horns* at the tip.
Compare Huntingdon and Belgian Elms (p.246). Only flowers and fruit (March–May) positively differentiate from Hybrid Elm variants.
Other trees American White Elm, *U. americana* (E and central North America; in odd collections) differs subtly: buds *ovoid*, often blunt; leaves broadest *above halfway*, less asymmetrical.

'GROENEVELD'   'DODOENS'   'PLANTIJN'   'LOBEL'

very narrow forking   neat habit

ORNAMENTAL HYBRID & WHITE ELMS 249

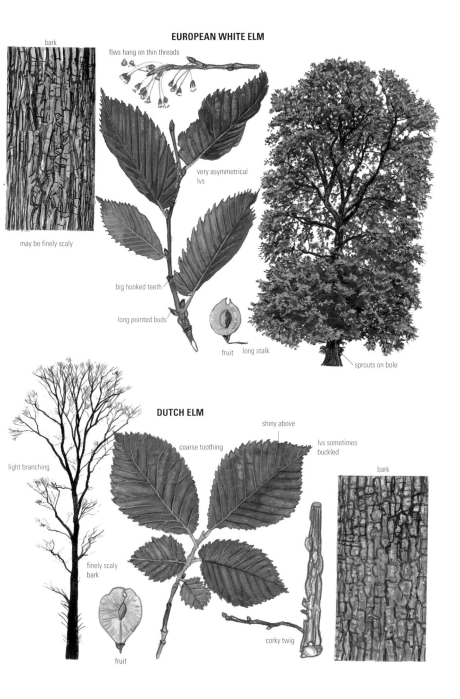

ELMS (ASIATIC)

Chinese Elm — *Ulmus parvifolia*

(Lace-bark Elm; *U. chinensis*) E Asia; S Japan. 1794. Rare. Often resists DED.
APPEARANCE **Shape** A beautiful tree, still green in late autumn. A dainty dome of narrow, blackish leaves; to 14 m. **Bark** Chocolate-brown, with coarse flaking ridges; or greyer and smoother between *orange scales* (cf. Chinese Zelkova, p.252). **Shoots** Minutely grey-felted. **Leaves** 2–6 cm, rounded at the base with about 2 mm of asymmetry, and with blunt, *simple* teeth; upper side glossy or slightly rough; some hairs under veins. **Flowers** *In autumn*.
OTHER TREES *U.* 'Regal' (USA, 1983 – involving 'Commelin' and Siberian Elm) is rare: a narrow dome, with long, spreading 'ostrich-plume' shoots; the bark is smooth and *silvery* until criss-cross grey-brown ridges develop. Leaves (about 8 cm, on conspicuously *long stalks*) are slender, nearly hairless and almost symmetrical, with blunt, often simple teeth.

Siberian Elm — *Ulmus pumila*

(*U. microphylla*) N and E Asia. *c.*1860. Rare, but usually DED-resistant so now a little more planted.
APPEARANCE **Shape** *Broad, irregular and untidy*: curving branches from a short, usually slanted bole (cf. 'Sapporo Autumn Gold'); long, lax, often weeping shoots, like ostrich feathers (cf. Lock Elm, p.244, and 'Regal', above). To 20 m. *Fresh or pale green* healthy foliage in summer. **Bark** Willow-like: a *very coarse network* of scaly brown ridges. **Shoots** Soon hairless. **Buds** Small, glossy brown. **Leaves** Small (6 cm), slender and *hairless* (sometimes with tiny tufts under vein-joints), and more or less *symmetrical at the base*; up to 3 or 4 secondary teeth on each jagged main tooth; stalk 1 cm, finely downy.
VARIANTS *U.* 'Pinnato-Ramosa' (*U. pumila* var. *arborea*) has *persistently hairy shoots* and longer-pointed leaves, carried in neat fish-bone patterns; a more gracefully domed, rare tree, to 20 m.

Japanese Elm — *Ulmus japonica*

Japan (where always rare); in a few collections and beginning to be planted more widely for its DED resistance.
APPEARANCE **Shape** Usually seen in the UK – from the sapling stage – as a *very wide, low*, deep-green dome of long, straggling, ostrich-plume shoots. **Bark** Grey-brown; very scaly ridges. **Shoots** Pale, typically downy (hairless in some planted trees); sometimes corky-winged. **Leaves** 3–10 cm, typically scrubby above and downy beneath (but smooth, hairless and shiny in some planted UK trees), with asymmetrical bases; stalks 15 mm.

Ulmus 'Sapporo Autumn Gold'

(Japanese × Siberian Elms; Wisconsin, 1973) Frequent: the *generally planted* DED-resistant elm of the 1980s, though scarcely in commerce by 2000. The single clone is readily learnt by its jizz.
APPEARANCE **Shape** Asymmetrically and jaggedly broad-domed (to 19 m so far); *light branches rise from a very short, slightly slanted bole*; dainty and *fresh-green* from early spring, with many luxuriant, long but not drooping shoots. **Bark** Scaly criss-cross *brown* ridges; *orange* fissures. **Shoots** Finely downy. **Leaves** 4–9 cm (smaller and slenderer than the new hybrid elms' on p.248); *slightly rough* but glossy above, with some down underneath; only about 2 mm of basal asymmetry (cf. 'Lobel', p.248).

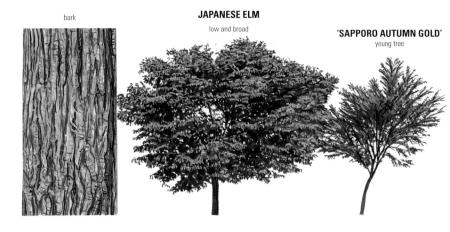

bark

JAPANESE ELM
low and broad

'SAPPORO AUTUMN GOLD'
young tree

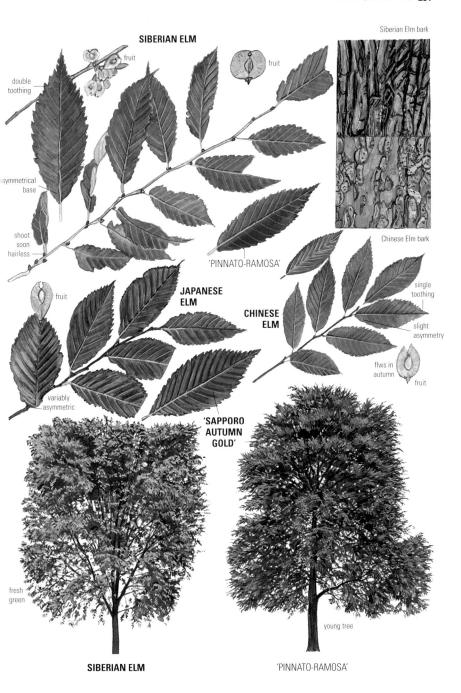

ZELKOVAS

Zelkovas (6 species) are trees related to the elms. They can also succumb to Dutch Elm Disease (see p.240). (Family: Ulmaceae.)

Things to Look for: Zelkovas

- Leaves: What shape are they at the base? What shape are their teeth? How long is the leaf-stalk?

Caucasian Elm *Zelkova carpinifolia*

(*Z. crenata*) Iran; Georgia; Armenia; E Turkey. 1760. Rather occasional; rarely naturalizing by its prolific suckers.
APPEARANCE Shape Usually a giant *'witch's broom'* *of steep stems from a 2 m bole with big, rounded flutings*, to a slender, slanted top (35 m), but sometimes 'tree-shaped' and occasionally bushy on sinuous, steep stems; rich, dark masses of little leaves. The 'witch's broom' shape (scarcely seen in the wild) makes big trees vulnerable to windthrow. **Bark** Buff-grey, remaining smooth but with a few *orange crumbling patches*. **Shoots** Slender, green/brown, hairy. **Buds** Small, blunt: elm-like, but a brighter dark red. **Leaves** Hard, to 10 cm; 9–11 big, rather *rounded* (but sharp) teeth each side; scattered scrubby elm-like hairs above and a softer down underneath, where stiff hairs also radiate from the main veins; stalks only 3–5 mm. **Fruit** (seldom seen) A green pea-sized nut.
COMPARE Chinese Zelkova (below); Macedonian Oak (p.222); Hornbeam 'Fastigiata' (p.194).
OTHER TREES Cut-leaved Zelkova, *Z.* 'Verschaeffeltii', is an obscure, daintily tree-shaped variant in a few big gardens, its leaves with *big outward-curving triangular lobes*. The bark, unlike other Zelkovas', may grow rugged dark brown ridges.
Cretan Zelkova, *Z. abelicea* (*Z. cretica*), is endemic to Crete and one of Europe's most endangered trees. Bushy plants have grown in a few UK collections since 1924: the leaves (crowded on downier shoots) are tiny (1–4 cm), with only 3–6 pairs of teeth.
Z. sicula was discovered in 1991 as a single population of 200 trees in SE Sicily: leaves with 6–8 pairs of sharper teeth.

Keaki *Zelkova serrata*

(*Z. acuminata*) Japan; Taiwan; Korea; NE China. 1862. Locally quite frequent as a quietly classy younger tree.
APPEARANCE Shape The commonly planted clone quickly grows a *broad but graceful*, light-limbed dome, on a short, straight, *smoothly rounded bole*, with hanging fresh-green foliage; amber/pink in autumn. To 26 m; rarely a giant bush. **Bark** Grey, with a few fine orange flakes; shaggy plates may develop after 80 years. **Shoots** Hairless by autumn. **Buds** As Caucasian Elm's. **Leaves** *Slender, long-pointed*, to 12 cm, with 6–13 pairs of big *curved triangular teeth*; hairy only under the main veins; stalk 5–10 mm. Many trees grow odd sprays of miniaturized foliage.
COMPARE Chinese Zelkova (below).
VARIANTS 'Green Vase' is a recent erect selection.

Chinese Zelkova *Zelkova sinica*

Central and E China. 1908. In some collections.
APPEARANCE Shape As Keaki; to 17 m. **Bark** Usually with *many bright orange scales* (cf. Chinese Elm, p.250). **Leaves** Small (6 cm), a hard dull green; *few shallow triangular teeth; downy beneath;* on downy *3 mm stalks*. The leaf-base (unlike Caucasian Elm's and Cretan Zelkova's) is *tapered*, and untoothed for the first 2 cm.

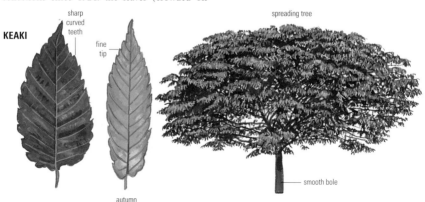

KEAKI — sharp curved teeth — fine tip — autumn — spreading tree — smooth bole

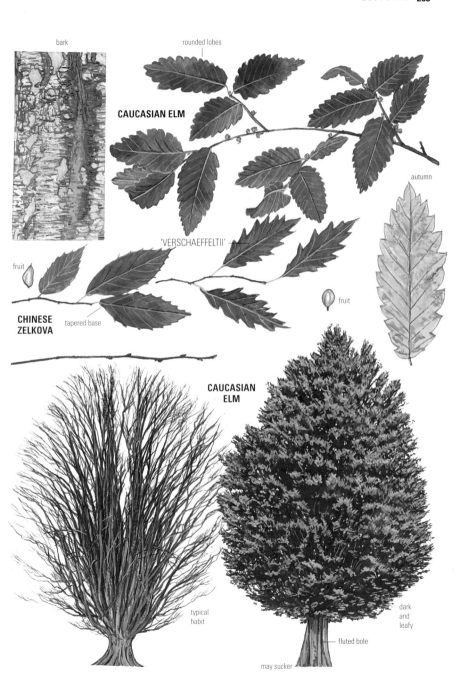

HACKBERRIES

Hackberries (70 species, mainly in tropical countries) have dry but edible berries. Their leaves are conspicuously 3-veined from the base (cf. Mulberries, pp.256–8). A fair variety lurk in UK collections, but gardeners have seldom shown much enthusiasm for them.(Family: Ulmaceae.)

Hackberry — *Celtis occidentalis*

(Nettle-tree) E North America: Manitoba to Alabama. 1656. Rare, but in the odd town park.
APPEARANCE Shape Broadly domed, to 16 m; rather untidy curving twigs in winter; in summer, dense, rich green. **Bark** Grey; smooth, then gradually developing *extravagantly knobbled and flanged ridges* (cf. Kentucky Coffee Tree, p.350). **Shoots** Slender, shiny brown, with some stiff hairs: downier in var. *cordata* (var. *crassifolia*), from W of the Appalachians, which is the common – and more vigorous – form in Europe. **Buds** White-hairy, long pointed; appressed. **Leaves** Hard and *rough but glossy above* (like some elms'), with stiff white hairs underneath, and *odd, irregular whiskered teeth* – sometimes absent all down one side; larger (to 15 cm) and often heart-shaped at the base in var. *cordata*. **Flowers** Small, green; 7–10 mm red/purple berries follow.
COMPARE Medlar-thorn (p.290): similar leaf-shape and texture.

Southern Nettle-tree — *Celtis australis*

S Europe (where it can live for 1000 years) E to Asia Minor. 1796. Very rare in the UK, needing summer heat but troubled by late frosts.
APPEARANCE Shape A low dome, to 14 m. **Bark** Grey, smooth and horizontally wrinkled (beech-like); a little rugged in old age. **Leaves** Like Hackberry's, but with *long twisted points* and *regular, jagged teeth*.

Fruit Slightly larger than Hackberry's (to 12 mm) and ripening black: probably the Homeric 'lotus'.

Caucasian Nettle-tree — *Celtis caucasica*

E Bulgaria through the Caucasus to N India. 1885. In a few collections.
APPEARANCE Shape A dense, slightly weeping dome; to 15 m. **Bark** Grey-brown; usually developing *wide, shallow, lime-like vertical ridges* between orange fissures. **Shoots** Hairy. **Leaves** Only 3–8 cm; blackish, coarsely toothed and harsh (like most elms'); downy on the *whitish-green* underside at first, and persistently so under the veins. **Fruit** Rufous, 1 cm.

Sugarberry — *Celtis laevigata*

(Mississippi Hackberry) S USA. *c.*1811. In some collections.
APPEARANCE Shape, Bark As Hackberry (above). **Shoots** *Hairless*. **Leaves** 8 cm; vivid green *on both sides* and *smooth above*, with a few hairs under the veins and tufts under the joints; sharply toothed only in var. *smallii*. **Fruit** 8 mm, slender, orange.
COMPARE In leaf, Osage Orange (p.256).
OTHER TREES *C. biondii* (central China, 1902; in some collections) has leaves sometimes toothed towards the tip, and downy beneath when young. *C. bungeana* (N China, 1882; in a few collections) has hairless leaves (except for tufts under the vein-joints) typically *matt beneath*; there are usually a few teeth near the tip. Its berries are *purple*.

coarse teeth
harsh

HACKBERRY
bark

flanged ridges

CAUCASIAN NETTLE-TREE

lime-like bark

under whitish green

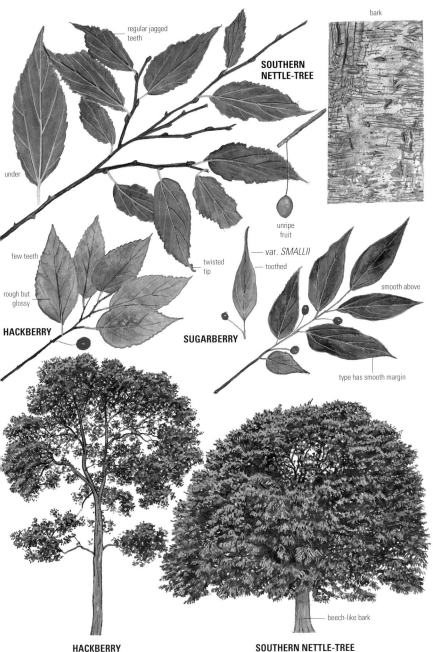

OSAGE ORANGE TO PAPER MULBERRY

Moraceae is a predominantly tropical family (including 2000 figs). Members of the family have a latex-like sap.

Osage Orange *Maclura pomifera*

(*M. aurantiaca*) S USA, from Arkansas to Texas, in moist fertile sites. 1818. Now very rare in the UK. **APPEARANCE Shape** A spiky, often luxuriant dome, to 15 m. **Bark** *Rufous*; rugged, *stringy* criss-cross ridges. **Shoots** Green, downy, then grey-brown; *one* 1 cm thorn stands by each small brown bud. (False Acacia, p.354, can have 2 thorns by each bud and Honey Locust, p.348, 3; the thorns of Hawthorn, etc, are spaced along the shoots while most 'thorny' trees have spine-tipped side-shoots instead.) **Leaves** To 12 cm, long-pointed, *untoothed*; dark and glossy above and slightly downy beneath, on downy stalks; they turn yellow in autumn. **Flowers** Dioecious. Little green-white clusters. **Fruit** (When male and female trees grow together, rarely in the UK) green then amber 13 cm 'oranges'. Inedible: a stringy, juicy pulp surrounds pale brown nuts. **COMPARE** Sugarberry (p.254). Cucumber Tree (p.260): similar jizz, but much larger leaves.

Fig *Ficus carica* ☠

E Mediterranean and W Asia, but long grown in the UK. The hardiest member of a huge tropical genus. Frequent, and naturalizing from pips in rubbish, reaching tree size in the warmest areas; abundant along some rivers in N England (notably the Don at Sheffield) where industrial use of cooling water has historically lifted water temperatures. The sap (in sun) is irritant, especially to the eyes.

APPEARANCE Shape Often bushy or leaning; very gaunt in winter, with sturdily upcurving twigs; sometimes a suckering mass. In London a few sturdy trees have reached 13 m. **Bark** Elephant-grey and smooth. **Shoots** Thick and knobbly, green/grey. **Buds** Yellow-green, long-pointed, to 15 mm. **Leaves** Very leathery, with a sour, minty aroma; to 30 cm, variably lobed (rarely unlobed); shiny but rough and hairy above; downy underneath. **Flowers, Fruit** Dioecious, but male trees ('caprifigs') are probably not grown in the UK. Female flowers are produced on the inside of the young fig (which is a modified shoot-tip with a hole at the end, and is pollinated in the wild by tiny wasps). Most named clones are self-fertile. In Britain the common one is 'Brown Turkey', ripening its figs in their 2nd year, given enough warmth.

Paper Mulberry *Broussonetia papyrifera*

China and Japan, where long grown for its fibrous bark, a source of paper and fine cloth. Rare: parks and gardens in warm areas.
APPEARANCE Shape A low dome, like Black Mulberry's (p.258); often bushy, yet to 15 m. **Bark** Pale grey-brown; criss-cross burry ridges. **Shoots** *Very woolly*. **Leaves** To 20 cm; varying more often from the heart-shaped norm than the mulberries' on p.258; often a *bright pale green, but very matt* with rough bristles above and dense wool below. **Flowers** Dioecious: male catkins (yellow and hanging, to 7 cm). **Fruit** 'Mulberries' ripening red but woody, inedible and seldom seen in the UK.

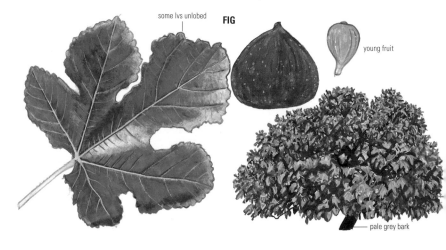

some lvs unlobed — **FIG** — young fruit — pale grey bark

OSAGE ORANGE TO PAPER MULBERRY

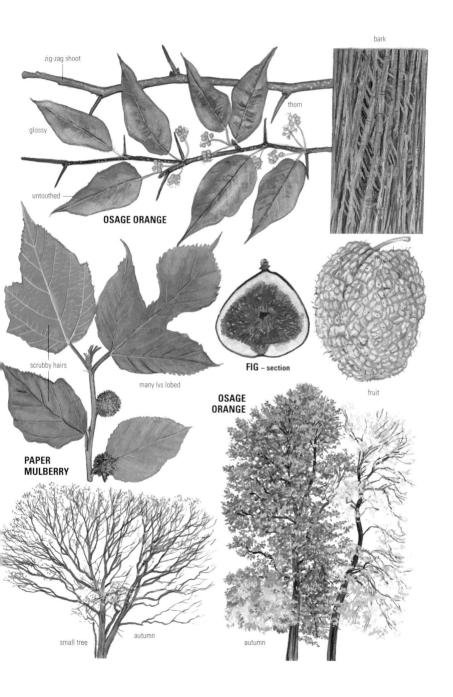

MULBERRIES & MAGNOLIAS

Black Mulberry — *Morus nigra*

Probably of W Asiatic origin, but long cultivated across Europe. Quite frequent in warm areas in cottage gardens, old parks etc; rarely naturalizing.
APPEARANCE Shape A low, dense, twiggy dome on twisty branches from a crooked trunk, to 12 m; late-flushing then singularly luxuriant through summer. Mulberries were often propagated by planting a 'truncheon' or large log – this sprouts but quickly rots from the sawn top, making young trees look 'ancient'. In fact there is no hard evidence that this is a long-lived tree; it certainly grows fast. **Bark** Orange-brown: scaly criss-cross ridges and some *bright, crumbling patches*; many big burrs. **Shoots** Stout, greyish, with some harsh hairs; strung with conspicuously *broad, sharp, purple buds* (cf. Wych Elm, p.242). **Leaves** 8–12 cm; thick; dark and *glossy but rough above* and hairy, especially beneath; heart-shaped, but variously dissected on sprouts and saplings (cf. Fig, p.256). **Fruit** Ripens in high summer: avidly consumed by anyone who does not have to worry about the incriminating blood-stains.
COMPARE Paper Mulberry (p.256); Italian Alder (p.192); Dove Tree (p.412). White Mulberry (below): jizz very different.

White Mulberry — *Morus alba*

China, but long cultivated in Europe: the favourite food-plant of the silkworm. Very occasional in warm areas of the UK.
APPEARANCE Shape An *open, upright dome of fine willowy shoots*; to 11 m. Often slender and straight-trunked. **Bark** Dull fawn-grey; a network of stringy but shallow ridges. **Shoots** Grey, soon smooth; small, sharp buds. **Leaves** *Pale fresh green*, flat and quite flimsy; nearly *smooth* and glossy above, but hairy under the veins. **Fruit** Pink/whitish: sweet but with little flavour.
COMPARE Crimean Lime (p.402); Hackberry (p.254). A tree (out of fruit) with no obvious affinities.
VARIANTS 'Pendula' is rather rare and makes a supremely picturesque glossy little igloo (to 4 m). 'Laciniata' (very rare) retains much-lobed leaves. 'Venosa' (very rare) has small, jagged leaves with exaggeratedly *broad, bright yellow veins*.
'Pyramidalis' (very rare) has stiffly erect branches.

Magnolias (about 80 species) have large 'primitive' flowers (landing pads for 'primitive' beetles of dubious aeronautical prowess). Their petals are not differentiated from sepals and are called 'tepals'. The leaves, always untoothed, are often enormous. Huge silky flower-buds adorn their shoot-tips; their berries dangle when ripe on silky strings from bright, knobbly 'cucumbers', before being blown free. Most are long-lived but of relatively recent introduction. (Family: Magnoliaceae.)

Things to Look for: Magnolias

- Bark: How rugged is it?
- Shoots: What colour? Aroma when bruised?
- Buds: Are the leaf-buds hairy?
- Leaves: Are they evergreen? Pointed? Wrinkled? How long are they? How downy beneath? How far up are they broadest?
- Flowers: Are they erect or nodding? Scented? How many tepals? and how wide? What colour?

Key Species

Southern Evergreen Magnolia (p.260): evergreen. **Cucumber Tree** (p.260): big leaves; flowers among them in early summer. **Campbell's Magnolia** (p.264): big leaves; flowers before them in spring. **Willow-leaved Magnolia** (p.268): small leaves; small white flowers before them in spring. **Saucer Magnolia** (p.270): bushy; tall, erect flowers before the leaves from an early age.

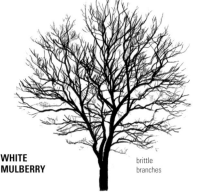

WHITE MULBERRY — brittle branches

'PENDULA'

MOSTLY EVERGREEN MAGNOLIAS

Michelia *Michelia doltsopa*

E Himalayas to W China. *c.*1918. Larger gardens in mild areas. The hardiest of 45 evergreen trees related to *Magnolia* but flowering from old wood, not at the shoot-tips.
Appearance Shape Densely and rather bushily upright, to 20 m. **Bark** Grey; a few cracks with age. **Leaves** To 18 cm; *quite shiny above*, like a lustre-finish photograph; *silver-bloomed beneath* (cf. Winter's Bark, p.274, and Korlinga, p.428) with fine down at first; rusty hairs persist under the veins. **Flowers** *Wreathing the shoots*, from rusty-silky buds, in early spring: creamy-white, to 10 cm; with a powerful, sweet, shaving-cream scent.

Southern Evergreen Magnolia
Magnolia grandiflora

(Bull Bay) Coastal SE USA. 1734. Quite frequent in warmer parts – often against old house walls.
Appearance Shape Irregularly and stiffly domed; to 12 m. **Bark** Grey; big shallow scales slowly develop. **Shoots** Fawn-woolly. **Leaves** Slender, to 25 cm; glossy above; orange wool coating the paler underleaf rubs thin through the year. **Flowers** Little by little from midsummer to late autumn. Richly scented; 9–15 huge tepals.
Variants Many older trees in the UK are 'Exmouth': upright, its narrow leaves only *sparsely* woolly and flowers with 18 tepals (such distinct clones developed because cuttings flowered after fewer years than plants raised from seed); 'Goliath' is bushy: *big broad leaves, almost hairless beneath*; flowers (from an early age) 30 cm wide.
Compare Gurass (p.428). Under-leaves readily distinguish the evergreen Magnolias.
Other trees Glossy Magnolia, *M. nitida* (SW China, Tibet; *c.*1920) is confined to big gardens in the mildest areas but is one of the most beautiful flowering trees: *brilliantly glossy, hairless* 10 cm leaves are finely silver-rimmed and flush coppery-red; scented creamy-white flowers early in spring.

Chinese Evergreen Magnolia
Magnolia delavayi

SW Yunnan, China. 1900. Rather rare in warmer, milder areas; sometimes against old house walls.
Appearance Shape A dense, wide, often bushy dome on twisting limbs; to 18 m. **Bark** Fawn-grey; *close, corky ridges*. **Leaves** Huge, broad and magnificent – *to 35 cm*; dull above and *silver-grey beneath*, with fine down (cf. Korlinga, p.428). **Flowers** Through late summer, to 20 cm wide – but they open at night and are at their best only for a few hours.

Cucumber Tree *Magnolia acuminata*

Ontario to Florida. 1736. Very occasional: older gardens in warmer parts.
Appearance Shape Conic; then a leafy dome to 25 m. **Bark** Orange-brown, soon with shallowly scaly *ridges*. **Leaves** To 22 cm; pointed and broadest below halfway; bright green above; pale and finely downy beneath. **Flowers** 5–10 cm, green-yellow, rather lost among the leaves at the start of summer. **Fruit** Erect 7 cm 'cucumbers', often deformed, shocking pink then red.
Variants Yellow Cucumber Tree, ssp. *cordata* (var. *subcordata*; SE USA; very rare to date), has flowers *bright yellow inside*. Usually bushy, with a more finely scaling bark and broader, darker, glossier leaves, long-hairy underneath.
Compare Epaulette Tree (p.432); Osage Orange (p.256). An enigmatic foliage tree; other summer-flowering deciduous tree-magnolias (p.262) have smoother grey barks and even bigger leaves.

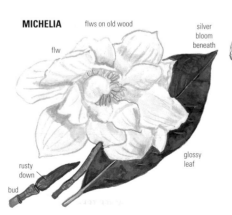

MICHELIA flws on old wood; silver bloom beneath; flw; glossy leaf; rusty down; bud

CUCUMBER TREE can grow tall

SUMMER-FLOWERING MAGNOLIAS

Big-leaf Magnolia *Magnolia macrophylla*

SE USA. 1800. Rare: big gardens in warm areas.
Appearance Shape Narrow and gaunt, to 16 m; or a bush in the UK. **Bark** Grey, developing odd scales and fissures. **Leaves** The *biggest of any hardy broadleaf* (to 1 m); *silvery* (cf. Chinese Evergreen Magnolia, p.260) and finely downy on the underside; there are usually *small rounded* auricles (cf. Fishtail Magnolia's wedge-shaped ones). **Flowers** Among the leaves: huge (to 30 cm) and fragrant, but rare in the UK.
Variants Ssp. *ashei* (NW Florida, 1949; very rare) carries absurdly large flowers even as a sapling.

Fishtail Magnolia *Magnolia fraseri*

(Fraser Magnolia) Virginia to Georgia. 1786. In some big gardens.
Appearance Shape Gauntly domed, to 15 m. **Bark** Grey; some distant scales with age. **Shoots** Brown; *dark purple* buds to 3 cm long. **Leaves** Fresh green; big and oval (to 45 cm), like Japanese Big-leaf Magnolia's (below), but big sharp auricles make the base *fishtail-shaped* (cf. Big-leaf Magnolia, above); *hairless*. **Flowers** Among the leaves in early summer – sweet-scented at a safe distance.

Japanese Big-leaf Magnolia *Magnolia obovata*

(*M. hypoleuca*) Japan; the Kuriles. 1865. Rather rare: big gardens in warmer areas.
Appearance Shape Gaunt: slender, wide limbs on a short trunk. **Bark** Grey, developing a few distant fissures. **Shoots** *Purple-brown, hairless*. **Leaves** Huge (to 40 cm), oval but broadest above halfway; glaucous and slightly hairy beneath. **Flowers** In early summer – singly among the leaves but 20 cm wide and intensely fragrant: sweet and creamy on the air, like sun-block; sickening and chemical at close range. **Fruit** The knobbly scarlet 'cucumbers' (to 18 cm) *taper to a (rounded) point*.
Compare Campbell's Magnolia (p.264): smaller leaves. Fishtail Magnolia (above); Cucumber Tree (p.260).
Other trees Chinese Big-leaf Magnolia, *M. officinalis*, is probably extinct in the wild (central China; trees were stripped of their bark, which contains the relaxant magnocurarine) but survives in some temple gardens. 1900: in a few big gardens in the UK; differs in *yellow-grey* young shoots and *flat-topped* fruits maturing purple-brown. 'Biloba' (1936) has *indented tips to many leaves* (cf. Sargent's Magnolia's smaller, narrower leaves, p.266). Watson's Magnolia, *M.* × *weiseneri* (*M.* × *watsonii*) is probably a hybrid with the white-flowered Japanese and Korean bush *M. sieboldii*: a stiffly bushy plant (to 9 m) in some big gardens, with smaller leaves (12–25 cm) and smaller (14 cm), whiter, equally scented flowers.

Umbrella Tree *Magnolia tripetala*

E USA: Indiana to Georgia. 1752. Rather rare: old gardens.
Appearance Shape Gaunt and rather bushy; to 14 m. **Bark** Grey; smooth, then some fissures/scales. **Leaves** Very big (to 50 cm), *long-tapering to the base* and more shortly to the tip; downy beneath when young. Even more than on other big-leaved magnolias, they grow *in whorls at the shoot-tips*, their shape creating almost solid 'umbrellas'. **Flowers** In late spring, among the leaves; to 25 cm wide, with an overpowering scent. **Fruit** 8 cm cerise 'cucumbers'; very showy.

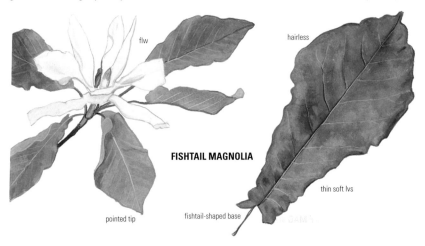

FISHTAIL MAGNOLIA

flw

hairless

thin soft lvs

pointed tip

fishtail-shaped base

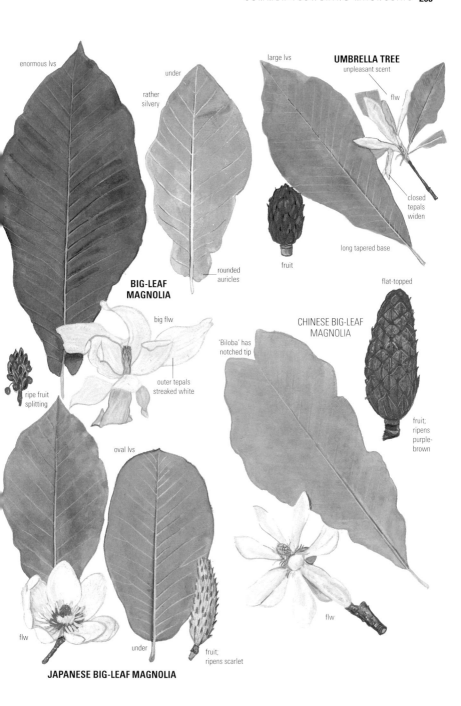

SPRING-FLOWERING MAGNOLIAS

Campbell's Magnolia *Magnolia campbellii*

(Pink Tulip Tree) Himalayas to W China. *c.*1870. Very occasional in milder areas: the most widely grown of the spectacular Asiatic spring-flowering tree species, but, as seedlings often take 20–30 years to begin to flower, hardly ever in smaller gardens.
APPEARANCE Shape Conic then openly domed, to 23 m, or on several straight trunks from the base; very wind-firm. The slender limbs run straight between sudden, sharp angles. **Bark** Typically grey with a few distant fissures; rarely rugged with close, buff, corky scales. **Leaves** Big (to 30 cm) and broad; rather matt; variably silky-hairy beneath. They are usually *oval with a small, sharp tip* (cf. Japanese Big-leaf Magnolia's even larger leaves, p.262), but can be long-tapered at the base like the other spring-flowering tree-magnolias' (p.266). **Flowers** Overwhelming the bare crown in *early* spring like a great flock of exotic birds – to 30 cm wide and typically a vivid *clear pink*. They remain more or less *erect* but at least the outer whorl of the *12–16 tepals* soon opens widely, flopping to the horizontal or beyond. The huge *ovoid* buds are vulnerable to frost-damage after their silky-hairy outer scales have been shed, and each bloom only lasts a few days before the tepals bruise and fall.
COMPARE Veitch's Magnolia (below). Sargent's Magnolia, Sprenger's Magnolia and Dawson's Magnolia (p.266): smaller/slenderer leaves; only Sprenger's Magnolia has similarly erect flowers.
VARIANTS f. *alba* has white flowers with a slight greenish cast: equally frequent in the wild but much less planted.
Ssp. *mollicomata* (Sikkim) is now more planted as a younger tree as it begins to flower usually at 13 years from seed, and *a week later*, when the blooms are less vulnerable to late frosts: its flowers (*hairy-stalked*; from *oblong pointed buds*) are usually a pale, imperfect *purplish* pink but are shapelier (the outer ring of tepals ultimately spreads horizontally while the inner still curves more regularly up, *like a cup on a saucer*).
'Lanarth' is a glorious wild-collected form of ssp. *mollicomata* with very *broad, thick leaves*. The 'cup-and-saucer' flowers open as early as the type's, an astonishing royal *magenta-purple*.
'Charles Raffill' (Kew, 1946) is the least rare of various garden hybrids between ssp. *mollicomata* and the type: the flowers combine the elegant 'cup-and-saucer' shape of the subspecies with the pure mid-pink colour of the best type trees, and are carried after about 13 years from seed.

Veitch's Magnolia
Magnolia × *veitchii* 'Peter Veitch'

A hybrid of Campbell's Magnolia (above) with Yulan (p.270); Exeter. 1907. Rather rare: larger gardens.
APPEARANCE Shape Conic then tall-domed; usually on a single trunk. Although Yulan is more of a bush, its hybrid has already made the *tallest magnolia* in the UK: to 27 m. **Bark** Grey; a few distant fissures with age. **Leaves** 15–30 cm; rather oval but broadest towards the sharp-pointed tip. **Flowers** A fortnight after Campbell's Magnolia, and held stiffly erect: the *9 tepals*, white with a strong purple-pink basal flush, remain *predominantly upright, in a vase shape* (cf. Saucer Magnolia, p.270).
VARIANTS 'Isca' is a rarer, broad-growing clone; its white tepals have a fainter basal flush.
'Alba' has ivory-white flowers; very rare.

VEITCH'S MAGNOLIA

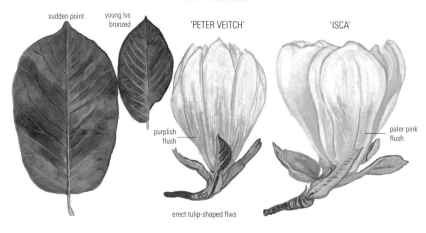

sudden point   young lvs bronzed   'PETER VEITCH'   'ISCA'

purplish flush   paler pink flush

erect tulip-shaped flws

SPRING-FLOWERING MAGNOLIAS 265

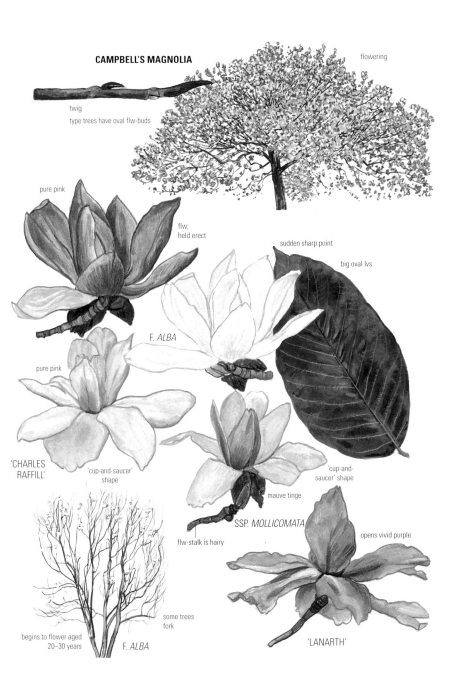

SPRING-FLOWERING MAGNOLIAS

Sargent's Magnolia
Magnolia sargentiana var. *robusta*

(*M. robusta*) SW China. 1910. Rare: a spectacular spring-flowering tree (its blooms opening just after Campbell's Magnolia, p.264) in the largest gardens in milder areas.
APPEARANCE Shape A sturdy tree (to 18 m), but often on several straight trunks. **Bark** Grey; usually remaining mainly smooth. **Leaves** Long, dark and *narrow*, to 20 × 8 cm, often with *indented tips* (cf. Chinese Big-leaf Magnolia, p.262); greyish hairs persist beneath. **Flowers** *Nodding at various angles*, to 30 cm wide; 10–16 *broad* blowsy tepals, pale pink to almost purple, soon spreading/hanging. Trees flower after about 13 years from seed.
COMPARE Dawson's Magnolia and Sprenger's Magnolia (below): slenderer tepals; shorter leaves. Campbell's Magnolia (p.264): *erect* flowers and broader leaves, never notched.
OTHER TREES The type is rarer: its broader leaves are seldom notched; the smaller flowers (10–13 tepals) are held at various angles, and begin after 25 years from seed. It is more often single-stemmed.

Sprenger's Magnolia *Magnolia sprengeri* var. *diva*

Central China. 1901. Rare: big gardens in milder parts.
APPEARANCE Shape Conic/domed; rather open and sometimes on several straight trunks; to 22 m. **Bark** Grey/buff, *often* developing small, ruggedly corky scales. **Leaves** To 17 cm, broadest near their pointed tip (cf. Kobushi, p.268, and Dawson's Magnolia, below), and *densely* downy under the veins. **Flowers** (First carried after about 20 years from seed, and usually a few days after Campbell's Magnolia's, p.264) *erect* and 20 cm wide; 12 quite broad, usually rich pink tepals soon spread widely.
VARIANTS Var. *elongata*, a small, erect tree (to 16 m) with leaves almost hairless beneath, has purple-stained *white* flowers; very rare.

Dawson's Magnolia
Magnolia dawsoniana

W China. 1919. Rare: big gardens in milder parts.
APPEARANCE Shape Short-stemmed, or *densely* tall-domed, to 18 m. **Bark** Usually buff; soon with small ruggedly corky scales. **Leaves** Shape and size as Sprenger's Magnolia's (above) but *darker* and slightly glossy, with a network of prominently *impressed veins*, and *hairless* except under the midrib; cf. Kobushi (p.268). **Flowers** Fragrant and drooping; the usually 9, soft-pink, ultimately hanging tepals are slenderer than Sprenger's Magnolia's, creating a spectacular but delicate show – almost a giant-scale *M.* × *loebneri* 'Leonard Messel' (p.268).
VARIANTS 'Chyverton Red', perhaps a hybrid with *M. sprengeri* var. *diva*, has almost crimson flowers.

Hybrid Tree Magnolias

Magnolias in cultivation hybridize promiscuously: many trees in big gardens cannot be specified. Hybrids which can grow into *shapely trees* include:
M. 'Galaxy' (*liliiflora* 'Nigra' × *sprengeri* var. *diva*, USA, 1963): nodding, blowsy, slightly scented flowers 20–25 cm wide, the 12 tepals *purple-red* outside (cf. 'Lanarth', p.264) and paler within.
M. 'Iolanthe' (involving *campbellii, sargentiana, denudata* and *liliiflora*; New Zealand, 1974): nodding, cup-and-saucer-shaped flowers 28 cm wide; 9 broad tepals clear pink outside, white within.
M. 'Star Wars' (*campbellii* × *liliiflora*; New Zealand, c.1970): typically inrolled outer tepals make the big bright pink, erect flowers rather star-shaped.

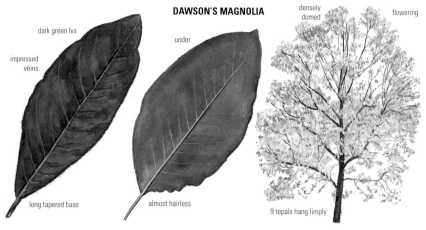

DAWSON'S MAGNOLIA

dark green lvs

impressed veins

long tapered base

under

almost hairless

densely domed

flowering

9 tepals hang limply

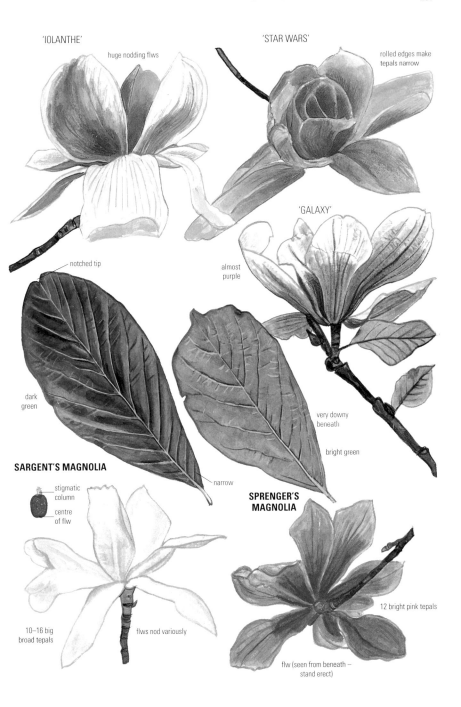

SPRING-FLOWERING MAGNOLIAS

Willow-leaved Magnolia
Magnolia salicifolia

Japan. 1892. Occasional: larger gardens in warmer areas.
APPEARANCE Shape A slender-limbed conic/rounded tree, to 17 m; dense crisp domes of foliage, *red-flushing*. **Bark** Usually buff; soon with small, ruggedly corky scales. **Shoots** *Slender, green*, hairless; *strong lemon-balm aroma when bruised*. **Buds** Flower-buds silky grey; *leaf-buds hairless*. **Leaves** The smallest of the tree-sized hardy magnolias': 8–14 cm and variably *slender, tapering* to a finely rounded tip; slightly *whitish* beneath, with a minute felt. **Flowers** Small (8 cm) snow-white nodding stars delicately overwhelm the tree a week after Campbell's Magnolia (p.264); 6 narrow, widely spreading main tepals; very sweetly scented. **COMPARE** Kobushi (below): differences emphasized. Yulan and white Saucer Magnolias (p.270): erect, *vase-shaped* flowers. Dawson's and Sprenger's Magnolias (p.266): rather larger leaves. Chinese Tupelo (p.412): similar foliage.
OTHER TREES *M.* × *kewensis*, the hybrid with Kobushi, is in some big gardens and can have more tepals. 'Wada's Memory', a densely upright clone named after Japanese nurseryman Koichiro Wada (though, in 1940, he was still alive), carries flowers *to 18 cm wide*, with *hanging, fluttering tepals*.

Kobushi
Magnolia kobus

(Northern Japanese Magnolia) N Japan; Cheju Do. 1865. Rather rare.
APPEARANCE Shape Typically sturdier then Willow-leaved Magnolia; to 18 m. **Bark** Like Willow-leaved Magnolia's; often coarser, corky ridges with age. **Shoots** *Pale brown, stoutish*; slight lemon aroma when bruised. **Buds** Flower-buds *and* leaf-buds *very downy*. **Leaves** Darker than Willow-leaved Magnolia's and *broad near their tips; green* beneath: most like Sprenger's Magnolia's (p.266). **Flowers** Like Willow-leaved Magnolia's, but a week earlier and usually carrying *a leaf-bud at the stalk-base*.

Star Magnolia
Magnolia stellata

(*M. kobus* var. *stellata*) Two sites in Japan. 1862. Abundant.
APPEARANCE Shape A twiggy bush (to 9 m). **Leaves** Slender, dark, undulant. **Flowers** Star-like, with *many* narrow white tepals; even on saplings.
OTHER TREES Loebner's Magnolia, *M.* × *loebneri*, is the hybrid with Kobushi (Germany; by 1910). Densely but delicately bushy, with stiff, straight twigs (*slight* lemon scent); now occasional (some small gardens) in various clones: solid in early spring, from an early age, with stars of many, spreading tepals. Leaves to 13 cm, from hairless buds.
'Merrill' (1939) is sturdily domed: its 15 cm flowers have up to 15 *broad*, pure white tepals.
'Leonard Messel', the most popular, is perhaps the loveliest magnolia: starry, spidery 12 cm flowers, the 12 narrow tepals flushed pink (almost *lilac* at a distance). Slow, slender and soon flat-topped.
'Ballerina' (Illinois, by 1970) has *up to 30* tepals, white with a basal pink flush.
M. × *proctoriana* (the hybrid with Willow-leaved Magnolia, 1928) is a rare, dainty, small-domed tree (to 9 m): slightly hairy leaf-buds; dark leaves slenderer than Loebner's Magnolia's; starry flowers with 6 (or sometimes up to 12) tepals, white except for a faint basal pink flush.

SPRING-FLOWERING MAGNOLIAS

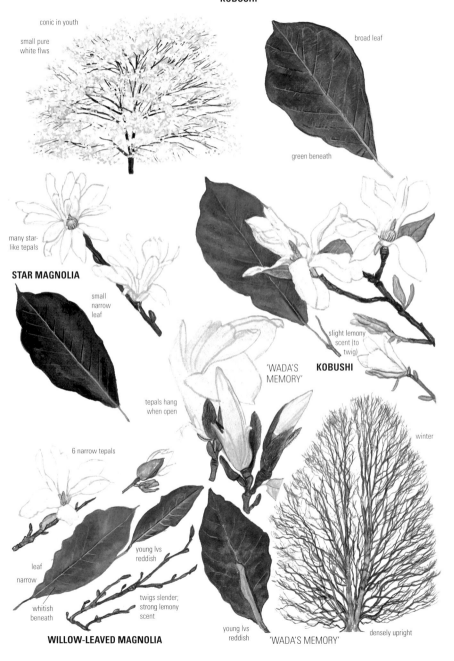

SPRING-FLOWERING MAGNOLIAS

Yulan — *Magnolia denudata*

(*M. heptapeta*) China. 1789. Now rare – larger gardens.
APPEARANCE Shape *Widely bushy, stiff but poised*; long-lived and growing slowly to 12 m. **Bark** Smooth, grey. **Leaves** To 15 cm, broadest near their pointed tips and finely downy beneath; they flush pale yellow for a week. **Flowers** Profuse but well spaced on the bare limbs: lemon-scented; erect and remaining *narrowly vase-shaped*; about 9 ivory-white tepals (which in China are eaten, fried, as a delicacy).
COMPARE Its hybrid, Veitch's Magnolia (p.264); Kobushi (p.268).

Saucer Magnolia — *Magnolia × soulangiana*

A hybrid of Yulan (above) with the more shrubby Japanese Lily-flowered Magnolia, *M. liliiflora*. The commonest magnolia, abundant in small gardens in a range of named and unnamed clones.
APPEARANCE Shape Untidily low-domed, to 13 m, or scraggy; usually a bush. **Bark** Smooth, grey. **Shoots** Grey-brown, with silky buds (leaf-buds 1 cm; flower-buds 2 cm). **Leaves** To 18 cm, broadest just above halfway; often downy only under the midrib; flushing yellowish. **Flowers** Erect and vase-shaped, from an early age; 9 white/pink/purple-stained tepals falling soon after they spread; spectacularly in early spring *then little by little through summer*.
VARIANTS Named clones include:
'Lennei': big flowers white inside, pale purple-pink outside, with a good autumn crop. Large, *broad* leaves, to 20 cm. 'Lennei Alba' is a seedling with pure white flowers.
'Brozzonii': large, *late* white flowers, purple-stained at the base; dark, matt leaves.
'Rustica Rubra': small flowers, mauve-red outside; dark, broad leaves; vigorous, rather *gaunt habit*.
'Picture': tall flowers, white within but with a central purple-pink stripe outside, and purplish at the base – found in a Japanese garden around 1930 by Koichiro Wada.
'San Jose' (1940, from California): creamy-white flowers, richly flushed pink.
'Verbanica': narrow tepals, usually a uniform mauve-pink outside; very late.

Smaller Hybrid Magnolias

Some now popular plants of compact (often scraggy) habit, bred to combine Saucer Magnolia's frost-resistant flowering, from an early age, with the flower-size and colours of the big tree species (see also p.266), include:
M. 'Heaven Scent' (× *veitchii* 'Rubra' × *liliiflora* 'Nigra'; USA, *c.*1960): tall very scented flowers with 12 narrow white tepals, dark pink at the base.
M. 'Peppermint Stick' (USA, 1962; same parentage): rather paler flowers with 9 tepals, particularly tall (11 cm) in bud.
M. 'Sayonara' (× *soulangiana* 'Lennei Alba' × × *veitchii* 'Rubra'; USA, 1966): white cup-shaped flowers with 9 broad tepals, flushed a subtle greenish pink.
M. 'Elizabeth' (*acuminata* × *denudata*: USA, 1978): first magnolia with profuse spring flowers to inherit Cucumber Tree's *rich cream colour*; 6–9 tepals.
M. 'Butterflies' (same parentage; USA, by 1988) has 10–16 *fully yellow* tepals.
M. 'Vulcan' (*M. campbellii* var. *mollicomata* 'Lanarth' × *M. liliiflora*; New Zealand, 1990): big, blowsy, vividly *crimson* flowers.

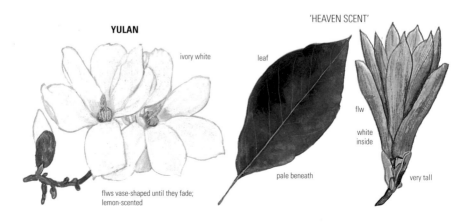

YULAN — ivory white; flws vase-shaped until they fade; lemon-scented

leaf; pale beneath

'HEAVEN SCENT' — flw; white inside; very tall

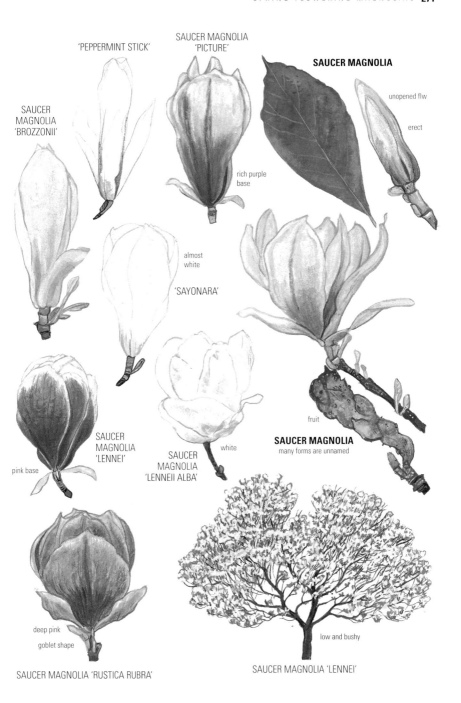

TULIP TREE TO SPUR-LEAF

Tulip Tree — *Liriodendron tulipifera*

(Yellow Poplar) E North America: Nova Scotia to Florida. Quite frequent in warmer areas.
Appearance Shape Towering, to 36 m; often rather twisted shoots and stems but sometimes starting off spire-shaped (like the best wild trees, which can have straight trunks 30 m long). Dense, rich-green summer foliage; reliably butter-yellow in autumn. Long-lived – one probable 1685 original thrives in Surrey – but easily broken by storms. **Bark** Pale grey: short, close, rather sharp criss-crossing ridges. Oldest trees more bronze and rugged; strange, ragged flanges. **Shoots** Greenish, bloomed lilac at first. **Buds** *Stalked and smoothly flattened, like a beaver's tail.* **Leaves** Deeply waisted on sprouts (like the adult leaves of Chinese Tulip Tree); occasionally with an extra pair of lobes; smooth and variably silvery beneath with a covering of minute waxy warts (papillae). (Juvenile plants have almost square leaves with no side lobes; the rare f. *integrifolium* retains this foliage.) **Flowers** Green-and-orange 5 cm tulips are abundant in June after a warm season but tend to be lost in the foliage. The 5 cm seed-heads last through the winter, biscuit-brown.
Variants 'Fastigiatum' is spire-shaped then rather broadly columnar, on twisting, *vertical branches*, and remains rare.
'Aureomarginatum' is rather rare but vigorous; the soft yellow leaf-margin fades to pale green through the summer and by autumn is not obvious. 'Aureopictum' (much rarer) has yellow blotches at its leaves' centres.
'Glen Gold' is a recent clone with uniformly yellow foliage in early summer.

Chinese Tulip Tree — *Liriodendron chinense*

E China to N Vietnam: discovered in 1875 as another example of a genus common to the 2 most diverse populations of temperate trees (E N America and China). 1901. Still rare, less vigorous in the UK than Tulip Tree (above); to 25 m.
Appearance Difficult to differentiate with confidence from Tulip Tree. **Bark** Can become brown and rather finely craggy at a younger age. **Shoots** More strongly white-bloomed. **Leaves** Flushing briefly *purplish; always* elegantly and strongly waisted (like Tulip Trees' sprout-leaves); a denser covering of papillae makes them somewhat whiter underneath. **Flowers** Generally greener and a few weeks later than Tulip Tree's.

Spur-Leaf — *Tetracentron sinense*

Central China, N Vietnam, N Burma, N India. 1901. A stylish tree in a few big gardens. This seems to be a very primitive species, with a rather conifer-like wood-structure. The only plant in its family, Tetracentronaceae.
Appearance Shape Flimsy, rather upright; to 13 m. **Bark** Grey-brown, scaly. **Leaves** Carried *singly on woody, alternate spurs* which grow year by year: mere lumps on the shiny red/green young shoots, but 2 cm long on interior twigs, and rough with annual growth-scars. Heart-shaped, thick, with sunken veins, to 15 cm; leaves flush deep pink and the colour persists for some time at the *blunt* tip of each fine but jagged tooth. **Flowers** Each spur can carry a very slender green catkin, to 20 cm long, from spring to late winter.
Compare Katsura (p.274): similar jizz but smaller, opposite leaves.

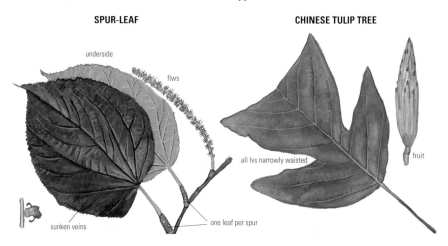

SPUR-LEAF — underside, flws, sunken veins, one leaf per spur

CHINESE TULIP TREE — all lvs narrowly waisted, fruit

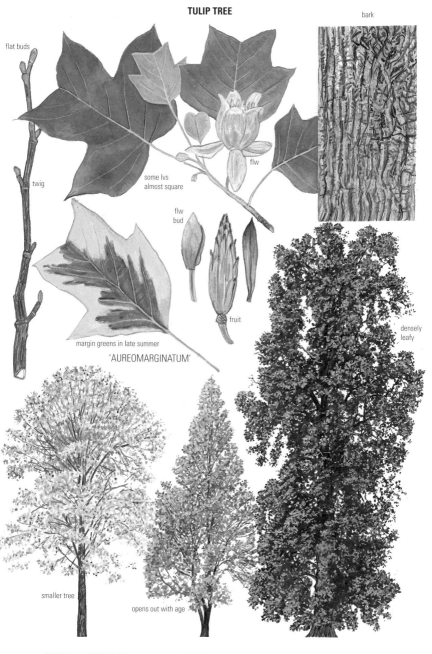

PRIMITIVE FLOWERING TREES

Katsura *Cercidiphyllum japonicum*

China; S Japan (an endangered giant). 1865. Rather occasional; happiest in mild, humid areas. (Family: Cercidiphyllaceae.)
Appearance Shape Daintily rounded-conic, to 25 m, often on several straight stems (Chinese trees, var. *sinense*, are more often single-trunked). Prominent paired buds; sweeping minor branches closely strung with neat little leaves, even inside the crown. **Bark** Pale brown/grey; long shaggy strips. **Leaves** In *opposite pairs* (alternate at the shoot-bases), finely round-toothed (the smallest scarcely so); to 10 cm; soon hairless. They flush pink and die off lemon and orange, with a *caramel scent* from high summer onwards which can lead you to distant trees. **Flowers** Dioecious: red tufts.
Compare Judas Tree (p.348): similar leaf-shape.
Variants 'Pendulum', long grown in Japan, weeps rather gauntly; very rare to date.
'Red Fox' is a new, *rich purple* clone (cf. Redbud 'Forest Pansy', p.348).

Wheel Tree *Trochodendron aralioides*

Japan; Korea; Taiwan. c.1894. Rare: big gardens. The only plant in its family, Trochodendraceae.
Appearance Shape Usually dense: bushy (to 17 m), with tiered evergreen foliage – like a tree of the unrelated Ivy. **Bark** Smooth; dark grey; aromatic. **Leaves** Glossy, leathery, hairless, to 12 cm; the *upper half* with distant teeth (cf. Bamboo Oak, p.230, and Peruvian Nutmeg, p.276); they droop on *stalks to 10 cm long*. **Flowers** Acid-yellow, in spring: wheels of radiating stamens, in 8 cm spires, backed by big, pale-pink bracts.

Winter's Bark *Drimys winteri*

S central Andes. 1827. Very occasional (mild areas). The bark was once chewed by sailors to combat scurvy: Captain William Winter brought it to England in 1580. Belongs to a 'living fossil' family, Winteraceae.
Appearance Shape Unusually, plants from the S of the range ('var. *latifolia*') are more tree-like: spires to 20 m, but soon wind-broken and with untidy, curving branches. **Bark** Smooth, pinkish brown. *Spicily aromatic* (more so than the leaves). **Shoots** Often brilliant crimson. **Leaves** 7–17 cm; glossy, leathery, hairless; variably *silvery* beneath; midrib prominent above. **Flowers** Big, blowsy, creamy-white heads, in late spring.
Compare Michelia (p.260): broader leaves; magnolia flowers. Silvery underleaf distinguishes Winter's Bark from most evergreens with long, untoothed leaves.

Chilean Firebush *Embothrium coccineum*

S central Andes. 1846. Very occasional: larger gardens in milder areas, on acid soils. The patterned timber is much in demand. (Family: Proteaceae.)
Appearance Shape *Slender*, to 20 m: erect/gauntly arching, densely clad stems. **Bark** Purplish brown; flaking with age. **Leaves** Hairless, flat; oval to lollipop-stick-shaped, 5–20 cm; slightly bluish beneath; shed in hard winters – leaving *long-pointed red end-buds*, to 15 mm. The common (and hardiest) clone is 'Norquinco Valley', with many narrow leaves. **Flowers** In early summer (sometimes again in autumn): spectacular, tropical-looking scarlet bunches wreathing the stems.
Compare Californian Laurel (p.276).

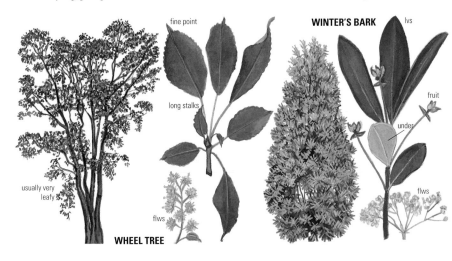

PRIMITIVE FLOWERING TREES

KATSURA

CHILEAN FIREBUSH

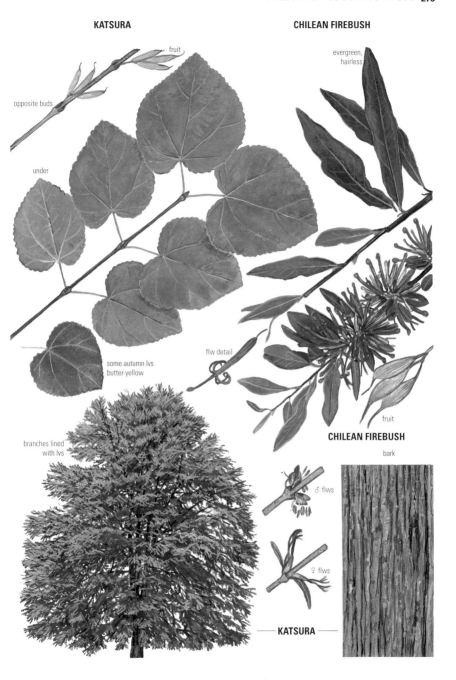

Sassafras
Sassafras albidum

(*Sassafras officinale*) Ontario to Florida, and W to Texas. In Spain by 1560. A rare but stylish tree, confined to acid soils in warm areas. (Family: Lauraceae.) **Appearance Shape** A rather narrow dome of twisting branches: to 18 m but sometimes shrubby and gaunt; suckering freely. **Bark** Grey; close sharp ridges. **Shoots** *Slender; leaf-green* for some years; *buds green*. **Leaves** Glossy green above, silvery under (downy in var. *molle*); *strongly vanilla-scented* (but carcinogenic?). Unique *range of oven-mitt shapes*, though in old trees unlobed leaves are the norm. In autumn, yellow, pinkish red and rich orange. **Flowers** Usually dioecious. **Fruit** Female trees carry black 1 cm berries on long red stalks. **Compare** In winter, Tupelo (p.412), Sweet Gum (p.278) and Sorrel Tree (p.430); green shoots distinguish Sassafras.

Californian Laurel
Umbellularia californica

(Headache Tree) California and Oregon. 1829. Rather rare; milder areas. A *headily and overwhelmingly aromatic* tree, used in Californian cuisine but quickly bringing on a sharp headache, or even rashes or unconsciousness. **Appearance Shape** A broad or irregular evergreen dome to 20 m; sometimes bushy; lower trunk often massively convoluted. **Bark** Grey, cracking with age into small squares. **Leaves** Untoothed, to 9 cm, round-ended, very *flat* and *fresh green each side*, the midrib conspicuously white beneath; soon hairless. **Fruit** 25 mm, ripening purplish – also aromatic. **Compare** Chilean Firebush (p.274): slenderer; leaves scarcely scented. Holm Oak (p.222): similar jizz.

Bay
Laurus nobilis

(Sweet Bay; Poet's Laurel) Mediterranean area. Long grown further N; abundant in milder areas; rarely naturalizing. 'Bachelor' (*baccalaurus*) derives from 'laurel-berry': the graduate was wreathed with Bay. **Appearance Shape** A dense evergreen spire to 20 m; bushier in colder parts. **Bark** Smooth, black-grey. **Leaves** 5–12 cm, finely long-pointed; *thin but hard with crinkled edges and the odd tiny tooth; hot, sharp fruity smell*. **Flowers** Dioecious; egg-yolk yellow in spring, from yellow globular buds prominent all winter. **Fruit** 12 mm black berries. **Compare** Holm Oak (p.222): leaves woolly beneath. Strawberry Tree (p.430): regularly serrated leaves. Other evergreens (eg. hollies) have thicker/glossier/less finely pointed leaves.
Variants Narrow-leaved Bay, 'Angustifolia' (undulant leaves less than 2 cm wide), is a remarkably distinctive but very rare foliage plant.
Golden Bay, 'Aurea', has gold leaves through winter/spring; rare, though hardier than the type.

Peruvian Nutmeg
Laurelia sempervirens

(Incorporating *L. serrata*) S central Andes. By 1868. Rare; confined to big gardens in mild areas. (Family: Atherospermataceae.) **Appearance Shape** A splendid evergreen to 20 m, rather resembling Bay but unrelated. **Bark** Dark grey; developing some large, shallow scales. **Leaves** In *opposite pairs*, to 10 cm, *glossy* but quite flimsy; coarsely and distantly serrated; yellow hairs often radiate under the midrib. Deliciously aromatic – *oranges and vanilla*. **Flowers** Clustered in the leaf-joints. **Fruit** 'Nutmegs' with fluffy seeds. **Compare** Wheel Tree (p.274): broader leaves not scented. Bamboo Oak (p.230): alternate leaves.

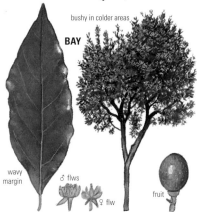

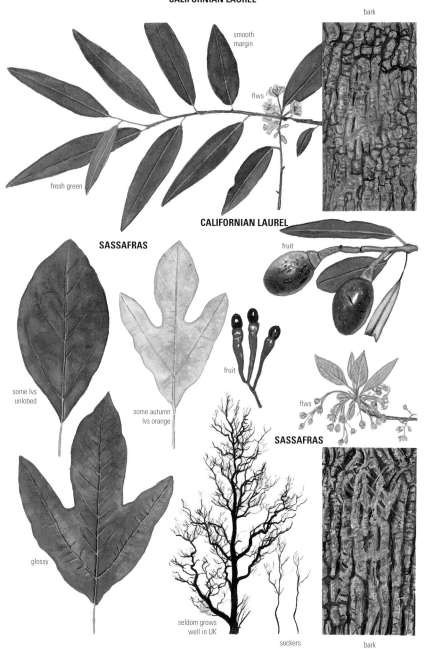

SWEET GUMS TO HARDY RUBBER TREE

Hamamelidaceae is a diverse family of often winter-flowering trees and shrubs.

Sweet Gum *Liquidambar styraciflua*

(American Sycamore) SE USA to Mexico – often in swamps; planted commercially for the balsam in its sap (the medicinal liquid storax). 1681. A frequent garden tree in warmer areas.
APPEARANCE Shape Stiffly conic, then domed or irregular, to 30 m, with twisting, upcurving, often broken branches and (in winter) a *spiky* outline (cf. Tupelo, p.412). **Bark** Grey-brown; knobbly, scaly ridges from the first. **Shoots** Young twigs (especially from suckers) can have corky wings (cf. English Elm, p.244; Field Maple, p.368). Saplings' short, spiky side-shoots with long-pointed buds suggest Aspen's (p.152) but are more forward-angled; buds are a clearer red/green. **Leaves** 5-lobed, finely toothed: maple-like but *alternate*; with small *tufts of hairs* under the vein-joints and scurf under the veins. Young trees can have 3–lobed leaves (cf. Chinese Sweet Gum, below). Fresh, *pale green* in summer; splendid autumn colours – lemon-yellows, crimsons, saturated purples. **Fruit** Hangs through winter – 3 cm spiky balls, rarely seen in the UK.
COMPARE Castor Aralia (p.424); Maple-leaved Crab (p.312).
VARIANTS 'Variegata' ('Aurea') is very rare: leaves yellow-splashed; 'Silver King' (young plants; rare; sometimes also sold as 'Variegata') has clear creamy-white leaf-margins.
'Lane Roberts' and 'Worplesdon' are autumn-colouring selections, seen occasionally as younger, grafted trees. 'Worplesdon' (crimson/yellow) tends to be bushy, with small, narrowly lobed leaves, but 'Lane Roberts' (blackish crimson) grows a good straight trunk of relatively smooth bark.
OTHER TREES Chinese Sweet Gum, *L. formosana* (China, Taiwan; 1884), is rare; to 23 m. Bark grey, usually *smooth*; leaves very *matt*, more or less 3–*lobed*; usually with some hairs across the underside. Autumn colours late, and less reliable.
Oriental Sweet Gum, *L. orientalis*, a large tree from Turkey and Rhodes (1750), remains bushy in the UK; very rare. Bright brown square-cracked bark; smaller matt leaves *completely hairless* beneath.

Persian Ironwood *Parrotia persica*

Forests S of the Caspian Sea. 1841. Rather occasional: town parks, larger gardens.
APPEARANCE Shape *Long, gaunt branches typically arch to the horizontal* from a dense central crown on a (very) short, muscularly fluted trunk. To 15 m. **Bark** Fine plates flake in *cream, grey and orange*: cf. Strawberry Dogwood (p.426) and Deciduous Camellia (p.408). **Shoots** With short, stiff hairs. **Buds** Purplish with short, stiff hairs. **Leaves** Glossy, to 12 cm; odd teeth/waves (like the shrubby witch-hazels'). Yellow and red autumn colours start early at the branch-tips. **Flowers** Maroon stamen-clusters in late winter (lacking witch-hazels' yellow petals).

Hardy Rubber Tree *Eucommia ulmoides*

(Gutta-Percha Tree) China: grown for its medicinal bark and 'rubber' but not known wild. 1896. Rare. The only plant in its family, Eucommiaceae.
APPEARANCE Shape Domed; upper limbs *untidily curling*. **Bark** Grey; deep, knobbly criss-cross fissures. **Leaves** Big, to 18 cm, hanging, *toothed*, glossy, with sunken incurving veins; downy when younger – unmemorable, but tear a leaf gently and the sap *hardens into strings* which hold the parts together (cf. only Dogwoods, p.424–6). **Flowers** Dioecious; studding the crown in late winter. **Fruit** On the rare females; elm-like.

CHINESE SWEET GUM — dark and matt

ORIENTAL SWEET GUM — hairless beneath

SWEET GUM 'SILVER KING'

'VARIEGATA'

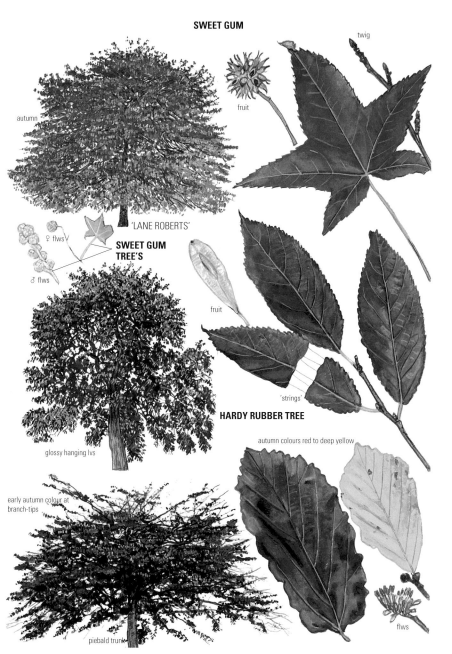

PLANES

The 6–7 species of plane originate in North America and Asia. (Family: Platanaceae.)

London Plane
Platanus × hispanica

(*P. × acerifolia*) Probably a fertile hybrid of Oriental Plane and the American Buttonwood, *P. occidentalis* (with clumsy, shallowly lobed leaves and scarcely grows in the UK; also called 'Sycamore' in North America, while in Scotland this book's 'Sycamore' is sometimes called 'Plane'), originating in Spain or S France (*c*.1650) but long planted abundantly in S England; very rarely naturalizing.
APPEARANCE Shape Long, straightish limbs rise from a buttressed trunk generally *clean for some metres*; smaller branches very twisting. To 44 m; oldest trees thriving at 320+ years. **Bark** Shallowly *scaling in greys and creams* – but see variants. **Shoots** Green/brown, scurfy at first, the end bud missing. **Buds** *1-scaled red/green horns*, hidden in the leaf-stalks in summer and in winter rimmed by their scars. **Leaves** Thick; maple-like but *alternate*; typically 3+ teeth on the shoulder of each main lobe. *Hard, deep green*; flushing late and buff-woolly. **Fruit** 3 cm balls of hairy seeds (usually 2–6 per string), breaking up in spring – can bring on asthma attacks.
VARIANTS Several clones can be distinguished, though few are named. One (mostly 18th-century plantings) has a tall, duller brown trunk and leaves with long, often untoothed triangular lobes; one has big leaves and greyish basal bark cracking in rectangles. Another (which could be called 'Baobab Plane', and probably carries a viral infection), has disproportionately small, very twisting branches on a short trunk *extraordinarily swollen with rough burrs*; its deeply cut leaves resemble Oriental Plane's but are darker, the lobes having *more* long teeth.

'Pyramidalis' (locally frequent) was christened before its crown had matured: it has straighter branches and is narrowly columnar at first, then broad and gaunt. The glossy leaves resemble Buttonwood's in being *shallowly 3-lobed* (smallest ones almost round) and the bark is soon *dull brown and furrowed, with many burrs*. The 45 mm fruit-balls, *1–2 per string*, are like Buttonwood's.
'Augustine Henry' is splendid but occasional: long bole and straighter branches clad in very clean, grey and creamy-white bark; sea-green leaves with drooping edges and *up to 5 regular long teeth* on the shoulder of each lobe; fruit-balls *1–3 per string*.
'Suttneri' is brightly white-variegated but very rare.

Oriental Plane
Platanus orientalis

SE Europe, with ancient planted trees E to Kashmir. Long grown in the UK; now rather occasional, in warmer areas.
APPEARANCE Shape Generally broader than London Plane: to 30 m: short, often massively burred bole and limbs which can bend to the ground and layer. Typically *fresh green*, delightfully *feathery* foliage; pale bronzy-purple in autumn. **Bark** As London Plane; on old trees pale brown and rugged (but almost white in hot climates). **Leaves** With deeper finger-like lobes than London Plane's (especially narrow in 'Digitata'); normally with only 1–2 teeth on the shoulder of each main lobe; often ('Cuneata') narrowly tapered to the base. *Sweet, balsam smell* from the foliage is stronger than in London Plane. **Fruit** Like a typical London Plane's but smaller: 3–6 balls per string.

ORIENTAL PLANE — spreading, long-lived, big burrs — 'CUNEATA'

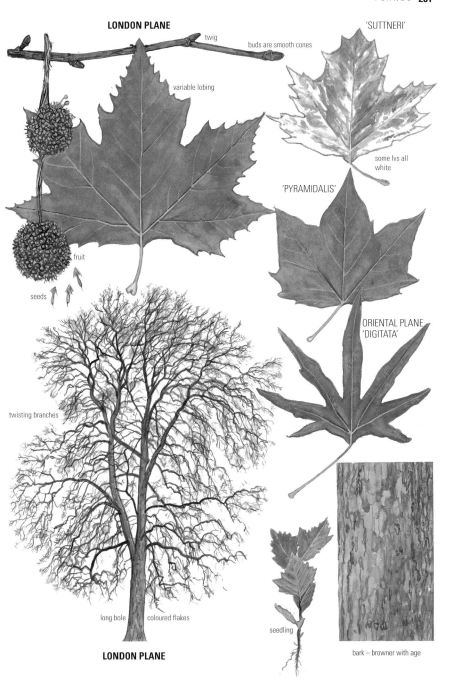

ROSE FAMILY TREES

Rosaceae is a family of 3000 herbs, trees and shrubs, many with edible fruits.

Himalayan Tree Cotoneaster
Cotoneaster frigidus

Himalayas. 1824. Quite frequent.
Appearance Shape Often multiple, leaning stems with *many vertical sprouts*; sometimes a sturdy tree, to 15 m. **Bark** Pale fawn-grey, harshly *scaly*. **Leaves** *Flat*, untoothed, 6–12 cm, falling through early winter; whitish beneath with a close felt reduced to sparse down by autumn; *blunt*, a tiny downward bristle at some tips. **Flowers** White: 5 cm heads in early summer. **Fruit** *Deep red* 7 mm berries can last till spring.
Compare Medlar (p.290). Hybrid Cockspur Thorn (p.290): serrated leaves.
Other trees *C.* × *watereri* covers a swarm of (occasionally naturalizing) hybrids with the evergreen *C. henryanus* and *C. salicifolius* (which have smooth grey barks); *veins impressed and leaf-margin downcurved*. *C.* 'Cornubia' (frequent; sometimes placed under *C. frigidus*) has the largest leaves (to 13 cm), almost hairless by late autumn, when they begin to drop; berries slightly larger (1 cm) and more vividly blood-red than Himalayan Tree Cotoneaster's. *C.* 'Exburiensis' has lemon-yellow and *C.* 'Rothschildianus' amber fruits; their smaller, more or less evergreen leaves remain woolly beneath.

Loquat
Eriobotrya japonica

(Japanese Medlar) China. 1787. Occasional in milder parts: in small gardens as often as big ones; rarely seeding.
Appearance Shape Rounded, often flat and bushy; surfaced with disproportionately *large, dark, drooping evergreen leaves* (to 30 × 12 cm; smaller with age). **Bark** Blackish; distantly cracked with age. **Leaves** Glossy; with *impressed veins* and beige felt beneath; *coarsely toothed*. **Flowers** Hawthorn-scented, creamy: on 15 cm heads in autumn. The sweet, yellow, pear-shaped 4 cm loquats ripen next summer (but seldom in the UK).
Compare Giant Photinia (below); Southern Evergreen Magnolia (p.260); Gurass (p.428).

Giant Photinia
Photinia serratifolia

(*P. serrulata*) China and Taiwan. 1804. Very occasional in bigger gardens in milder parts.
Appearance Shape A densely surfaced evergreen dome, to 15 m. **Bark** Grey; smooth; some reddish scales with age. **Leaves** Large (to 20 cm); leathery; very *finely serrated*; white down under the midrib at first, then hairless. They unfold glistening brownish red (this is one parent of the ubiquitous, much smaller-leaved bush, *P.* × *fraseri*). **Flowers** In 15 cm, white, hawthorn-scented heads among the bright spring leaves; 6 mm red berries follow.
Compare Cherry Laurel (p.344): distant teeth. Southern Evergreen Magnolia (p.260).

Deciduous Photinia
Photinia beauverdiana

W China. 1908. Rare; large gardens but deserving better. Not obviously akin to Giant Photinia.
Appearance Shape Lightly branched, elegant; domed, to 12 m. **Bark** Pale grey, smooth. **Leaves** *Glossy* and hairless but thin and deciduous; *minutely serrated*; to 12 × 4 cm (but *to 9 cm wide* in the commoner var. *notabilis*); veins sunken **Flowers** In 10 cm white heads; airy plates of thornlike red berries follow.
Compare Snowy Mespil (p.320): shorter, fresh-green leaves. Alder-leaved Whitebeam (p.294) recalls var. *notabilis* in habit, flower and fruit.

HIMALAYAN TREE COTONEASTER

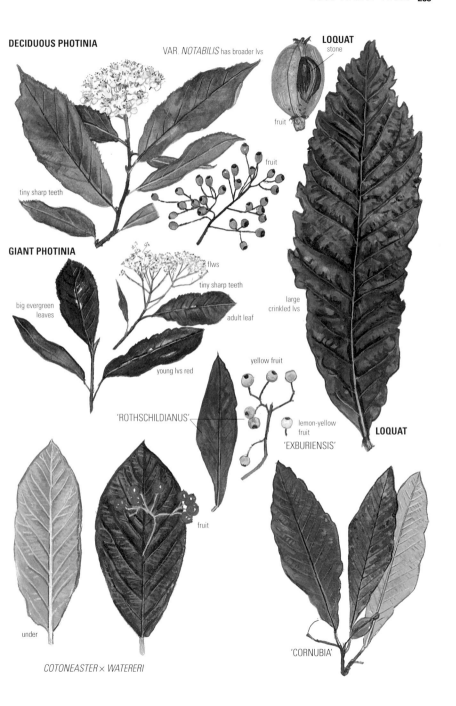

HAWTHORNS

Hawthorns (over 200 species) are small, usually thorny trees.

Things to Look for: Hawthorns

- Leaves: What shape? Are they downy above/beneath?
- Flowers: How many anthers? and what colour? How many styles?
- Fruit: How big? What colour? How many pips?

Key Species

Common Hawthorn (below): deeply lobed leaves.
Scarlet Thorn (p.288): big, shallowly lobed leaves.
Broad-leaved Cockspur Thorn (p.288): unlobed leaves.

Common Hawthorn — *Crataegus monogyna*

(Quickthorn; Whitethorn; May) Europe (including Britain and Ireland) to Afghanistan. Dominant in most hedgerows; abundant everywhere except on poor sands.
APPEARANCE Shape A twiggy mass of stiff branches, to 15 m. Can be long-lived. **Bark** Brown; shallow, often spiralled, scaling ridges. **Shoots** Glossy green- or red-brown; many *1–2 cm rose-like thorns*. (The 'spines' of other 'thorny' trees only *tip bud-bearing side-shoots*; Hawthorns have these too.) **Leaves** (To 6 cm) cut *at least halfway to the midrib* by lobes *toothed at their tips*; main veins *curve outwards*, with *tufts under the joints*. **Flowers** White (rarely fading red); *1 style.* **Fruit** 'Haws' with *1 pip*.
COMPARE Midland Hawthorn (below). Oriental Thorn and Tansy-leaved Thorn (p.286): hairy leaves. Large-leaved Thorn (p.286): leaves to 15 cm.
VARIANTS 'Fastigiata' is occasional: as nearly vandal-proof as any street tree, but no beauty. 'Pink May' has delicate pink flowers but is much rarer than the red cultivars of Midland Hawthorn.

('Paul's Scarlet' is colloquially called 'Pink May'.) Glastonbury Thorn, 'Biflora' (very rare), is said to have sprung from the staff of Joseph of Arimathaea, planted on Christmas Day at Glastonbury in front of a sceptical heathen congregation; its *premature secondary crop* of leaves and flowers is seldom as early as December.
OTHER TREES *C. calycina* (France E to Greece and Russia; in some 'native' plantings in Britain) has leaf-lobes *toothed almost to their bases*, and slenderer, oval haws.

Midland Hawthorn — *Crataegus laevigata*

(*C. oxyacantha*; *C. oxyacanthoides*) W Europe, including S/central England: ancient woods/old hedges *on clay*.
APPEARANCE Shape Bushy in the wild, but odd park trees are as large as Common Hawthorns. **Leaves** Dark, slightly glossy, and lobed *less than halfway to midrib*; almost hairless; main veins straight or *curved upwards*. **Flowers** *2–3 styles.* **Fruit** Haws with *2–3 pips*: hybrids with Common Hawthorn (*C. × media*: as widespread as the species) have 1–3. Some wild trees (f. *rosea*) open pink.
COMPARE Common Hawthorn (above): differences emphasized. Grignon's Thorn (p.290).
VARIANTS Many probably derive from *C. × media* (leaves lobed at least halfway to the midrib). 'Punicea Flore Plena' (frequent) has mauve-pink double flowers; some may be reversions from 'Paul's Scarlet' – an abundant sport (1858) of this old form, whose *double crimson* flowers fade whitish inside. 'Masekii' (rare) has double soft pink flowers. 'Plena Alba' has double white flowers (fading pink; occasional). 'Punicea' (very occasional) has single flowers of crimson petals and a white heart; 'Crimson Cloud' is similar.

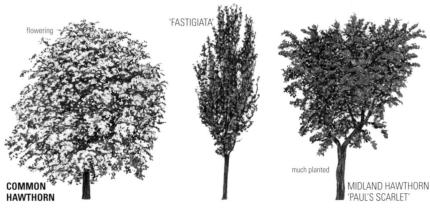

COMMON HAWTHORN (flowering); 'FASTIGIATA'; MIDLAND HAWTHORN 'PAUL'S SCARLET' (much planted)

HAWTHORNS

Large-leaved Thorn
Crataegus pinnatifida var. *major*

N China. c.1880. Odd trees in parks and older gardens.
APPEARANCE Shape Shapely, to 10 m; seldom thorny. Bark Scaly, with orange/yellow tints; sometimes brown and more square-cracked. Leaves Shaped much like Common Hawthorn's but dark and *huge* (to 15 cm); slightly hairy along the veins. (The type – scarcer – has smaller, slightly hairier, more deeply lobed leaves.) Flowers Showy. Fruit 20 mm, deep red, with 3–4 pips and a dimpled tip.

Oriental Thorn
Crataegus laciniata

(*C. orientalis*) SE Europe; Sicily; Spain. 1810. Very occasional: parks, gardens.
APPEARANCE Shape Low, picturesque: twisting, almost thornless branches. Bark Coarsely scaly; often with *orange-pink* tints. Shoots Woolly at first. Leaves Much more narrowly and deeply lobed than Common Hawthorn's, the base tending to sweep down the short stalk; dark *grey*-green from *fine hairs on both sides* (cf. Tansy-leaved Thorn). Flowers In early summer. Fruit Large (18 mm), bright brick-red haws; 5 pips.
OTHER TREES The following European species, in the odd UK garden, also have downy, sharply lobed leaves.
Azarole, *C. azarolus* (Mediterranean areas E to Iran), is a thornier tree with less hairy leaves, closer to Common Hawthorn's in shape. Its haws, *yellow and 22 mm wide* (red/white in some cultivars), *taste like apples* (cf. Tansy-leaved Thorn) and are used in S Europe for jams and liqueurs; they have 1–4 pips.
C. heldreichii (mountains of Albania, Greece and Crete): differs from Azarole in haws with only 1–3 pips.
C. schraderiana (mountains of N Greece; the Crimea): differs from Azarole in its dull, *dark* red haws with 2–4 pips.
C. pentagyna (E Europe) has small *purple-black* haws with 4–5 pips (cf. Hungarian Thorn); leaves often larger than Azarole's.
Hungarian Thorn, *C. nigra* (Danube valley), has rather *large* (4–8 cm), dark, shallowly lobed leaves, downy on both sides (cf. Scarlet Thorn, p.288, and Red Thorn, below), and *black* haws with 4–5 pips. Red Thorn, *C. sanguinea* (Russia), has 5–8 cm, shallowly lobed, downy leaves (cf. Scarlet Thorn, p.288) and red 1 cm haws with 2–5 pips; a poor bush in the UK.

Tansy-leaved Thorn
Crataegus tanacetifolia

Asia Minor to Syria. 1789. Rather rare.
APPEARANCE Differences from Oriental Thorn (above) as follows: Bark Often *pear-like*, black, square-cracked. Leaves Grey-woolly; main teeth *tipped with tiny knobbly glands*. Fruit Yellowish, 22 mm haws (with 5 pips), which taste like apples (cf. Azarole, above).
OTHER TREES *C. × dippeliana* is a putative hybrid with Spotted Thorn in some town parks etc: a picturesque tree with very twisted branches and dark grey bark often growing *interlacing ridges*. Leaves to 8 cm, almost hairless above by autumn; lobed *less than halfway to the midrib* and scarcely at all on weak shoots; haws 15 mm, dull red.

Spotted Thorn
Crataegus punctata

E North America. 1746. Locally very occasional in town parks.
APPEARANCE Shape Vigorous and handsome, to 12 m; *broad, rather tabular branching* (cf. Hybrid Cockspur Thorn, p.290); thorns 7 cm. Bark Orange-brown; scaly. Leaves Tapering to the base, scarcely lobed; *sunken, parallel veins; matt* and almost hairless above, finely downy beneath. Flowers In massed, overpoweringly scented plates. Fruit Oval haws, 20 mm, with 5 pips; dull red (but yellow in f. *aurea*), and speckled with *pale dots*.

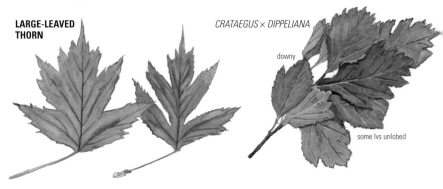

LARGE-LEAVED THORN

CRATAEGUS × DIPPELIANA

downy

some lvs unlobed

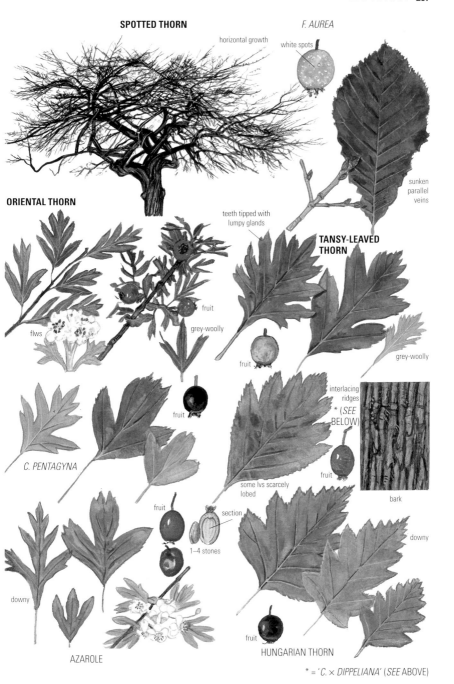

HAWTHORNS (AMERICAN)

Scarlet Thorn — *Crataegus mollis*

E USA. One representative of a huge group of similar E North American thorns, occasional in some variety in older gardens and parks (and rarely naturalized). Most of these species can only be differentiated in flower.
Appearance Shape Twiggy and seldom thorny; sometimes bushy but often with a strong central axis. **Bark** Grey-brown; rather scaly. **Leaves** *Big* (to 12 cm) and sparse, a dull pale green; *softly downy beneath* and rather *scrubbily* hairy above; base rounded/*heart-shaped*. **Flowers** On very woolly stalks; about *20 stamens* with *pale yellow anthers*. **Fruit** The red, downy haws, to 25 mm long, have 4–5 pips.
Compare Hungarian Thorn and Red Thorn (p.286). Service Tree of Fontainebleau (p.296): thicker leaves, grey-woolly beneath. Yunnan Crab (p.308); Italian Crab (p.312); Turkish Hazel (p.198).
Other trees The following similarly large-leaved allies are sometimes found in gardens.
C. submollis differs in having flowers with *10 stamens*.
C. coccinioides has yellowish green, *almost hairless* leaves with rounded/heart-shaped bases. Its flowers have *20 stamens with red anthers* (cf. *C. chrysocarpa*); fruits bright red, 15 mm.
C. chrysocarpa is one of the commoner, more vigorous species. Leaves with broadly *tapering bases*, glossy and *hairless except under the veins* and, at first, across the upper surface. Flowers (on hairy stalks) have *10 stamens with pale yellow anthers*; haws (red/yellowish) are only 12 mm long.
C. pedicellata has leaves rather scrubby above and hairless beneath, with broadly tapering bases. Flowers (on stalks only slightly hairy) have *10 stamens with pale pink anthers*; the rather pear-shaped, rich scarlet haws are 18 mm long. Out of flower, the finely scrubby upper leaf surface distinguishes from *C. chrysocarpa*.
C. ellwangeriana has leaves with broadly tapering bases, finely scrubby above and softly downy beneath. Flowers have *10 stamens with pale pink anthers*; haws, like those of *C. pedicellata*, are very showy. (Differs from *C. submollis* in tapering leaf-base and pink anthers.)

Broad-leaved Cockspur Thorn — *Crataegus persimilis* 'Prunifolia'

(*C. prunifolia*) A selection (by 1797) of another North American thorn: the *glossy oval leaves* turn brilliant orange-red in autumn. Frequent in parks, streets and gardens.
Appearance Shape Broad and twiggily domed; to 9 m. **Bark** With very scaly, often spiralling ridges. **Shoots** Hairless; thorns 2 cm. **Leaves** To 8 cm, *never lobed; shiny*, but finely hairy under the midrib. **Fruit** Dark red haws, 15 mm, adding (briefly) to the autumn spectacle.
Compare Snowy Mespil (p.320): leaves less glossy. Tartar Maple (p.376): opposite leaves). Other thorns with scarcely lobed (but narrower) leaves include Spotted Thorn (p.286), Cockspur Thorn (below), and Hybrid Cockspur Thorn (p.290).

Cockspur Thorn — *Crataegus crus-galli*

NE North America. 1691. Now rare.
Appearance Shape A low tree bristling with *purple 8 cm thorns*, like cocktail sticks or *branched*. **Bark** Greyish, only *finely* cracking. **Shoots** Hairless. **Leaves** *Completely hairless from the first and very glossy*; narrow, to 8 cm long; *orange* in autumn. **Flowers** On *hairless* stalks. **Fruit** 1 cm, deep red haws; they hang on through winter.
Compare Hybrid Cockspur Thorn (p.290).

BROAD-LEAVED COCKSPUR THORN

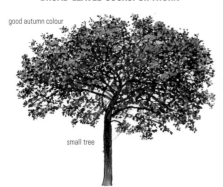

good autumn colour

small tree

COCKSPUR THORN

good autumn colour

fruits persist through winter

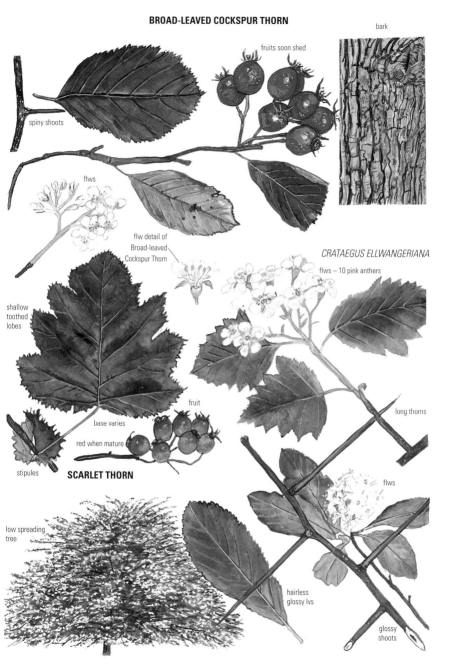

HYBRID THORNS TO QUINCE

Hybrid Cockspur Thorn
Crataegus × *lavallei* 'Carrièrei'

(*C* × *carrierei*) A hybrid probably of Cockspur Thorn (p.288) with the semi-evergreen Mexican *C. stipulacea*; by 1880. Frequent in parks and streets.
Appearance Shape Picturesque: rather level, *dense* branch-systems; to 12 m. **Bark** Grey; very scaly. **Shoots** With long hairs but few thorns. **Leaves** *Long* (to 10 cm, the lowest third untoothed), glossy, *blackish*; some stiff hairs above and down underneath: they flush late, then look almost evergreen until they drop, still green, early in winter. **Flowers** Showy, in early summer, on woolly stalks. **Fruit** Dull scarlet, 18 mm; lasting until spring.
Other trees Grignon's Thorn, *C.* × *grignonensis*, is another, rather rarer hybrid of *C. stipulacea* (*c*.1873), more upright and untidy. Shoots hairless; leaves much shorter (to 6 cm), paler and less glossy, with *variable large teeth/small forward-pointing lobes* – on strong shoots they can resemble Midland Hawthorn's (p.284). Flowers in late spring; haws (some lasting all year) *bright* red.

Medlar-thorn
× *Crataemespilus grandiflora*

A hybrid of Medlar and probably Midland Hawthorn, long known but quite rare (town parks).
Appearance Shape Picturesque; rather tabular; to 10 m. **Bark** Brown, very scaly. **Shoots** Downy. **Leaves** Long (to 11 cm) and slender; dull dark green and very *downy on both sides*; finely toothed, and variously *lobed towards the tip*. **Flowers** To 3 cm, in 2s/3s in early summer. **Fruit** 2 cm, with a medlar's wide sepal-ring but a haw's mealy taste.
Other trees Bronvaux Medlar, + *Crataegomespilus dardarii*, is a chimaera of Medlar and Common Hawthorn, found on a grafted Medlar in a garden at Bronvaux near Metz around 1895; a *weak, twiggy plant* (to 7 m) in a few collections. Leaves (with *odd round teeth*) smaller than Medlar-thorn's; flowers in heads of up to 12; fruits rather medlar-like but *clustered*. 'Jules d'Asnières' (from a second branch of the Bronvaux tree) tends towards hawthorn, with short, dull, lobed leaves and brown 12 mm haws (cf. Cutleaf Crab, p.308).

Medlar
Mespilus germanica

SE Europe to Iran; long grown in the UK and (rarely) well naturalized in SE England; occasional in old gardens. Grafted often on hawthorn and sometimes on pear.
Appearance Shape Low and *tangled*; spiny. (Old fruiting cultivars – 'Nottingham', 'Royal', 'Dutch' – are thornless.) **Bark** Grey-brown; long, scaly oblong plates. **Shoots** Densely white-hairy at first; prominent pale grey lenticels. **Leaves** To 15 cm, on *5 mm stalks, untoothed*; dull and wrinkly; dense hairs beneath. **Flowers** White, to 6 cm. **Fruit** In Britain the 5 cm medlars are edible – half pear, half date – only when they have gone 'sleepy' in late autumn ('bletted').
Compare Himalayan Tree Cotoneaster (p.282).

Quince
Cydonia oblonga

(*Pyrus cydonia*) Long cultivated and rarely naturalized in the UK (probably W Asian). Now very occasional (but a rootstock for many pears).
Appearance Shape Low; often bushy. **Bark** Grey, *smooth;* then with some big, thin, rufous scales. **Leaves** Broad, to 10 cm, *untoothed*; woolly beneath when young. **Flowers** White/pink, 5 cm. **Fruit** The quinces (Portuguese *marmelo*, whence marmalade) are overwhelmingly fragrant, but inedible raw.
Compare Big-leaved Storax (p.434).

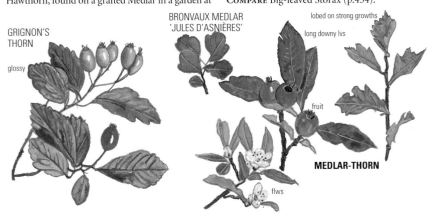

GRIGNON'S THORN — glossy

BRONVAUX MEDLAR 'JULES D'ASNIÈRES'

lobed on strong growths
long downy lvs
fruit
MEDLAR-THORN
flws

HYBRID THORNS TO QUINCE

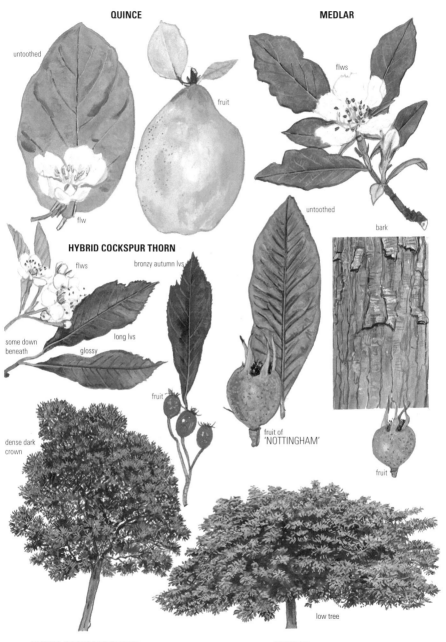

WHITEBEAM & BOLLWYLLER PEAR

The Sorbus *genus comprises 100 trees and shrubs including whitebeams (simple leaves) and rowans (compound leaves).*

Things to Look for: *Sorbus*

- Bark: What is it like?
- Buds: Are they hairy? Sticky?
- Leaves: What shape? Are they hairy beneath?
- Fruit: What colour? How big? How many seeds?

Key Species

Common Whitebeam (below): oval leaves, white-downy beneath. **Alder-leaved Whitebeam** (p.294): oval leaves, soon hairless. **Swedish Whitebeam** (p.298): lobed leaves, grey-downy beneath. **Wild Service** (p.296): lobed leaves, soon almost hairless. **Bastard Service** (p.298): some leaflets at the leaf-base. **Common Rowan** (p.300): fully compound leaves.

Common Whitebeam *Sorbus aria*

S and central Europe, including S England and Co. Galway; Morocco. Locally frequent in the wild on chalk scarps and light sands; abundantly grown everywhere.
Appearance Shape A narrow dome on stiff, steep branches; to 23 m. Wild trees are often multi-stemmed. **Bark** Grey; smooth, then with odd fissures. **Shoots** With white wool quickly shed; *brick-red in sun, grey-green in shade*. **Buds** Conic, to 15 mm; *brown and green* down-tipped scales. **Leaves** To 12 cm, irregularly toothed/slightly round-lobed; 8–13 pairs of *close-set* parallel veins; stalks *slender*. They unfold erect and silvery, like magnolia blooms; the hairs above are soon shed but the under-side stays startlingly white-woolly. **Flowers** In 7 cm-wide heads. **Fruit** Dull red fruit with 2 seeds, which birds soon pillage.
Compare Other whitebeams (p.294); *S. mougeotii* (p.298); *S. croceocarpa* (p.296); Pillar Apple (p.308).
Variants 'Lutescens' (1892) is the most planted clone; erect then neatly narrow-domed to 14 m. The leaves (broad but not long) retain some *pale mealy down above* until late summer, giving the crown a steel-grey cast; berries grey-woolly until half-ripe.
'Majestica' ('Decaisneana') is a group of frequent clones, to 20 m: long/broad leaves (to 15 cm), *dark and glossy above*, with up to 15 close vein-pairs (cf. 'Wilfrid Fox' and Mitchell's Whitebeam, p.294).
Golden Whitebeam, 'Chrysophylla', is rare: slender leaves pale yellow above, fading to olive.
Other trees Cliff Whitebeam, *S. rupicola*, is the least rare of several British and Irish microspecies: limestone cliffs in Scotland, Wales, Ireland, the Pennines and Devon (also in Scandinavia and Estonia). Leaves *small (3–7 cm)*, the lower third narrowly tapered and untoothed, with very small, sharp lobes only round the *almost fan-shaped* top; fruits broader than long. Subtly distinct are *S. eminens* (Wye Valley and Avon Gorge), *S. hibernica* (central Ireland), *S. lancastriensis* (S Cumbria), *S. porrigentiformis* (N Devon, the Mendips, S Wales), *S. vexans* (N Devon) and *S. wilmottiana* (Avon Gorge). All are bushy in their typical habitats, and confined as planted trees to a few botanic gardens. *S. graeca* (Sicily E to Iraq; N Africa) covers a complex of similarly small-leaved trees; *S. umbellata* (Balkans to Crimea) covers another group with particularly squat, fan-shaped leaves, intensely white beneath and with only 5–8 vein-pairs.

Bollwyller Pear *× Sorbopyrus auricularis*

The hybrid of Common Whitebeam with Common Pear was found at Bollwyller (the Alsace) in 1619 and has been propagated by grafting, but is now confined to a few collections.
Appearance Shape Rather stiffly oval; untidy with big apple-like spur-shoots. Out of fruit it is most like Orchard Apple (p.306). **Bark** Roughly *square-cracked*. **Leaves** Narrower than apple's (to 9 × 5 cm) and *sometimes heart-shaped* at the base; closely grey-woolly beneath. **Flowers** Like apple-blossom: white, from cerise buds. **Fruit** Sweet, red, pear-shaped; to 3 cm. The type has produced a few seedlings over the centuries, including 'Malifolia', with shorter, broader leaves and larger (5 cm) yellow fruit.

BOLLWYLLER PEAR

flws in heads

shiny dark green lvs

pear-shaped fruit

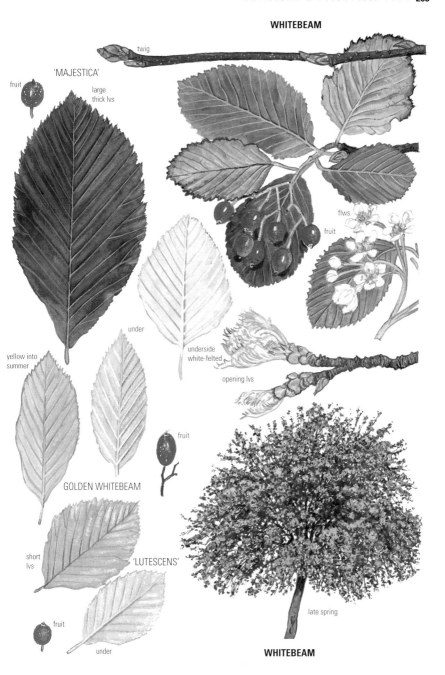

WHITEBEAMS (ASIATIC)

Himalayan Whitebeam — *Sorbus vestita*

(*S. cuspidata*) Himalayas. 1820. In many big gardens (some may be hybrids).
Appearance Shape To 20 m; intensely steely-grey. **Bark** Slightly rougher than Common Whitebeam's: papery, grey-purple ridges. **Shoots** Thick, woolly-white. **Buds** Pink/green, blunt. **Leaves** Long (to 20 cm; cf. Mitchell's Whitebeam, below) and variably broad, stout-stalked and thick, yet with *only 6–12 distant vein-pairs* (cf. *S. croceocarpa*, p.296), the white felt beneath *plastered flat* as if with a tiny iron. **Flowers** Hawthorn-scented. **Fruit** *20 mm*, russet, with 4–5 seeds and ripening in *early winter*.
Other trees *S.* 'Wilfrid Fox' is probably a hybrid (*c.*1920) with Common Whitebeam, named after the creator of the Winkworth Arboretum in Surrey and very occasional. Stiffly slender at first; to 14 m. Shoots stout, soon dark brown; leaves oblong, inheriting Himalayan Whitebeam's *stout stalks/midribs* and brilliant felt, with 12–15 pairs of *slightly incurving* main veins. (The *closer* veins of the Common Whitebeam 'Majestica', p.292, tend to curve outwards.) Fruit *to 20 mm, dull orange-brown*.
S. hedlundii, in some collections, has *12–17* pairs of veins with (on adult trees) *orange* hairs beneath them (in contrast to the white felt).
S. pallescens (China 1908; in some collections, to 20 m) has pointed, narrowly oblong leaves, to 12 × 5 cm (cf. Folgner's whitebeam), grey-green underneath by autumn *as the white felt wears thin*; veins in 10–13 pairs; fruits with 2–5 seeds. The bark may be finely scaly.

Mitchell's Whitebeam — *Sorbus thibetica* 'John Mitchell'

(*Sorbus* 'Mitchellii') A now-frequent clone of a very rare Himalayan/W Chinese whitebeam. Named after a curator of Westonbirt Arboretum, Gloucestershire, where the 1938 original reached 17 m.
Appearance Shape Sturdily broad-domed, or shapeless. **Bark** Purple-grey; shallow, scaly ridges develop. **Shoots** Woolly-white for a year. **Leaves** *Huge and broad*, to 20 × 17 cm; fluffily but brilliantly white-felted beneath and blackish-green above; stalk/midrib *3 mm thick*. Minor veins make a *lattice between the rather distant main veins*. In winter, still white-backed, the leaves litter the ground like paper plates from an old picnic. **Fruit** With 2–3 seeds (cf. 4–5 in Himalayan Whitebeam and *S. hedlundii*, above).

Folgner's Whitebeam — *Sorbus folgneri*

Central and W China. 1901. A beautiful tree in some big gardens, with brilliant late-autumn oranges and crimsons.
Appearance Shape Graceful, to 18 m; arching or weeping slender branches. **Leaves** Only to 9 × 4 cm, usually *narrowly tapered to each end* (cf. *S. pallescens*, above); blackish-green and *wrinkled* on top, and intensely silver-felted beneath. **Fruit** 1 cm, reddish (yellow in 'Lemon Drop'); sepals shed.

Alder-leaved Whitebeam — *Sorbus alnifolia*

E Asia. 1892. A rare but very elegant, puzzling tree.
Appearance Shape Lightly domed, to 17 m. **Bark** Grey, smooth. **Shoots** Slender, white-woolly at first. **Buds** 6 mm, copper-brown. **Leaves** Small (4–8 cm), thin and crisp, with impressed, closely parallel veins; *oblong, finely and sharply doubly toothed* or slightly lobed; *soon hairless beneath*. Scarlet and orange in autumn. **Flowers** Showy among the fresh young leaves. **Fruit** In airy crimson/cerise heads; sepals shed.
Compare Deciduous Photinia (p.282); Snowy Mespil (p.320); Callery Pear (p.318).

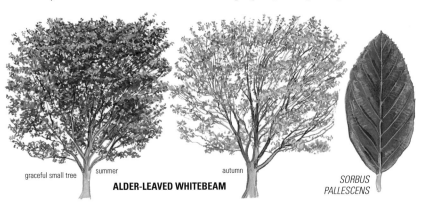

graceful small tree — summer — autumn
ALDER-LEAVED WHITEBEAM
SORBUS PALLESCENS

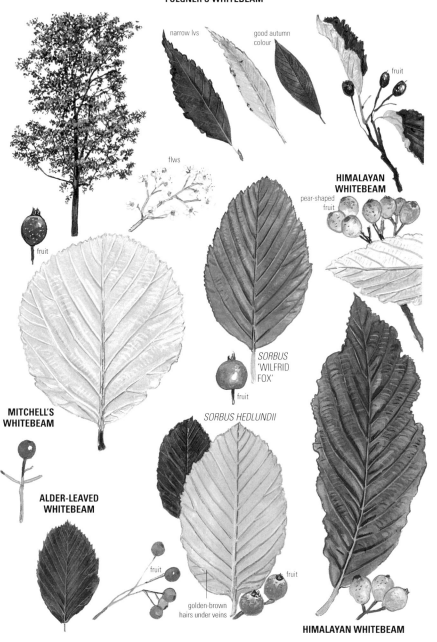

SERVICE TREES

Wild Service *Sorbus torminalis*

(Chequers Tree) S and central Europe, including England N to Cumbria; Algeria; the Caucasus; Syria. Locally frequent on heavy soils, reproducing in Britain from suckers much more often than from seed (summers are often too cool for seeds to ripen; plants raised from seed can be S. × *vagensis*, below); recently more widely planted.
APPEARANCE Shape A large tree (to 28 m); straight twigs on heavy, twisting limbs. **Bark** Grey, then black-brown with close, scaly oblong plates. **Shoots** Shiny grey-brown, soon hairless. **Buds** *Green, round, like peas*. **Leaves** Resemble only Italian Crab's (p.312) but are larger and more sharply lobed; shiny beneath, with hairs soon confined to the veins. Sometimes taken for a maple – but leaves alternate, the main veins not radiating from the base. Often exotically bright red autumn colours. **Flowers** In showy creamy heads in late spring. **Fruit** The brown, 1 cm berries taste of dates when over-ripe, and used locally to be brewed into an alcoholic drink (whence, it is assumed, 'The Chequers' as a pub name).
COMPARE Scarlet Thorn (p.288); Turkish Hazel (p.198); Cut-leaved Alder (p.190). Maple-leaved Crab (p.312): similar jizz. In winter very like True Service (p.298). Wild Pear (p.316): twiggier; bark more closely square-cracked.

Service Tree of Fontainebleau
Sorbus latifolia

Probably originating as a hybrid of Wild Service, and confined in the wild to the Ile de France, but the most widely planted in the UK of this group of microspecies. Very occasional.

APPEARANCE Shape Broad, on sturdy horizontal branches; to 20 m. **Bark** Purple-grey; wide *papery-scaling* ridges. **Leaves** Glossy, with a *dull green-grey felt* beneath; broadly tapering at the base and *almost as broad as long*; small but consistent *triangular lobes tip the distant main veins*. **Fruit** Dull red-brown, 12 mm.
COMPARE Swedish Whitebeam (p.298): rounded lobes. Common Whitebeam (p.292): leaves less lobed, bright white beneath; smoother grey bark, and closer main veins.
OTHER TREES Several comparable wild microspecies (planted only in a few botanic gardens) reach tree size in Britain and Ireland:
French Hales, *S. devoniensis*, locally frequent in Devon, E Cornwall and SE Ireland, has *rather narrower* and smaller leaves (9 × 6 cm).
Bristol Service, *S. bristoliensis* (Leigh Woods and Clifton Down near Bristol), is small-growing: leaves only 5 × 9 cm, whiter beneath, with *narrow-tapered bases* and veins crowded towards the rounded tip; fruit *bright orange*. Bigger orange-berried trees with narrow but longer, dully felted leaves, in a few town parks, are probably *S. decipiens* – variously understood as a Continental hybrid of Common Whitebeam and Wild Service, or a German microspecies (Burgberg). Exmoor Service, *S. subcuneata* (cliffs on Exmoor), shares narrow leaves tapered at the base, but has browner berries.
S. × *vagensis* is a similar, infertile, broad-leaved hybrid of Common Whitebeam and Wild Service – wild in the Wye Valley and the Bleal in Kent.
S. croceocarpa (*S.* 'Theophrasta') is a rare park tree (naturalized on Anglesey); broad leaves *scarcely* lobed (rougher bark, darker crown, rounder leaves with fewer main veins than Common Whitebeam); fruit most like Service Tree of Fontainebleau's.

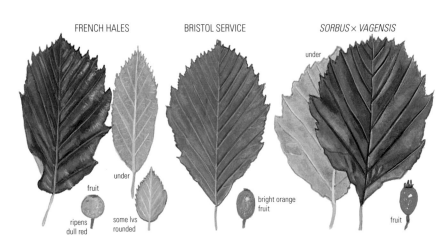

FRENCH HALES — under — fruit — ripens dull red — some lvs rounded

BRISTOL SERVICE — bright orange fruit

SORBUS × VAGENSIS — under — fruit

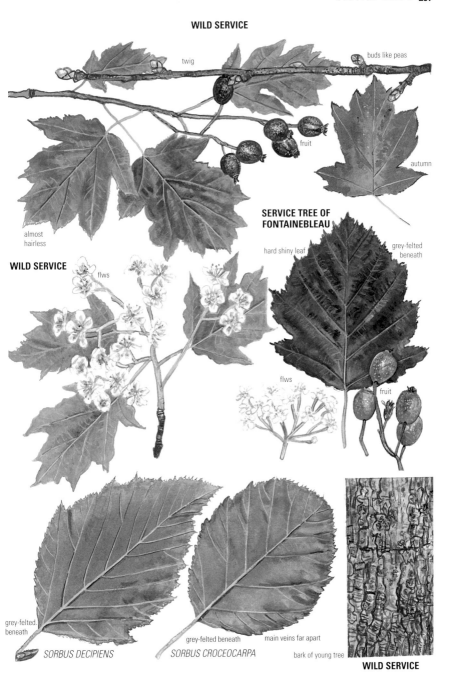

SERVICE TREES

Swedish Whitebeam — *Sorbus intermedia*

(*S. scandica*) Baltic region; long grown in the UK and sometimes seeding. Abundant: a singularly tough street tree.
Appearance Shape A lollipop dome of spreading, twisting branches; to 20 m. **Bark** Grey, remaining largely *smooth*. **Leaves** Sessile Oak-shaped (but the round lobes are serrated, with *very narrow sinuses*), grey-woolly beneath; 6–9 vein-pairs. **Flowers** In particularly showy creamy-white heads. **Fruit** Scarlet berries soon falling or eaten.
Compare Service Tree of Fontainebleau (p.296): leaves with triangular lobes. Common Whitebeam (p.292): leaves, rarely more shallowly lobed, whiter beneath.
Other trees *S. mougeotii* (Pyrenees to Austria) grows in a few collections: *more numerous small, close, rounded lobes*, often *untoothed on inner side*; *10–12* vein-pairs. *S. anglica*, from limestone areas of SW England, Wales, Co. Kerry, is a similar but shrubby microspecies (specimens in some botanic gardens have reached tree size). *S. austriaca* (Austria to the Balkans) has leaves whiter beneath. Arran Whitebeam, *S. arranensis* (N Arran; bushy), has leaves with much deeper basal lobes.

Bastard Service — *Sorbus thuringiaca* 'Fastigiata'

Wild (infertile) hybrids of Common Rowan and Common Whitebeam occur rarely; this erect variant (grown since the late 18th century) is now quite a frequent street tree.
Appearance Shape A neat *balloon of close, erect branches*; to 18 m. **Bark** Grey, smooth. **Leaves** Grey-woolly beneath, with *2–4 free leaflets* on stronger shoots (to 14 leaflets in the rare clone 'Decurrens').
Fruit Red, in heads of 10–15.
Other trees Finnish Whitebeam, *S. hybrida* (*S. fennica*; a group of stable hybrids of Common Rowan and Cliff Whitebeam), from S Scandinavia, makes *wide-spreading* trees in some gardens/streets, with shorter leaves. Arran Service, *S. pseudofennica*, is a native microspecies of this group, restricted to the N of the island. *S. meinichii* (W Norway) has 4–6 free leaflet-pairs.

True Service — *Sorbus domestica*

(*Cormus domestica*) S Europe to the Caucasus; N Africa. The True Service's native status in Britain was disputed until the discovery of wizened trees perhaps 1000 years old on cliffs near Cardiff in 1984; one mysterious individual in Wyre Forest, ancient by 1678, had finally been burnt down by a tramp 184 years later. As a planted park tree it remains very occasional.
Appearance Shape To 20 m; may be massively but gracefully domed. Dark yellowish crown of *drooping* leaves. **Bark** Black-brown; crisply *criss-cross, square-cracking ridges*. **Shoots, Buds** As Wild Service (p.296). **Leaves** As Common Rowan (p.300) but slightly bigger; softly hairy beneath. **Flowers** Cream; big heads in late spring. **Fruit** *To 3 cm*, green-brown; apple-shaped in some trees (f. *pomifera*), pear-shaped in others (f. *pyrifera*); edible when over-ripe. The variants are equally common and produce seedlings of either form.
Compare Black Walnut (p.178): dark bark but much larger, longer-pointed leaflets. Wild Service: very similar in winter (bark usually slightly paler and scalier). Wild Pear (p.316): a stockier, twiggy tree in winter.

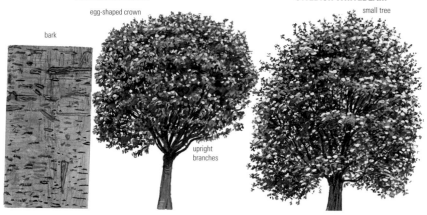

BASTARD SERVICE
egg-shaped crown
bark
upright branches

SWEDISH WHITEBEAM
small tree

ROWANS

Common Rowan — *Sorbus aucuparia*

(Mountain Ash – a name shared by a Eucalypt, p.422.) Europe, including Britain and Ireland; N Africa; Asia Minor. Abundant on light, acid soils, especially in Highland Scotland. Planted everywhere in streets, parks and gardens.
APPEARANCE Shape Roughly and openly spire-shaped at first; branches always cleanly ascending. Exceptionally to 25 m. **Bark** Smooth and silvery-grey, horizontally streaked with dark lenticels; scarcely rugged in age. **Shoots** Purplish grey with hairs soon shed. **Buds** To 15 mm: *purple, non-sticky scales are edged with long grey hairs* – like the conic abdomen of some big spider. **Leaves** Typically with 15 rather *rectangular, often yellowish* leaflets each 5 cm long and serrated to within 1–2 cm of base; dense hairs beneath are normally shed during summer. Autumn colour orange – best in north. Saplings can have deeply toothed/lobed leaflets (cf. Cut-leaved Rowan, below). **Flowers** Big, flat, creamy-white heads in late spring. **Fruit** Scarlet 1 cm berries soon eaten by birds.
COMPARE Other rowans: a highly ornamental group with berries of many colours and with a gamut of additional Asiatic species in big gardens. Most Common Rowans' rather oblong, dark-yellowish, matt leaves are distinctive. True Service (p.298): closest in foliage, but a large, craggy-barked tree. Others with alternate, serrated, compound leaves (eg. Black Walnut, p.178) have bigger foliage. False Acacia (p.354): untoothed leaflets.
VARIANTS 'Sheerwater Seedling' is the most frequent of several *narrowly upright* selections – a shapely, conic, straight-trunked, very floriferous tree, with rather deeply toothed leaves.

'Beissneri' (1899) is quite frequent: a small, upright plant with yellowish green leaflets sometimes jaggedly lobed (cf. Cut-leaved Rowan) and remarkable bark, *dull pinkish-orange* and waxy, shining when wet but fading as algae colonize it. Autumn colour yellow; berries sweeter (cf. Edible Rowan).
Cut-leaved Rowan, 'Aspleniifolia', is rare: a tree with normal bark and habit but leaflets more downy beneath, and (even when adult) variously and *jaggedly lobed*/double-toothed (cf. 'Chinese Lace', p.302).
Golden Rowan, 'Dirkenii', is upright; *bright yellow foliage* fades to soft green through summer. Rare.
Golden-fruited Rowan, 'Fructu Luteo' ('Xanthocarpa'; 1893), is rather occasional and has *golden-orange* fruits (cf. 'Joseph Rock', p.304).
Edible Rowan, 'Edulis', a selection from central Europe (occasional?), has widely spaced narrow leaflets, almost hairless beneath and toothed only on the upper half (cf. Hubei Rowan, p.304). The scarlet fruits are unusually large and *edible* (sharp but not bitter). Another rare cultivar with palatable fruits (large and dark red) is 'Rossica Major', whose broad leaflets (on purplish main stalks) remain particularly downy beneath.

American Rowan — *Sorbus americana*

E North America. 1782. Very rare – large gardens.
APPEARANCE Shape As Common Rowan (above); to 10 m. Differences as follows: **Buds** Redder, the scales slightly *sticky* (cf. Japanese Rowan and Sargent's Rowan, p.302) and hairless except at their tips. **Leaves** Leaflets soon hairless beneath; often slightly larger and more slender-pointed. **Fruit** A richer red.

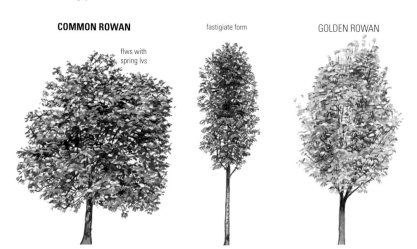

COMMON ROWAN flws with spring lvs

fastigiate form

GOLDEN ROWAN

ROWANS (ASIATIC)

Japanese Rowan — *Sorbus commixta*

Korea, Japan, Sakhalin. *c.*1890. Rather occasional. (Sometimes labelled 'S. *matsumarana*', 'S. *discolor*' or 'S. *serotina*'.)
APPEARANCE Habit, bark and aspect much as Common Rowan (p.300). **Buds** *Crimson* (or green and red); variably *sticky* (cf. American Rowan, p.300, and Sargent's Rowan). **Leaves** Rather deep glossy green; often held densely above the horizontal. Leaflets *longer-pointed* than Common Rowan's; hairless except sometimes for brownish down under the midrib (extensive in var. *rufo-ferruginea*; cf. Ghose's Rowan); each to 7 cm. *Purple* and crimson autumn colours. **Flowers** A week after Common Rowan's; sweet-scented. **Fruit** Bright red, in great heads.
VARIANTS 'Embley' is a (slightly commoner) selection: smaller, *narrower, close-set*, 5 cm leaflets (cf. 'Joseph Rock', p.304); *scarlet* then deep crimson in autumn.

Kew Hybrid Rowan — *Sorbus* × *kewensis*

Rare grafted trees (to 17 m) are probably hybrids of Common Rowan with the NW Chinese *S. pohuashanensis* and/or *S. essertauana* (the nomenclature is confused).
APPEARANCE **Shape** As Common Rowan. **Leaves** Leaflets slightly larger, often bluish green; woollier beneath, on woollier mauve common stalks and shoots; stipules (miniature fan-shaped leaves, cf. Sargent's Rowan, and Ladder-leaf Rowan, p.304) persist at the base of *many more leaf-stalks*/flower-heads. **Fruit** *Bright red*.
VARIANTS 'Chinese Lace' is a small, upright clone whose shoot-tip leaves at least are even more *deeply and jaggedly lobed* than in the Common Rowan 'Aspleniifolia' (p.300); rare.
OTHER TREES *S. essertauana* (including trees labelled '*S. conradinae*') and *S. pohuashanensis* are probably confined to collections: bushier, sparsely branched; larger, dark green, slightly leathery leaflets, *all the upper leaves* (and the flower-heads) *with large, toothed stipules*; huge heads of scarlet berries (yellow in *S. essertauana* 'Flava').

Ghose's Rowan — *Sorbus* 'Ghose'

Probably a hybrid (by 1960) of an unnamed E Himalayan Rowan (KW 7746) with Common Rowan. In some large gardens.
APPEARANCE **Shape** An upright, leafy tree of great panache. **Shoots** Covered in brown/white wool. **Leaves** Leaflets quite large (6 cm), pointed; matt *dark green* and serrated in their upper halves; *brown-white wool persists underneath until autumn*. **Flowers** Fan-shaped, toothed stipules back the flower-heads. **Fruit** Lavish clusters of *pinkish cerise* berries.

Sargent's Rowan — *Sorbus sargentiana*

W Sichuan, China. 1910. Still rather rare.
APPEARANCE **Shape** Often gauntly bushy; to 15 m. **Shoots** Stout, dark brown. **Buds** Big, *crimson and very sticky* (cf. Horse Chestnut's opposite buds p.393). **Leaves** With semicircular stipules at the base; 9–11 *long* (to 13 cm), pointed leaflets, softly downy beneath and each with 20–25 pairs of sunken veins; they flush dark brown and turn brilliant gold and crimson in autumn. **Fruit** Tiny (6 mm), but bright red and in heads of 200–500.
COMPARE Chinese Varnish Tree (p.360): unserrated leaves. Japanese Big-leaved Ash (p.440): opposite leaves.

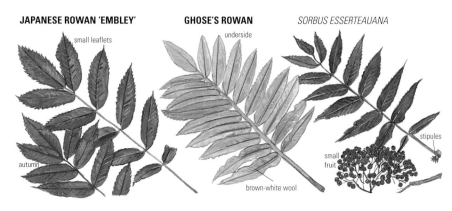

JAPANESE ROWAN 'EMBLEY' — small leaflets, autumn
GHOSE'S ROWAN — underside, brown-white wool
SORBUS ESSERTEAUANA — stipules, small fruit

ROWANS (ASIATIC)

Joseph Rock's Rowan *Sorbus* 'Joseph Rock'

(*S. rockii*; Rock 23657) A clone probably found wild in NW Yunnan (China); now an occasional garden tree.
Appearance Shape Upright, with spreading branches, to 12 m; dainty. **Shoots** Soon hairless. **Leaves** *Small* (10–18 cm): usually with 19 pointed, *closely set* 3–4 cm leaflets, fresh green and rather glossy. **Fruit** *Primrose-yellow* (white on the shaded side), in big heads – gorgeous among the crimson-and-purple autumn foliage.
Compare *S. commixta* 'Embley' (p.302): longer, darker leaflets; red fruit.

Ladder-leaf Rowan *Sorbus scalaris*

W Sichuan (China). 1904. Rare.
Appearance Shape Broad and open. **Shoots** Woolly-white when young. **Leaves** Small (to 20 cm), with semicircular basal stipules and up to *33 very narrow, closely set leaflets* like the rungs of a ladder; a blackish, rather glossy green above and cobwebbed with white down beneath; the downcurved rims are toothed only near the pointed tips. They flush red-brown, and turn gold and red in autumn. **Fruit** Small (6 mm), but in huge bright red clusters.

Vilmorin's Rowan *Sorbus vilmorinii*

(*S. vilmoriniana*) Yunnan (China). 1904. A particularly dainty, ferny-leaved rowan; very occasional.
Appearance Shape Broad, on a trained stem or bushy. **Shoots** With rusty hairs at first. **Leaves** Small (12 cm): up to 29 tiny, round-tipped leaflets, dark green above and greyish beneath, wrinkled and almost hairless; they unfold pale brown and turn deep red in autumn. **Fruit** 1 cm, *deep maroon-pink then ripening to pink-flushed white*.

Compare Kowhai (p.352): evergreen. Hubei and Kashmir Rowans (below), also with white/pink fruit, have far fewer, bigger leaflets.
Other trees *S. prattii*, in collections, is a very small, open tree from W China with *white* berries; *S. koehneana* is similar, but has *hairless* shoots.

Hubei Rowan *Sorbus glabrescens*

Yunnan, China. 1910. Occasional in parks and gardens; generally labelled *S. hupehensis* – an ally (from Hubei) scarcely grown in the UK.
Appearance Shape Strongly though narrowly *domed*; to 17 m. **Bark** Grey-brown, developing *slightly corky ridges*. **Shoots** Sturdy, soon hairless. **Buds** Green then red. **Leaves** With 11–17 rather blunt, *glaucous* leaflets, silvery-grey beneath with tiny waxy warts but almost hairless; finely toothed only in the upper half (cf. Common Rowan 'Edulis', p.300). They unfold bronze, and turn lambent *pink* and red in autumn. **Flowers** In open, creamy-white puffs. **Fruit** *White, flushed pink*.
Other trees *S. oligodonta* (also Yunnanese and often grown as *S. hupehensis* var. *obtusa*; *S. pseudohupehensis*) is equally widespread: narrow, straggling crown of *fine shoots* (rarely to 15 m), *very glaucous* leaves whose top leaflets are largest, and, in the usual clone(s), 'Rufus'/'Rosea', long-lasting *rich magenta berries*.

Kashmir Rowan *Sorbus cashmiriana*

W Himalayas. *c.*1932. Very occasional.
Appearance Shape A low, open plant, to 6 m. **Leaves** Most like Common Rowan's (p.300) but without the yellowish tinge (they turn pale yellow in autumn). **Flowers** In large, pale *pink* heads. **Fruit** *Marble-sized* and *pure white* (tending to bruise brown).

HUBEI ROWAN

VILMORIN'S ROWAN

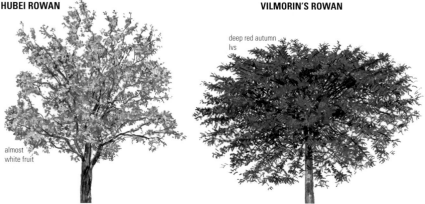

almost white fruit

deep red autumn lvs

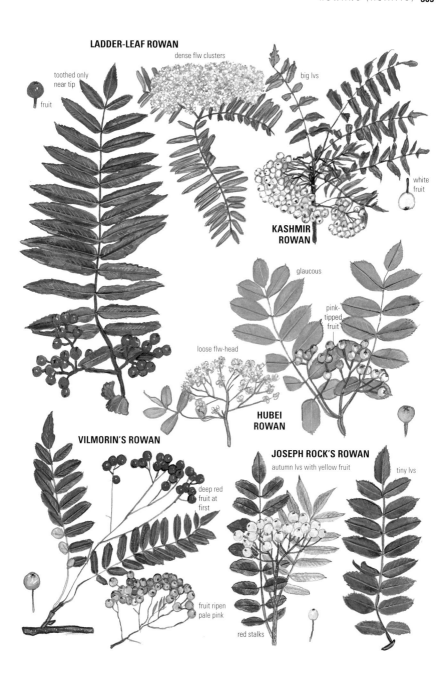

APPLES

Apples (about 30 species and several thousand hybrid cultivars, of which only the commoner ornamental forms are covered here) have berry-sized fruits, or 'apples'. Their sepals (which make the brown star at the bottom of eating apples) may be shed as they mature. Most trees flower together in late spring: Japanese Crab (p.308), Cherry-crab (p.312) and Purple Crab (p.314) are usually first, with Orchard Apple a week later.

Things to Look for: Apples

- Leaves: How broad? Are they hairy beneath?
- Flowers: Are they double? What colour are they?
- Fruit: What shape is it? (rounded? conic? lemon-shaped?) How big is it? Are the sepals retained?

Key Species

Orchard Apple (below): big, oblong leaves and 'apples'. **Japanese Crab** (p.308): slender leaves; 1 cm yellow fruit. **Hubei Crab** (p.310): longer, glossy leaves and scaly bark. **Purple Crab** (p.314): purplish leaves. **Maple-leaved Crab** (p.312): deeply lobed leaves.

Orchard Apple *Malus domestica*

Fruiting apples have been bred over millennia from central Asian stock (not from Wild Crab); abundant in gardens and orchards and as 'wildings' in scrub and by roads or railways.
APPEARANCE Shape Low and wide, with many short, fruiting 'spur' shoots and vigorous, straight extensions; not spiny. **Bark** Shallowly scaly; greys, browns, some purples. **Shoots** Stout, grey; slightly *hairy*. **Buds** More or less *grey-woolly*. **Leaves** Thinly *woolly beneath* (cf. Pillar Apple, p.308) and on the stalk; *oblong*, large (to 12 cm), with small, irregular teeth; dark, rather matt and variably crumpled. **Flowers** From rich pink buds: white with soft pink shading. **Fruit** *At least 4 cm wide*; most are edible.
COMPARE *M.* 'Dartmouth' (p.312); *M. domestica* 'Elise Rathke' (p.314); Wild Crab (below); Bullace (p.340). Many 'wild' apple trees are clearly hybrids and can be tall-domed, to 18 m: '*M. pumila*' is a name sometimes used to cover them. Other crabs have narrower/more pointed, less downy leaves; Plum-leaved and Chinese Crabs (p.310) are most similar. Bollwyller Pear (p.292): leaves sometimes *heart-shaped* at the base.
OTHER TREES Danube Apple, *M. dasyphylla* (SE Europe; scarcely grown in the UK) is a wild species with similarly downy foliage and 4 cm apples.

Wild Crab *Malus sylvestris*

Europe including Britain and Ireland; frequent in old woods and hedges on heavier soils but until recently hardly ever planted.
APPEARANCE Shape Irregular; to 17 m; sometimes spiny when young. **Bark** Purplish brown; closely scaly ridges. **Shoots** *Soon hairless* and glossy green/brown. **Buds** Brown, downy only at their pointed tips. **Leaves** Almost *hairless*, quite glossy; oval, to 6 cm; rather folded; with small, rounded-triangular teeth. **Flowers** White, from pink buds. **Fruit** Crab-apples yellow-green, hard and *very acid*, to 4 cm, dropping in winter and often carpeting the ground.
COMPARE Orchard Apple (above). 'Wild' apples with pinker flowers, larger fruit or leaves downy beneath are hybrids. Snowy Mespil (p.320): leaves flatter and paler. Wild Plum (p.340): leaves more wrinkled. Wild Pear (p.316): similar winter shoots; blacker bark; leaves flatter and glossier.

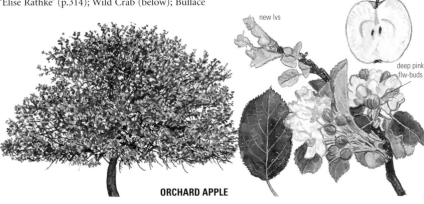

ORCHARD APPLE

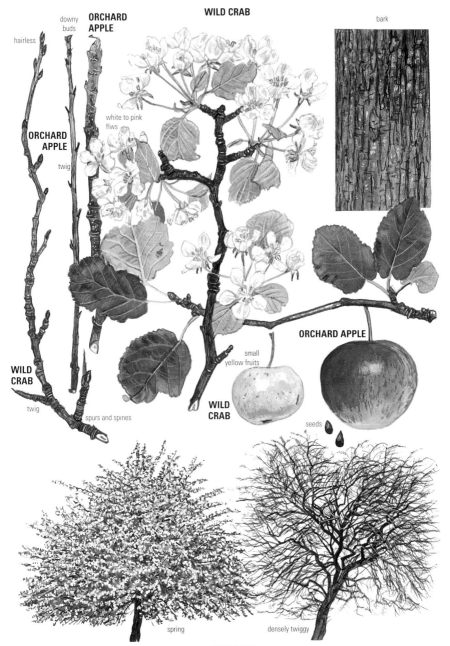

FLOWERING CRABS

Japanese Crab — *Malus floribunda*

Japanese gardens. 1862 (probably a long-cultivated hybrid). Abundant: a most dazzling flowering crab.
APPEARANCE Shape A low, particularly *tangled* dome of zig-zag branches, to 10 m; sparsely leaved by late summer. **Bark** Dull dark brown, *closely fissured* into knobbly oblongs. **Leaves** *Small, quite narrow*, pointed, 7 × 3 cm (cf. Myrobalan Plum, p.340), with the odd big lobe on strong growths (cf. Siebold's Crab); dull dark green above and finely downy beneath; stalk finely hairy. **Flowers** From rich carmine buds, ultimately white: very long-stalked, and so profuse they *totally smother the very early leaf-flush*. **Fruit** 1 cm: dull yellow; sepals shed; seldom conspicuous.
COMPARE Siberian Crab (p.310); Hall's Crab (p.314); Cherry-crab (p.312); *M.* × *scheideckeri* 'Excellenz Thiel' (p.314).

Siebold's Crab — *Malus toringo*

(*M. sieboldii*) Japan. 1856. Rather rare.
APPEARANCE Shape Lower and more spikily tabular than Japanese Crab. **Leaves** More frequently with the *odd massive lobe*. **Flowers** Spectacular, but a fortnight later; narrow, *starry* white/pink petals, from rich pink buds. **Fruit** *Tiny* (5 mm; sepals shed); yellow/brownish red.
OTHER TREES Cutleaf Crab, *M. bhutanica* (*M. toringoides*; W China, the Himalayas; 1904), is a rare taller tree whose lobed leaves (*predominant*; scarce on some mature plants) tend to be *hawthorn-like* (cf. Bronvaux Medlar, p.290). Flowers late, creamy white, less showy; fruit spectacular: bright yellow with scarlet cheeks (10–15 mm; sepals shed).
M. transitoria (NW China, 1911; very rare) has hairier leaves than Cutleaf Crab, *most of them* very deeply and narrowly lobed; fruit similar.

Pillar Apple — *Malus tschonoskii*

(Chonosuki's Crab; *Eriolobus tschonoskii*) Japan. 1897. Rare in the wild but now very frequent in parks and streets, principally for its vibrant autumn *golds and scarlets*; short-lived.
APPEARANCE Shape *Conic on steep branches*, to 17 m. **Bark** Grey; *smooth*. **Leaves** *Large*, to 12 × 8 cm, leathery; *grey-felted beneath*. **Flowers** White, among the *silvery* young leaves – restrained. **Fruit** 25 mm, yellowish with a purple cheek; sepals persistent. Seldom abundant.
COMPARE Bollwyller Pear (p.292); Grey Alder (p.192). Chanticleer Pear (p.318): another common erect street tree with silvery young leaves.
OTHER TREES Yunnan Crab, *M. yunnanensis* (W China, 1908), is rare: *wide* crown; greener leaves with more distinct triangular lobes (cf. Sweet Crab). Flowers showy, strongly azalea-scented; fruits smaller (12 mm), *deep red with white dots*. Pratt's Crab, *M. prattii* (W China, 1904), has dotted fruit like Yunnan Crab's; its similarly large leaves, soon almost *hairless*, may be narrower.

Sweet Crab — *Malus coronaria*

(Garland Tree) E North America. 1724. Very rare.
APPEARANCE Shape Very untidy. **Leaves** Oval to 3- or 5-lobed; *large* (to 11 × 9 cm); soon *hairless*. **Flowers** *In early summer*: big (4 cm), pink-flushed white; violet-scented. **Fruit** To 4 cm, yellow-green.
VARIANTS 'Charlottae' has semi-double, almost rose-like flowers. A plant of poor constitution.
OTHER TREES Prairie Crab, *M. ioensis* (Central USA), has woolly shoots and downy underleaves. *M.* 'Red Tip', a hybrid with a Purple Crab (1919; rare), has *red-brown young leaves*, broad and lobed on strong shoots, and rich pink flowers; fruits *45 mm*, green and red.

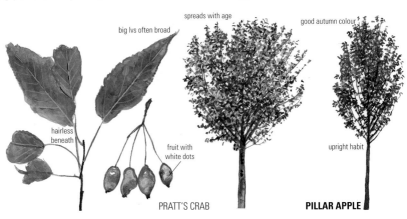

big lvs often broad
hairless beneath
fruit with white dots
PRATT'S CRAB

spreads with age

good autumn colour
upright habit
PILLAR APPLE

FLOWERING CRABS

Hubei Crab — *Malus hupehensis*

(*M. theifera*) Central and W China (with endangered populations in Japan and Taiwan). 1900. Quite frequent: a tree which would be grown for its luxuriant foliage (used as tea in China) even if it did not flower spectacularly.
APPEARANCE Shape A *luxuriant* wide dome; to 17 m. **Bark** *Long, scaling, often spiralling plates*, with *orange/pink tints*. **Leaves** Typically *long* (to 10 × 5 cm), boat-shaped; flushing purplish then usually a *vivid, shiny green*; downy under the main veins; stalk woolly. **Flowers** White (pink in f. *rosea*) from pink buds, a few days after Japanese Crab's, petals *overlapping*; usually 3 styles. **Fruit** 1 cm, *dull* yellow/red; sepals shed.
COMPARE Siberian Crab (below): differences emphasized.

Siberian Crab — *Malus baccata*

NE Asia. By 1784. Occasional: streets and bigger gardens.
APPEARANCE Shape Lower: more twiggy and sprouty than Hubei Crab (above); to 15 m. **Bark** Usually browner and more closely cracking. **Leaves** Generally smaller, more matt; often soon *hairless*. The stalk is hairy only in the eastern var. *mandshurica*, which is common in the UK. (Its rare clone 'Jackii', from Seoul, has leaves as big and glossy as Hubei Crab's, but *blackish-green*.) **Flowers** Pure white from pink buds; petals *not overlapping*; usually 5 styles. **Fruit** *Bright* yellow/red, 1 cm; sepals shed and the stalk-end slightly sunken.
VARIANTS 'Lady Northcliffe', of uncertain parentage, has flowers darker pink in bud and dull, brownish fruits; rare.
COMPARE Hubei Crab; Cherry-crab (p.312).

Plum-leaved Crab — *Malus prunifolia*

An apple of obscure, probably Chinese, origin; occasional in parks and gardens.
APPEARANCE Shape Quite dense and upright, to 10 m; dull, slightly greyish cast (like Snow Pear, p.320). **Bark** Grey-brown, often rather regularly fissured. **Leaves** Rather oblong, 6 cm; downy beneath. **Flowers** White, fragrant; less spectacular and delicate than Siberian Crab's. **Fruit** Profuse, greenish yellow to reddish; *lemon-shaped* (to 3 cm; cf. 'John Downie', p.312); sepals retained on top of the 'nipple'.
COMPARE Chinese Crab (below); Orchard Apple (p.306); Cherry-crab (p.312); *M.* × *zumi* 'Golden Hornet' (p.312).

Chinese Crab — *Malus spectabilis*

A cultivated N Chinese flowering crab, now only occasional in parks and gardens.
APPEARANCE Shape Tall, *open* and untidy; to 15 m. **Bark** Very shallowly scaly: a large, *pale grey, brown and purplish* patchwork. **Leaves** *Broad*-oval, to 8 cm; glossy above and hairy beneath at first. **Flowers** *Very large* (5 cm), with a few extra petals; rich pink from red buds; clustering near shoot-tips amid fresh-green leaves. **Fruit** 2 cm, yellow; sepals retained; never showy.
OTHER TREES Magdeburg Crab, *M.* × *magdeburgensis*, is a low, spreading tree; now rare. Flowers much smaller (25 mm), but clothing the branches in late spring; up to 12 petals, deep pink outside and paler within.
M. 'Van Eseltine' (1930) is a now popular hybrid, *funnel-shaped*, with *fully double* pink flowers.

MAGDEBURG CRAB — spring; small tree
PLUM-LEAVED CRAB — compact flw-clusters; fruit; sepals retained on 'nipple'
CHINESE CRAB — big broad lvs

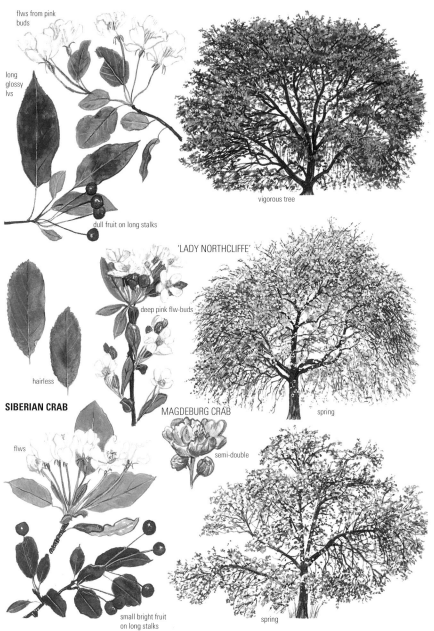

FRUITING CRABS; MAPLE-LEAVED CRAB

Cherry-crab — *Malus × robusta*

A group of hybrids of Siberian and Plum-leaved Crabs, very frequent in small gardens.
APPEARANCE Shape Spiky; frequently bushy. **Leaves** Quite large, coarse, oblong; downy beneath; early flushing. **Flowers** White from red buds (recalling Japanese Crab's (p.308), but duller and less massed). **Fruit** Hanging *until next spring*: cherry-like, *many with a star of sepals, some without*; golden in 'Yellow Siberian', crimson in 'Red Siberian' and the popular 'Red Sentinel'.
COMPARE Siberian Crab (p.310): smaller fruits; sepals always shed. Plum-leaved Crab (p.310): lemon-shaped fruit, always with sepals.
OTHER TREES Some other comparable crabs have spectacular fruit, often lasting into winter:
M. 'John Downie' (by 1891; frequent) has rather narrow leaves (6 × 3 cm), glossy green on a hairy stalk; white flowers from pale pink buds in late spring. Fruits bright orange/red, to 4 cm (cf. 'Neville Copeman', p.314), *tapering*: they look glorious on the tree but put owners in a quandary as they taste delicious and make the best jelly.
M. × *zumi* 'Golden Hornet' (1949; abundant) has perhaps the richest fruit display of all. In summer, *gauntly spiky* and resembling Plum-leaved Crab (p.310); can be vigorous (to 9 m). White flowers from pink buds lead to great *wreaths of yellow, lemon-shaped 20 mm fruits*, with sepals, that rot brown on the tree.
M. 'Winter Gold' (by 1947) has glossy-green leaves, occasionally 3-lobed like Japanese Crab's (p.308). White flowers from pink buds; the very showy yellow, 12 mm, orange-shaped fruits shed their sepals.
M. 'Crittenden' (1921) makes a low tree with twisting branches (cf. 'Red Jade', p.314). Flowers pale pink; fruits 2 cm, rather lemon-shaped, with sepals, very glossy deep red, at least on the sunny side.
M. 'Butterball' (1961) has white flowers from pink buds, and yellow 25 mm fruit, flushed orange.
M. 'Dartmouth', an old American variety, resembles Orchard Apple (p.306), but is grown for its white flowers and big, yellow-and-crimson fruits, bloomed purple, *5 cm wide* (cf. 'Wisley Crab', p.314, which has purple-tinged leaves).

Maple-leaved Crab — *Malus trilobata*

(*Eriolobus trilobatus*) NE Greece to the Holy Land. 1877. Rare in the wild and now very occasional in the UK; a singularly handsome, unmistakable tree.
APPEARANCE Shape Sturdy, upright; to 15 m. **Bark** Pear-like: black-brown with close, rugged, square-cracking ridges. **Shoots** Very downy. **Leaves** Like Field Maple's (p.368), but alternate; shiny blackish green above and downy below; bright red in autumn. **Flowers** In woolly, hawthorn-like heads; **Fruit** Usually 1–3 together; red/yellow, to 2 cm.
COMPARE Wild Service (p.296): similar jizz.
OTHER TREES Italian Crab, *M. florentina* (*Malosorbus florentina*; N Italy to N Greece, *c.*1877) is another rare one-off: a sturdily conic tree to 10 m, its leaves cut like Wild Service's (p.296), but smaller (3–8 cm) and with more rounded lobes (cf. Hungarian Thorn, p.286). The bark – almost Quince-like – sheds big, pale, shallow scales; fruits red, 12 mm; sepals shed.

'DARTMOUTH' CRAB — small flws — large apples

ITALIAN CRAB — flws — autumn — fruit on long stalks

PURPLE & WEEPING CRABS

Hall's Crab — *Malus halliana*

A cultivated Chinese and Japanese crab. 1863. Rather rare.
APPEARANCE Shape Open, twiggy habit, to 9 m. **Shoots** Dark purple, soon hairless. **Leaves** Narrow, *long-pointed* (5 × 2 cm), blackish glossy green; *fine red margin* and often crimson veins create a purplish cast. **Flowers** Rich pink; purple-stalked; up to 8 petals (to 15 in 'Parkmanii'). **Fruit** *Tiny (5 mm), pear-shaped, purple*; sepals shed.
OTHER TREES *M.* × *hartwigii*, the hybrid with Siberian Crab, has similar if greener foliage. Red buds open to blush-white flowers with a few extra petals; 12 mm dark red fruits follow. Its cultivar 'Katherine' (1928; rare; often bushy) has splendid flowers, 55 mm wide, with up to 20 tiered petals, pink in bud then white; tiny (6 mm) bright red-and-yellow fruits. *M.* 'Snowcloud' (a seedling of 'Katherine', by 1978) combines the double flowers (from pink buds) with an upright habit and bronzy young leaves.

Purple Crab — *Malus* × *purpurea*

An abundant garden hybrid (by 1900).
APPEARANCE Shape Very untidy; often almost leafless by late summer. To 10 m. **Bark** Purplish grey, shallowly scaling/cracked. **Shoots** Black-purple. **Leaves** Unfolding glossy purple and fading to dark mauvish grey-green. To 6 × 3 cm; rarely 3-lobed (cf. 'Red Tip', p.308). **Flowers** Smouldering red-purple – quite dazzling in late sun – then mauve. **Fruit** Purple-red, round, 2 cm; sepals retained.
VARIANTS 'Lemoinei' (1922; probably now the commonest clone) has a denser crown, richer purplish green in summer, leaves to 10 cm (seldom lobed), and larger (5 cm) flowers with odd extra petals; fruits small, very dark purple. 'Aldenhamensis' (1912) is upright and open, flowering later (with some extra petals); only the newest leaves are maroon in summer. Bark paler, shallowly scaly; fruits purple, *tangerine-shaped*; sepals shed. 'Eleyi' (by 1920; rare) has dark crimson flowers and *conic* purple 25 mm fruits; sepals often shed. A *much leafier* reddish green tree. 'Neville Copeman' (1952; occasional) is a poor, sparse plant, but its *round 4 cm brilliant orange-crimson fruits* (sepals retained) are briefly spectacular (cf. 'John Downie', p.312).
OTHER TREES *M.* 'Wisley Crab' (1924) has *large* (6 cm) purple-red fruits, with sepals.
M. 'Royalty' (1958), open and sprawling, keeps the *darkest foliage* (among which its purple flowers are rather lost); bark with coarse vertical ridges; fruits 20 mm, rich purple, sepals retained.
M. × *moerlandsii* 'Profusion' (Holland, by 1938; 'Lemoinei' × Siebold's Crab) has crimson flowers *soon fading to soft mauve* among purplish leaves which mature a shiny, healthy, rather reddish green. Its *bright red* fruits are *small* and round (15 mm), their sepals shed. 'Liset' (a darker-flowered sister tree) is scarcer.

Weeping Purple Crab — *Malus* × *gloriosa* 'Oekonomierat Echtermeyer'

A garden hybrid (1914), frequent.
APPEARANCE Shape Shoots *hang to the ground* from low zig-zag limbs; the purplish young leaves fade to grey-green. **Flowers** Purplish mauve. **Fruit** Purple-red, 25 mm.
VARIANTS *M.* 'Royal Beauty' (1980) is a healthier, leafier improvement.
OTHER TREES Other weeping crabs include *M.* × *scheideckeri* 'Excellenz Thiel' (1909), effectively an untidily igloo-shaped Japanese Crab (p.308); the vigorous *M.* × *scheideckeri* 'Red Jade' (1935) with pink-flushed cup-shaped white flowers and cherry-sized *brilliant red* crabs (sepals shed); and *M. domestica* 'Elise Rathke' (1886), wide-spreading, with large, yellow, *edible apples*.

MALUS × HARTWIGII

PURPLE CRAB

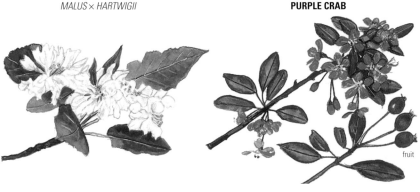

PEARS

Pears (about 30 species) have gritty fruits.

Things to Look for: Pears

- Bark: What is it like?
- Leaves: What shape are they? Are they downy beneath? Are they toothed?
- Fruit: How big is it? Are the sepals retained?

Common Pear
Pyrus communis

Fruiting pears (var. *culta*; many hundreds of clones) have been bred over millennia in Europe/W Asia from the wild species. Abundant in parks, gardens and, locally, old orchards.
APPEARANCE Shape Often gaunt and spiky; dense with fruiting spurs; very upright, strong shoots. Sometimes massively domed with age: clones like 'Pitmaston Duchess' reach 20 m and live for 300 years. **Bark** Black-brown, split – often as if by a knife – into *small knobbly oblongs*. **Shoots** Shiny brown, sometimes downy; rarely spiny. **Buds** Brown, pointed. **Leaves** Variable – often *heart-shaped* at the base. Rounded, narrowly oblong or slightly triangular, 3–8 cm long; *minutely round-toothed* (rarely untoothed). More leathery than apple leaves and glossier *blackish green* above; often hairless beneath but sometimes woolly (especially at first). **Flowers** Creamy-white heads overwhelm the tree a fortnight before apple-blossom. **Fruit** Pear-shaped in varying sizes, edible; sepals persistent.
COMPARE Wild Pear (below); Wild Crab (p.306); Broad-leaved Cockspur Thorn (p.288). Himalayan and Callery Pears (p.318): berry-like fruit. Sand Pear (p.318): whisker-toothed leaves. Almond-leaved Pear (p.318): narrow leaves. Snow Pear (p.320): silver-woolly.
VARIANTS 'Beech Hill' is used as a younger street tree. *Strong, vertical shoots and younger branches make a gauntly funnel-shaped but ultimately broad crown (to 14 m)* – much less shapely than the commoner Chanticleer Pear (p.318).
OTHER TREES Wild Pear, *P. pyraster* (Choke Pear; *P. communis* var. *pyraster*), is less distinct from Common Pears than Orchard Apples are from Wild Crab, but Neolithic pear charcoal has been found in England, and (wild) pear trees feature often in Anglo-Saxon charters. Very rare plants, in old hedgerows and ancient woods, have tall, domed crowns, spiny young shoots, small leaves (4 cm long), and small rounded fruits (to 4 cm; sepals retained), yellow-green and *hard as wood* even when dropping.
Austrian Pear, *P. austriaca* (central Europe), has leaves woolly beneath.
Poirier Sauger, *P. salvifolia* (continental Europe; rarely planted in the UK), may be a hybrid with Snow Pear (p.320): slender, untoothed leaves, woolly beneath and at first above.

Plymouth Pear
Pyrus cordata

(*P. communis* var. *cordata*) W France, Iberia and a few hedges around Plymouth and Truro – one of England's rarest wild trees (young plants are in some botanic gardens).
APPEARANCE Shape Shrubby and spiny. **Bark** Like Common Pear's. **Leaves** Small (4 cm), neat, rounded; seldom heart-shaped at the base. **Flowers** Among fresh-green young leaves; crimson stamens give a *pinkish* cast. **Fruit** *Like marbles*; sepals soon *shed*; edible when over-ripe.
Compare Himalayan Pear (p.318).
OTHER TREES *P. magyarica* (Hungary) and *P. rossica* (central European Russia) are similar local species.

PLYMOUTH PEAR

small hard fruit

small lvs

sturdy tree WILD PEAR

small rounded lvs

tiny round fruit

sepals are soon shed

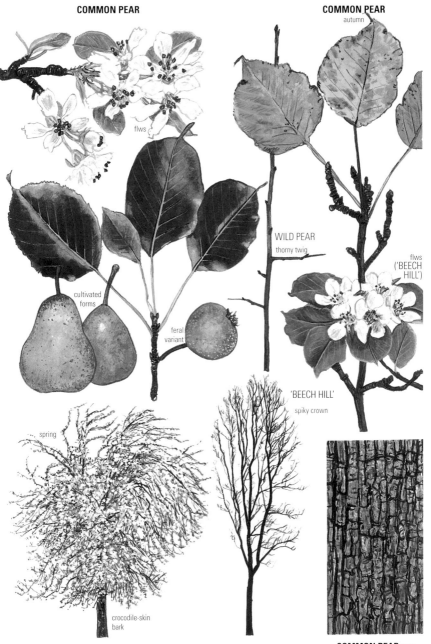

PEARS

Himalayan Pear *Pyrus pashia*

Afghanistan to W China. 1908. A very rare but handsome tree.
APPEARANCE **Shape** Tall (to 20 m); rather *open*. **Bark** Like Common Pear's (p.318). **Leaves** Quite small and often narrower than Common Pear's; can be 3-lobed on vigorous (sometimes spiny) shoots. **Flowers** In very dense heads, with rounded petals; crimson stamens give a *pale pink* cast. **Fruit** Like *Plymouth Pear's* (p.316).

Callery Pear *Pyrus calleryana*

Central and S China. 1908. A handsome pear with brilliant very *late* autumn colour; confined to collections.
APPEARANCE **Shape** Rather upright; to 15 m. **Bark** *Pale* grey-brown; rugged, largely vertical ridges. **Buds** Pinkish, woolly. **Leaves** *Brilliant green* and hairless when mature, finely toothed; sharply pointed; to 7 cm. **Flowers** In *very early spring* among silvery unfolding leaves. **Fruit** Tiny: like Plymouth Pear's (p.316).
VARIANTS Chanticleer Pear, 'Chanticleer', a North American selection, became one of the most deservedly popular street and park trees of the late 20th century: *neatly spire-shaped* (egg-shaped in age or if the trunk forks), with short spreading twigs; to 15 m so far (cf. Pillar Apple, p.308).
'Bradford', an older selection (1914), is rare: steep branches make a rougher, rounded/funnel shape.

Sand Pear *Pyrus pyrifolia*

W and central China. 1909. Much bred for eating pears in China, but rare in Europe.
APPEARANCE **Shape** Domed. **Bark** Grey-brown; rugged, *scaly, vertical ridges*. **Leaves** Large (to 10 cm), dark and glossy; rather *triangular* and beautifully *fringed with fine whisker-tipped teeth*; hairy beneath at first. Brilliant red autumn colours. **Fruit** In the wild species, brown, rounded and hard, 3 cm; sepals shed as they ripen.
OTHER TREES Kieffer Pear, *P.* × *lecontei*, is the hybrid with Common Pear; rare. Leaves *minutely* whisker-toothed and tending to the triangular; pears large and tasty.
P. ussuriensis (NE Asia, 1855; in some collections) differs from Sand Pear in greater vigour (to 16 m), black-brown often square-cracked bark, nearly hairless often *oblong* leaves, prolific flowers *pink* in bud and pears which *retain* their sepals.

Almond-leaved Pear *Pyrus amygdaliformis*

Italy to Bulgaria. 1810. Rare, but in some parks.
APPEARANCE **Shape** Peculiarly spiky; broad, often slightly weeping. A dark, greyish-olive tree, to 12 m, sometimes with spines. **Bark** Dark grey-brown, ruggedly cracked. **Leaves** *Oblong, variably narrow* (40 × 15 to 70 × 20 mm): silky-hairy when new almost hairless by midsummer, shiny above; minutely and *distantly round-toothed*. **Fruit** Flattened, 25 mm, brownish yellow; sepals persistent.
OTHER TREES Michaux's Pear, *P.* × *michauxii*, in some gardens, is probably a Levantine hybrid with Snow Pear (p.320). Leaves *untoothed*, and retaining silvery hairs beneath: a darker, more irregular tree than Snow Pear.
P. bourgaeana (Iberia; Morocco) has regularly round-toothed leaves; *P. caucasica* (E Greece to the Caucasus) has untoothed, long-tipped leaves. These are confined to a few collections in the UK.

CALLERY PEAR 'BRADFORD'

ALMOND-LEAVED PEAR

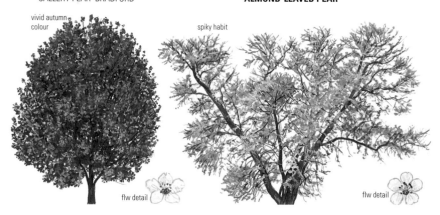

vivid autumn colour

spiky habit

flw detail

flw detail

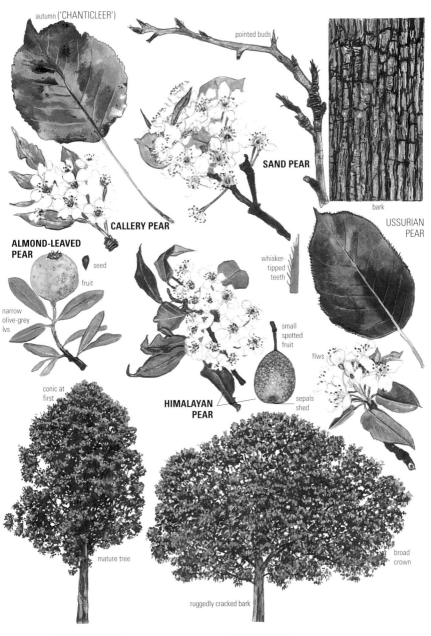

SILVER PEARS & SNOWY MESPIL

Willow-leaved Pear *Pyrus salicifolia*

Caucasus to N Iran. 1780. Very frequent: intensely silvery in spring, fading to iron-grey.
APPEARANCE Shape A low, tangled dome, to 13 m, the thin outer branches *weeping* except in shade. (Most grafts in gardens are probably the cultivar 'Pendula'.) **Bark** Black; ruggedly fissured. **Shoots, Buds** Silky-white. **Leaves** *Untoothed; silvery-hairy*, the upper side slowly turning shiny dark green; 4 × 1 cm to 9 × 2 cm. **Flowers** White, in well-spaced heads among the silvery young leaves. **Fruit** Pear-shaped, 3 cm; sepals persistent.
COMPARE Oleaster (p.412): silver *scales* under the leaves. Common Osier (p.170).

Snow Pear *Pyrus nivalis*

Italy to Romania (orchards in France). 1800. Very occasional: streets, parks.
APPEARANCE Shape Spikily branched but often lollipop-shaped; to 12 m. **Bark** Dark grey-brown; rather square-cracked. **Leaves** Unfolding silvery-hairy; the hairs fall quickly from the upper side and *slowly thin below* so that the crown becomes dark and steely and finally dull olive-grey. To 8 × 4 cm, narrowly rounded and more or less *untoothed*. **Flowers** In dense, white, showy heads among the grey young foliage. **Fruit** Yellow-green, 4 cm, with persistent sepals; round and hard – used for perry-cider and, sometimes in town parks, as projectiles.
COMPARE Poirier Sauger (p.316).
OTHER TREES *P. elaeagrifolia* (SE Europe, Crimea, Tunisia; 1800) is confined to collections: spiny, with slenderer leaves (intensely grey in early summer) and 1–2 cm purplish green fruits.

Snowy Mespil *Amelanchier lamarckii*

(*A.* × *grandiflora*; Serviceberry; Juneberry) Abundant; commonly naturalized on sandy soils in SE England. Probably a stable hybrid of the E North American Shadblow (*A. laevis*, as which many specimens are grown but which has *hairless young leaves*): the nomenclature is much confused. Garden trees are also sometimes labelled *A. canadensis*, which is properly a North American suckering plant with shorter petals.
APPEARANCE Shape A low, twiggy but crisp dome, to 13 m, the trunk often very twisted. **Bark** Smooth, rather silvery-grey; neat shallow spiralling or criss-cross ridges can develop in old age. **Shoots** Slender, hairless. **Buds** Long-pointed and coppery – almost beech-like. **Leaves** *Flat but flimsy*, moss-green, unfolding coppery and with *silky hairs*, but soon hairless except on the stalk; oval, but with neatly toothed rather *parallel sides* half-way up. They glow rich orange and red in autumn, and individual jewel-like coloured leaves can be found throughout the season. **Flowers** With soft white, narrow, starry petals, in small heads above the young leaves in mid-spring: one of the most delicately beautiful of flowering trees. **Fruit** 9 mm berries ripen red then purple-black in midsummer, and are soon eaten by birds.
COMPARE Wild Crab (p.306); Callery Pear (p.318); Alder-leaved Whitebeam (p.294). Deciduous Photinia (p.282): glossier leaves with finer teeth. Broad-leaved Cockspur Thorn (p.288).
OTHER TREES *A. rotundifolia* (*A. ovalis*; S Europe, E Asia, N Africa) is a bush with leaves hairless from the first and round-petalled, less graceful flowers; little grown in the UK.

SNOWY MESPIL

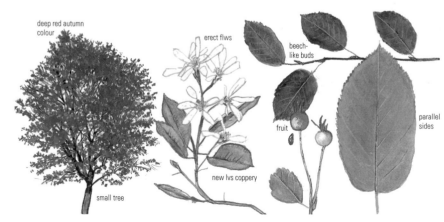

SILVER PEARS & SNOWY MESPIL

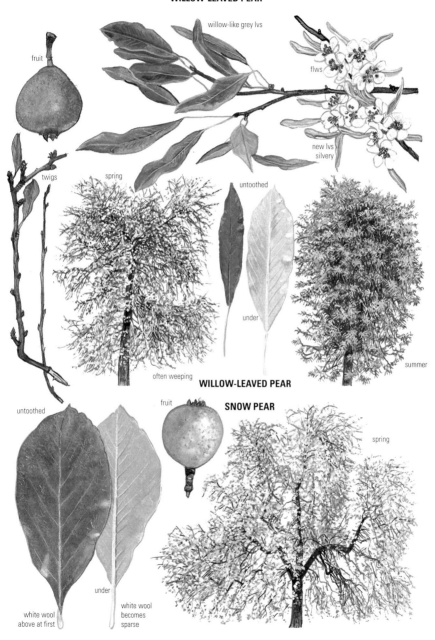

CHERRIES

Cherries (400 species) grow as trees or shrubs. The leaf-stalk normally has glands near the top (cf. willows, pp.164–171, Hybrid Black Poplars, p.156, and Idesia, p.408): these are nectaries, attracting ants to combat leaf-eating insects. Most are best learnt in flower.

Things to Look for: Cherries

- Shape? Bark? What are they like?
- Leaves: How long are they? How hairy beneath? Are their teeth single or double (and whisker-tipped?)? Are the leaf-stalks hairy?
- Flowers: What colour(s) are they? How early? How many petals? How many styles, and of what shape? Are they among young leaves (of what colour?), or do they precede them? Are the sepals (what colour?) serrated?

Key Species

Wild Cherry (below): big long leaves (with coarse teeth), and flowers in bunches: see Key 'Flowering Cherries', below. **Bird Cherry** (p.342): big long leaves (with tiny teeth), and flowers in stiff tails. **Almond** (p.338): long very narrow leaves. **Tibetan Cherry** (p.324): showy bark. **Plum** (p.340): smaller oval leaves. **Apricot** (p.338): almost round leaves. **Portugal Laurel** (p.344): big long evergreen leaves.

Key Flowering Cherries

Wild Cherry: jaggedly toothed leaves (odd hairs below). **Cheal's Weeping Cherry** (p.332): a weeping dome. **'Japanese Cherries'** (pp.326–8): whisker-toothed, hairless leaves. **Sargent's Cherry** (p.330): hairless leaves with sharp but scarcely whisker-tipped teeth. **Yoshino Cherry** (p.332): similar leaves on *downy* stalks. **Fuji Cherry** (p.332): distinctly downy leaves and stalks.

Flowering sequence of commoner *Prunus*:
Timings, throughout spring, vary from region to region. 'Kanzan', the commonest, is typically at its best in London in mid-April, and in Belfast or Edinburgh in early May. The early sequence varies more from warm to cold years.
Hubei Cherry (single pale pink); **'Okame'** (single dark pink); **Myrobalan Plum** (single pure white); **Pissard's Plum** (single pale pink); **Double Cherry-plum** (double deep pink); **'Accolade'** (double rich pink); **'Pandora'** (single soft pink); **Fuji Cherry** (single white/pale pink); **Weeping Spring Cherry** (single pale pink); **Rose-bud Cherry** (single rich pink); **Almond** (single dusky pink); **Yoshino Cherry** (single pale pink); **'Spire'** (single mid pink); **Sargent's Cherry** (single rich pink); **Wild Hill Cherry** (single white/pink); **Seagull Cherry** (single white); **Blackthorn and plums** (single off-white); **'Shirotae'** (semi-double pure white); **Great White Cherry** (single pure white); **Wild Cherry** (single off-white); **Naden** (double pale pink); **'Ichiyo'** (double pale pink); **Kiku-shidare Zakura** (double deep pink); **'Jo-nioi'** (single white, scented); **'Ukon'** (semi-double yellow-white); **Chinese Flowering Cherry** (double pure white); **'Amanogawa'** (semi-double pale pink); **Double Gean** (double pure white); **'Kanzan'** (double deep pink); **'Pink Perfection'** (double rich pink); **'Fugen-zo'** (double rich pink); **'Shirofugen'** (double pale pink); **'Shogetsu'** (double pure white); **Bird Cherry** (white stiff tails); **Rhex's Cherry** (double white).

Wild Cherry   *Prunus avium*

(Gean; Mazzard; *Cerasus avium*) Europe, including Britain and Ireland; N Africa; W Asia. Frequent on rich/heavy soils, forming suckering stands. Abundantly planted everywhere: a favourite in amenity woodlands for its blossom and great early vigour.
APPEARANCE Shape Spire-shaped when growing fast, with annual branch-whorls. Old trees, *to 30 m*, spikily domed. **Bark** *Purplish* grey: horizontally peeling papery strips and rough lenticel-bands (craggily fissured in age; cf. Downy Birch, p.182). **Shoots** Brown, bloomed grey; hairless. **Buds** Long, rufous (cf. English Oak, p.216: cherry-buds are clustered *only on flowering spurs*). **Leaves** Big (to 11 × 6 cm), dull green, with *deep* (2–4 mm), *blunt, irregular but simple teeth* (cf. 'Okame', p.336); finely hairy under the main veins. In autumn, gold and scarlet-pink. Stalk hairless, with typical 2–5 knobbly crimson glands near the top. **Flowers** Ivory-white, in posies (stalks don't branch), flooding the tree in mid-spring just as leaves unfold fawn-bronze. **Fruit** The cherries in mid-summer are sweet/bitter (not sour) but accessible only to birds.
COMPARE Sour Cherry and Schmitt's Cherry (p.324); Japanese Apricot (p.338).
VARIANTS These include hundreds of relatively low, broad, fruiting cherries, abundant in cottage gardens and (locally, in drier climates) in orchards. (Others – Duke Cherries, *P.* × *gondouinii* – are hybrids with Sour Cherry, p.324.)
Double Gean, 'Plena' (by 1700), is abundant: brilliant white, *double flowers, 2 weeks after the type's*, hang on long stalks under green young leaves (cf. 'Shogetsu', p.328). Bigger than other 'flowering cherries' (*to 20 m*); often identified by its broad but open, spiky-topped crown and elaborately *fluted*, rough-barked bole; its slender leaves tend to hang. Weeping Gean, 'Pendula', is very rare: a parasol of hanging shoots from high, haphazard branches.

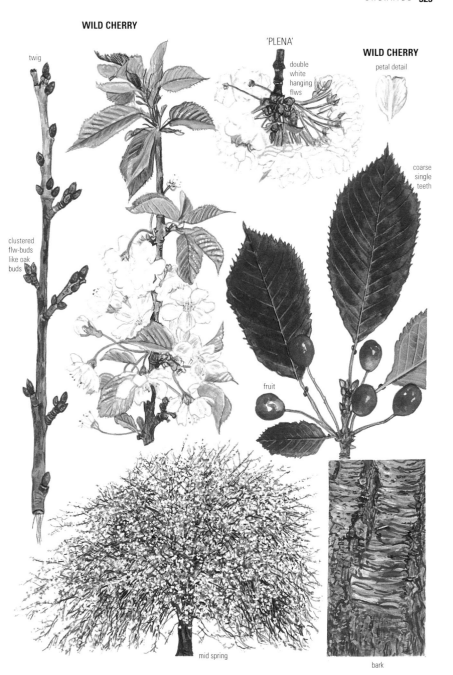

CHERRIES

Sour Cherry *Prunus cerasus*

(*Cerasus vulgaris*) Naturalized scarcely across Europe; probably not known wild.
APPEARANCE Shape Bushy; *abundantly suckering*. **Bark** Duller brown than Wild Cherry's. **Leaves** Smaller (to 8 × 4 cm; cf. Plum, p.340), a glossier sea-green above; usually *hairless; neat teeth finely rounded*, and rather *double*. **Fruit** Acid rather than bitter: Morello and Kentish Cherries are cultivars.
VARIANTS Rhex's Cherry, 'Rhexii', grown since the 16th century, is very occasional, carrying small (3 cm) double white pompom-like flowers at the end of spring among dark green young foliage. Leaves are smaller and darker than in other similar small 'flowering cherries' except the sharp-toothed 'Pandora' (p.336).
All Saints' Cherry, 'Semperflorens' (1623; almost extinct), continues to flower little by little until autumn on twigs in *congested masses*.

Schmitt's Cherry *Prunus × schmittii*

Wild Cherry crossed with Greyleaf Cherry (*P. canescens*: a shrub in collections with narrow, downy leaves and a bark almost like Tibetan Cherry's). 1923. Now a frequent street tree.
APPEARANCE Shape Unmistakable: *long, slender, erect, closely leafy branches* make a narrow, open balloon, pointed in youth. To 20 m; short-lived. **Bark** Finely peeling, *lustrous red-brown horizontal strips* between close rough lenticel-bands. **Leaves** Most like Wild Cherry's but shorter; coarsely toothed. **Flowers** Pale pink; small and fleeting among olive unfolding leaves.
COMPARE Tibetan Cherry. Sargent's Cherry 'Fastigiata' (p.330): more open; bark dull brown.

OTHER TREES Dawyck Cherry, *P. × dawyckensis* (1907; probably another hybrid of Greyleaf Cherry, to 8 m in some big gardens), shares the bright bark; its sparse, *gauntly drooping shoots* are strung with deeper pink, showier flowers.

Tibetan Cherry *Prunus serrula*

W China. 1908. Occasional in gardens and newer street-plantings – entirely for its unique bark; regenerating freely.
APPEARANCE Shape A finely twiggy dome of hanging, greyish leaves on rather stiff branches; to 15 m. **Bark** *Crimson, satin-smooth* between rough brown bands; shredding shaggily with age; ultimately cracked, dull and sprouty. Conscientious owners scrub their trees with toothbrushes. **Shoots** Shortly downy. **Buds** Long, slender, chestnut. **Leaves** *Narrow*, to 11 × 3 cm; usually silky-hairy under the veins. **Flowers** Yellow-white, tiny and fleeting among the foliage. **Fruit** Cherries red, oval, 15 mm long.
COMPARE Schmitt's Cherry (above). Dull-barked old trees can even recall Almond (p.338).

Manchurian Cherry *Prunus maackii*

NE Asia. 1910. Rather rare.
APPEARANCE Shape A vigorous spire; then untidily open-domed, to 18 m. **Bark** Smooth, *honey-gold*, with horizontal lenticel-bands; greyer and fissured with age. **Shoots** Orange, downy. **Buds** Appressed. **Leaves** *Finely toothed*; downy on the stalk and under the main veins. **Flowers** White, fragrant: in short (to 7 cm), stiff tails *on last year's twigs* (cf. St Lucie Cherry, p.342).
COMPARE Bird Cherry (p.342): similar leaves, but almost hairless.

MANCHURIAN CHERRY

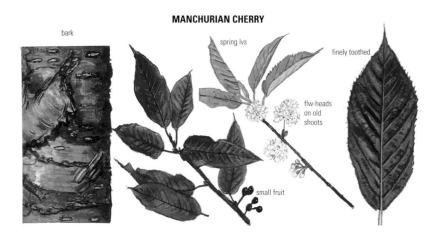

bark · spring lvs · finely toothed · flw-heads on old shoots · small fruit

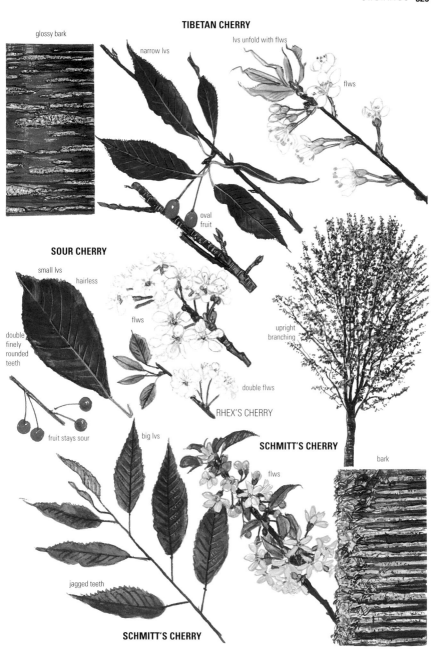

JAPANESE CHERRIES

Japanese Cherries

'Japanese Cherries' (*Sato-zakura*) were bred in Japan (and China) over many centuries from wild species (though 'Pink Perfection' and the still-rare 'John Baggeson' are seedlings of 'Kanzan' raised in England); sometimes corralled as cultivars of *Prunus serrulata* (p.330). All are short-lived in the UK.

APPEARANCE Shape Generally low; customarily grafted at head-height on a stem of Wild Cherry, which outgrows the branches above it like a pillar box. (In 'Sheraton' cherries, roots of Wild Cherry and the head of a Japanese Cherry are grafted onto a stem of Tibetan Cherry, p.324, with its bright bark.) **Bark** Typically pewter-grey, with brown lenticel-bands. **Leaves** Large, *sparse; hairless and on hairless stalks; with more or less simple but strongly whisker-tipped teeth*; autumn colours rich ambers and crimson-pinks. **Flowers** On variably long, *branching stalks* (cf. only Wild Hill Cherry, p.330): larger and generally later than those of other 'flowering cherries', often double, and opening when the (often red) young leaves are *half-expanded*.

🔑 *Prunus* 'Kanzan' (c.1913) is the most abundant and has given its whole clan a bad name. (The Japanese have the sense to use it sparingly). Briefly gorgeous as magenta buds swell under red baby leaves; less so as the 23–28 petals fade to hard pink and the leaves expand olive-green. Funnel-shaped when young; vigorous (to 14 m); heavy limbs soon curve untidily to the horizontal. Leaves large, very sparse; tomato-red growth-tips through early summer.

OTHER TREES *P.* 'Fugenzo' ('James Veitch'; 1892) is occasional: purer pink double flowers a week later contrast more subtly with orange-brown young leaves. The sepals (as in 'Shirofugen', p.328, of which this is a colour variant) are *serrated in their middle third* – untoothed in 'Kanzan'.

P. 'Royal Burgundy' (2000) has flowers as 'Kanzan', among smaller leaves which *remain rich purple*.

P. 'Pink Perfection' (Surrey, 1935; frequent), has bright but mottled pink double flowers, timed with 'Kanzan' but hanging among *quickly green* young leaves. In summer a rather *globular* tree with *drooping outer shoots* and *short*, dark leaves.

P. 'Hokusai' ('Uzuzakura'; 1866) is vigorous but becoming rare. Soft pink flowers with *10–15 petals*, 2 weeks before 'Kanzan', have a centre of crimson stamens (and 1 style; 'Kanzan' has 2–3), among coppery-brown young leaves (cf. 'Taoyame', p.328; Naden, p.334).

P. 'Ito-kukuri' is rare: soft mauve-pink flowers with 15–24 petals in *dense pompom clusters* among *bronze-green* young foliage, 10 days before 'Kanzan'; the 1 style has leaf-like flanges. A rounded tree with rather shiny leaves, yellow in autumn; their teeth have particularly long, crimped whiskers (cf. 'Shirotae', p.328).

OTHER TREES *P.* 'Yae-murasaki-zakura' (less crowded flowers with 11–14 slightly darker petals and 1 *perfectly formed* style) is also rare.

P. 'Amanogawa' (Lombardy Poplar Cherry) is planted abundantly for its *vertical* shape rather than its flowers, which (opening with those of 'Kanzan') are also in erect posies, the 6–15 pale pink petals a poor contrast to bronzy-green young leaves. Old trees are as broad as tall but still have twisting, stiffly *upright* twigs.

'KANZAN'

branches arch over in old trees

'FUGENZO'

very late

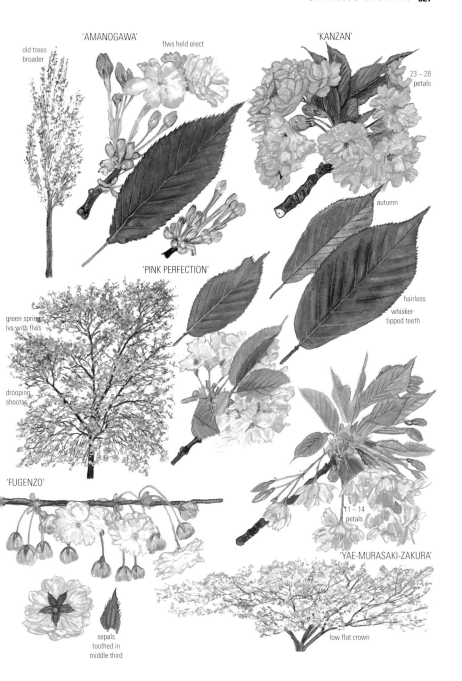

JAPANESE CHERRIES

Prunus 'Ichiyo' is a magnificent but only locally frequent Japanese Cherry. Flowers 10 days before 'Kanzan', among young leaves which are soon *fresh green*; pink in bud fading to *creamy* blush: a flat, neat ring of 16–22 petals *like a ballerina's tutu* surrounds green/red eyes (with 1–2 styles) that stare at you from all over the tree. In summer, a vigorous cherry (often on its own stem) with rising then spreading limbs, and relatively *small*, dull, narrow leaves; its bark develops *rugged vertical fissures*, orange at the base.

P. 'Taoyame' is gorgeous but rare: *some flowers single, some with up to 15 petals*, set off by *rich maroon young leaves* (cf. 'Hokusai', p.326), and by rather purple stalks and *purplish sepals whose tips curl upwards*. Leaves *taper towards the base and are broadest above halfway*.

Great White Cherry, *P.* 'Tai Haku', was found growing in a Winchelsea garden in 1923 by the great cherry-breeder Collingwood Ingram after having been assumed lost in Japan for two centuries; descendants of this one half-dead tree are now very frequent. *Huge* (6–8 cm) single snow-white flowers, a few days after 'Shirotae', are complemented by *copper* young leaves, *very large* and dark green in summer. A funnel-shaped then widely *low-domed* tree.

OTHER TREES *P.* 'Ojochin' (rare) has 5–8 large, very pale pink petals, opening a few days later among bronzy-green leaves which in summer are large, dark, coarsely toothed and rather oblong: *many have a blunt end* (cf. *a few* leaves of 'Shirofugen', 'Ichiyo', 'Tai Haku' and 'Ukon').

P. 'Ukon', very frequent, has 9–15 petals *flushed greenish yellow*, a week later; young leaves coffee-brown. Summer foliage as 'Tai Haku' (habit – a straggling funnel, as 'Kanzan' – and paler growing tips will often differentiate).

P. 'Shirotae' ('Mount Fuji'; *c.*1905) is very frequent: the first 'Japanese Cherry' to flower: dazzling white, honey-scented blossoms with 6–11 petals hanging under green baby leaves (cf. Chinese Flowering Cherry, p.330). Unique in summer: strong *very horizontal* branches (taller and gaunt in shade) carry bright rich green leaves, their teeth with *extra-long* crimped whisker tips.

P. 'Jo-Nioi', rather rare, has delicate billowing masses of relatively small (4 cm), single, white, *gorse-scented* flowers, a few days before 'Kanzan', and rather ahead of the golden-brown young leaves. Later and laxer than Seagull Cherry (p.336).

P. 'Shogetsu' ('Shimidsu Zakura'; 'Longipes') is frequent: pink buds open a week after 'Kanzan' into pure white flowers with 20–28 petals which hang under yellow-green young leaves. Small, rounded, slightly weeping crown, often cankered and dying back.

P. 'Shirofugen', the last to flower, is very frequent and unsurpassed: almost white, double flowers hang from pink buds under *maroon* young leaves, then *fade soft pink* as the leaves turn green. In summer as 'Fugenzo' (p.326) – a vigorous, spreading cherry: medium-sized, dark, often oblong leaves, the whisker-tips of the teeth generally appressed.

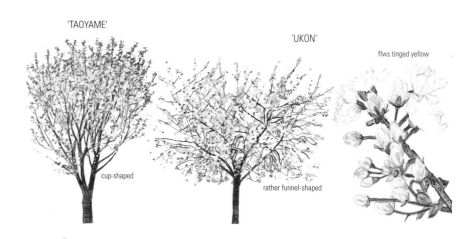

'TAOYAME'
cup-shaped

'UKON'
rather funnel-shaped

flws tinged yellow

FLOWERING CHERRIES

Sargent's Cherry — *Prunus sargentii*

Mountains of Korea, Sakhalin and N Japan (where it makes a huge timber tree). 1893. Very frequent.
APPEARANCE Shape Rounded; shapely, with light, ascending branches; to 15 m; *densely clad* in *dark yellowish*, rather matt, hanging leaves. **Bark** Purple-brown, with orange lenticel-bands: a *browner* cast than other 'flowering cherries'; more rugged in age. (Garden trees are often grafted on a stem of Wild Cherry.) **Leaves** Hairless and on hairless stalks, the sharp teeth scarcely double *and scarcely whisker-tipped*; rather broadly oblong (to 14 × 8 cm) and generally *convex* (the tip down-curving); whitish beneath. Autumn colours *early and scarlet*. **Flowers** Brilliant mauve-pink, single but quite large (4 cm), *daintily half-filling the crown* a week before the first Japanese Cherry, on non-branching stalks among *ruby-red baby leaves which are still dark bronze as the last petals drop*.
COMPARE Wild Hill Cherry (below). Yoshino Cherry (p.332): hairy leaf-stalks. Hubei Cherry (p.334): glossier leaves. Sour Cherry (p.324) and 'Spire' (p.336): similarly hairless but shorter leaves. The (hairless) leaves of 'Japanese Cherries' (pp.326–8) are always sparser (and whisker-toothed).
VARIANTS 'Fastigiata' is occasional: vigorous, *gaunt, erect stems*, to 16 m (cf. the denser, compact, pale-pink 'Spire', p.336). 'Rancho' (1961; rare?) is more compactly columnar.
OTHER TREES Judd's Cherry, *P.* × *juddii*, is a probable hybrid with Yoshino Cherry (1914). Flowers as large but much paler (cf. 'Hillieri', p.336); its young leaves (hairless but *downy-stalked*) are less reddish.

Wild Hill Cherry — *Prunus jamasakura*

(*P. serrulata* var. *spontanea*) Foothills in S Japan. c.1914. Surpassed by few if any trees in its combination of blossom and natural grace, but confined to big gardens; rarely seeding.
APPEARANCE Shape Dainty: light limbs and masses of hanging leaves; *to 18 m*. **Bark** Smooth pewter-grey, with prominent horizontal lenticel-bands; finally rugged. **Leaves** Hairless and on hairless stalks; slenderer than most Japanese Cherries', and whiter beneath, the usually simple sharp teeth scarcely whisker-tipped. In autumn, brilliant yellow/red. **Flowers** Single, silvery-white to clear pink, timed with Sargent's Cherry's but on the Japanese Cherries' *branching stalks*: vivid clouds overtaken by usually red young leaves.
OTHER TREES Korean Hill Cherry, *P.* × *verecunda* (*P. serrulata* var. *pubescens*; 1900: odd garden gardens), has leaves with coarser teeth, *downy beneath and on the stalks* (cf. Fuji Cherry, p.332), pink flowers, and duller young leaves.

Chinese Flowering Cherry — *Prunus serrulata* 'Albi-Plena'

As highly bred as the Japanese Cherries, this tree reached the UK first (1822; the Japanese 'Ichihara-toro-no-o' is very similar). Occasional.
APPEARANCE Shape Stylish; very flat (cf. 'Shirotae', p.328), and recognized by its *exaggeratedly lumpy spur-shoots*. **Leaves** Like Japanese Cherries', but whiter beneath. **Flowers** Quite small (38 mm), late, hyacinth-scented, with 18–21 white petals (like Old English roses as they expand, from red buds), on branching stalks among green young foliage.
COMPARE Rhex's Cherry (p.324).

CHINESE FLOWERING CHERRY

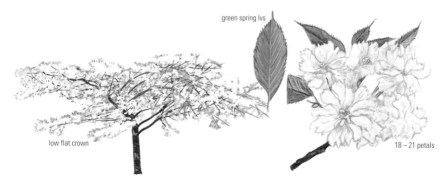

green spring lvs

low flat crown

18 – 21 petals

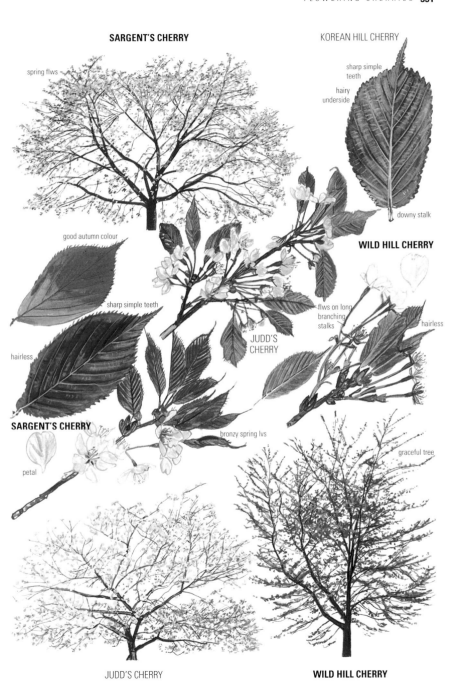

FLOWERING & WEEPING CHERRIES

Fuji Cherry — *Prunus incisa*

Mountains of central Japan. 1910. Very occasional: mostly big gardens.
Appearance Shape Gracefully bushy or to 10 m; twiggy, with dense, small leaves. **Bark** Smooth, purplish grey; brown lenticel-bands. **Leaves** *Hairy above*, under the veins, and on the stalks (which have *stalked* glands), oblong and only to 6 × 3 cm; *deeply double-toothed* (cf. 'Spire', p.336). **Flowers** Silvery/pale pink, small (22 mm) and *nodding*, before the leaves ('Praecox' and 'February Pink' are among the first cherries to flower): not super-abundant, but uniquely delicate.
Compare Naden (p.334), Korean Hill Cherry (p.330), and 'Kursar' (p.336): *larger* downy leaves.
Variants Most distinctive are f. *yamadae* ('Midori-zakura': flowers with *pale green* (not reddish) eyes, ivory *en masse*); 'Oshidori-zakura' (double rich pink flowers hang on 4 cm stalks); and 'Kojo-no-Mai' (dwarf; twigs *kink* at each bud).

Yoshino Cherry — *Prunus* × *yedoensis*

Now the most abundant garden cherry in Japan. c.1910; frequent in the UK.
Appearance Shape Usually low and dense (slightly weeping at the edges), on *heavier, more zig-zag limbs* than other 'flowering cherries'; to 15 m. **Bark** Pewter-grey, smooth; brown lenticel-bands. **Leaves** Large, dark, olive-green, rather hanging; whisker-toothed; downy under the main veins *and on the stalk*. **Flowers** Pale blush, on unbranched stalks, slightly scented and *densely wreathing* the graceful shoots *ahead of its leaves*.
Compare Other 'flowering cherries' with hairs largely confined to the leaf-stalks: Winter and Rose-bud Cherries (p.334); 'Pandora', 'Kursar', 'Accolade' and Seagull Cherries (p.336).
Variants Swallow Cherry, 'Tsubame', is rare: long *horizontal* then drooping low branches and nearly white flowers.
The frequent 'Shidare Yoshino' ('Perpendens'; 'Pendula') makes a *neat, dense igloo* of hanging shoots (*P.* 'Hilling's Weeping' is a *white* variant); 'Ivensii' (by 1929; also white) *weeps from widely arching limbs*.
Other trees Weeping Cherries, whose branches arch to the ground from a high graft, also include the following.
Cheal's Weeping Cherry, *P.* 'Kiku-shidare-zakura', is an abundant Japanese Cherry (see p.326; 'Geraldinae' ('Asano') is its much rarer funnel-shaped equivalent. Mauve-pink, very double flowers, a week before 'Kanzan', wreathe the *sparse, gaunt branches* in solid pompom clusters, among olive-green young foliage. The slender leaves (scarcely whisker-toothed) mature *glossy and puckered*.
Weeping Spring Cherry, *P. pendula* (*P. subhirtella* var. *pendula*; 1894), is a Weeping-Willow-shaped selection of a wild Japanese tree (see p.334) in a few large gardens, its branches *zig-zagging up* before the shoots swoop down. Flowers small, early, pale pink; leaves small and *narrow* (to 8 × 3 cm), hanging, often double-toothed; hairy on their stalks *and at least under the main veins*. Three low, *wispily igloo-shaped* cultivars are occasional: 'Pendula Rosea' ('Pendula') has very *pale pink* single flowers; 'Pendula Rubra' ('Beni-shidare') has rich, smoky-pink *single* flowers; 'Pendula Plena Rosea' ('Yae-beni-shidare') has rich pink *double* flowers.

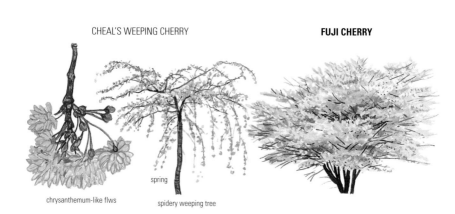

CHEAL'S WEEPING CHERRY — spring — spidery weeping tree — chrysanthemum-like flws

FUJI CHERRY

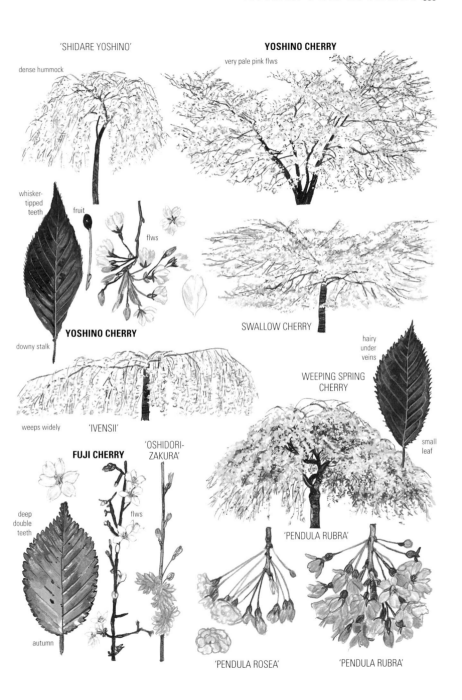

FLOWERING CHERRIES

Rose-bud Cherry
Prunus pendula var. *ascendens* 'Rosea'

('Beni Higan Sakura') A selection of the wild variety of *P. pendula* (see p.332). *c.*1916. Rare.
Appearance Shape Billowing; untidily twiggy, as Winter Cherry (below). **Leaves** To 8 × 4 cm, often double-toothed; hairy at least under the main veins and on the stalks (cf. Yoshino Cherry, p.332). **Flowers** Bright pink (darker than 'Pandora', p.336); a week before Sargent's Cherry's and ahead of the leaves; in abundant clouds but *small* (20 mm across).
Variants 'Fukubana' (1927, rare) has small, semi-double rich-pink flowers (cf. 'Accolade', p.336). 'Stellata' ('Pink Star'; 1955) is low and rare: larger, crowded flowers have *narrow, star-like, deep-pink petals*.

Winter Cherry
Prunus × *subhirtella* 'Autumnalis'

('Jugatsu-zakura') Japanese gardens. *c.*1900. Very frequent.
Appearance Shape Very twiggy and untidy; spreading, to 9 m. *Bare lengths of thin, twisting shoots* with small leaves at the tips (cf. 'Accolade', p.336); rather *pale moss-green foliage*. **Bark** Grey-brown, smooth; brown lenticel-bands. **Leaves** Small (7 × 4 cm), often down-curling near the tip; doubly toothed though not so deeply as Fuji Cherry's (p.332); hairy under the main veins and on their stalks. **Flowers** Small, semi-double, *almost white* (crimson stamens make them look pink); opening little by little *from October* until mid-spring when there is a final, quiet flourish at the branch-tips, among pale green young leaves.
Variants 'Autumnalis Rosea' has *soft pink* flowers and is slightly less common.
Other trees 'Fudan Zakura' is a rare small Japanese Cherry (with the typical foliage; see p.326); its bigger (4 cm), white, single, short-stalked flowers are spectacular in early spring (among copper young leaves) but *start in winter*.

Naden
Prunus × *sieboldii* 'Takasago'

('Caespitosa') An old Japanese selection; introduced by 1864 and once one of the commonest Japanese cherries in the UK, but now distinctly rare: *takasago* means 'good health and long life', neither of which it enjoys.
Appearance Shape Rather horizontal; sometimes dwarfish, with stiff shoots. **Leaves** *Dark* green: most like Yoshino Cherry's (p.332) but finely *velvety on both sides* as well as on the stalks. **Flowers** 4 cm, *hairy-stalked*: 10–13 pale pink petals, in the middle of the Japanese Cherry season (cf. 'Hokusai', p.326), among bronze young leaves.
Compare Fuji Cherry (p.332); Korean Hill Cherry (p.330). The leaves, Japanese Cherry-sized but velvety, are very distinctive.

Hubei Cherry
Prunus hirtipes

(Incorporating *P. conradinae*) W China. 1907. Rare.
Appearance Shape A graceful tree, with the habit of Wild Hill Cherry (p.330), but smaller-growing in the UK (to 10 m). **Bark** As Wild Hill Cherry. **Leaves** Broadly oval (to 12 × 7 cm); doubly but finely toothed. *Glossier* than most 'flowering cherries'; there are odd hairs under the veins and sometimes above. **Flowers** Very pale pink, small (25 mm), fragrant; scarcely stalked. They string the shoots in dollops almost like Yoshino Cherry's (p.332), but in *late winter* – one of the year's first great displays.
Variants 'Semi-plena' (1925; flowers with a few extra petals) is now the commoner form.

NADEN

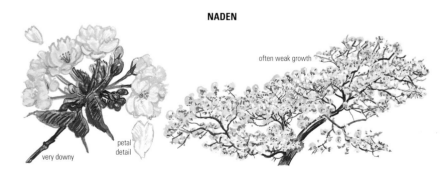

often weak growth

petal detail

very downy

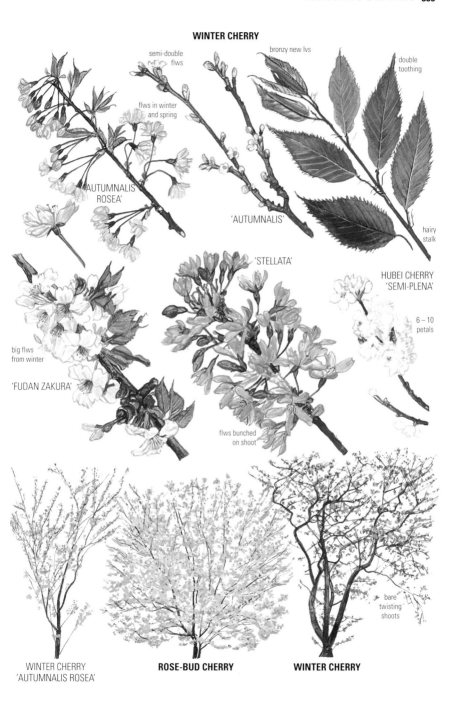

HYBRID FLOWERING CHERRIES

Prunus 'Spire'

(*P.* × *hillieri* 'Spire') A 1930s hybrid of Fuji Cherry; locally abundant as a street tree.
APPEARANCE **Shape** *Erect* thin limbs usually from a *very* disproportionate Wild Cherry pedestal; densely leafy, *compact but irregular*; finally as broad as tall (to 12 m). **Leaves** To 8 × 5 cm, *almost round*, with a sudden long point (cf. 'Umineko'), and *deep, rounded* double teeth; slightly downy beneath; scarlet in autumn. **Flowers** 4 cm; blush-pink clouds, among red baby leaves; a week after 'Pandora'.
OTHER TREES *P.* 'Hillieri' is a rare, spreading sister-tree (cf. Fuji Cherry, p.332, and *P.* × *juddii*, p.330).

Prunus 'Pandora'

A complex garden hybrid (by 1940); quite frequent.
APPEARANCE **Shape** Funnel-shaped, then broad; quite graceful. **Bark** Dull grey-purple; rugged with age. **Leaves** *Small* (7 × 3 cm), *glossy, blackish, neatly and sharply double-toothed*; almost hairless; they hang rather sparsely by late summer. **Flowers** 4 cm: 5 petals, pale pink but darker near their margins: early and long-lasting, overtaken at last by bronze young leaves.
COMPARE Other small-leaved forms – 'Okame' (below); Fuji Cherry (p.332); Tibetan and Sour Cherries (p.324).

Seagull Cherry   *Prunus* 'Umineko'

Fuji Cherry crossed with Oshima Cherry (*P. speciosa*). 1948. Frequent.
APPEARANCE **Shape** Dense; neatly upright at first (like a seagull with raised wings); branches *arch over at 7 m*. *Dark, vivid green*; clear orange in autumn. **Leaves** Downy-stalked; *almost round* with a sudden long point, to 9 × 6 cm; *deep*, neat, double teeth: cf. 'Spire' (above: with blunt toothing) and Fuji Cherry (p.332: small, narrower leaves). **Flowers** Similar to Wild Cherry's (p.322) but snowier and earlier; slowly overtaken by green young leaves. Understated but very lovely.
OTHER TREES 'Snow Goose' is a rare sister tree: wider, gaunter habit and slightly larger leaves.

Prunus 'Okame'

Fuji Cherry crossed with the red-flowered Bell Cherry (*P. campanulata*). 1947. Rather occasional.
APPEARANCE **Shape** Usually dwarfishly globular; *stiffly twiggy and untidy*. **Leaves** Small, dark, *narrow* (6 × 3cm), with odd hairs above at first; *jaggedly and irregularly* toothed (even lobed on saplings). **Flowers** Small, *dark, smoky magenta-pink clouds from almost crimson buds, a month before* the leaves.
OTHER TREES 'Kursar', another Bell Cherry hybrid, is rarer: taller, shapelier; flowers downy-stalked; leaves (downy beneath; on finely hairy stalks) larger and regularly toothed. Its 1979 seedling, 'Collingwood Ingram', is even darker and rather upright; 'Shosar', also upright and almost as dark (very rare), is a fortnight later.

Prunus 'Accolade'

Sargent's Cherry × *P.* × *subhirtella*. 1952. Frequent.
APPEARANCE **Shape** Spreading: *very untidy*, tangled, hanging, *twisting bare shoots*, like Winter Cherry's (p.334). **Leaves** Bigger (to 11 × 5 cm) and darker; hairy-stalked. **Flowers** The tree's redeeming feature: 4 cm; *about 12 brilliant then pale pink petals* (cf. Winter Cherry): superabundant strings and clusters, well before the soft-green leaves and often starting late in winter.

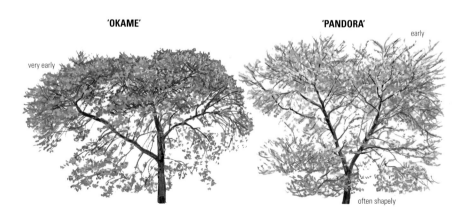

'OKAME' — very early

'PANDORA' — early, often shapely

HYBRID FLOWERING CHERRIES

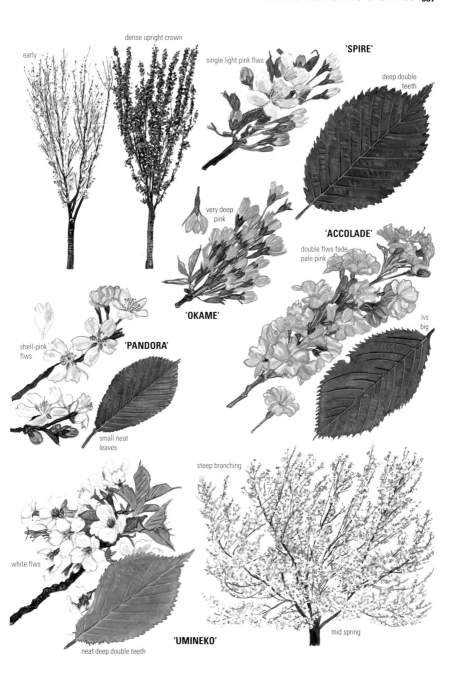

ALMONDS TO APRICOTS

Almond
Prunus dulcis

(*P. communis*; *P. amygdalus*; *Amygdalus dulcis*) Mediterranean regions; long grown in the UK. Flowering selections are frequent in small gardens.
APPEARANCE Shape Gauntly globular, open and rather erect, to 11 m; wild trees can be spiny. **Bark** *Blackish; soon finely but ruggedly cracked.* **Shoots** Like Plum's (p.340), but with slightly downy buds. **Leaves** Willow-like, to 12 cm; dark and glossy but crumpled (often folded up at the base); hairless; finely serrated; on stalks *to 25 mm,* which carry the typical cherry glands. **Flowers** Big (to 5 cm), mauve-pink; sparsely single or paired on very short stalks; well before the leaves. **Fruit** Like small greenish peaches, to 5 cm; finally dark brown. Chemicals in the almonds of unselected trees (var. *amara*) can release dangerous quantities of cyanide – 50 may be fatal.
COMPARE Peach (below); Black Cherry (p.344); Bay Willow (p.166).
VARIANTS 'Alba' has white flowers; very rare.
Double Almond, 'Roseoplena', has double pink flowers and a rather weeping habit. Rare. (cf. Cheal's Weeping Cherry, p.332.)
'Praecox' opens its pink flowers in late winter; rare. Sweet Almond, 'Macrocarpa', bears edible almonds in fruits to 8 cm wide; scarcely grown in the UK.
OTHER TREES *P. webbii* (Mediterranean region) is spiny and has small (4 cm) leaves and 2 cm white flowers; scarcely grown in the UK.

Peach
Prunus persica

(*Amygdalus persica*) N China. Long cultivated in S Europe; occasional in the UK.
APPEARANCE Shape Bushy, often trained against a wall or bursting out of glasshouses. **Leaves** Slightly narrower than Almond's (to 15 × 3 cm), and on *shorter stalks* (12 mm). **Flowers** Smaller, paler and later. **Fruit** The peaches need plenty of warm sunshine to ripen well.
VARIANTS Flowering cultivars, with double/red flowers, are all now rare. Nectarine, var. *nucupersica* (var. *nectarina*), has hairless fruit.

Apricot
Prunus armeniaca

(*Armeniaca vulgaris*) N China; cultivated and naturalized in S Europe but rare outdoors in the UK.
APPEARANCE Shape Gauntly globular, on twisting branches. **Bark** Soon rugged; paler grey than Almond's. **Shoots** (Like the plums') *lacking a big end bud.* **Leaves** *Rounded* with a sudden twisted tip (particularly broad in var. *ansu* from E Asia); round-toothed; hairless or with tufts under the vein-joints. **Flowers** Rather small (25 mm) and early; white/pink-flushed. **Fruit** Apricots of unselected trees are only 3 cm wide.
COMPARE St Lucie Cherry (p.342): smaller leaves. A puzzling plant at first sight, though with typical cherry glands at the top of the leaf-stalk.
OTHER TREES Japanese Apricot or Mei, *P. mume* (1844), is rare and bushy in the UK. Leaves narrower; sharply, often doubly, toothed; hairy at least when young. The 3 cm apricots are scarcely edible; various cultivars carry white/pink, often double, scented flowers in late winter, of almost camellia-like beauty.
Briançon Apricot, *P. brigantina* (Alpes Maritimes), has doubly toothed leaves downy beneath. The stones of the small yellow apricots are used locally to make a fragrant oil, *huile de Marmotte.*

ALMOND — often dusky pink — rugged black bark

APRICOT — fruiting tree

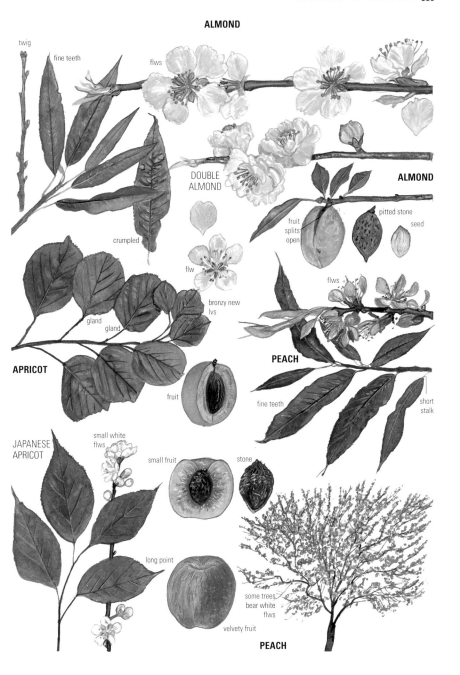

PLUMS

Plum — *Prunus domestica*

Long grown in the UK; probably a Caucasian hybrid of Blackthorn and Myrobalan Plum. Abundant: gardens, some orchards; suckering and naturalizing.
APPEARANCE Shape Low, dense; to 10 m; *not* spiny. **Bark** Purple; finely roughened then widely fissured. **Shoots** Red-purple, soon *hairless*. **Buds** Longer and sharper than the other plums' on this page (all *lack a big end-bud* and can have 2–3 side-buds at each leaf). **Leaves** Wrinkled (cf. Goat Willow, p.168); to 8 cm, broadest above halfway; downy under the veins; stalks with the typical cherry glands. **Flowers** Off-white, in mid-spring. **Fruit** At least 3 cm long.
COMPARE Sour Cherry (p.324); Apricot (p.338); St Lucie Cherry (p.342); Orchard Apple (p.306).
OTHER TREES Bullace, *P. insititia* (*P. domestica* var. *insititia*), abundantly naturalized in hedges, is often bushy. Fine purple-grey shoots, *downy for a year*, are sometimes spiny; leaves smaller (but broader than Blackthorn's), with fine *down on both sides*. Fruits round, 2–3 cm; purple ('Black Bullace') or yellow ('Shepherd's Bullace'); *sweet*. Damsons (sweeter oval purple plums), Mirabelle Plums (round yellow plums) and Greengages (yellow-green oval plums) are rarely naturalized selections.
Naples Plum, *P. cocomila* (S Italy; S Balkans; scarcely grown in the UK), has 4 cm yellow plums.

Blackthorn — *Prunus spinosa*

(Sloe) Europe, including Britain and Ireland; N Asia. Dominant in scrub; hedges on heavier soils.
APPEARANCE Shape Generally a suckering *bush*. **Bark** Purple-black, finely roughened. **Shoots** Finely downy then *almost smooth by winter; gleaming purple* or grey-bloomed (green in shade); *many side-shoots end in vicious spines*. **Buds** 1–2 mm. **Leaves** Small, *slender* (to 5 × 2 cm), broadest above halfway; wrinkled; downy when young. (Their stalks often *lack* glands.) **Flowers** Off-white, in mid-spring but before the leaves. Sloes purple then black, 15 mm, still intensely *sour* by winter.
VARIANTS 'Purpurea', has purple leaves (smaller and paler than Pissard's Plum's) and pink flowers. 'Plena' has double white flowers. Both are very rare.

Myrobalan Plum — *Prunus cerasifera*

(Cherry Plum) Long grown in the UK; abundantly naturalized. (Wild trees – Balkans to central Asia – may be distinguished as *P. divaricata*.)
APPEARANCE Shape Larger than other plums: *to 15 m*. **Bark** Dark *grey*; widely fissured with age. **Shoots** Soon *hairless; green*. **Leaves** Fresh green; slender, often broadest *below halfway*; downy under veins. **Flowers** Whiter than Blackthorn's and *much earlier* (weather-dependent). **Fruit** Sweet, 4 cm, round, red/gold plums, ripe *late in summer*.
VARIANTS Pissard's Plum, 'Pissardii' ('Atropurpurea'; Iran, 1880), is abundant: untidily upswept, with very pale pink flowers, shiny purple shoots and *heavy purple foliage*. 'Nigra', replacing it as a more graceful plant (scarcely separable in summer) has *deep pink flowers* a few days later.
'Hessei', low and twiggy, has blotchy *bronze-red* leaves and white flowers; rare but much prettier. 'Lindsayae' (Iran, 1937) is rare: pink flowers.
OTHER TREES Double Cherry-plum, *P.* × *blireana* 'Moseri' (Pissard's Plum × Japanese Apricot, 1895), is quite frequent: lower, with broader red-purple leaves and *double* 4 cm flowers, rich pink.

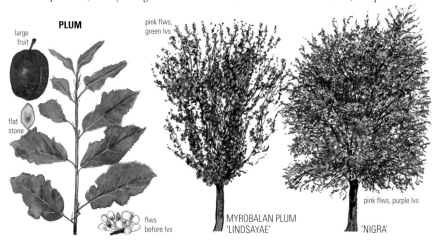

PLUM — large fruit, flat stone
pink flws, green lvs
flws before lvs
MYROBALAN PLUM 'LINDSAYAE'
pink flws, purple lvs
'NIGRA'

BIRD CHERRIES

St Lucie Cherry — *Prunus mahaleb*

(*Cerasus mahaleb*) Belgium E to Central Asia. 1714. Now rare but once planted for game-cover so very occasionally naturalized on sandy soils.
Appearance Shape A low, neat dome of fine, sometimes weeping twigs; to 8 m. **Bark** Dark brown; soon rugged. **Leaves** Oval, even *heart-shaped* at the base; small (5 × 3 cm); glossy and finely toothed, with hairs under midrib. The stalk has 2 cherry glands. **Flowers** Late; in close-set stumpy *sprays* of 6–10 year-old twigs (cf. Manchurian Cherry, p.324): pure white and *scented* but only 18 mm wide. **Fruit** Tiny (8 mm) black cherries.
Compare Apricot (p.338), Sour Cherry (p.324) and Plum (p.340): bigger/longer leaves. Wild Pear (p.316) and Wild Crab (p.306): more leathery leaves, without glands on their stalks.

Bird Cherry — *Prunus padus*

(Hawkberry; Hagberry; *Padus racemosa*) N Eurasia, including hilly, limestone parts of Ireland and of Britain S to Norfolk; locally frequent. The type has seldom been grown in gardens, but suckers from grafts may naturalize.
Appearance Shape Conic and whorled, then openly domed (to 14 m): straight, rising limbs and fine, sweeping twigs. **Bark** Dull grey; very finely roughened but *never fissured/peeling*. **Shoots** Soon hairless; dark, dull green-brown with pale lenticels and sharp 1 cm buds – like Aspen's (p.152), but lacking perpendicular side-shoots. **Leaves** Dull green, and hairless except for tufts under the vein-joints; much *finer, sharper serrations* (1 mm deep) than Wild Cherry's (p.322). The hairless stalks have the typical cherry glands. **Flowers** In well-separated *stiff tails 8–15 cm long* from the shoot-tips in late spring. **Fruit** 8 mm, bitter, black cherries.
Compare Black Cherry (p.344); Manchurian Cherry (p.324).
Variants 'Watereri' (1914), the common garden clone until recently, *grows to 25 m*. Very open and untidy with curving branches; larger (15 cm) leaves, always tufted beneath; much *longer* (to 20 cm), horizontally spreading flower-heads.
'Albertii' is occasional as a younger tree: neatly and narrowly ascending branches make a cone/egg-shape, at least in youth. Flowers a few days earlier, in short but dense, broad heads.
'Colorata' is occasional as a gawky young plant, strikingly beautiful in flower when *soft pink* tails stand among pale *bronze-purple* young leaves. In summer the shoots are purple and the leaves dull grey-green, with a purplish cast beneath (cf. Canada Red Cherry, below).
'Plena' has large, double, long-lasting flowers, in typical spikes; very rare.
Var. *commutata* (Siberia, Manchuria; in some big gardens) flowers/flushes *a month before the type*.

Choke Cherry — *Prunus virginiana*

N North America. 1724. In a few collections, but widely naturalized in continental Europe.
Appearance Shape Bushy, suckering Bird Cherry. **Leaves** Glossy, 6 × 3 cm (cf. Black Cherry, p.344). **Bark** *Smooth*, brown/grey. **Fruit** Often red cherries.
Variants Canada Red Cherry, 'Shubert' (1950), is rare: flowers in stiff white tails, among glossy-green young leaves which turn *brownish purple by early summer* (cf. Bird Cherry 'Colorata', above).

BIRD CHERRY 'ALBERTII'

young trees
conic

'WATERERI'

longer flws;
bigger lvs

BLACK CHERRY TO CHERRY LAUREL

Black Cherry
Prunus serotina

(Rum Cherry) Nova Scotia to Florida and Arizona. 1629. Very occasional in gardens, but sometimes well naturalized on sandy soils.
Appearance Shape A somewhat shapeless but quite dense, evergreen-looking tree (late into leaf and gold late in autumn); outer branches sometimes very weeping. To 20 m. **Bark** Black-brown; soon *fissured and peeling harshly*, and rugged in age. **Shoots** Slender; glossy red-brown, with 4 mm yellowish buds, usually blunt. **Leaves** To 12 × 4 cm (but some much smaller); rather leathery; dark and glossy above and smooth beneath but with *dense orange/white hairs bristling from the prominent lower midrib*; fine *incurving* teeth. Stalk to 15 mm only, carrying the typical cherry glands. **Flowers** In tails like Bird Cherry's, in early summer. **Fruit** 8 mm crimson then purple-black cherries, which were used to flavour rum and brandy.
Compare Portugal Laurel (below); Choke Cherry (p.342); Almond (p.338); Sorrel Tree (p.430).

Portugal Laurel
Prunus lusitanica

Iberia. By 1648. Abundant in hedges, shrubberies and game-coverts; thoroughly naturalized in the UK in woods on heavy soils, where it spreads by layering and may smother the native vegetation. Ironically it is threatened in the wild by an increasingly dry climate.
Appearance Shape Bushy or on a sturdy, irregular trunk, to 18 m; densely and handsomely evergreen. **Bark** Dark grey; finely roughened. **Leaves** Very glossy; thinly leathery, flat and hairless; finely serrated and broadest below halfway; to 12 cm. **Flowers** In narrow, arching, fragrant tails to 25 cm long, turning the tree creamy-white in early summer. **Fruit** 1 cm bitter purple 'cherries'.
Compare Cherry Laurel (below); Giant Photinia (p.282). *Rhododendron ponticum* (p.428), which misbehaves in the UK in the same ways, is sometimes confused, but has scaly reddish bark and untoothed leaves.
Variants 'Variegata' (brightly but narrowly white-margined leaves, flushing yellow) is rare.

Cherry Laurel
Prunus laurocerasus

Asia Minor to Iran; Bulgaria; Serbia. 1576. An abundant evergreen in hedges, shrubberies and old game-coverts, where on moist heavy soils it colonizes woodlands even more aggressively than Portugal Laurel (above). Used sparingly in herbal medicine ('Cherry Laurel water'), but chemicals in the leaves can release dangerous quantities of cyanide: clipping and burning it are best avoided.
Appearance Shape Bushy (exceptionally to 18 m): sprawling, layering stems and branches gauntly clothed with big (to 20 cm) leaves. **Bark** Blackish; very finely roughened. **Leaves** Bright deep green; shiny, thickly leathery and hairless; broadest above the middle and slightly *convex*, with indented veins and *distant* tiny teeth. **Flowers** In erect white tails to 12 cm long in mid-spring. **Fruit** 15 mm blackish 'cherries' (toxic in bulk).
Compare Portugal Laurel (above); Loquat (p.282); Southern Evergreen Magnolia (p.260); Tarajo (p.366).
Variants 'Latifolia' ('Magnoliifolia'; a rare garden selection) has rich green leaves *to 30 cm long*.
'Camelliifolia' has leaves curling curiously in hoops, as if treated with weedkiller. Rare.
Several dwarf, pretty forms, such as 'Otto Luyken' (1940), are now very popular ground-cover plants.

CHERRY LAUREL

black smooth bark

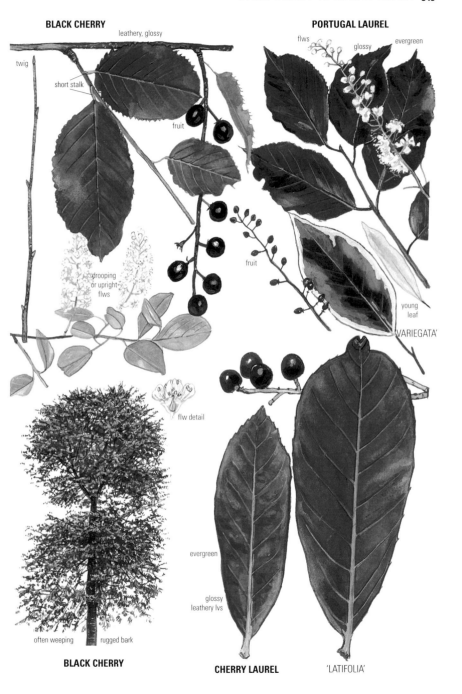

MIMOSA TO PINK SIRIS

Leguminosae is an enormous family (1200 species in the Acacia *genus alone) that is mostly herbaceous. Their leaves are usually compound and seldom toothed; buds are often hidden in the bases of leaf-stalks; and their fruits come in bean-like pods. Nitrogen-fixing bacteria in root-nodules allow many species to colonize poor, sandy soils.*

Mimosa — *Acacia dealbata*

(*A. decurrens* var. *dealbata*; Silver Wattle) SE Australia and Tasmania. 1820. The hardiest *Acacia*. Locally frequent in the mildest areas (Ireland; Cornwall; inner London); cut to the ground in the occasional hard winter in SE England but regrowing with huge vigour; easily blown down. Naturalized in Mediterranean Europe.
APPEARANCE Shape Often conic with a tilted, wispy tip and light limbs, to 23 m; sometimes broad and leaning with sinuous branches. Very silver trees (var. *alpina*) are hardiest. **Bark** At first glaucous-green; then smooth coppery-brown and minutely wrinkled. Older trees blackish, with fluted trunks. **Leaves** Feathery, doubly compound, to 15 cm; evergreen; individual leaflets only 3.5 × 0.5 mm. **Flowers** Creamy buds stand above the foliage from autumn and open from late winter to spring as dazzling 10 cm branching strings of tiny yellow balls of stamens. Seed-pods to 10 cm; blue-white.
COMPARE Pink Siris, Cape Wattle (below).
OTHER TREES Cootamundra Wattle, *A. baileyana* (New South Wales; *c.*1888), is more tender and confined to very mild areas. Leaves much smaller (*to 5 cm*) and vivid, like silver filigree. Flowers arranged like clusters of miniature laburnum-chains, smothering the crown. Its cultivar 'Purpurea' has pretty, *purple-brown foliage*.

Blackwood — *Acacia melanoxylon*

SE Australia and Tasmania. 1808. Confined to mild areas. Occasionally naturalized in S Europe.

APPEARANCE Shape Vigorous and slender, to 25 m; sometimes regrowing with several sinuous trunks. **Bark** Brown; fissured with age. **Leaves** Mimosa-like on saplings (the leaflets fewer and larger). On older plants these true leaves are almost entirely replaced by evergreen, *matt, hanging 'phyllodes'* (enlarged, flattened leaf-stalks), crescent-shaped with a rounded tip and to 14 × 3 cm. These have parallel veins so rather resemble the leaves of Snow Gum (p.420). **Flowers** In sparse, small, pale yellow heads along the shoots.

Pink Siris — *Albizia julibrissin*

(Silk Tree) Probably of Chinese origin, but long cultivated W to Europe. 1745. Very occasional; a few street trees in London. Despite appearances it is quite hardy, but needs maximum summer warmth to grow well.
APPEARANCE Shape A low, bushy plant in the UK. **Bark** Smooth, grey. **Leaves** Doubly compound, like Mimosa's but *deciduous* and larger (to 45 cm), the individual leaflets up to *3 mm wide* and dark greyish green; they fold up overnight. **Flowers** Exotic-looking, large cream/pink tufts (magenta in 'Rosea'), through the summer.
OTHER TREES Cape Wattle, *Albizia lophantha* (Plume Albizia; *Paraserianthes lophantha*; *A. distachya*), from Australia, is very rare to date and marginally hardy in the UK: a vigorous, upright plant with fine, *vivid but dark evergreen foliage, velvety shoots/main stalks* and some short bottle-brushes of creamy flowers through spring and early summer.

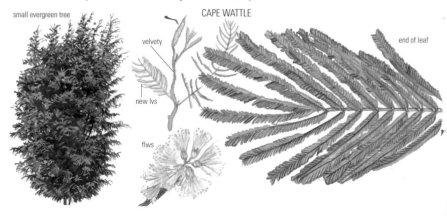

small evergreen tree · CAPE WATTLE · velvety · new lvs · flws · end of leaf

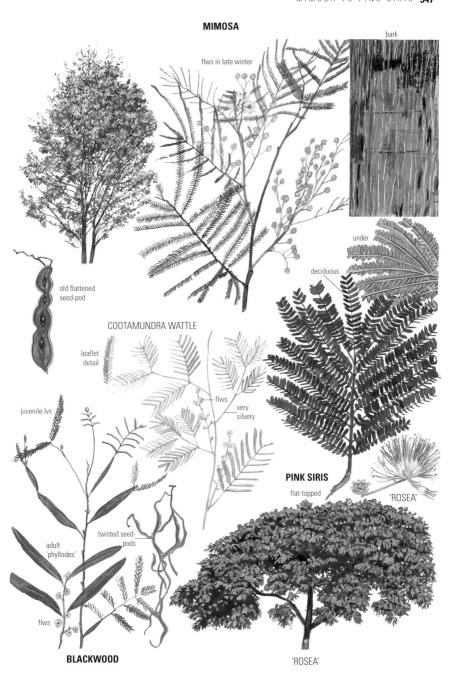

JUDAS TREE TO HONEY LOCUST

Judas Tree *Cercis siliquastrum*

E Mediterranean; long grown in the UK and quite frequent in parks and gardens in warmer parts.
APPEARANCE **Shape** Low (to 14 m); twisted branches and dense grey-tinged foliage; sprawling and layering given the chance. **Bark** Dark grey; smooth then with *close vertical corrugations*; small, rugged ridges in age. **Shoots** Red-brown. **Buds** Sharp, red; hidden in the bases of the leaf-stalks then rimmed by their grey scars. **Leaves** Kidney-shaped, to 10 cm wide, *untoothed and grey-green*, hairless. **Flowers** Like Sweet Peas, magenta-pink, in late spring with the young leaves; budding endearingly from the bark of the branches and trunk itself. 10 cm brown pea-pods follow and last into winter.
COMPARE Katsura (p.274): leaves of similar shape in mostly *opposite pairs* and more or less serrated.
VARIANTS 'Alba' has off-white flowers. Curiously ineffective and very rare.
OTHER TREES *C. canadensis* 'Forest Pansy' is a cultivar of the Redbud, from the E/central USA – *pointed*, heart-shaped leaves (sometimes downy beneath), and tiny (12 mm) but brilliant flowers. This rare clone has *purple young leaves* (fading to bronze; cf. Katsura 'Red Fox', p.274).

Mount Etna Broom *Genista aetnensis*

Sardinia, Sicily. Rather occasional.
APPEARANCE **Shape** The largest broom: curving branches *to 12 m* on a short, slanted trunk. **Bark** Green, then pale brown and shallowly ridged. **Leaves** Almost absent. The fine green photosynthesizing shoots make an open crown – like a winter willow. **Flowers** Large and golden-yellow in high summer; pods 12 mm.
COMPARE Tamarisk (p.410).

Honey Locust *Gleditsia triacanthos*

Central North America. 1700. Locally frequent in warm parts.
APPEARANCE **Shape** Slender and open on a long, sinuous stem, to 27 m; dainty and fresh-green in summer but very gaunt through winter, with thickly curling twigs. One of the last trees into leaf. **Bark** Purplish grey; developing wide, flanged ridges. Trunks of typical trees – rare here – carry *clumps of vicious 20 cm spines*. Thornless forms (f. *inermis*), often selected for straight trunks (eg. 'Shademaster'), have naturally been preferred in streets and parks. **Shoots** With 3 spines by each 1 mm orange bud in the thorny form. **Leaves** To 20 cm, *singly or doubly compound on the same tree*; the small (2–4 cm) glossy leaflets have *wavering margins* (rarely toothed). The doubly compound leaves have up to 4 pinnae (each like the illustrated leaves but usually smaller) on each side of their green central stalks. **Flowers** In thin, inconspicuous tails, followed (on non-sterile clones) by huge black-brown pods, like squashed bananas, which clatter in the wind. The 'honey' is their pulp, from which beer was once brewed in America.
COMPARE False Acacia (p.354).
VARIANTS 'Sunburst' (1954) is frequent: *fresh yellow* leaflets fade to lime-green; thornless. A less brilliant but subtler tree than Golden Robinia (p.354).
'Ruby Lace' (1961) has *purple-brown* young foliage, fading to a strange greeny-grey-brown which recalls a badly pressed leaf; rare so far.
'Nana' (origins obscure) is rare but distinct: a short, smooth (but spiny) *dark-grey* trunk carries a neat *egg-shape of close, steep, twisting limbs*, to 18 m.
OTHER TREES Chinese Honey Locust, *G. sinensis* (a few collections) has larger leaflets (each to 8 cm).

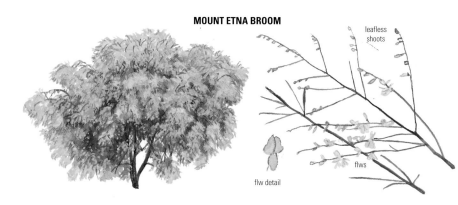

MOUNT ETNA BROOM

leafless shoots

flws

flw detail

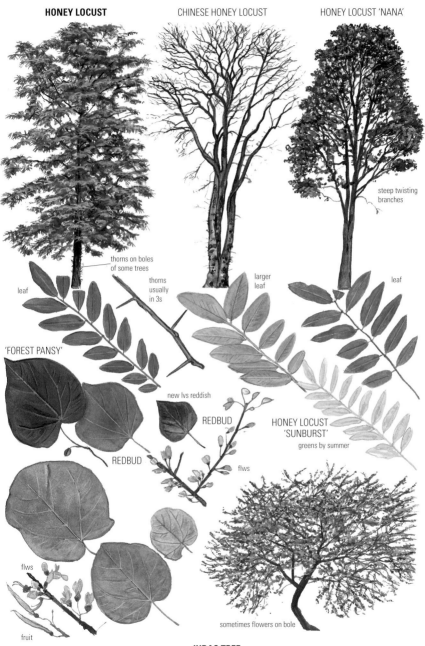

KENTUCKY COFFEE TREE & LABURNUMS

Kentucky Coffee Tree *Gymnocladus dioica* ☠

(*G. canadensis*) E and central USA. By 1748. Rather rare; confined to warmer areas.
Appearance Shape An open dome, to 17 m; very gaunt in winter: few, thick, twisting shoots, knobbly and bloomed pink. **Bark** Grey-brown; rugged and often *flanged* ridges (cf. Hackberry, p.254). **Leaves** Doubly compound and up to 1 m long; fresh green; unfolding late and pink, and turning yellow in autumn. (The illustration shows one leaf, not a leafy shoot.) Unlike most Leguminosae, the leaf-margins are (usually) serrated. **Flowers** Dioecious; rare in the UK. The waxy seeds, in pods to 25 cm long, were once roasted to make an ersatz coffee (they are poisonous in the raw).
Compare Golden Rain Tree (p.398): similar jizz.
Other trees Japanese Angelica Tree, *Aralia elata*, from the Far East, is unrelated (*Araliaceae*) but has similarly enormous, doubly compound leaves (darker and glossier). Occasional: usually a spinily grey-stemmed suckering bush; whitish flowers in huge (to 60 cm) airy heads in late summer followed by great towers of berries.

Common Laburnum
Laburnum anagyroides ☠

(Golden Chain Tree) Central and S Europe. Long grown in the UK: abundant as an older plant and rarely naturalizing. The hard, dark, greenish timber is valued in cabinet-making, but the whole tree is highly poisonous.
Appearance Shape Irregular, or rather weeping; to 10 m. **Bark** Greenish/coppery brown; smooth for many years, then distantly fissured or slightly corky. **Shoots** Grey-green with *silky hairs.* **Buds** Silvery-hairy. **Leaves** *Sparse*; the three 6 cm untoothed leaflets emerge silvery and remain *silky-hairy beneath*. **Flowers** In strings to 25 cm long in late spring; abundant, twisted seed-pods follow.

Variants Golden-leaved Laburnum, 'Aureum', is very rare. (Sickly trees of the type often show premature autumn gold.)
'Pendulum' makes an igloo of hanging shoots; rare. 'Quercifolium' has 3–5 *round-lobed* leaflets, on winged stalks. Handsome but very rare.
Other trees Scots Laburnum, *L. alpinum* (damper places and higher mountains in S Europe; 1596), is occasional (least scarce in Scotland); a taller, upright, rather funnel-shaped tree, to 12 m. Its browner bark becomes *ruggedly plated*; shoots are *soon hairless*; leaves rather larger and darker, a deeper green beneath and *almost hairless*. Flowers a fortnight later: longer strings (to 40 cm), though less packed with blossom; the seed-pod's upper seam has *1 mm flanges*.
Voss's Laburnum, *L.* × *watereri* 'Vossii', is the form of the hybrid of these two tree-species which is *now the commonly planted laburnum*: flower-strings as long as Scots Laburnum's and as crowded as Common Laburnum's. Bark and habit are intermediate; young shoots are soon hairless, the dense leaves are very slightly hairy beneath; the toxic pods *scarcely develop* (a major recommendation wherever they may tempt children).

Adam's Laburnum
+ *Laburnocytisus adamii* ☠

A chimaera of Common Laburnum and a dwarf purple broom (*Cytisus purpureus*) – two closely related but very different-looking plants – originating in Jean Louis Adam's nursery at Vitry near Paris (1825). Now very rare.
Appearance Shape An extraordinary, sickly tree with bits of both 'parents' breaking out all over it. **Leaves** Smaller than Common Laburnum's, nearly hairless; there are *random patches* of tiny broom foliage (leaflets 2 cm long). **Flowers** Laburnum-like but purple-flushed – and mingled with ordinary laburnum and broom flowers.

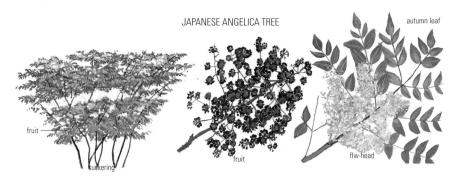

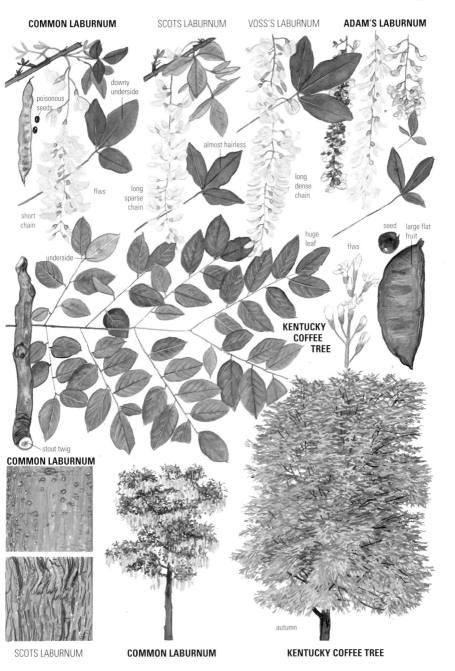

PAGODA TREE TO YELLOW-WOOD

Pagoda Tree *Sophora japonica*

(Scholar's Tree; *Stryphnolobium japonicum*) China, Korea (long grown in Japan). Quite frequent in warmer areas. 1753 original still thriving – horizontally – at Kew.
Appearance Shape Roughly domed, on heavy twisting limbs: a big, broad tree, to 25 m, sometimes recalling Huntingdon Elm (p.246) in winter; rich green in summer with yellow growth-tips. **Bark** Grey-brown; *criss-cross ridges*, less rugged and irregular than False Acacia's (p.354). **Shoots** *Dark green.* **Buds** 1 mm, hidden in the leaf-stalks then rimmed by their scars. **Leaves** To 25 cm: *9–15 untoothed, finely pointed* 3–6 cm leaflets. **Flowers** On trees 30 or more years old, showy after hot summers: white (mauve in 'Violacea'), *big branching heads in early autumn*. Seed-pods, to 8 cm, are rare in the UK. Flowers and fruits are much used in Chinese herbal medicine.
Compare False Acacia and *Robinia × slavinii* (p.354): differences emphasized. Chinese Yellow-wood (below) has more oblong leaflets.
Variants 'Pendula' is a rare but singularly picturesque weeping tree, to 10 m; it does not flower. 'Variegata' (very rare) has white-margined leaves. (The young leaves of type trees are conspicuously haloed with grey down.)

Kowhai *Sophora tetraptera*

New Zealand, Chile and Tristan da Cunha: the necklace-like seed-pods (4-ridged; to 20 cm) float, and are viable for 3 years, allowing dispersal by water. 1772. Very occasional in warmer areas.
Appearance Shape Rather stiff and gaunt; generally a multi-stemmed bush, to 12 m. **Bark** Blackish, smooth. **Leaves** Tiny: silky-hairy, dark, evergreen, convex, untoothed leaflets 12–35 mm long. **Flowers** Spectacular mustard-yellow clusters encrust the crown as the old leaves drop in late spring.
Compare Ladder-leaf Rowan (p.304).

Maackia *Maackia amurensis*

NE Asia and (var. *buergeri*) Japan. 1864. Very rare.
Appearance Shape Low (to 10 m); gracefully spreading. **Bark** Brown, cherry-like: *flaking finely and horizontally* between rough lenticel-bands. **Leaves** Most like Pagoda Tree's: leaflets rounder and darker (downy beneath in var. *buergeri*); buds *not fully hidden* by the swollen stalk-base. **Flowers** Ivory-white: *stiffly upright 12 cm heads* in high summer.
Compare Chinese Yellow-wood (below): smooth grey bark.

Yellow-wood *Cladrastis kentukea*

(*C. lutea*; *C. tinctoria*) Indiana to Carolina. 1812. Rare: large gardens. The timber, gold when freshly cut, yields a yellow dye.
Appearance Shape Delicately wide-spreading; to 15 m. **Bark** *Dark grey; finely* roughened. **Buds** *Scale-less,* 4 mm; hidden in the leaf-stalks, then rimmed by their scars. **Leaves** With *large* (to 12 × 7 cm), matt fresh-green, untoothed leaflets, well spaced more or less *alternately* up the central stalk and with silky hairs beneath. One of the few Leguminosae with good autumn colour – gold. **Flowers** Fragrant but infrequently seen here: large (20–40 cm), hanging heads in early summer.
Compare Walnut (p.178): dark, leathery leaves.
Other trees Chinese Yellow-wood, *C. sinensis* (1901), is very rare and has up to 17 *narrowly oblong leaflets* (to 10 × 3 cm), with long parallel sides and *rusty hairs* under midrib; flower-heads showy and *erect* (cf. Maackia, above, with papery bark).

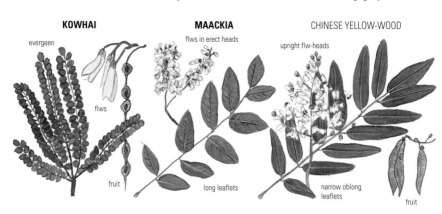

KOWHAI — evergreen, flws, fruit

MAACKIA — flws in erect heads, long leaflets

CHINESE YELLOW-WOOD — upright flw-heads, narrow oblong leaflets, fruit

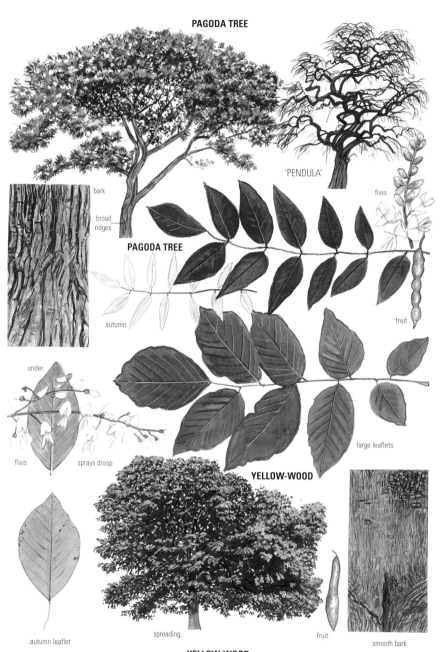

FALSE ACACIAS

False Acacia *Robinia pseudoacacia*

(Black Locust) E USA. By 1630. Abundant in warm areas, naturalizing by suckers and, more rarely, seedlings.
APPEARANCE Shape Gaunt, to 28 m: twisting, much-shattered branches on often straight, slanting stems. **Bark** Grey-brown; soon *very craggy* with long, *deep fissures*. **Shoots** Dark red, ribbed: stronger ones with *2 spines* by each tiny, scale-less bud (hidden in summer by the base of the leaf-stalk, and then rimmed by its scar). **Leaves** With 9–23 *oval*, untoothed leaflets (each 4 × 2 cm), fresh/bluish green, with a little, soft bristle at the *round*, minutely notched tip; soon hairless. **Flowers** White, scented: laburnum-like cascades at the start of summer after hot years. Dark brown, 10 cm seed-pods hang on in bunches.
COMPARE Pagoda Tree (p.352). Honey Locust (p.348): smaller, often doubly compound leaves.
VARIANTS Golden Robinia, 'Frisia' (1935), is now the most abundant and perhaps the *brightest* golden broadleaf in gardens in warm areas: large leaflets, clear yellow all summer, darken to gold in autumn. (In very hot summers they can fade to pale green.) Slender, to 18 m; short-leaved. ('Aurea', by 1864 and now very rare, is broader: yellow leaves mature lime-green. Type trees flush briefly yellow.)
'Bessoniana' (by 1871) is a locally frequent street-tree: a good *straight* trunk (typically 5 m) carries a *rounded crown* of heavy, twisting limbs, to 14 m; spineless (except from below a graft) and shy-flowering.
Mop-head Robinia, 'Umbraculifera' (c.1820; often mis-named 'Inermis'), is frequent: a *frizz of often thin, twisting stems* from a graft makes a *dense* but fragile low dome, to 9 m; very seldom flowering.
'Rozynskiana' (1903; rare) has *long, drooping, elegant leaves* with particularly *distantly set leaflets*.
'Pyramidalis' ('Fastigiata'; 1839) is rather rare: *Lombardy Poplar-shaped*, with few flowers.
Single-leaved Robinia, 'Unifoliola' ('Monophylla'; c.1855), is also rare. Large (10 × 4 cm) *simple leaves* are mixed with trifoliate ones (whose side-leaflets are relatively much smaller than laburnums').
'Monophylla Fastigiata' is confined to collections: columnar, its leaves often with 5 leaflets. There is also a semi-weeping 'Monophylla Pendula'.
'Tortuosa' has twisted, steep branches (cf. 'Umbraculifera'), rather *contorted leaves* and short, dense tails of blossom; rare.
OTHER TREES Clammy Locust, *R. viscosa* (Virginia to Alabama, 1797: rare in the wild, and in a few big gardens, to 15 m), has *sticky hairs* on its young shoots/leaf-stalks; flowers *pink*, scentless. Its hybrids with False Acacia (*R.* × *ambigua*) are very occasional: the *pale pink* 'Decaisneana' (1863) is hard to tell from False Acacia out of flower (shoots very slightly sticky); 'Bella-Rosea' (c.1860; deeper pink) is less vigorous, with stickier young growths.
Another pink-flowered hybrid seen occasionally (often as the 1933 clone 'Hillieri', blossoming showily as a sapling) is *R.* × *slavinii* (False Acacia × the American bush *R. kelseyi*; 1914), which has *only 9–11 well-spaced, slender leaflets with long-tapered tips*. These (and its *dainty habit*) suggest Pagoda Tree (p.352), but are quickly hairless and carry the robinias' small, soft bristle at their tip.

MOP-HEAD ROBINIA

dense dome of frizzy branches

'PYRAMIDALIS'

'FRISIA'

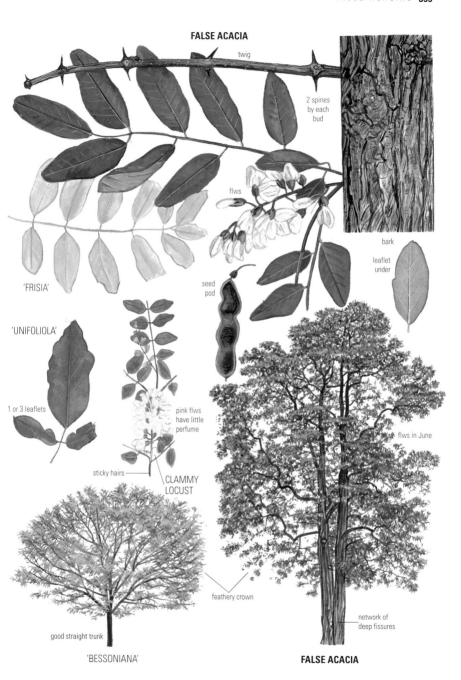

EUODIA & CORK TREES

*The 1700 plants of the family Rutaceae include lemons and oranges (*Citrus*) from SE Asia, but they are now widely grown in frost-free parts of S Europe.*

Euodia   *Tetradium daniellii* ☠ ❧

(*Euodia hupehensis, E. velutina*) China, Korea. 1907. Rare, but sometimes in town parks.
APPEARANCE Shape A broad dome on light, clean, ascending branches; to 24 m. **Bark** *Dark* grey, brown-speckled; smooth. **Shoots** Velvety; slender for a tree with compound foliage. **Buds** *Scale-less*, silvery/red, 3 mm. **Leaves** To 35 cm, in opposite pairs: 5–9 almost toothless, cupped leaflets, glossy above with odd hairs; whitish beneath with more down; heavily and sourly *lemon-scented*. **Flowers** Dioecious; showy white heads (to 15 cm) in *high summer*. **Fruit** Purplish, in heavy clusters.
COMPARE Manna Ash (p.438): scaled buds and earlier, superficially similar blossom. Phellodendrons (below). Chinese Varnish Tree (p.360): alternate leaves.

Phellodendron is a genus of 10 NE Asian trees, often with corky barks.

Things to Look for: *Phellodendron*

- Bark: Is it corky?
- Leaves: What shape are the leaflets at the base? How downy beneath?

Japanese Cork Tree   *Phellodendron lavallei* ❧

(*P. amurense* var. *lavallei*) Japan. 1865. Very occasional in warmer parts: big gardens; odd town parks.
APPEARANCE Shape A wide and dense or gaunt dome on twisting, ascending branches; to 17 m. **Bark** *Pale* grey-brown, soon with shallow *corky* ridges; much used in Chinese herbal medicine. **Buds** Broad, *apple-green*; almost hidden by the leaf-stalk's base then rimmed by its thin grey scar. **Leaves** To 40 cm, in opposite pairs: 7–13 untoothed leaflets tapering/rounded at their base, glossy above and with white hairs *under the main veins* and round the rim; they flush fawn-yellow. **Flowers** Dioecious: inconspicuous yellow-green heads to 15 cm wide. **Fruit** Shiny, black.
COMPARE Other Cork Trees; Euodia (above). The bark, though much shallower than Cork Oak's (p.224), helps distinguish from trees with *alternate* compound leaves (Chinese Cedar, p.358; Black Walnut, p.178).
OTHER TREES Amur Cork Tree, *P. amurense* (NE mainland Asia, 1885), is probably rarer here (older labels seldom distinguish the two): leaflets downy *only* at the rim and under *the base of the midrib*.

Japanese Phellodendron
                         Phellodendron japonicum ❧

Japan. *c.*1863. Rare.
APPEARANCE Shape A small, broad tree, often gaunt; to 14 m. **Bark** Grows criss-cross but *scarcely corky* grey-brown ridges. **Leaves** With broader, yellower leaflets than Japanese Cork Tree's; *level or heart-shaped* at the base, and duller above; *velvety-white on the central stalk and underneath*.
COMPARE Other Cork Trees; Euodia (above). With its less-distinctive bark this can be a puzzling species; its opposite, untoothed leaves differentiate it from Chinese Varnish Tree (p.360) or Yellow-wood (p.352).
OTHER TREES *P. chinense* (central China; 1907) is confined to collections and has *taper-based* leaves, thinly downy beneath. Female trees carry their fruit in a dense head, as *broad as high*.

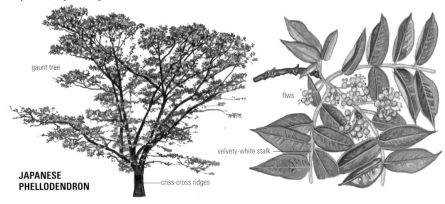

gaunt tree

flws

velvety-white stalk

JAPANESE PHELLODENDRON   criss-cross ridges

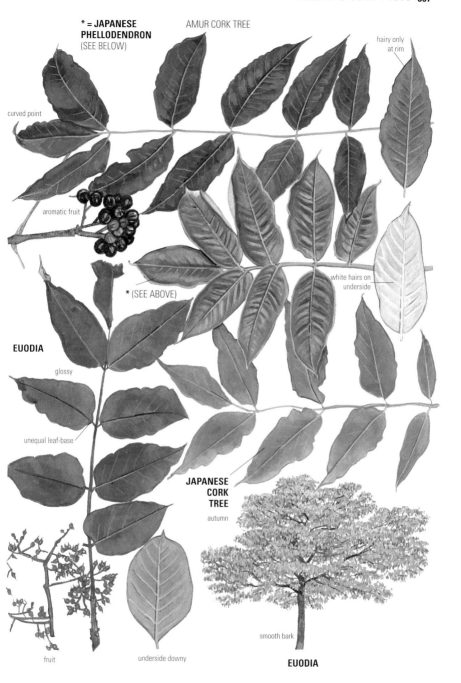

BOX TO CHINESE CEDAR

Box *Buxus sempervirens*

Europe (including SE England); N Africa; W Asia. Very locally abundant on chalk scarps (place-names suggest a wider former distribution); naturalized more widely and grown everywhere in garden hedges. The valuable, bone-like timber can sink in water. (Family: Buxaceae.)
APPEARANCE Shape Generally densely bushy, on steep sinuous stems; to 12 m. **Bark** Grey-fawn; close rugged ridges. **Leaves** Opposite: small (2–3 cm), convex, untoothed, often notched at the tip; glossy, leathery and evergreen, with an odd sweet smell. **Flowers** Yellow, tufting the shoots from late winter.
COMPARE Phillyrea 'Latifolia' (p.442); Small-leaved Azara (p.408).
VARIANTS Most are bushy. 'Aurea Pendula' (to 8 m; occasional) weeps, with gold-margined leaves.
OTHER TREES Balearic Box, *B. balearica* (also from SW Spain and Sardinia; 1780), is rare: *larger, flatter, matt yellowish leaves* (to 5 cm); showier, scented flowers.

Tree of Heaven *Ailanthus altissima*

N China. 1751. (A Moluccan member of the genus is called 'Ailanto', 'a tree reaching the skies'.) Scarce in the wild but aggressively naturalizing in US cities and S Europe. Frequent in warm parts: one of the toughest trees for dry, polluted urban sites. (Family: Simaroubaceae.)
APPEARANCE Shape Tall (to 28 m); stout, twisting branches from a short, column-like trunk. **Bark** Smooth, dark grey, with *fine white vertical 'snakes'*; sometimes slightly corky ridges in old age. **Shoots** Thickly tapering, finely velvety. **Buds** (Above big, pale, heart-shaped leaf-scars) tiny, reddish, domed. **Leaves** Huge, malodorous; flushing late and *red* (cf. Common Walnut, p.178); 11–21 glossy leaflets lightly downy at first underneath, and around the margin, which is untoothed except for *1–6 big teeth at the base of each leaflet* (glands on these ooze nectar, attracting ants to combat leaf-eating insects). **Flowers** Usually dioecious. Males in stinking, cream plumes in mid-summer. Females, understandably, are the norm, sometimes as grafts: fruits like ash-keys in huge bunches, glamorous and tropical-looking in late summer when – given enough warmth – they ripen apricot, scarlet or ('Erythrocarpum') crimson.
COMPARE Chinese Cedar, Varnish Tree (p.360): both lack the basal teeth. Butter-nut (p.180): fully serrated leaves.
OTHER TREES Downy Tree of Heaven, *A. vilmorniana* (W China,1897; collections) looks alike but has *odd soft spines* on its young shoots and *very downy leaves*; the red main stalk is sometimes bristly. The bole may be sprouty.

Chinese Cedar *Toona sinensis*

(Chinese Mahogany; *Cedrela sinensis*) N and W China. 1862. Rare; warm parts. (The aromatic timber of Australian Red Cedar, *T. australis*, gave this genus its perplexing name. Family: Meliaceae.)
APPEARANCE Shape Gaunt, open, to 27 m; suckering freely. **Bark** Grey; long, shallow plates becoming *shaggy*. **Shoots** *Pale orange*, with green buds. **Leaves** To 70 cm, generally minus an end leaflet (cf. Black Walnut, p.178, with a similar jizz); almost hairless when mature; with distant teeth *or none*. They *smell strongly of onions* (young shoots are eaten in China). **Flowers** Fragrant, open, white plumes, hanging to 50 cm, in mid-summer.
COMPARE Varnish Tree (p.360).
VARIANTS 'Flamingo' has shocking-*pink young leaves*, fading to green via cream. Very rare.

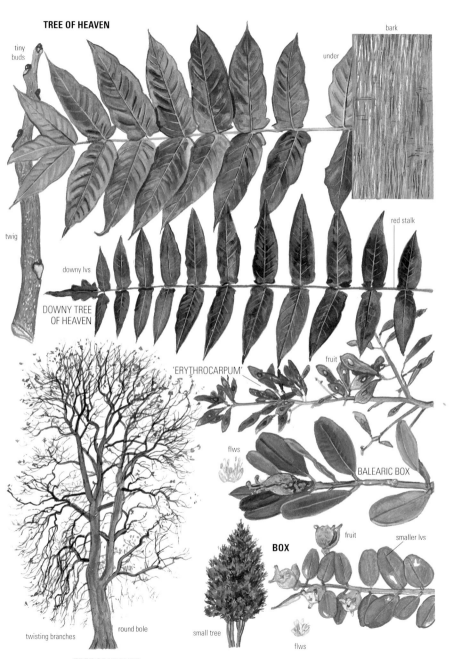

SUMACHS

Anacardiaceae is a mainly tropical family. The Rhus *genus (200 species) includes Poison Ivy (R. radicans) – some people are allergic to these trees' sap.*

Stag's-horn Sumach — *Rhus typhina*

E North America. *c.*1610. Abundant in small gardens, and suckering aggressively.
APPEARANCE Shape A *miniature, short-trunked dome* (to 7 m). **Bark** Brown, finely roughened. **Shoots** *Like stags' antlers in velvet.* **Buds** 4 mm hairy, orange, scale-less domes; hidden in summer by the base of the leaf-stalk. **Leaves** To 60 cm: up to 25 *serrated*, long-pointed leaflets. They droop then fall individually, often crimson, but may hang into winter. **Flowers** Dioecious. Most garden trees are female: dense, crimson-furred fruiting spires last until spring.
VARIANTS Cut-leaved Sumach, 'Dissecta' ('Laciniata'), has very feathery foliage; occasional (cf. Cut-leaved Walnut, p.178).

Varnish Tree — *Rhus verniciflua*

Himalayas, China. 1874. Rather rare. Its (toxic) sap is tapped in Japan for lacquer production.
APPEARANCE Shape Gaunt, open, to 21 m: often whorled branches from straight stems. **Bark** Dark grey; roughened with black diamond-marks then small, curving scales. **Shoots** *Pale grey*, freckled orange. **Buds** Shiny brown, the end one *beaked, 1 cm.* **Leaves** To 70 cm: 7–19 *untoothed* leathery leaflets glossy above, velvety beneath; smelling faintly of balsam (don't crush them) and red in autumn. **Flowers** Often dioecious. Whorls of sparse off-white sprays to 25 cm long in midsummer. **Fruit** The glossy yellow-brown berries are pressed for their oil in China.
COMPARE Chinese Cedar (p.358). Tree of Heaven (p.358): *large teeth* at base of leaflets.

Chinese Varnish Tree — *Rhus potaninii*

N China. 1902. Rare: big gardens.
APPEARANCE Shape Wide (rarely to 24 m): twisting twigs on level limbs. **Bark** Grey, growing *oak-like orange fissures*. **Buds** Scaleless, white-hairy, and hidden in the leaf-stalks in summer; the end one missing. **Shoots** Green, often finely velvety at first. **Leaves** With 7–11 glossy, sometimes toothed leaflets, the tip one *projecting level*; pink and red in autumn. **Fruit** Bright red, downy berries.
COMPARE Maackia (p.352). Euodia (p.356): leaves in opposite pairs.

Smoke-bush — *Cotinus coggygria*

(Venetian Sumach; *Rhus cotinus*) Central and S Europe to China. *c.*1656. The type is now rare.
APPEARANCE Shape Bushy; to 9 m. **Bark** Pale grey-brown; closely ridged. **Leaves** *Untoothed*, hairless; to 7 × 5 cm, rather racquet-shaped; conspicuous parallel veins and lambent autumn reds. **Flowers** In *smoky heads* of largely barren, hairy stalks.
COMPARE Alder Buckthorn (p.398); Snowbell Tree (p.434).
VARIANTS f. *purpureus* has purple-pink flower-heads; 'Foliis Purpureis' has purplish leaves and flower-heads; the abundant 'Notcutt's Variety' and 'Royal Purple' have purple leaves (green beneath) and flower-heads (but crimson autumn colour).
OTHER TREES Chittam Wood, *C. obovatus* (*C. americanus*; SE USA, 1882), is rare in the wild and (as hybrids?) in old gardens: a *domed tree* (rather mulberry-like in winter) with criss-cross orange-brown bark-ridges, and narrower leaves than Smoke-bush's (to 12 × 7 cm), more *wedge-shaped at the base*; seed-heads open and not showy.

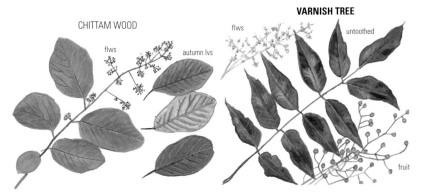

CHITTAM WOOD — flws — autumn lvs

VARNISH TREE — flws — untoothed — fruit

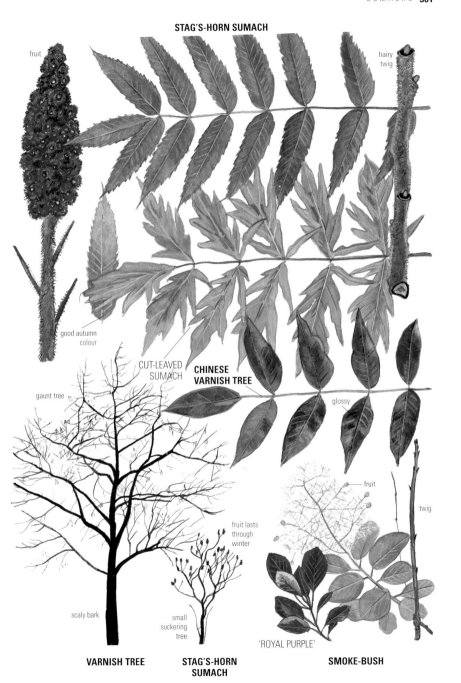

HOLLIES

Aquifoliaceae is a widely distributed family dominated by the hollies.

Common Holly *Ilex aquifolium* 🌲 ☠

Europe (including Britain and Ireland); W Asia. Abundant on non-chalky, free-draining soils away from the coldest areas: woods, old commons and hedges, and even exposed moorlands and shingle ridges. Planted everywhere: the sole European member of this genus of 400 diverse species has some claim to be the most ornamental of all.

APPEARANCE Shape Spire-shaped, then irregularly upright, to 23 m; often with densely cascading lower shoots; suckering and sometimes widely layered. **Bark** Brownish grey, finely roughened and frequently with small, round warts. **Leaves** Evergreen, with a high-gloss finish above (pale and matt beneath); massively spined on young plants. The spines deter browsing mammals but are expensive to produce: tall trees have almost entirely unspined, plane leaves and can look puzzling. **Flowers** In late spring. Males have a cross of 4 yellow-headed stamens; females (on separate trees) have a single style in the middle. **Fruit** Satiny-crimson, toxic berries (whose abundance hinges on how many pollinating insects were on the wing at flowering time).

COMPARE Highclere Hollies (p.366). Holly-leaved Osmanthus (p.442) and sometimes Phillyrea (p.442): similarly spined but *opposite* leaves. Japanese Evergreen Oak (p.230) and Chinese Tree Privet (p.442): glossy, always unspined leaves. Californian Live Oak (p.222), Cork Oak (p.224) and young Holm Oaks (p.222): softer spines and different barks.

OTHER TREES American Holly, *I. opaca*, has long been grown here but is now very rare: leaves very similar but *matt* (cf. Himalayan Holly, p.366).

Madeira Holly, *I. perado* (with varieties in the Azores and Canaries), is hardy only in the mildest areas. Its *broad*, glossy leaves have a few *slender forward-pointing spines*, or none. (The Highclere Hollies, p.366, are much more widespread.)

VARIANTS Clones of Common Holly are either male or female, but trees can be sexed with confidence only when in flower, as unpollinated females will not berry. Some trees, marketed before they had ever flowered, have names that belie their sex. Golden Holly, f. *aureomarginata*, is popular in a range of clones: leaves normally shaped, but margined with deep yellow (see p.364 for further variegated hollies with odd-shaped foliage).
'Golden Queen' is frequent and distinctive. Male; sturdy, with *much-pimpled bark* and broad leaves, *pure gold on the numerous sprouts*.
'Madame Briot' is – happily – female: purple shoots; broad, very spiny leaves *mottled* as well as margined with deep gold.
Egham Holly, 'Heterophylla Aureomarginata', is female. Its broad leaves are *unspined even on young plants*. (Cf. 'Belgica Aurea', p.366.)
'Watereriana' is male and has a *dense, neat habit* and quite broad, usually unspined leaves with a narrow but neat rich gold margin (cf. 'Laurifolia Aurea', p.364).
'Golden Milkboy' is male and has a dark gold splash *in the middle* of each leaf (cf. Golden Hedgehog Holly and 'Crispa Picta', p.364). Prone to revert, so very rarely seen as an older tree. 'Golden Milkmaid' is the female equivalent.

COMMON HOLLY — a female tree

GOLDEN HOLLY — some lvs fully yellow

MADEIRA HOLLY — broad lvs

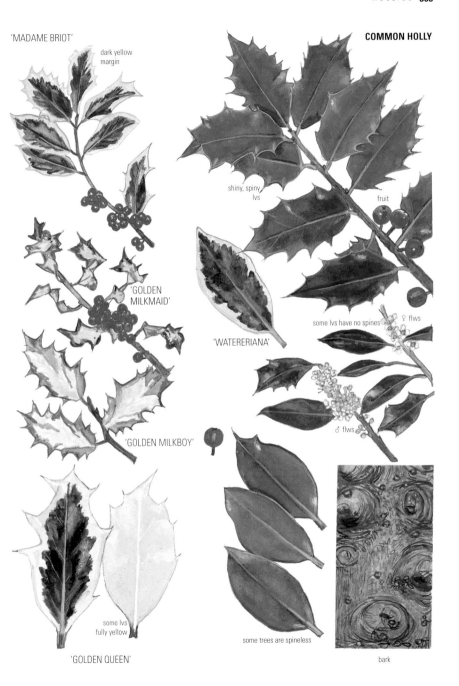

COMMON HOLLY CULTIVARS

Silver Holly, *Ilex aquifolium* f. *argenteomarginata*, is as common as Golden Holly (p.362) in various clones: its normal-shaped leaves have white/pale creamy-yellow margins.
'Argentea Marginata' is a popular tall-growing and freely fruiting female clone with very bright, white leaf-margins.
'Handsworth New Silver' is female and has *purple shoots* and rather slender leaves with curiously massive spines *in a flat plane*, the centre mottled in grey and the silver margin brilliant.
'Silver Queen' is male, with *purple* shoots and bright silvery leaf-margins.
'Silver Milkboy' – this one actually *is* male – has a cream splash *in the middle* of the leaves and easily reverts (cf. 'Golden Milkboy'/'Golden Milkmaid', p.362).

Hollies with unusual leaf-shapes include Hedgehog Holly, 'Ferox', grown since Stuart times and now very occasional. The small leaves have a stubble of spines *across the upper face* towards the tip: a dark, rather gaunt male tree (to 12 m only) which is only obvious at close range.
Silver Hedgehog Holly, 'Ferox Argentea', has similarly aggressive leaves, *margined* in pale yellowy cream (not really silver); more vigorous than 'Ferox', and occasional. Golden Hedgehog Holly, 'Ferox Aurea', has leaves with a *central old-gold splash*; rare and bushy.
'Crispa' is extraordinary (though hardly ornamental) and now rare. Thick, leathery leaves on purple shoots are highly convoluted and have few or no spines except for a downward prickle at the tip. Male (cf. Kohuhu, p.410.) 'Crispa Picta' is a variant whose leaves are splashed yellow in the centre.

Leather-leaf Holly, 'Crassifolia', is an even more extreme form, now very rare; the narrow leaves, *as thick as leather*, have small, flat spines; female.
Myrtle-leaved Holly, 'Myrtifolia', is occasional and very distinct: its *tiny narrow leaves* (to 50 × 15 mm) are flat, with neat, small spines. Male; rather spire-shaped, to 12 m. 'Myrtifolia Aurea' has similar if slightly broader leaves, margined yellow.
Laurel-leaved Holly, 'Laurifolia', is a vigorous, scarce male tree with purple shoots and *narrow*, mostly unspined leaves (to 8 × 3 cm). 'Laurifolia Aurea' has leaves with neat, narrow gold margins (cf. the broader-leaved 'Watereriana', p.362).
Pyramid Holly, 'Pyramidalis', is a now-popular female selection with a neat spire-shape, generally *unspined leaves* and heavy crops of berries. Trees almost as fine are sometimes seen in the wild (f. *heterophylla*).

Other unusual forms include Moonlight Holly, 'Flavescens', with entirely *brilliant yellow young leaves*, which fade slowly to dark green so that the new growths stand out against the interior. Rare (wild trees can have dull yellowish leaves when sickly; their leaves also die off individually yellow, mostly in spring).
Yellow-berried Holly, f. *bacciflava* ('Fructu-Luteo'), is occasional but recognizable only when carrying its *bright yellow berries*.
Weeping Holly, 'Pendula' (rather rare), makes a dome of foliage in *dense curtains* (except in dense shade) from a high graft; female.
Perry's Weeping Holly, 'Argenteomarginata Pendula', is one of the best hollies, fruiting freely: the low dome is of brightly silver-edged leaves. The equally attractive golden-variegated equivalent, 'Aurea Pendula', is strangely very rare.

WEEPING HOLLY

'LAURIFOLIA AUREA'

LAUREL-LEAVED HOLLY

most lvs unspined

COMMON HOLLY CULTIVARS

HOLLIES

Tarajo *Ilex latifolia*

Japan. 1840. Rare; large gardens in milder areas.
Appearance Shape A puzzling, slender, giant-leaved tree, to 12 m. **Bark** Rather silvery. **Shoots** Woolly. **Leaves** *Very leathery* and glossy (hairless); convex and long-pointed, to *20 cm*; sometimes toothed but not spined. **Flowers, Fruit** Like Common Hollies'.
Compare Giant Osmanthus (p.442); Giant Photinia (p.282); Southern Evergreen Magnolia (p.260); Cherry Laurel (p.344).

Highclere Holly *Ilex × altaclarensis*

The tender Madeira Holly (p.362) used to be popular in conservatories; in spring the pots were wheeled into the air and Common Hollies sometimes fertilized the females. Their descendants are particularly vigorous, pollution-tolerant, trees, with flattened shoots and broader leaves than Common Holly's, but a similar bark and habit. Some cannot be named with confidence. Forms include the following.
Hodgins' Holly, 'Hodginsii' (by 1836), is now occasional: leaves to 10 × 8 cm, thick and glossy, with irregular, small, level spines. Male, its shoots purple in sun. 'Wilsonii' (a female variant, now rare) has paler leaves with small, more or less regular spines.
'Camelliifolia' has longer, slenderer, finely pointed, high-gloss leaves (to 13 × 6 cm); female, with violet-based petals. A narrow tree (to 20 m), its foliage dense and slightly weeping; the leaves vaguely suggest Japanese Evergreen Oak's (p.230), and even Tarajo's. Glamorous but rather occasional.
Henderson's Holly, 'Hendersonii' (very occasional), is female, with *matt* leaves, almost as broad as those of 'Hodginsii' and often spineless; 'Mundyi' (rare) is similar but male, its leaves rather wrinkled. 'Golden King' (female, of course; *c.*1870) is a sport of 'Hendersonii', to which unhappily it readily reverts. Quite frequent, and often densely bushy; its leaves have a few small spines and are matt, but *brightly gold-margined*; 'Lawsoniana' (*c.*1865; occasional; also female) instead has a *central marbling of yellow and pale green*.
'Belgica Aurea' (by 1908) is more vigorous than 'Golden King': its broad, *glossy* leaves have pale creamy-yellow margins. Female; apt to revert and now very rare.

Himalayan Holly *Ilex dipyrena*

E Himalayas to W China. 1840. Rare: large gardens.
Appearance Shape To 16 m, but often bushy. **Bark** *Brown*-grey, finely roughened. **Leaves** *Narrow*, to 12 × 3 cm; dark bluish and *matt above* but a shinier green beneath; untoothed or with fine, *appressed* yellow spines. **Fruit** *Dull* red berries, not densely clustered.
Compare Henderson's Holly (above); Giant Osmanthus (p.442).
Other trees The hybrid with Common Holly, *I. × beanii*, is very rare: shorter, broader dull leaves (never with the big upwards-and-downwards spines of Common and American Hollies, p.362).

Perny's Holly *Ilex pernyi*

Central China to E India. 1900. Rare.
Appearance Shape A gaunt, spiky little plant (to 9 m). **Leaves** More or less stalkless, almost *triangular*; crowded as if on a cocktail stick (smaller and shorter than the *opposite* leaves of Holly-leaved Osmanthus, p.442).

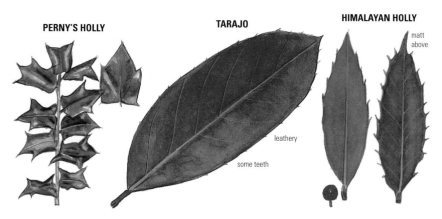

PERNY'S HOLLY   **TARAJO**   **HIMALAYAN HOLLY**

matt above

leathery

some teeth

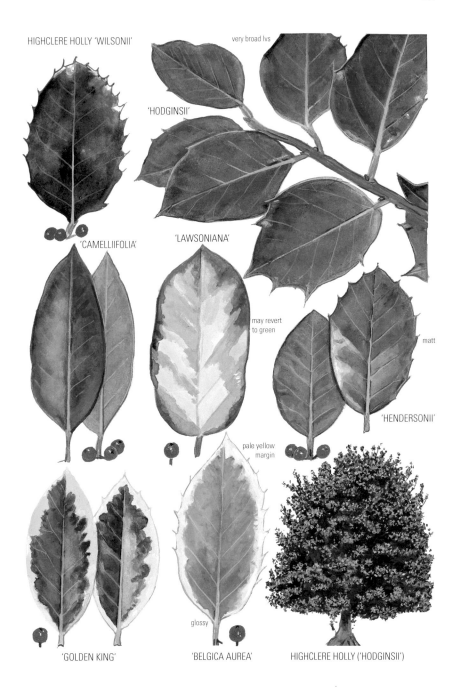

MAPLES

Aceraceae is a family of Northern-hemisphere trees and shrubs, all but 2 being maples. Maples (150 species) have leaves in opposite pairs that are very varied in shape, but most typically have 5 veins radiating neatly at 45° from the base to 5 lobes. (Other large hardy trees with 'maple-shaped' leaves – Planes, p.280, Sweet Gums, p.278, and Castor Aralia, p.424 – all bud alternately.) Trident Maple (p.374), Lime-leaved Maple (p.382), Tartar Maple (p.376) and Hornbeam-leaved Maple (p.384) have generally unlobed leaves of particularly abnormal shapes. Flowers occasionally have showy petals; fruits are bunches/strings of paired 'keys'.

Things to Look for: Maples

- Shape: What shape is the tree?
- Bark: What is it like?
- Shoots: What colour are they?
- Buds: What colour are they? Are they stalked?
- Leaves: What shape? Are they hairy or glaucous beneath? Are their teeth whisker-tipped?
- Flowers: What colour are they? Are the flowers/fruits bunched (erect/drooping?) or in tails?

Key Species

Field Maple (below): 5-lobed leaves with few, rounded teeth. **Norway Maple** (p.372): 5-lobed leaves with few, long-pointed teeth. **Sycamore** (p.370): 5-lobed leaves with many coarse jagged teeth. **Smooth Japanese Maple** (p.384): 5- or 7-lobed leaves with many fine teeth. **Cappadocian Maple** (p.374): 5- or 7-lobed leaves, more or less untoothed. **Downy Japanese Maple** (p.386): leaves with up to 11 lobes. **Silver Maple** (p.378): leaves very glaucous beneath. **Père David's Maple** (p.380): retains a green bark with white stripes. **Paper-bark Maple** (p.388): trifoliate leaves; red bark. **Box Elder** (p.390): 3–7 leaflets; grey bark. **Syrian Maple** (p.376): glossy, semi-evergreen leaves.

Field Maple *Acer campestre*

(English Maple) Europe, including England and Wales; SW Asia; N Africa. Abundant on rich/heavy soils; frequently planted everywhere.
APPEARANCE Shape Densely twiggy (often almost solid in winter); usually domed when free-standing; to 25 m. **Bark** Pale rather bright brown, with close ridges from the first: slightly corky (cracked into small squares in age) and feeling warm in cold weather. **Shoots** Thin; pale brown, with small grey-hairy buds. They are wrinkly by their second year then may develop corky wings (cf. English Elm, p.244; Sweet Gum, p.278). **Leaves** *Small* (to 10 cm wide); very dark and slightly shiny. The 5 (3) neat lobes have just a few big *rounded* teeth. Rich yellow autumn colour (rarely red). The sap, like that of Norway Maple, Oregon Maple, Miyabe's Maple and Cappadocian Maple and its allies, is *milky* not clear (snap a leaf-stalk and squeeze). **Flowers** In little erect yellow-green posies as the leaves emerge; the green/crimson keys have *horizontal* wings. Some trees are consistently male or female.
COMPARE Montpelier Maple and Balkan Maple (p.376): both easily mistaken but with consistently different leaf-shapes. Miyabe's Maple (p.386). Oregon Maple (p.378): much bigger leaves. Maple-leaved Crab (p.312): alternate leaves.
VARIANTS Fastigiate Field Maple, f. *fastigiatum*, is planted occasionally in streets and parks in a number of grafted clones such as 'Elsrijk', though wild trees (especially when crowded) can have closely ascending limbs. No clone is strikingly narrow for long, and all broaden with age.
'Compactum' has a small, globular crown of tiny leaves; rare.
'Pulverulentum' has leaves densely white-stippled. A very bright, ghostly tree, slow and spiky, but rare and tending to revert.
'Carnival' is an exciting recent clone: brilliant thick white margins to leaves which unfold red-pink.
'Postelense' (1896) has gold young leaves; green by summer except for late growths. Very rare.
'Schwerinii' (by 1899) has maroon leaves, maturing reddish green. Very rare. In 'Red Shine', a recent improvement, the younger leaves (and shoots) are a shining coppery-red. (Wild trees cut in hedges can also have coppery new growths.)
OTHER TREES Spanish Maple, *A. granatense* (mountains of SE Spain and N Africa; scarcely grown in the UK), is a small tree with smooth grey bark and leaves less plane and more toothed (cf. Balkan Maple, p.376); its sap is *clear*.

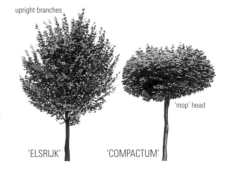

upright branches

'mop' head

'ELSRIJK'   'COMPACTUM'

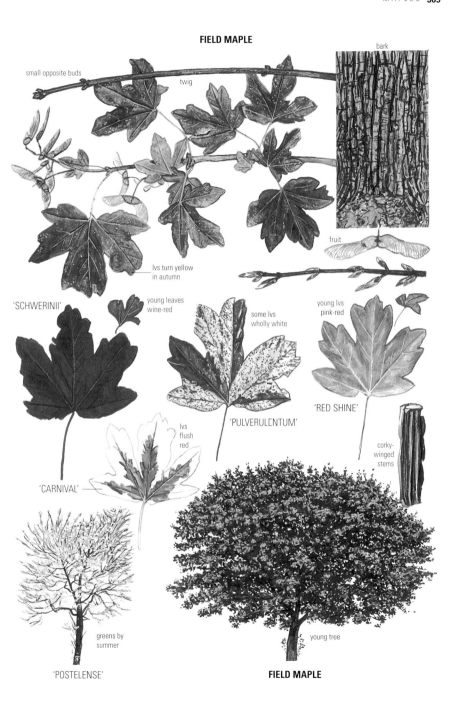

MAPLES

Sycamore — *Acer pseudoplatanus*

(Great Maple and – in Scotland – 'Plane') Europe N to Paris; long grown in the UK and fully naturalized: seedlings quickly dominate on rich/heavy soils. Much hated and much hacked: the heavy leaf-fall smothers ground flora, though it does support a high insect biomass – so plenty of birds. Strangely tough (considering its origins): highly valued for its shelter in uplands and on coasts.
APPEARANCE Shape A huge, heavy dome, *to 38 m* where it does best (eg. S Scotland; Kent); twisting, short twigs on straight limbs. **Bark** Pinkish grey: smooth, then by 80 years *shaggy* with small pale grey plates. **Shoots** Greenish grey-pink, stout. **Buds** Big, *green*. **Leaves** Large – to 18 × 26 cm on young trees – with many *coarse, round-tipped teeth*; dull dark green; some tawny down beneath. They flush orange-brown and, in lowlands, drop early, much blackened by Tar-spot Fungus (*Rhytisma acerinum*). **Flowers** Yellow-green, on 6–12 cm *tails*; the keys (green; red in f. *erythrocarpum*) hang from the tail's central stem.
COMPARE Downy Maple (p.388) and Horned Maple (below): similar foliage. Heldreich's Maple (p.388); Italian Maple (p.376); Deep-veined Maple (p.386).
VARIANTS Purple Sycamore, f. *purpureum*, is frequent in gardens and the wild: bark pink-tinged; leaves *green above but mauve/royal purple beneath* – a novel maroon-brown as the sun shines through them. The darkest ('Atropurpureum') are grafts. 'Brilliantissimum' is quite frequent: a low, dense dome (ultimately 15 m); small, sharp-lobed leaves unfold *pink-white, fading to yellow-green* by summer. 'Prinz Handjery' (1883) is more local: its darker leaves stay *mauve* beneath (cf. 'Nizetii', below).
Variegated Sycamore, f. *variegatum*, has leaves *radially splashed* yellow-white; quite frequent and brightly yellowish at a distance; bark usually pale grey. 'Nizetii' (occasional; rather narrow) has leaves also *mauve beneath*. Clones include 'Leopoldii' (*c*.1860; yellowish pink/purple splashes in spring), and 'Simon-Louis Frères', (darker leaves flush pink and stay pinkish creamy-green beneath).
Golden Sycamore, 'Worleei', has small leaves with triangular lobes, *bright yellow* then fading slowly to lime-green except at the growth-tips; occasional. Corstorphine Plane, 'Corstorphinense' (1600), has leaves of normal size and shape, fading *from gold to deep green in 6 weeks*; now rare.

Horned Maple — *Acer diabolicum*

Japan. 1880. Rare. Readily taken, out of flower, for a delicate little Sycamore.
APPEARANCE Shape Rounded, on straight, fine branches, to 15 m. **Bark** Grey; rather smooth. **Shoots** Green, purple-bloomed. **Buds** *Brown*. **Leaves** Slightly brighter, smoother and *harder* than Sycamore's (Plane-like); *fringed with hairs* (cf. Oregon Maple, p.378) and downy under the veins. Seldom much autumn colour. **Flowers** Yellow bells, in clusters hanging under the unfolding leaves to recall *open parachutes*. Many trees are f. *purpurascens* (Japanese gardens, 1878): *dusky-red* bells prettily set off by fresh green foliage. **Fruit** With stinging bristles; the 2 styles persisting between the wings suggest devil's horns.
COMPARE Downy Maple (p.388): similar if less showy flowers.

MAPLES

Norway Maple — *Acer platanoides*

Europe (but not Britain/Ireland); the Caucasus. 1683. Abundant; sometimes well naturalized.
APPEARANCE Shape A neat leafy dome; to 30 m. **Bark** Pale grey; *closely corrugated* with small regular ridges; oldest trees more rugged. **Shoots** Shiny brown; broad, red-brown buds. **Leaves** Elegant, plane; *hairless* except for tufts under the vein-joints; the *few long teeth mostly* (until tattered by weather) *have whisker-tips*. Yellow autumn colours (rarely purplish then red). Squeezed stalks ooze *milky* sap. **Flowers** In acid-yellow *branching erect posies,* with the leaves (cf. Italian Maple, p.376).
COMPARE Sugar Maple (below): differences emphasized. Lobel's Maple (p.374); Miyabe's Maple (p.386).
VARIANTS 'Drummondii' (1903) is one of the brightest and commonest silver-variegated trees (yellow as its leaves open); rather globular and slow, but speeding up as reversions take over. 'Goldsworth Purple' (1936) has *heavy purple* foliage (greener beneath and greenish black by autumn); abundant, to 17 m so far. Winter buds saturated purple; *purple* bracts and stalks make the flowers look orange-brown. 'Crimson King' (1946) and 'Faassen's Black' (1936) are hard to separate; 'Crimson Column' is newer and erect.
'Schwedleri' (1870), frequent as a *large* older tree, to 28 m, starts off purple-*red* (with *red* flower bracts/stalks); its leaves turn mauvish *grey-green* by high summer, but show purple autumn colour. 'Reitenbachii' (rare?) is *lower,* and *often burry*: slenderer leaves, more *reddish* green in summer.
'Palmatifidum' ('Lorbergii', 'Dissectum') is a complex of very occasional clones, densely low-domed, whose leaves are *cut almost to the centre*. Eagle's Claw Maple, 'Laciniatum', is rare: *slender, erect; deep jagged lobes curl down* at their tips.
'Cucullatum' (by 1880) is rather rare: erect, then to 25 m, on steep limbs; leaves rounded and *fringed with many tattered lobes*.
'Columnare' is occasional as a younger street tree: leaves short-lobed; steep branches make a narrow, *dense, often irregular funnel/column,* to 25 m (cf. Black Maple 'Temple's Upright'). 'Olmsted' (1952) and 'Almira' are broader but shapelier (rare?). 'Globosum' makes a dumpy lollipop of a tree; rare.

Sugar Maple — *Acer saccharum*

(*A. barbatum*) E Canada to Georgia. 1735. The maple on the Canadian flag, and the one most often tapped for maple syrup. Very occasional.
APPEARANCE Shape Easily taken for Norway Maple. To 25 m. **Bark** Grows slightly coarser, *shaggier* ridges. **Shoots** Olive, often with a purple-red band by each (*finely downy*) bud-pair. **Leaves** Flimsy, variably *downy beneath*; the main lobe-tips *but not the* (few) *teeth* whisker-tipped; gold and scarlet in autumn. Squeezed stalks ooze *clear* sap. **Flowers** Yellow, *dangling on clusters of fine threads.*
VARIANTS Black Maple, ssp. *nigrum* (*A. nigrum*; similar habitats; 1812), is very rare: *dull,* dark leaves with drooping sides to 3 (5) shallow lobes; bark purplish; shoots orange; 'Temple's Upright' is a rare fastigiate clone of this ssp. (cf. Norway Maple 'Columnare', above, and Red Maple 'Scanlon', p.378).

SUGAR MAPLE

whiskers tip main lobes
downy beneath
fruit
autumn
slender straight shoots
NORWAY MAPLE 'CUCULLATUM'

MAPLES

Cappadocian Maple
Acer cappadocicum

(Coliseum Maple; *A. laetum*) Caucasus region, extending E to China (as ssp. *sinicum*). 1838. A frequent street and park tree, closest in appearance to Norway Maple (p.372) but the commonest species whose leaves have consistently *untoothed* lobes; rarely naturalizing by its *suckers*.
APPEARANCE Shape Densely broad-domed on a usually short, often knobbly bole; to 25 m. **Bark** Pale grey: *smooth* then very shallowly fissured. **Shoots** Crimson on strong growths/suckers, bloomed grey; then smooth and *green for several years*, and finely striped white (cf. Snake-bark Maples, p.380, and Lobel's Maple). **Leaves** With 5 (7) whisker-tipped lobes; tufted under vein-joints. They expand briefly crimson, and are invariably butter-yellow in autumn. **Flowers** Yellow: in erect, branching heads among the leaves in late spring.
COMPARE Lobel's Maple (below).
VARIANTS 'Aureum' has brilliant whitish yellow younger foliage; this fades slowly to dark olive-green but is still sprinkled with the yellow stars of late leaves when the crowns of type trees become sprinkled with early-colouring foliage. Occasional, to 22 m.
'Rubrum' is a confused name, sometimes applied to the common clone in cultivation but more often to rare grafts with consistently crimson young shoots and leaves retaining pinkish margins.
The Chinese ssp. *sinicum*, in collections, has smaller, more deeply and narrowly lobed leaves, a slightly rougher bark, and bright red keys.
OTHER TREES *A. mono* (*A. pictum*; E Asia, 1881) is in many collections, to 18 m. Differs in often *shaggy* whitish bark and twigs *grey-brown and fissured by their second year*; the leaves are downy beneath in the rare f. *ambiguum*.

Shandong Maple, *A. truncatum* NE China, 1881) is also confined to collections, and differs from *A. mono* in its delicate, fan-shaped small leaves which are usually *cut flat across at the base*, and never show more than tufts of hair underneath; they tend to flap in the breeze on their very long stalks.

Lobel's Maple
Acer lobelii

S Italy. 1838. Quite frequent as a younger street tree.
APPEARANCE Shape Differs from Cappadocian Maple (above) most obviously in its non-suckering, naturally *columnar habit* with stiffly erect branching, making it ideal for street planting. To 28 m: the oldest trees are broader and irregularly domed. Cappadocian Maple suckers can surround grafted examples. **Bark** Fawn-grey; shallowly ridged with age. **Shoots** *Green for several years* with faint white stripes (like Cappadocian Maple's), but *bloomed purple* in their first year. **Leaves** Dark green; sometimes with *odd big teeth* on their wavy margins (but fewer than Sugar Maple, p.372); held level, the points of the 5 (3) lobes *twisting up or sideways* (cf. Eagle's Claw Maple, p.372).
COMPARE Norway Maple 'Columnare' and Black Maple 'Temple's Upright' (p.372).

Trident Maple
Acer buergerianum

(Tang Maple) China. 1896. A characterful little maple in some big gardens.
APPEARANCE Shape An often dense but irregular dome; to 16 m. **Bark** Bright brown; soon *very shaggily ridged*. **Leaves** With *3 forward-angled main veins*, usually running to 3 untoothed lobes; oblong on some trees, or with odd rounded teeth; shiny and leathery; glaucous beneath. Rich crimson autumn colours.
COMPARE *A. pycnanthum* (p.378): toothed lobes.

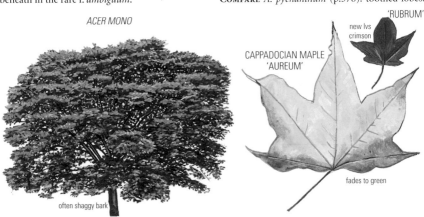

ACER MONO

often shaggy bark

CAPPADOCIAN MAPLE 'AUREUM'

'RUBRUM'
new lvs crimson

fades to green

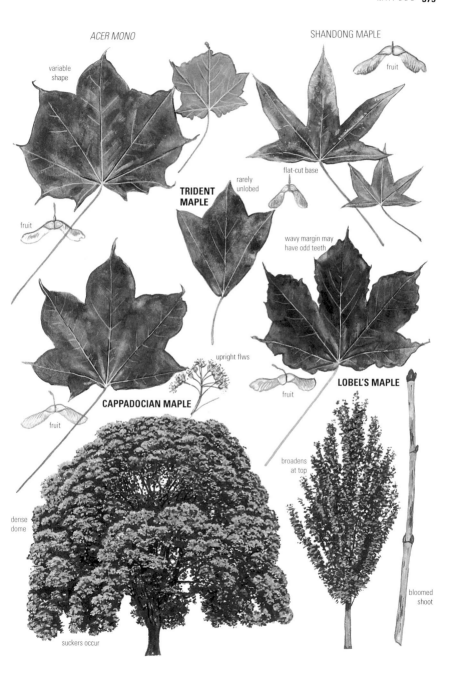

MAPLES

Zoeschen Maple — *Acer × zoeschense*

Cappadocian × Field Maple. By 1870. Rare.
APPEARANCE Shape Usually densely domed, to 24 m. **Bark** Grey-brown; soon shallowly ridged. **Leaves** Held flat: narrower triangular lobes than Cappadocian Maple's (p.374) and the odd tooth at their 'shoulders'; downy beneath at first. *Red-purple* stalks and major veins usually create a *dark, maroon cast* (cf. Norway Maple 'Reitenbachii', p.372, with many more teeth).

Tartar Maple — *Acer tataricum*

Austria E to the Caucasus. 1759. Very rare.
APPEARANCE Shape A low dome; to 8 m. **Bark** Grey-brown; smooth then cracking rather shaggily. **Leaves** Bright green, usually oval (some narrow triangular lobes on strong growths); downy along the sunken veins. **Flowers** In *creamy, globular, almost hawthorn-like heads* among spring-green leaves; keys ripen deep red.
COMPARE Broad-leaved Cockspur Thorn (p.288).
VARIANTS Amur Maple, var. *ginnala* (*A. ginnala*; NE Asia, Japan; 1860), is more widespread: an open tree to 12 m in some small gardens, its darker leaves much more often *jaggedly, irregularly lobed* (typically a long central lobe and 2 small, narrow basal ones); brief early autumn crimsons.

Montpelier Maple — *Acer monspessulanum*

S and central Europe; SW Asia; N Africa. 1739. Rather rare, but in odd town parks.
APPEARANCE Shape Easily taken for a Field Maple (p.368: differences emphasized): irregularly but very densely and twiggily domed; to 15 m. **Bark** *Dark grey*; closely rugged, square-cracked ridges and burrs. **Leaves** Dark and rather shiny: *3 lobes, never with teeth* (except in saplings). **Flowers** Yellow, drooping, in showy heads before the leaves; the key-wings *converge*.

Syrian Maple — *Acer obtusifolium*

(Cyprus Maple; *A. syriacum*) E Mediterranean. 1752. Very rare.
APPEARANCE Shape Twiggy; usually bushy, to 12 m. **Bark** Grey; rather *smooth*. **Leaves** *Evergreen*, hard; shiny on both sides; unlobed/3-lobed (serrated on younger plants); 4–6 cm.
OTHER TREES Cretan Maple, *A. sempervirens* (*A. creticum*; *A. orientale*; E Mediterranean; slightly less rare), has semi-evergreen *tiny* leaves *(1–4 cm)*, with odd teeth/lobes. Twiggy and often bushy, with *rugged* brown bark – in summer recalling Midland Hawthorn (p.284).

Italian Maple — *Acer opalus*

Mountain forests from the Pyrenees to the Caucasus, and in N Africa. 1752. Very occasional.
APPEARANCE Shape A sturdy Sycamore-like dome; to 22 m. **Bark** Grey, then developing shaggy scales with more pink, yellow and orange lights than old Sycamores' (p.370). **Shoots** *Dark brown*; buds a little browner than Sycamore's. **Leaves** With 3 or 5 *shallow rounded lobes*, irregularly or not toothed (cf. Montpelier Maple). Dark green; variably grey- or white-woolly beneath (cf. Downy Maple, p.388). **Flowers** Daffodil-yellow, in spectacular *nodding* bunches before the leaves – much showier than Norway Maple's (p.372).

Balkan Maple — *Acer hyrcanum*

SE Europe to Iran. 1865. In some big gardens.
APPEARANCE Shape A low, twiggy dome. **Bark** Brown, with yellows and greys; scaly ridges. **Leaves** Small (5 × 7 cm) on notably *long* (to 9 cm) pink/yellow stalks; some down beneath. The 5 (3) stubby lobes have neatly *parallel* sides and rather few, rounded teeth (cf. Spanish Maple, p.368). **Flowers** Yellowish green, in nodding bunches.

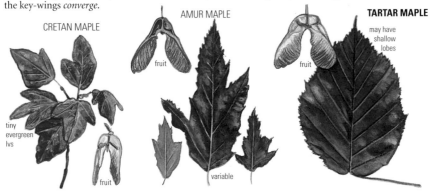

CRETAN MAPLE — tiny evergreen lvs — fruit

AMUR MAPLE — fruit — variable

TARTAR MAPLE — may have shallow lobes — fruit

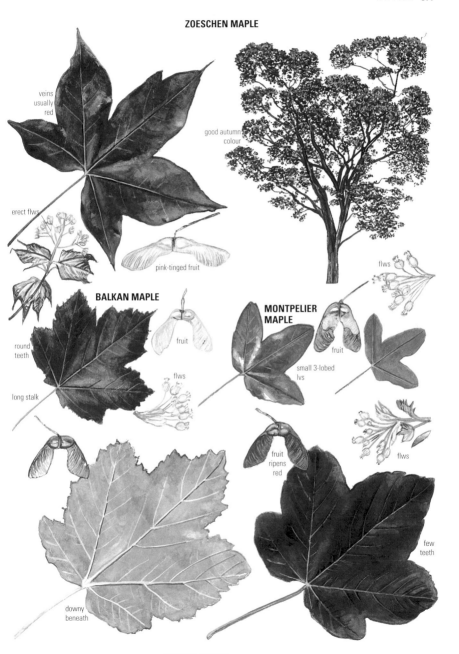

MAPLES (AMERICAN)

Silver Maple — *Acer saccharinum*

(*A. dasycarpum*) Quebec to Florida. 1725. A very frequent municipal tree.
APPEARANCE Shape An irregular, *airy, willowy* dome: steep limbs and often weeping outer branches from a generally leaning trunk; to 30 m. Vigorous but fragile and short-lived. **Bark** Grey, smooth; then, by 60 years, with shallow but *shaggy* cream-grey plates; often sprouty. **Shoots** Green then red-brown, with clear red/green 8 mm buds. **Leaves** *5-lobed*, jaggedly toothed; *bright silvery-grey underneath* with some fine down; flushing orange-brown; pale yellow and pink-red in autumn. **Flowers** Wreathed along the shoots before the leaves, *dull* ochre or reddish.
COMPARE Red Maple (below). Sometimes confused with *A. saccharum* (p.372), thanks to the similar name.
VARIANTS Cut-leaved Silver Maple, f. *laciniatum* (including the 1873 clone 'Wieri'), is now the most planted form: leaves *deeply, jaggedly cut*; crown often fountain-like.
'Lutescens' has soft-yellowish foliage (like some sickly or chlorotic trees of the type); very rare.

Red Maple — *Acer rubrum*

Newfoundland to Texas. 1656. Much less frequent than Silver Maple.
APPEARANCE Shape Denser, often on a better stem and twisting branches; to 26 m. **Bark** Usually darker and more closely cracked. **Shoots** Slender, reddish. **Buds** *Small* (2–4 mm), red. **Leaves** Bright silver-grey beneath with some fine fluffy down; *3 (5) lobes* angled forwards and *shallower* than Silver Maple's. Gold and rich red in autumn. **Flowers** In *shining crimson* clusters before the leaves: quietly spectacular against a dark background.
COMPARE Trident Maple (p.374): untoothed lobes.
VARIANTS 'Scanlon' is occasional as a younger tree: *neatly* and densely conic-columnar, to 17 m so far (cf. Black Maple 'Temple's Upright', p.372).
'Fastigiatum' is vigorous and gaunt: *long, erect shoots* and sinuous, vertical branches; rare.
'October Glory' is an autumn-colouring selection of good oval shape, now much planted.
OTHER TREES *A. pycnanthum* is a threatened tree from the mountains of N Honshu (Japan), now in many collections: slender, with more shallowly lobed (even unlobed, but toothed) leaves, glaucous beneath but *hairless* except near the main veins.

Oregon Maple — *Acer macrophyllum*

(Bigleaf Maple) W North America, where it makes the world's largest maple. 1826. Parks and gardens; very occasional.
APPEARANCE Shape A sturdy, upright dome on strong, arching limbs; to 27 m; sometimes suckering. **Bark** Dark brown; soon criss-crossed with rather craggy ridges. **Shoots** Purple/green for some years; buds red/green, almost hidden in summer by the *swollen base* of each leaf-stalk. **Leaves** Dark and sometimes huge (to 30 cm), with *narrow-based, floppy lobes*; hair-fringed (cf. Horned Maple, p.370), and tufted under the vein-joints. **Flowers** *In Sycamore-like tails*, but longer (15 cm); the large keys – wings 5 cm across – have stinging bristles.
COMPARE Norway Maple (p.372): similar jizz but with whisker-tipped lobes. Miyabe's Maple (p.386): a rare little tree with smaller leaves, downy above. Van Volxem's Maple (p.388): also with huge leaves.

OREGON MAPLE

RED MAPLE

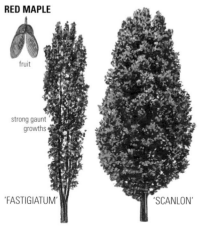

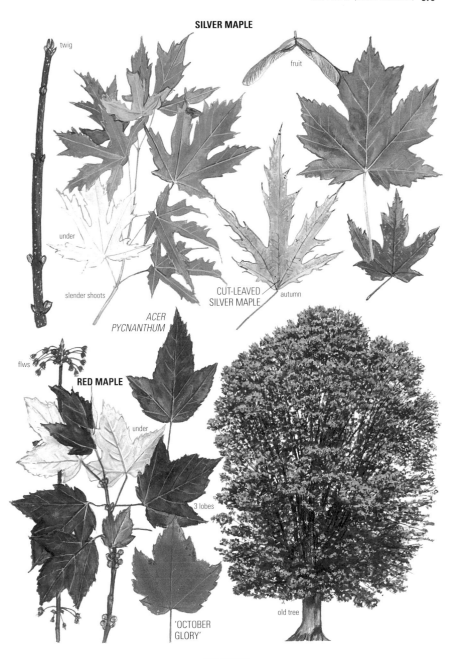

SNAKE-BARK MAPLES

'Snake-bark maples' have indistinctly *stalked buds*; the *leaf-green bark, with vertical streaks*, may be lost with age. (Spindle, p.448, has dark green bark with fawn 'snakes' as a younger bush.)

Red Snake-bark Maple   *Acer capillipes*

(Kyushu Maple) S Japan. 1894. The most frequent snake-bark as a younger planting.
APPEARANCE Shape Airily funnel-shaped, on light limbs arching out; to 16 m. The leaves seldom hang. **Bark** Bright; lasting well. **Buds** Purple-red. **Leaves** Long, with little regular side-lobes and *sunken parallel veins*; soon almost hairless; 1 mm 'pegs' persist under vein-joints. Rich mixed autumn colours. **Flowers** Yellow-green.
COMPARE Grey Snake-bark Maple (below); Hers' Maple (p.332).
OTHER TREES Forrest's Maple, *A. forrestii* (S China; Upper Burma; 1906), is in a few big gardens. To 15 m: arching then *cascading* branches closely clad in *pink/scarlet-stalked* fresh-green leaves with *long sharp side-lobes* but little autumn colour; the dull yellow flowers have red eyes.

Père David's Maple   *Acer davidii*

China; occasional in larger gardens and some municipal plantings. The commonest clone is 'George Forrest' (Yunnan, 1921–22):
APPEARANCE Shape Light, clean branches rising from a short bole make a wide dome; elegant *tiers of hanging foliage*. Taller and laxer in shelter: to 17 m. **Bark** Vivid (but ultimately brown-grey and finely chiselled, or even oak-like). **Leaves** Quite large (15 cm); *unlobed* or with a hint of side-lobes; *glossy blackish green* between red sunken veins, on dark red stalks; pale beneath with rusty hairs under the veins only at first; curious 1 mm white 'pegs' persist under the vein-joints, as in many Snake-bark Maples. Flushes dark orange but shows *no autumn colour*. **Flowers** Yellow-green, in arching 6 cm tails.
COMPARE Hers' Maple and Lime-leaved Maple (p.382). Red (above) and Grey Snake-bark Maple (below): regular side-lobes.
VARIANTS 'Ernest Wilson' (1879; rare) has much narrower, folded leaves, grey beneath and not hanging (cf. Uri Maple, p.382): they flush green but turn a good red in autumn.
'Madeline Spitta', in some collections, is tall, lax and open (to 20 m), with bright green bark lasting well and leaves most like those of 'George Forrest' but turning orange in autumn.
OTHER TREES *A. × conspicuum* 'Silver Vein' is a 1961 hybrid of 'George Forrest' and the Moosewood 'Erythrocladum' (p.382) – whose *yellow-and-red younger bark* and stronger side-lobes it inherits; very rare to date.

Grey Snake-bark Maple   *Acer rufinerve*

(Honshu Maple) Japan. 1879. Rather rare.
APPEARANCE Shape Open, to 14 m, on steep, arching limbs. **Bark** Green with white or *grey with pink stripes*; dull and cracked in age. **Shoots** Green, *bloomed lilac-white*. **Buds** 2 green outer scales *brightly bloomed grey*. **Leaves** Matt, *as broad as long*, with *spreading* side-lobes; rusty hairs under the base are *slowly* shed (*no* 1 mm 'pegs' in the vein-joints). Good red autumn colours.
COMPARE Red Snake-bark Maple. Moosewood (p.382): massive side-lobes. Hers' Maple (p.382).
VARIANTS 'Hatsuyuki' (f. *albolimbatum*; it often comes true from seed) has leaves finely splashed white, and whiter shoots/buds. Now very rare; *A.* 'Silver Cardinal' (a seedling of Moosewood, by 1985) has broad leaves *brightly splashed* with white/pink; rather weeping, with crimson shoots.

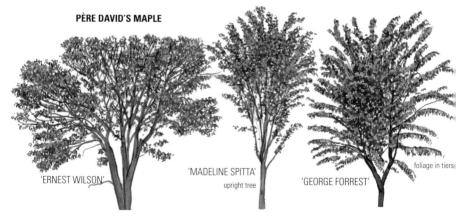

PÈRE DAVID'S MAPLE

'ERNEST WILSON'

'MADELINE SPITTA' upright tree

'GEORGE FORREST' foliage in tiers

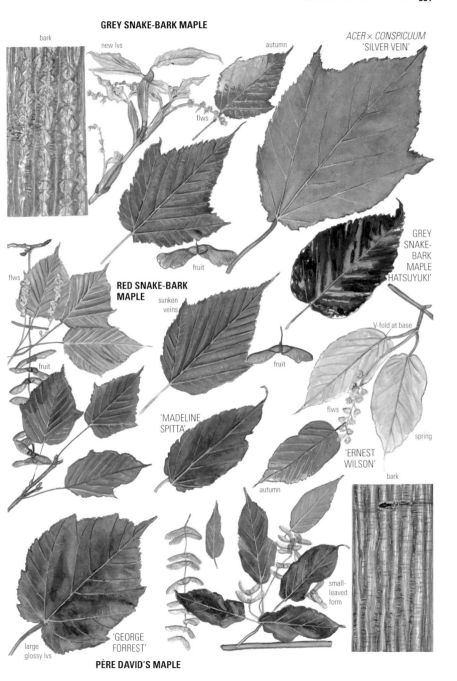

SNAKE-BARK & LIME-LEAVED MAPLES

Hers' Maple — *Acer grosseri* var. *hersii*

(Green Snake-bark Maple; *A. hersii*) China. 1924. Very occasional: a snake-bark maple often *green* in every part.
Appearance Shape Open, on steep then arching, closely clad branches; to 15 m. **Bark** Olive; boldly white-striped at least in youth. **Shoots** Olive-green (pale pink on the strongest growths). **Buds** *Green* (unbloomed). **Leaves** Green-stalked; thick and rubbery, quite small (10 cm) and *rather round*; with stubby side-lobes (weak/absent in the type, which is in a few collections); the fine toothing is rather blunt. Rusty hairs underneath are soon shed; 1 mm white 'pegs' persist in the vein-joints. Autumn colours are more yellow than red. **Flowers** Yellow-*green*, on 12 cm arching tails.
Compare Père David's Maple (p.380): glossier leaves with finer toothing. Birch-leaved Maple (below).

Moosewood — *Acer pensylvanicum*

(Linnaeus' mis-spelling has to be retained.) Nova Scotia to Georgia. 1755. The only American snake-bark maple; it has typical bark. Very occasional.
Appearance Shape Light branches, rising and arching; to 17 m. **Shoots** Green. **Buds** Red/brown. **Leaves** Often *large* (to 22 cm), matt; with *big, usually forward-pointing side-lobes*; rusty hairs underneath are *slowly* shed (*no* 1 mm 'pegs' in the vein-joints); yellow autumn colours. **Fruit** Abundant keys, the nuts *flattened*.
Compare Grey Snake-bark Maple (p.380): often smaller, spreading side-lobes; rounded nuts.
Variants 'Erythrocladum' has *pale, clear crimson* winter shoots, and gold younger bark with *red and white 'snakes'* (cf. its hybrid 'Silver Vein', p.380); unfortunately a sickly tree, and now very rare.

Uri Maple — *Acer crataegifolium*

(Hawthorn-leaved Maple) Japan. 1879. Very rare: large gardens.
Appearance Shape Slender: tiers of foliage on *twisting, maroon*/green shoots. **Bark** Green, white-striped; soon fading to grey-brown. **Buds** Purple/green. **Leaves** Small (to 7 cm), dark but *pink-tinged; slender* and variably lobed; odd fawn hairs but no 1 mm 'pegs' under the vein-joints. **Flowers** Yellow-white, in *erect*, 4 cm heads.
Compare Birch-leaved Maple (below); Père David's Maple 'Ernest Wilson' (p.380).

Birch-leaved Maple — *Acer stachyophyllum*

(Incorporating *A. tetramerum*) E Himalayas and W China. 1901. Rather rare: big gardens.
Appearance Shape Weak but steep branches and arching shoots, to 15 m; sometimes a suckering bush. **Bark** Faintly striped, then burnished greenish ochre, with pale lenticels. **Shoots** Green. **Leaves** Flimsy; rather triangular with *rounded* bases and the odd lobe, on long (6 cm) green/*scarlet stalks*; fine *white down covers the shiny underleaf*. **Flowers** Dioecious; in short, sometimes branching tails.
Compare Hers' Maple (above): leaves soon hairless.

Lime-leaved Maple — *Acer distylum*

Japan. 1879. In a few big gardens on acid soils.
Appearance Shape Broad, on arching limbs; to 9 m. **Bark** Green, striped orange; soon rougher and brown. **Buds** Red, downy; long-pointed (to 1 cm); *never stalked*. **Leaves** Large (to 17 cm), pale green and soon hairless; distinctively *heart-shaped*; flushing pink, then gold in autumn. **Flowers** In *stiff, branching tails* to 10 cm long.
Compare Père David's Maple (p.380). Limes, etc all have alternate leaves.

HERS' MAPLE — all parts green
autumn
MOOSEWOOD
'ERYTHROCLADUM'

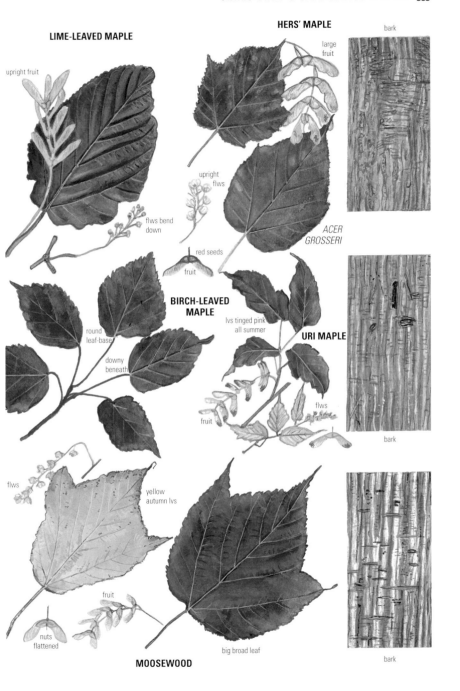

MAPLES (EAST ASIATIC)

Smooth Japanese Maple
Acer palmatum

('Acer') China, Korea and Japan, where long cultivated in many generally bushy cultivars. 1820. A singularly elegant plant, quite frequent in mostly larger gardens; seedlings spring up freely.
Appearance Shape Rounded, to 15 m: light limbs from a short bole and dainty horizontal sprays of dense, star-like, moss-green leaves. **Bark** Smooth and brown-grey: faint white snakes at first, then some muscular ripples. **Shoots** Slender, bright red/green, *all tipped by 2* crimson/green buds (cf. Downy Japanese Maple, p.386). **Leaves** 4–7 cm (but to 12 cm in ssp. *amoenum* and the widely grown var. *heptalobum*); hairless except for tufts under the vein-joints; the 5 or 7 (rarely 9) deep lobes are *finely and sharply* but seldom doubly serrated. **Flowers** Deep red, in small spreading heads, turning trees amber as the green leaves flush. **Fruits** Pale red; erect/drooping clusters.
Compare Deep-veined Maple (p.386).
Variants 'Osakazuki' (a Japanese selection of var. *heptalobum*; 1861) has larger leaves (seldom 5-lobed), flushing deep pink so that the crown is maroon-tinged through summer, with reliably *blood-red autumn colours* (type trees usually turn pale orange); fruits *rich* red. Quite frequent.
Purple Japanese Maple, f. *atropurpureum*, is quite frequent and crops up among seedlings. Leaves bronze, purplish, or a superb deep crimson in selections like 'Bloodgood'; paling to scarlet in autumn.
Coral-bark Maple, 'Sango-kaku' ('Senkaki'), has stiffly rising *brilliant red* winter shoots; the small, slightly yellow leaves turn gold in autumn.
Cut-leaved Japanese Maple, f. *dissectum*, has finely feathery foliage and usually grows as a miniature mushroom; to tree height in shade. Very frequent in many selections (cf. cut-leaved Norway Maples, p.372, and the Downy Japanese Maple 'Aconitifolium', p.386). 'Dissectum Atropurpureum' has soft, silvery-purple foliage.
'Hagoromo' ('Sessilifolium'; rare) was once considered a different species: its leaves (*very shortly stalked*) often have *3 or 5 jaggedly lobed leaflets*. A slender, rather upright tree, to 14 m.
'Linearilobum' is rare: deep, narrow *finger-like* lobes (not themselves dissected); 'Linearilobum Atropurpureum' ('Atrolineare') is the purple version. 'Shishigashira' ('Ribesifolium'), *narrow, stiff and very dense*, has *much-twisted and cut dark leaves*. 'Aureum' has soft yellow leaves (gold in autumn); very rare.
'Albomarginatum' has delicately white-margined leaves; rare; 'Kagiri-Nishiki' ('Roseomarginatum') has 5-lobed leaves with fine pink margins (fading creamy-white); tall-growing but readily reverting.
Other trees *A. oliverianum* is a sometimes similar-looking central Chinese tree in some collections: its 3- or 5-lobed leaves, often with only a few teeth, are stiffer, darker and downy along the veins on both sides.
A. ukurunduense (mountains of Japan and NE Asia; in collections), has downy *stalked* buds (like snake-bark maples', pp.380–2) and big leaves with 5 or 7 shallower, *doubly toothed* lobes; all the veins are downy beneath (cf. Deep-veined Maple, p.386).

Hornbeam-leaved Maple
Acer carpinifolium

Japan. 1879. Rare.
Appearance Shape A broad, bushy plant (to 12 m), suggesting anything but a maple. **Bark** Smooth, grey. **Leaves** Elegant, fresh-green, hanging, with *parallel veins in 18–23 pairs*; doubly toothed. Silky-hairy under veins; gold in autumn. **Flowers** Dioecious, rich green, on very slender, hanging 10 cm tails.
Compare Hornbeams (pp.194–6): *alternate* leaves.

SMOOTH JAPANESE MAPLE   **HORNBEAM-LEAVED MAPLE**

dainty foliage

smooth grey bark

often bushy

MAPLES

Deep-veined Maple — Acer argutum

Mountains of Japan. 1881. In some big gardens.
APPEARANCE Shape Weak, steep, purplish stems; often bushy. **Bark** Dark green, fading to brown; smooth. **Shoots** Green, finely hairy. **Buds** Green/pink, *stalked* (cf. snake-bark maples, pp.380–2). **Leaves** Dark and hard; downy under the prominent veins, the 5 lobes *doubly* toothed (cf. *A. ukurunduense*, p.384) and with long, *jaggedly serrated tips*. Bright yellow autumn colour. **Flowers** Dioecious; on thin, hanging tails.
COMPARE Smooth Japanese Maple (p.384): simple toothing.
OTHER TREES *A. acuminatum* (W Himalayas; one of several similar maples in collections) has 3 forward-angled lobes (plus odd little lobes at the base), with comparably long, jagged tips.
Mountain Maple, *A. spicatum* (E North America; very rare and rather bushy) has 3 or 5 spreading, double-toothed lobes with less attenuated tips; flowers in erect, 15 cm tails.

Downy Japanese Maple — Acer japonicum

(Full Moon Maple) Japan. 1864. Rare: large gardens.
APPEARANCE Shape Usually bushy; sinuous, steep branches with open, straggly tops. To 13 m. **Bark** Grey, smooth. **Shoots** Red/green, quickly hairless, *all tipped by 2 buds* (cf. Smooth Japanese Maple, p.384; they are hidden in summer by the base of the leaf-stalks). **Leaves** Broad (to 15 cm) and dark, with 7–11 shallow lobes; unfolding hairy on hairy stalks (the hairs confined to veins by late summer). **Flowers** Purple, in long-stalked bunches.
VARIANTS 'Vitifolium', a vigorous autumn-colouring selection (gold then dark red) is the most grown clone.
'Aconitifolium' has much-divided lobes running to 1 cm from the leaf's centre. Rare; rather bushy.
OTHER TREES *A. shirasawanum* 'Aureum' (often called *A. japonicum* 'Aureum' but now considered to derive from this Japanese ally which has *only tufts of down* under its leaves) is very occasional in larger gardens. Bushy, to 8 m: gold then *fresh yellow-green leaves* in crisp layers.
Keijo Maple, *A. pseudosieboldianum* (Korean Maple; E Asia, 1903), is in some collections. Bushy; leaves with *9 or 11* slightly deeper lobes; teeth very *slender-pointed*; autumn colour *crimson with orange and purple shades*. Flowers purple.
Siebold's Maple, *A. sieboldianum* (Japan, *c*.1880), is still rarer: *yellow* flowers and autumn colours; *grey-downy shoots* and leaf-stalks.
Vine Maple, *A. circinatum* (W North America, 1826), is also rare. Often bushy: the branches layer and make thickets. Smaller leaves (to 10 × 13 cm), with only *7 or 9 lobes*, turn *orange* and red in autumn. White petals and purple sepals make a pretty combination.

Miyabe's Maple — Acer miyabei

Japan. 1895. Very rare: big gardens.
APPEARANCE Shape A broad and rather willowy dome; to 16 m. **Bark** Brown; orange and grey scales. **Buds** Red-brown, mostly hidden in summer by the base of the leaf-stalks. **Leaves** *Downy on both sides*, at least near the veins, but not hair-fringed; often very heart-shaped at the base; yellow in autumn. The stalks, snapped and squeezed, ooze milky sap.
COMPARE Horned Maple (p.370); Sugar Maple (p.372); Oregon Maple (p.378).

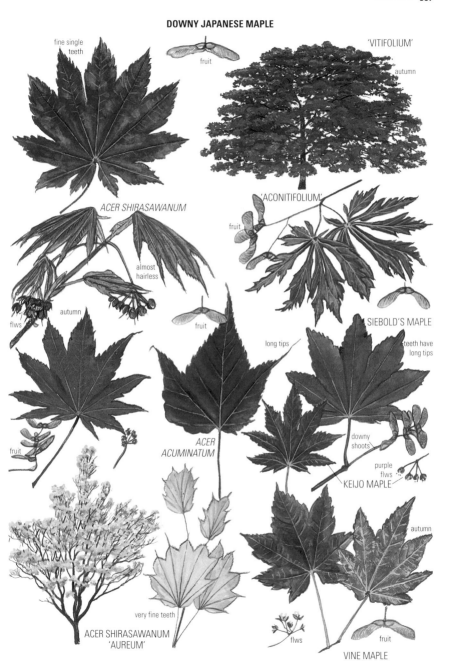

MAPLES

Downy Maple — *Acer velutinum*

Caucasus, Iran. 1873. Rare. Easily passed over as a Sycamore (p.370), so may be overlooked in neglected, Sycamore-infested gardens (differences emphasized; best told in flower/fruit).
APPEARANCE Shape Densely domed, to 20 m; often *rich, fresh green*. **Bark** *Smooth dark grey*, even on old trees; prominent rings round branches. **Shoots, Buds** *Browner* than Sycamore's. **Leaves** Unfolding late; Sycamore-shaped on often red stalks; variable amounts of velvety brownish down beneath. **Flowers** Pale green, in a big *upright bunch*. Keys *in bunches* not on tails; wings spread widely.
COMPARE Trautvetter's Maple (below); Horned Maple (p.370).
VARIANTS Van Volxem's Maple, var. *vanvolxemii* (the Caucasus) is the most grown form: *large* leaves (to 25 cm) on long, *yellow-pink* stalks; the down underneath is soon *confined to the veins*.

Trautvetter's Maple — *Acer trautvetteri*

(Red-bud Maple; *A. heldreichii* ssp. *trautvetteri*) Caucasus to Iran. 1866. Rare.
APPEARANCE Shape Narrow-domed, to 19 m. **Shoots, Buds** *Bright* red-brown (especially in winter). **Leaves** Glossier and with narrower, more finger-like lobing than Sycamore's (but less deeply divided than Heldreich's Maple's, below; greyer beneath, with denser, orange hairs under the veins). **Flowers** Yellow, in erect heads. **Fruit** Seeds have showy, *rosy-pink*, incurving wings.
COMPARE Sycamore (p.370). Downy Maple (above): leaves with more forward-angled serrations; seed-wings widely spread.

Heldreich's Maple — *Acer heldreichii*

Greece; the Balkans. *c.*1879. Very rare.
APPEARANCE Shape Openly narrow-domed on rising branches; to 23 m. **Bark** Remaining smooth; pinkish grey. **Shoots** Brown. **Buds** *Sharply pointed, many-scaled*, red-brown. **Leaves** Slightly Sycamore-like but green-flushing, glossier and less toothed; each lobe is narrowly cut *to within 15 mm of the base*, above which there is 1 tuft of hair. Yellow and red in autumn. **Flowers** Yellow, in erect heads (cf. Sycamore's long tails). **Fruit** The seed-wings curve inwards and may overlap.
COMPARE Trautvetter's Maple (above).

Paper-bark Maple — *Acer griseum*

Central China. 1901. Rather occasional, but in many gardens and some parks for its unique bark. The seeds often mature without fertilization and are infertile, and the difficulty of raising seedlings helps to explain why this gorgeous tree is still not commoner.
APPEARANCE Shape A low, neat, finely twiggy dome (gaunt when unhappy); to 15 m. **Bark** *Papery* scrolls from the first: *rich cinnamon-red* and chocolate-brown, sometimes blue-bloomed. Odd trees can develop harder curling scales, a dull orange-brown. **Leaves** With *3 leaflets*, each roundly toothed/lobed; dark grey-green above and hairy; dense blue-white down beneath, and on the pink/red stalks. They flush very late, pale orange; usually orange and crimson autumn colours. **Flowers** Usually in 3s.
COMPARE Other trifoliate maples (p.390): bark distinguishes these most easily. Chinese Red-barked Birch (p.188) and Chinese Stuartia (p.408) have comparably bright, papery barks.

PAPER-BARK MAPLE

MAPLES

Chosen Maple — *Acer triflorum*

(Rough-barked Maple) N China, Korea. 1923. Very rare – needing plenty of summer heat to grow well. **Appearance Shape** A dense, low dome, like Paper-bark Maple (p.388). **Bark** Pale grey-brown; *coarsely shredding vertical ridges*. **Leaves** Like Paper-bark Maple's, though the leaflets have fewer, shallower lobes/teeth; crimson in autumn. **Flowers** In 3s.

Nikko Maple — *Acer maximowiczianum*

(*A. nikoense*) Central China; Japan. 1881. Rather rare (not to be confused with *A. maximowiczii*, a very rare snake-bark maple). **Appearance Shape** Domed; level sprays of heavy, un-maple-like foliage; to 15 m. **Bark** Dark grey; very smooth. **Leaves** Larger and darker than Paper-bark Maple's, on stout pink/green hairy stalks; red and yellow in autumn. **Flowers** In 3s among the young leaves.

Vine-leaved Maple — *Acer cissifolium*

Japan. 1875. Rare: bigger gardens. **Appearance Shape** A *flat, wide dome*; rarely to 15 m. **Bark** *Pale* grey; rather smooth then irregularly ridged. **Leaves** With *jagged* teeth/lobes and *long wire-thin red stalks;* each of the 3 leaflets has its own red stalk. *Sparsely* hairy on both sides; pale yellow in autumn. **Flowers** In long (10–16 cm) *drooping tails*.
Other trees *A. henryi* (central China, 1903; collections) is bushier. Leaves *glossy*, often untoothed; flower-stalks (off the 'tail') *only 1 mm long* (3–6 mm in Vine-leaved Maple). Spectacular when decked in its red fruits.

Box Elder — *Acer negundo*

(Ash-leaved Maple; *Negundo aceroides*) Across North America; sometimes tapped for maple syrup. By 1688. Very frequent. **Appearance Shape** Broad, *untidy*: leaning stems and steep younger limbs. To 18 m. **Bark** Pale grey; burred, sprouty; *irregular networks of ridges*. **Shoots** Rich *plastic-green*, variously bloomed. **Buds** Silky-white. **Leaves** With *3–5* (rarely 7 or 9) *leaflets*, each deeply/haphazardly lobed; velvety on downy shoots in var. *californicum* (W coast); little autumn colour. **Flowers** Dioecious: in showy hanging plumes before the leaves.
Compare Vine-leaved Maple (above); Smooth Japanese Maple 'Hagoromo' (p.384). Larger and much coarser than the (consistently) trifoliate maples.
Variants var. *violaceum* covers mid-Western trees with *strongly purple-bloomed* but hairless shoots. Foliage *darker* and heavier; males with spectacular salmon-pink flowers. Occasional.
'Variegatum' (1845; female) has brightly white-variegated leaves and white-winged fruit. Much planted in the mid-20th century but quickly reverting: today seen generally as a graft with *white-bloomed shoots*, yellowish, *slightly blotchy-green* foliage and the odd milk-white sprout.
'Elegans' has glossy, boldly yellow-margined leaves and blue-bloomed shoots; probably also reverting quickly as no big trees are now known.
'Flamingo' has white-variegated leaves unfolding pink; rather rare as a small tree, but less unstable.
'Auratum' (1891) has golden then pale green leaves; rather rare. A *female* clone, prone to revert.
'Kelly's Gold' is a bright, small, recent *male* clone.

NIKKO MAPLE   **VINE-LEAVED MAPLE**   *ACER HENRYI*

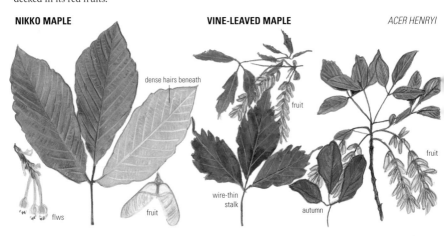

dense hairs beneath · fruit · flws · fruit · wire-thin stalk · autumn · fruit

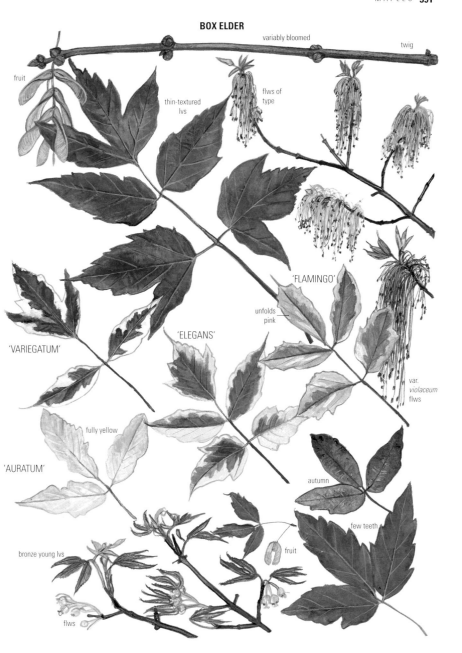

HORSE CHESTNUTS

Hippocastanaceae is a small family dominated by 25 horse chestnuts/buckeyes.

Things to Look for: *Aesculus*

- Bark: Does it remain smooth?
- Leaves: How many leaflets? Are they stalked? Hairy beneath? Doubly toothed?
- Flowers: Colour? Are the petals hairy? Sticky?
- Fruit: Are the conker-husks prickly?

Key Species

Horse Chestnut (below): usually 7 unstalked leaflets. Indian Horse Chestnut (p.394): usually 7 stalked leaflets. Red Horse Chestnut (p.394): 5 scarcely stalked leaflets. Yellow Buckeye (p.396): 5 distinctly stalked leaflets.

Horse Chestnut *Aesculus hippocastanum*

Growing freely from conkers in the UK and abundantly planted since 1616, but confined in the wild to mountains in N Greece/Albania. Its name probably derives from its role in horse-medicine and its planting by smithies (and from association with the unrelated Sweet Chestnut, whose fruits have a similar design).
Appearance Shape An often narrow dome of short, twisting twigs. The main limbs shatter easily, particularly when rain weighs down the dense foliage. To 39 m; potentially very long-lived. **Bark** In youth, smooth and pink-grey; red-brown and coarsely scaly by 80 years. **Shoots** Thick, red/grey. The pale leaf-scars are horseshoe-like (3 'nail-holes' one side, 4 the other, where sap ran to each leaflet). **Buds** Deep red-brown, *sticky* (cf. Californian Buckeye, p.396; Sargent's Rowan, p.302); end one huge. **Leaves** With 7 (5 or 6) *stalkless* leaflets, jaggedly/double-toothed; unfolding early and by late summer often browned by fungal rusts. **Flowers** The showiest of any tall tree's; the yellow basal blazes can turn red after pollination. **Fruit** Many conker-husks on each tree bear *sharp spines*.
Compare Japanese Horse Chestnut (below); Dallimore's Chestnut (p.394); Ohio Buckeye (p.396).
Variants 'Baumanii' (1820) has whiter, stumpier candles of *double* flowers, so no conkers; quite frequent. It grows almost vertical central limbs, *densely clothed* in rather small, cupped leaves.
'Pyramidalis' is rare: *stiffly upright* branches make a neat, narrow tree with a pointed top.
'Hampton Court Gold' has evenly yellow-green foliage (many sickly type trees are blotchily yellow); 'Honiton Gold' is taller: both are very rare.
'Digitata' has very *narrow* leaflets (often only 3) on a flattened, leaf-like main stalk; f. *laciniata* has *feathery*, cut leaves; both are feeble and very rare.

Japanese Horse Chestnut
Aesculus turbinata

Mountains of Japan. *c.*1880. Rare.
Appearance Shape Domed; to 25 m. **Bark** Pink-grey; *white streaks* when young, later a *few* fissures. **Shoots** Pink. **Buds** Rufous, sticky. **Leaves** Like Horse Chestnut's (usually 7 stalkless leaflets), but often even larger; slightly silvered beneath, with *orange tufts* in the vein-joints (Horse Chestnut has browner tufts); margins with *regular (not double) small teeth*. **Flowers** White, blotched red, a fortnight after Horse Chestnut's. **Fruit** The conker-husks *lack any spines*.

JAPANESE HORSE CHESTNUT — regular small teeth; orange tufts under midrib
HORSE CHESTNUT 'DIGITATA' — winged stalk
'HAMPTON COURT GOLD' — yellow-green lvs

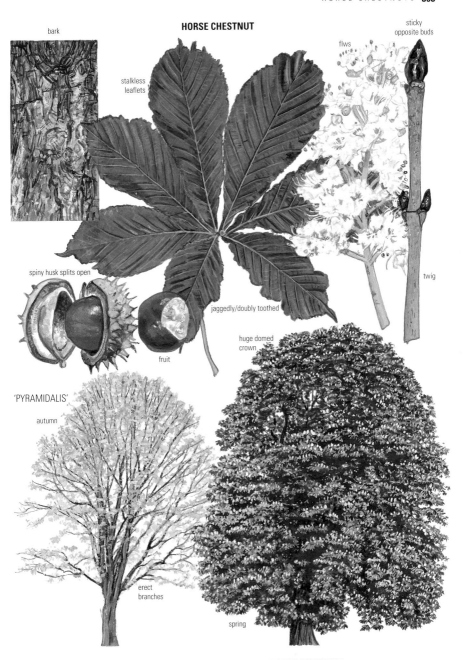

HORSE CHESTNUTS

Indian Horse Chestnut *Aesculus indica*

NW Himalayas. 1851. A splendid tree which is still distinctly occasional; often seeding abundantly.
APPEARANCE Shape A fine dome on straightish limbs, to 26 m; sometimes a giant bush. **Bark** Smooth, pink-grey; distantly scaling with age. **Buds** Green/pinkish red and sticky; weaker shoots end in a pair. **Leaves** With 5–9 (usually 7) *slender, finger-like* leaflets each on *1 cm reddish stalks*; glossy and *very dark* above; pale grey beneath but *hairless*. They open briefly *tomato-red* (cf. Sunrise Horse Chestnut, p.396) but – the tree's one limitation – show little autumn colour. **Flowers** Pale pink (with yellow blazes turning crimson), in long, elegant candles, from late spring into *early summer*. 'Sydney Pearce' (1928; rare; grafted trees) has many rich red and even *purple* blazes. **Fruit** Conkers black-brown, in leathery, spineless husks.
COMPARE Californian Buckeye (p.396). Japanese Horse Chestnut (p.392): stalkless leaflets.

Red Horse Chestnut *Aesculus × carnea*

A cross of Red Buckeye (p.396) with Horse Chestnut (by 1818), which has since doubled its chromosomes so breeds true. An abundant plant of rather endearing ugliness.
APPEARANCE Shape Low (rarely 20 m); very twisting, often weeping branches. **Bark** Smoother for longer (with prominent lenticels) then less closely/shaggily ridged than Horse Chestnut's; often much cankered. **Buds** Greyish, *scarcely sticky*; weaker shoots end (like the buckeyes') in a pair of buds. **Leaves** Dark, *crumpled* but slightly glossy; 5 (6 or 7) leaflets, each *jaggedly* toothed and scarcely stalked; usually smaller than Horse Chestnut's. **Flowers** *Dull crimson* ('carnea' means 'meat-coloured', which is a bit unfair), in dumpy rather radiating candles. **Fruit** The conker-husks have few or no spines.
COMPARE Ohio Buckeye (p.396): much more elegant foliage.
VARIANTS 'Briotii' (1858) is now the commonly planted clone: a shapelier dome (to 25 m); *glossier, less-crinkled* leaves; candles variably *bright* red-pink (often one flower with a yellow blaze next to one with a magenta blaze).
'Plantierensis' (*A.* × *plantierensis*), locally occasional, is a back-cross with Horse Chestnut: a low-domed tree whose long leaflets are shaped like Horse Chestnut's but crinkled and shiny and usually in 5s, with *soft apricot and salmon-pink flowers* in huge, loose candles – the finest *Aesculus* of all in blossom. It *does not fruit*.

Dallimore's Chestnut

Aesculus + dallimorei

This chimaera was found in Kent in 1955 by William Dallimore, a retired director of the Royal Botanic Gardens at Kew: a branch of Yellow Buckeye had fused with tissues of the Horse Chestnut onto which it was grafted. Three independent sports are now known, all arising in Kent. Showy, but still almost confined to collections.
APPEARANCE Shape A vigorous, dense, leafy dome. **Leaves** The usually 7 crinkled leaflets suggest Horse Chestnut's (p.392) but are *longer-pointed* and rather white-hairy beneath (though with brownish hairs like Horse Chestnut's along the midrib). **Flowers** *Creamy yellow* (or whiter and blotched red), in great candles like Horse Chestnut's but a fortnight later.

INDIAN HORSE CHESTNUT

short bole

RED HORSE CHESTNUT

few spines
fruit
dull red flws

'PLANTIERENSIS'

apricot to pink

BUCKEYES

Californian Buckeye — *Aesculus californica*

California. *c.*1850. Very occasional.
Appearance Shape A broad, low dome, to 14 m, or bushy. **Bark** Smooth and grey-pink; finely scaly with age. **Buds** Sticky (unlike the other buckeyes'). **Leaves** *Small*: 5–7 dark glossy-metallic leaflets 5–10 cm long, downy only when young, often *spread in a full circle*. **Flowers** Fragrant; much showier than other buckeyes': heavy, stumpy, radiating candles, white/soft pink, in *early summer*. **Fruit** Conker-husks roughened but without any sharp spines.
Compare Indian Horse Chestnut (p.394).

Yellow Buckeye — *Aesculus flava*

(Sweet Buckeye; *A. octandra*) E USA. 1764. Rather occasional: parks and gardens in warmer areas.
Appearance Shape Narrowly domed, to 26 m, or irregular; often *twisting* branches. **Bark** Pink-grey; smooth (prominently lenticellate), then usually with big, harsh, curved scales. **Buds** Pale pink-brown, non-sticky; weaker shoots end in a pair. **Leaves** With 5 (3 or 4) *elegant*, smooth, slightly glossy, fresh-green leaflets each on a *1 cm stalk*, and remaining *more or less downy beneath*. A tree grown firstly for its foliage; bright orange-red in autumn. **Flowers** In meagre yellow candles among the fresh leaves – piquant, not spectacular (red in the rare f. *virginica*, which is confined in the wild to W Virginia). The tubular, *long-hairy* petals lack Red Buckeye's sticky glands, and (unlike Ohio Buckeye's) are long enough to hide the stamens. **Fruit** The leathery conker-husks always *lack spines*.
Other trees Red Buckeye, *A. pavia* (SE USA, 1711), is very occasional: usually *bushy* with straight, radiating limbs, or rather weeping branches from a high graft. Its bark typically *remains* smooth and grey. The small, dark, glossy leaves (often crimson-stalked) are less downy beneath; its rich crimson (rarely yellow) petals form a very narrow, coal-scuttle shape and are *hairless, but sticky with glands* (which make a minute stubble round their margins).
Hybrid Buckeye, *A.* × *hybrida* (by 1815), is the occasional cross with Yellow Buckeye: often narrowly columnar to 22 m on straight, steep, smoothly grey-barked limbs, with darker foliage than Yellow Buckeye's; the yellow/red petals are both *hairy and sticky with tiny glands*.

Ohio Buckeye — *Aesculus glabra*

E USA. *c.*1809. Very rare.
Appearance Shape Often low and irregular, on twisting limbs; to 18 m. **Bark** Pale brown, strong-smelling; *rugged, corky ridges* develop (smoother but *almost white* in var. *leucodermis* from S Missouri/Arkansas). **Leaves** With 5–7 elegant, slightly stalked leaflets; darker than Yellow Buckeye's, the hairs under them soon *confined to the main veins*. **Flowers** In stumpy green-yellow candles, their petals *all shorter than the stamens*. **Fruit** The conker-husks (unlike other buckeyes') have some *short prickles/sharp warts*.

Sunrise Horse Chestnut — *Aesculus* × *neglecta* 'Erythroblastos'

A sport of an E North American wild hybrid buckeye: grown in some gardens since 1933 for the colour of its young leaves.
Appearance Shape A stunted, irregular dome (to 10 m). **Buds** With *spreading scale-tips*. **Leaves** *Shrimp-pink* when young, fading over a fortnight to bright yellow-green (cf. Indian Horse Chestnut's reddish flush, p.394) then maturing dull green with *yellow near the midrib*. **Flowers** As Yellow Buckeye (above).

CALIFORNIAN BUCKEYE

SUNRISE HORSE CHESTNUT

often a bush

fades to pale green

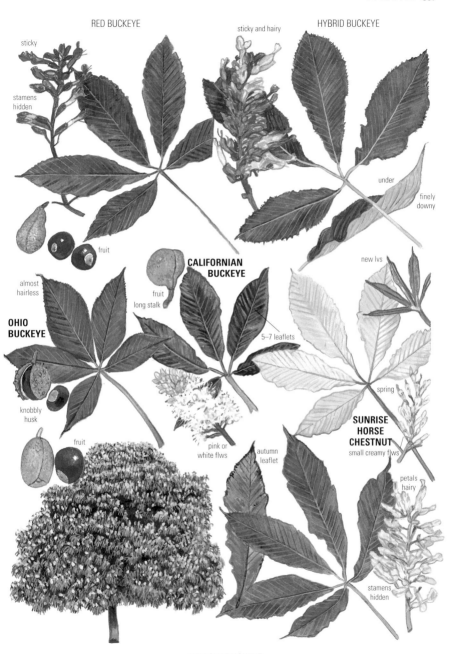

GOLDEN RAIN TREE & BUCKTHORNS

Golden Rain Tree *Koelreuteria paniculata*

(Pride of India) China, Korea, Japan. 1763. Occasional in warmer parts; sometimes seeding freely. (Family: Sapindaceae.)
APPEARANCE Shape Densely domed, on thick, rising and twisting branches; to 15 m. Matt, very dark foliage. **Bark** Brown; close, ultimately craggy, interlacing ridges. **Shoots** Pale coppery-brown, the short, sharp buds standing above *raised black-rimmed leaf-scars*. **Leaves** Variously subdivided but unmistakable; flushing garish red then amber-pink. **Flowers** Dramatic, 30 cm mustard-yellow plumes festoon the crown in high summer. The equally eye-catching *pinkish bladders*, like Chinese lanterns, each hold 3 pea-sized seeds.
COMPARE Kentucky Coffee Tree (p.350): superficially alike, at least in winter.
VARIANTS 'Fastigiata' is very rare: the twisting branches are *vertical* and make an exceedingly narrow tree, to 14 m.

Purging Buckthorn
Rhamnus cathartica

Europe (including England, Wales and Ireland); W and N Asia. Frequent but inconspicuous in scrub and woodland *on calcareous soils*; scarcely planted. This and Alder Buckthorn, its counterpart on acidic soils, are the food-plants of the Brimstone Butterfly, the sulphur-yellow males often being the most conspicuous evidence of the plant's proximity. Its bark and fruits (skin irritants) were once used to make a yellow dye. (Family: Rhamnaceae.)
APPEARANCE Shape Twiggy, often bushy, on a short but sturdy stem; exceptionally to 15 m. Sideshoots often end in Blackthorn-like spines. **Bark** Dark brown; soon *shaggily scaly*. **Shoots** Slender, straight, grey-brown. **Buds** Conic, black-brown; *appressed; usually in opposite pairs but sometimes staggered*. **Leaves** Dark, about 6 cm long: with incurving veins like Common Dogwood's (p.424) but *finely round-toothed*. Usually hairless; yellow in autumn. **Flowers** Small yellow-green stars, clustered at the base of new shoots. **Fruit** 6 mm black berries (powerfully purgative and once widely used as a kill-or-cure medicine).
COMPARE Spindle (p.448): narrower leaves. Wild Crab (p.306): similar jizz, but consistently alternate foliage.

Alder Buckthorn *Frangula alnus*

(*Rhamnus frangula*) Europe (including England, Wales and Ireland), E to Siberia; N Africa. Locally frequent in wet heathlands and swamps, or as a straggling plant in *acidic* woodlands; scarcely planted. A close ally of Purging Buckthorn, though it looks very different. Its timber – a startling lemon-yellow – once provided the finest charcoal for gunpowder. The sap is intensely bitter and irritant.
APPEARANCE Shape Spire-like when young, on elegant, straight, light stems, but scarcely to tree size. **Bark** Smooth, dark grey. **Shoots** Straight, very slender; purple-brown; elongated lenticels make *fine white streaks*. **Buds** Alternate, *scale-less* – 3 mm tufts of orange fluff. **Leaves** Small (about 5 cm), flat; matt fresh green; *untoothed*. They are tapered at the base but *blunt-tipped* (except on strong growths), like those of Common Alder (p.190), with which it often grows. **Flowers** In tiny green clusters. **Fruit** Red berries, ripening purple; toxic.
COMPARE Smoke-bush (p.360): similar foliage.

PURGING BUCKTHORN

ALDER BUCKTHORN

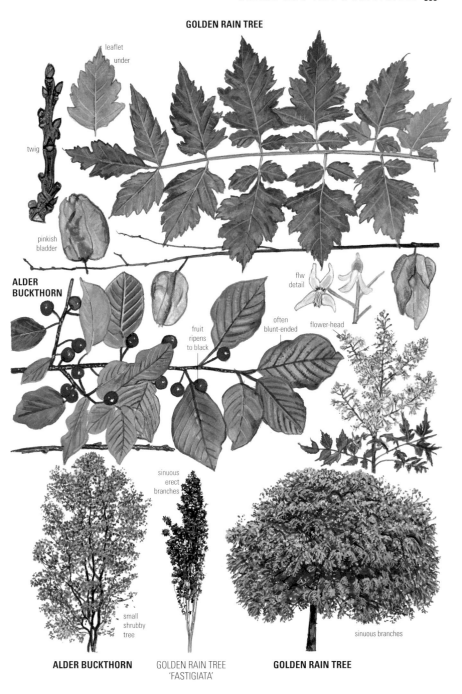

400 LIMES

Tiliaceae is a diverse family including one main group of trees – the limes. Limes (30 species) have heart-shaped leaves that are typically bulged more on one side of the slanting stalk than the other; there is no big end-bud. Their flowers, with showy and aerodynamic bracts, are deliciously scented. Leaf-aphids often drip sticky 'honeydew' through the summer.

Things to Look for: Limes

- Shoots: Are they hairy/felted? How red are they?
- Leaves: Are they glossy? Downy/felted beneath? Are there tufts (what colour?) under the vein-joints? Is the leaf-stalk downy? How long is it?

Broad-leaved Lime   *Tilia platyphyllos*

Europe (including England and Wales); SW Asia. A rare native from the Pennines, Wye Valley, Cotswolds and SE downland scarps. Planted abundantly for some centuries, though less than Common Lime.
Appearance Shape Tall-domed; to 42 m. **Bark** Greyish, with often clean criss-cross ridges; seldom sprouty. **Shoots** Grey-green (redder in sun; clear red in the rare 'Rubra'/'Corallina', which has dense, bright foliage); *fine hairs* wear off through winter. **Buds** Fat, the *3 scales* grey/dull red, with sparse hairs. **Leaves** Often *dull, dark* green and *softly furry*; on *hairy stalks* in N European trees (ssp. *cordifolia*); hairy only under the veins in the central European ssp. *platyphyllos*; nearly hairless in the E European ssp. *pseudorubra*. To 15 × 15 cm; the sides can droop. **Flowers** Hanging, in mid-summer; 3–6 per bract. **Fruit** *Often strongly 5-ribbed*, downy.
Compare Common Lime (p.402). Silver Lime (p.404): *furry* buds in winter.
Variants Cut-leaved Lime, 'Laciniata' ('Aspleniifolia'), has small, very variably lobed leaves; a narrow tree (to 22 m), flowering profusely.

'Fastigiata', very occasional as a young tree, has close, steep limbs making a narrow, pointed dome; 'Orebro' (1935; rarer) is more gracefully conic.

Small-leaved Lime   *Tilia cordata*

(Pry Tree) Europe (including England N to Cumbria, and E Wales); the Caucasus. Locally abundant in old woods and hedges. Pollen deposits show that 5000 years ago this was the dominant tree in the NW European lowland 'Wildwood'; but it is vulnerable to grazing by stock and needs summer heat to ripen its seeds, and is now absent in the wild from many counties. Some giant coppice stools, surviving from pre-history, may be England's oldest 'trees'. Occasional as an older planted specimen, but now in fashion.
Appearance Shape Domed, to 38 m. **Bark** Grey/buff; craggier than Broad-leaved Lime's; often very sprouty. **Shoots** Quickly *hairless* and (in sun) shining red. **Buds** Fat, hairless, with 1 big and 1 small scale – like boxing-gloves. **Leaves** Small (8 × 8 cm); usually sturdily *flat-surfaced*; hairless except for *rufous* tufts under vein-joints. Underside matt and slightly silvered (can be glossier green in shade/on saplings). **Flowers** Profuse: *spreading at all angles* in high summer, so that the crown turns creamy yellow; 5–11 per bract. **Fruit** *Hairless; scarcely ribbed*.
Compare Common Lime (p.402): fawn tufts under leaves; drooping flowers. Mongolian Lime (p.404).
Variants 'Rancho' and 'Greenspire' are 1961 clones with steeply rising branches; rare so far.

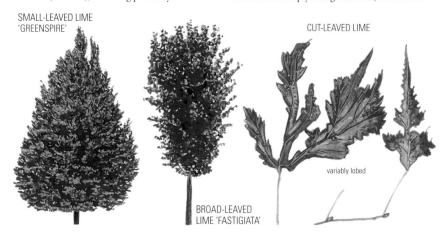

SMALL-LEAVED LIME 'GREENSPIRE'

CUT-LEAVED LIME

variably lobed

BROAD-LEAVED LIME 'FASTIGIATA'

LIMES

Common Lime
Tilia × europaea

(*T. × vulgaris*) A hybrid of Broad- and Small-leaved Limes, very rare in the UK in the wild (more widespread on the Continent), but from the early 17th to the mid-20th centuries the most planted lime: the liberal sprouts made propagation cheap. In many high streets, submits cheerfully to annual lopping.
APPEARANCE Shape Domed or, in one common clone, *columnar with close, vertical limbs*. The tallest wild broadleaf, to 46 m; the small leaves near the top make it look even bigger. **Bark** Pale grey-brown, irregularly ridged; some clones envelop themselves in sheaths of sprouts, ideal for nesting birds. **Shoots** Red in sun; soon hairless. **Buds** Fat, like boxing gloves: 1 big and 1 small reddish scale *fringed* with fine hairs. **Leaves** Typically 10 × 10 cm, *flimsy*, soon more or less hairless except for *buff* tufts under the vein-joints (cf. Small-leaved Lime's more *rufous* tufts). Underside usually shiny pale green, but matt and slightly silvered on the shoots (in sun) of one clone; stalks soon hairless. **Flowers** Hanging, in high summer; 4–10 per bract. **Fruits** *Downy, but only faintly ribbed*.
COMPARE The parents (p.400); American Lime and Crimean Lime (below).
VARIANTS Golden Lime, 'Wratislaviensis' (*c.*1890), is strangely rare: flushing green, then *gold*; later dark but still sprinkled with yellow late growths.

American Lime
Tilia americana

(Basswood) New Brunswick to Kentucky. 1752. Locally occasional in warm areas; easily passed over as a well-groomed Common Lime.
APPEARANCE Shape Narrow-domed, to 24 m, or gaunt, or rather weeping; dark, vivid, *yellowish green*. **Bark** Dark grey; rather vertical ridges. **Shoots, Buds** Green-brown/reddish, hairless. **Leaves** Common Lime-sized (but to 30 cm long on odd gaunt trees); rather *oblong* with the base slashed obliquely across; *large, elegant, yellow-tipped teeth* (elongated in the clone 'Dentata'). Under-leaf slightly *darker* grey-green than Common Lime's, with *tiny* whitish tufts in the vein-joints (sparsely downy in the northern var. *vestita*, in some collections); stalk hairless. **Flowers** Often *10–12 per bract*. **Fruit** *Smooth, hairless*.

Crimean Lime
Tilia × euchlora

(Caucasian Hybrid Lime) *T. cordata × dasystyla*? – in the Crimea by 1860. Frequently planted: the *glossy foliage* is not aphid-friendly.
APPEARANCE Shape The common clone is soon unmistakably ugly: a *straight trunk* degenerates halfway up into a *tangle of curling limbs*, above which fine branches make a *narrow*, dense dome to 20 m. **Bark** Dark grey; smooth; widely fissured with age. **Shoots** *Lime-green* (amber-red in sun by winter), finely downy. **Buds** Green/reddish. **Leaves** Plane, elegantly pale-toothed, on hairless stalks; *big brown tufts* under the vein-joints. They yellow, one by one, early in autumn. **Flowers** Late; rich gold and headily scented. **Fruit** Downy, shallowly 5-ribbed.
COMPARE Common Lime (above). Silver Pendent Lime (p.404): slightly similar branching.
OTHER TREES The Crimean *T. dasystyla* (as which Crimean Lime is sometimes grown) and its Caucasian ssp. *caucasica* are in some collections; elegant leaves downy under the main veins (cf. ssp. *pseudorubra* of Broad-leaved Lime, p.400).

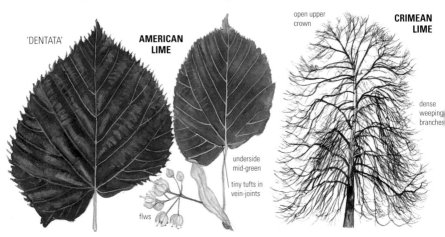

SILVER LIMES; MONGOLIAN LIME

Silver Lime *Tilia tomentosa*

Hungary to SW Russia and NW Turkey. 1767. Quite frequent in warmer parts, especially as a younger tree.
APPEARANCE Shape Precisely domed, usually on straight, steeply radiating limbs, to 32 m. The heavy, sombre foliage, in *level plates*, is lit up by the silvery leaf-backs. **Bark** Dark grey-brown; coarse but shallow criss-cross ridges. Old specimens are often grafted. **Shoots, Buds** *Finely but closely grey-downy*. **Leaves** With large teeth (and rarely pointed lobes at the 'shoulders'); *white-woolly* beneath. The minutely silver-scurfy stalk is *less than half as long as the leaf*. **Flowers** Late, 7–10 per bract; very fragrant, with a hint of soap-powder; a sugar in the nectar intoxicates and can kill bumble bees. **Fruit** 5-ridged, white-hairy.
VARIANTS Silver Pendent Lime, 'Petiolaris' (*T. × petiolaris*; of obscure origin, by 1842), is slightly more frequent and possibly the most ornamental (and vigorous) lime. Fountain-like crown, to 33 m, with *weeping side-branches from heavy, crooked main limbs*; leaf-stalk (much) *more than half as long as the leaf*. Always grafted (usually on the sprouty Common Lime) and awkwardly outgrowing the stock; good yellow autumn colours.
OTHER TREES *T.* 'Moltkei' (Germany; by 1880), may be a cross of 'Petiolaris' with American Lime: slightly weeping crown (to 25 m), *hairless* shoots and buds, and *big leaves* (to 25 cm), often finely downy above and *very thinly felted* beneath (no distinct tufts under the vein-joints). Rare, but more likely to be found as an old park tree than is the (smaller-leaved) White Basswood (below). *T.* 'Spectabilis' (Silver × American Limes?; now almost extinct), with smaller but similar leaves, has finely downy shoots and buds.

White Basswood, *T. heterophylla* (E USA), is in some collections: its leaves, shaped like American Lime's (p.402), have smaller teeth and *buff tufts* under the vein-joints, which can be more or less hidden by the under-leaf's fine white/grey felt. Its shoots and leaf-stalks are very finely woolly.

Oliver's Lime *Tilia oliveri*

Central China. 1900. A very stylish lime, still almost confined to the biggest gardens.
APPEARANCE Shape Domed, on clean limbs; to 25 m. **Bark** *Smooth* and soft grey; dark folds cap the branch-scars. **Shoots** *Hairless*, green/pinkish; buds hairless except at their tips. **Leaves** *Flat* and very elegant: fresh matt green, brilliantly silvery-felted beneath, with small, neat, white-tipped, rather distant teeth; the apple-green stalks are *quite hairless* except for a little down at each end. **Flowers** Small – but up to 20 per bract.
COMPARE Silver Lime (above) and its allies: woolly shoots *or* leaves only dully and sparsely felted.

Mongolian Lime *Tilia mongolica*

N China; E Russia. 1904. In some collections and now a little more widely planted (town parks etc).
APPEARANCE Shape Neatly domed, to 20 m. **Bark** Grey; rather vertical fissures. **Shoots, Buds** Shiny and soon hairless; crimson in sun. **Leaves** Small and neatly flat-surfaced, like Small-leaved Lime's (p.400; often whitish beneath but not downy), but with *few and coarse teeth; sharply lobed* at least on younger trees. **Flowers** Showy: sometimes up to 30 per bract.
COMPARE Silver Lime (above): occasionally lobed leaves silver-woolly beneath. Swedish Birch (p.182); Birch-leaved Maple (p.382).

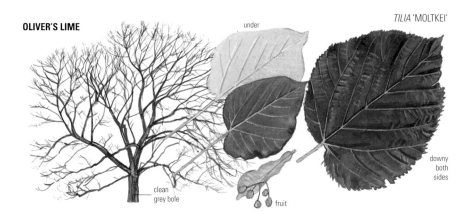

OLIVER'S LIME
under
TILIA 'MOLTKEI'
downy both sides
clean grey bole
fruit

SILVER LIMES; MONGOLIAN LIME

SILVER PENDENT LIME

steeply ascending branches

clean trunk

SILVER LIME

grafted

under

short stalk

flws

under

fruit

long stalks

SILVER PENDENT LIME

small dainty lvs

MONGOLIAN LIME

autumn

WHITE BASSWOOD

EUCRYPHIAS

The 5 or 6 Eucryphias *(in their own family, Eucryphiaceae) have opposite leaves and philadelphus-like flowers, with 4 white petals that stud their crowns in late summer.*

Rose-leaved Eucryphia
Eucryphia glutinosa

(*E. pinnatifolia*) Central Chile. 1859. Very rare in the wild as a rainforest understorey, but in many larger gardens: the hardiest eucryphia, though it demands an acid soil.
APPEARANCE **Shape** Roughly, spikily domed, with glittering foliage; to 13 m but often a bush. **Shoots** Pale brown, with scattered hairs. **Buds** Green, long-pointed, to 1 cm. **Leaves** Glossy; *3–7 finely toothed leaflets*. *Deciduous*: oranges and reds in autumn. **Flowers** Earlier in summer than the other species'. VARIANTS 'Plena' arises frequently, but its double flowers are no improvement on the type's.

Ulmo
Eucryphia cordifolia

Rainforests of central Chile. 1851. Rare: large gardens in humid areas. Out of flower a puzzling tree.
APPEARANCE **Shape** *Naturally fastigiate*, to 22 m; blackish green. **Leaves** Densely set in perpendicular pairs; *simple*, rather oblong, to 8 × 4 cm; dark and slightly glossy but roughened above and *very downy beneath*. Thick, like card, with downcurved edges; toothed in young plants; later with wavy margins.
OTHER TREES Tasmanian Leatherwood, *E. lucida*, is much less hairy, with slenderer (1–2 cm wide), very glossy, often pointed leaves, *silvered beneath*; honey from its flowers is much prized. *E. milliganii* is a bushier Tasmanian ally (small, sometimes notched leaves; flowers only 18 mm wide). Both grow in a few big gardens.
E. 'Penwith' is an intermediate hybrid of Ulmo and Tasmanian Leatherwood, in some large gardens – the pointed leaves are slightly silvery underneath. Tobira (*Pittosporum tobira*) is a slightly similar, tall E Asian bush with *alternate* evergreen leaves, also confined to large gardens in mild areas.

Nymans Eucryphia
Eucryphia × nymansensis 'Nymansay'

Ulmo and Rose-leaved Eucryphia crossed at Nymans, Sussex, in 1914 and this form (originally called 'Nymans A', to distinguish it from another promising seedling, 'B') is now the only eucryphia to be found – occasionally, and in milder areas – away from large gardens.
APPEARANCE **Shape** Naturally *fastigiate*: many close stems, often from the base, with dense upswept blackish evergreen foliage and spiky tops dying back in exposure; to 23 m. **Bark** Smooth, dark grey. **Shoots** Green; very hairy. **Leaves** On densely hairy stalks: the larger trifoliate, the smaller simple (cf. Rostrevor Eucryphia); *oblong*, finely round-toothed, and slightly downy on both sides.

Rostrevor Eucryphia
Eucryphia × intermedia 'Rostrevor'

Rose-leaved Eucryphia and Tasmanian Leather-wood hybridized at Rostrevor, Co. Down, early in the 20th century. The descendants are very occasional: larger gardens in milder areas.
APPEARANCE **Shape** Dense, narrow, billowing *domes*, to 17 m. **Leaves** Trifoliate *or* (smaller ones) simple (cf. Nymans Eucryphia); slightly hairy *only on the stalks and margins*; teeth tiny, irregular. *Rich green* but less glossy than Nymans Eucryphia's; slightly glaucous beneath.

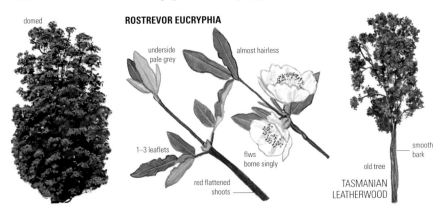

ROSTREVOR EUCRYPHIA
domed
underside pale grey
almost hairless
1–3 leaflets
flws borne singly
red flattened shoots
smooth bark
old tree
TASMANIAN LEATHERWOOD

STUARTIAS TO SMALL-LEAVED AZARA

Deciduous Camellia   *Stewartia pseudocamellia*

(*S. koreana*) Japan, Korea. 1874. Rather rare: large gardens on acid soils in S England. Named in honour of John Stuart, Earl of Bute; Linnaeus' misspelling is normally retained in the botanical name. A tree of the Tea family (Theaceae).
APPEARANCE Shape Elegant, slender and open, to 15 m, the foliage in crisp plates. **Bark** *With long, thin flakes of orange, cream and purple-grey* (cf. Persian Ironwood, p.278, and Strawberry Dogwood, p.426). **Shoots** More or less *hairless*. **Buds** Shiny green/red; flattened. **Leaves** Often glossier beneath than above; slightly crumpled, hair-fringed and with tiny distant teeth; yellow to dark red in autumn. **Flowers** Camellia-like, in midsummer: 6 cm wide, but a little lost among the leaves. The 2 green bracts, which back the flowers/fruit, are *shorter* than the densely grey-hairy sepals. **Fruit** About 22 mm long.
OTHER TREES *S. monadelpha* (S Japan, Cheju Do; *c*.1903) is rarer. A slender little tree, often with a brighter orange, more finely peeling bark, it has *hairy* young shoots and its 2 leafy flower-bracts are much *longer* than the silky sepals. Flowers only 3 cm wide and fruits small (1 cm long).

Chinese Stuartia   *Stewartia sinensis*

Central and E China. 1901. Rare: a plant of unearthly beauty in some large gardens.
APPEARANCE Shape Elegant: lightly domed, to 15 m, with crisp sprays. **Bark** Stone-cold and silky-smooth (squirrel-proofing the tree): the surface layer darkens from cream through flesh-pink to dull purple before peeling off during autumn like sunburnt skin (and like only a few tree-rhododendrons, such as *R. barbatum*). Odd trees have a finely flaking, more orange bark (cf. *S. monadelpha*). **Shoots** Deep red, downy at first. **Leaves** With odd teeth; lightly hairy, especially above; almost smooth by autumn, when they colour orange and scarlet. **Flowers** 4–5 cm wide, fragrant: the green bracts behind them are *about as long* as the silky sepals. **Fruit** 2 cm.

Idesia   *Idesia polycarpa*

E Asia. 1864. Very rare. (Family: Flacourtiaceae.)
APPEARANCE Shape Open, on wide, level branches; to 20 m. **Bark** *Yellow-pink*; finely roughened, with prominent lenticel-bands. **Leaves** In opposite pairs, and to 30 cm: Catalpa-like (see p.444), but with many *hooked teeth*. Glaucous beneath, and downy in var. *vestita*. The *scarlet stalks* have 2 nectar-bearing glands near the top (cf. Hybrid Black Poplars, p.156; poplars have *alternate* leaves). **Flowers** Dioecious; yellow and fragrant, in 25 cm plumes in mid-summer. **Fruit** Deep red, 8 mm, hanging like grapes in hot years on female trees when (as hardly ever happens in the UK) males grow nearby.
COMPARE Chinese Necklace Poplar (p.162).

Small-leaved Azara   *Azara microphylla*

S central Andes. *c*.1861. The hardiest member of its genus and occasional in warmer areas.
APPEARANCE Shape Often bushy (to 11 m); daintily rising then cascading stems densely strung with shining, blackish, evergreen leaves. **Bark** Buff, finely scaling. **Leaves** 2 cm; alternate, but with *a 6 mm leaf-like stipule* (or a pair of them) *opposite each stalk*; odd tiny teeth. **Flowers** Little chocolate-scented yellow clusters string the twigs late in winter. **Fruit** Red berries.
COMPARE Black Beech (p.202); Box (p.358).

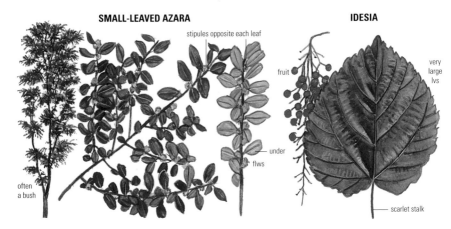

SMALL-LEAVED AZARA — stipules opposite each leaf; fruit; under flws; often a bush

IDESIA — very large lvs; scarlet stalk

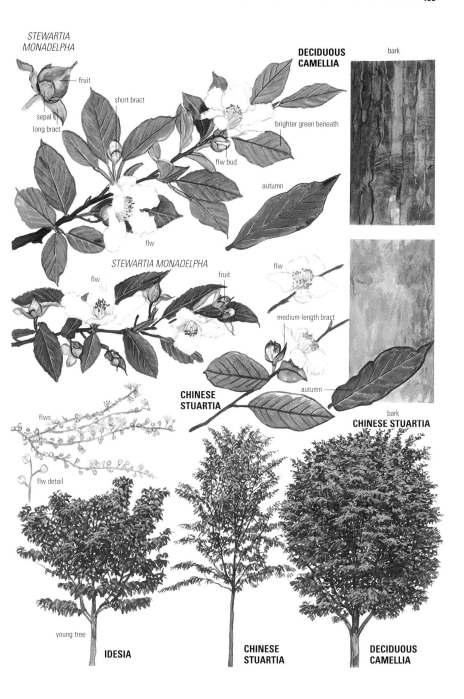

Kohuhu — *Pittosporum tenuifolium*

New Zealand. *c.*1850. Quite frequent in mild areas; sometimes seeding. (Family: Pittosporaceae.)
APPEARANCE Shape To 17 m but densely bushy in colder parts, on many slim straight rising stems. **Bark** Dark grey-brown; smooth. **Leaves** Foliage a favourite in floristry: purple-brown shoots carry thin, softly sea-green, evergreen 5 cm leaves, as undulant as crisps (cf. Holly 'Crispa', p.364). **Flowers** Purple-brown, honey-scented, 7 mm bells stand by the leaf-stalks in late spring.
VARIANTS 'Silver Queen' has misty-grey leaves finely margined in silver; frequent. 'Warnham Gold' has soft yellow foliage; rare. 'Purpureum' has green young leaves, maturing a saturated blackish purple; occasional.
OTHER TREES Also from New Zealand and with equally undulant evergreen leaves (thinly *grey-felted* beneath) is Akiraho (*Olearia paniculata*), a broad, bushy tree of the Daisy family (Compositae), confined to the mildest areas. Bark brown and *very stringy*; heads of off-white, scented daisies open late in autumn.

Tamarisk — *Tamarix gallica*

(*T. anglica*) N France to N Africa, near the Atlantic. Long naturalized in the UK (possibly native) and abundant in gardens on coasts; occasional inland. (Family: Tamaricaceae.)
APPEARANCE Shape Broad and bushy, on a gnarled, sprouty stem; to 8 m. **Bark** Brown; stringy with close vertical ridges. **Shoots** Red; willow-like but knobbly with close, sharp buds. **Leaves** In light green sprays 1 mm thick – Cypress-like, but with spiralling not paired scale-leaves, and shed in winter. **Flowers** Pink (rarely white), spraying *from the young shoots in summer*.
COMPARE Mount Etna Broom (p.348).
OTHER TREES *T. africana* (Atlantic coasts; now more planted, but a wispy bush) normally *flowers from spring on year-old twigs*. 6 more S European species (*T. ramosissima*: S Russia, Asia Minor; *T. canariensis*: W Mediterranean and Portuguese coasts; *T. smyrnensis*: SE Europe; *T. dalmatica*: E Mediterranean, in marshes; *T. parviflora*: Balkans/Aegean; *T. tetrandra*: mountains of SE Europe) are hard to separate with confidence.

Elaeagnaceae is a family of 50 silver-scaly trees and shrubs. Their roots, like alders and many pea-family trees, carry nodules of nitrogen-fixing bacteria allowing them to grow in poor soils.

Sea Buckthorn — *Hippophae rhamnoides*

Eurasia (including British coasts from the Scottish Borders S to Dungeness). Planted and aggressively suckering elsewhere on sand-dunes; a frequent, *intensely silver* garden plant.
APPEARANCE Shape Very spiky and spiny, to 10 m (but columnar in parts of its range); black and stocky in winter. **Bark** Dark grey-brown; shaggy, willow-like ridges. **Shoots** *Silver-scaled*, closely strung with conspicuous orange buds. **Leaves** To 70 × 7 mm; dull green above with a frosting of the tiny silver scales that also *completely coat the underside*. **Flowers** Dioecious. Tiny clusters in spring; female trees (with longer, pointed buds) are wreathed from September to February in clusters of *orange berries* (too acid for most birds to eat) whenever a male is close enough to pollinate them.
COMPARE Oleaster (p.412); Silver Willow (p.164).
OTHER TREES *H. salicifolia* (the Himalayas) is in a few big gardens: leaves to 70 × 12 mm, finely white-*felted* rather than scaly beneath. A sturdier tree, to 12 m, with *pallid yellow* berries.

TAMARISK

SEA BUCKTHORN

female tree

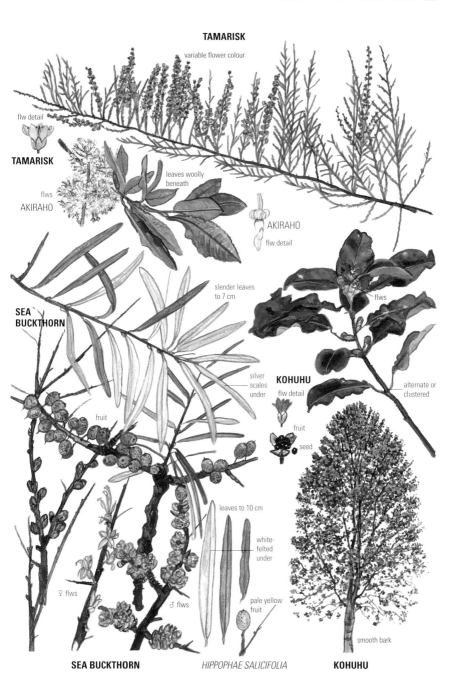

OLEASTER TO DOVE TREE

Oleaster *Elaeagnus angustifolia*

(Russian Olive) W Asia; naturalized in S Europe and long grown further N; very occasional in warmer parts.
APPEARANCE Shape Broad and spikily twiggy on a short leaning bole, to 11 m; occasionally spiny. **Bark** Black-grey; shaggy with superimposed criss-cross ridges. **Shoots** Silver-scaled; buds inconspicuous. **Leaves** Larger than Sea Buckthorn's (p.410) – to 80 × 18 mm, but with the same silver scales. **Flowers** Hyacinth-scented bells (white outside, yellow inside), in early summer. **Fruit** 12 mm, yellow, with silver scales; sweet.
COMPARE Olive (p.442): *evergreen, opposite* leaves; larger fruit, inedible raw.
OTHER TREES *E. umbellata* (the Himalayas, China, Japan. 1830) can also reach tree size: leaves *rounded-oblong*, to 10 × 4 cm, *fresh green above* but with the characteristic silver scales beneath; its 1 cm fruits ripen from silver to red.

Tupelo *Nyssa sylvatica*

(Black Gum) Wet land from Ontario to NE Mexico. Rather rare: larger gardens in warm parts. (Family: Nyssaceae.)
APPEARANCE Shape Conic/domed, to 22 m; very leafy; level, slightly *curved, spiky side-shoots*. **Bark** Grey; soon with rugged, triangular ridges. **Shoots** Green/brown, soon hairless; buds finely pointed, green/red-brown. **Leaves** Variable but more or less *untoothed*: dark or yellowish green (pale green beneath), oval/boat-shaped; usually *hairless and glossy*. Late-leafing; autumnal reds and golds as intense as any tree's. **Flowers** Dioecious; in tiny, *long-stalked*, greenish heads.
COMPARE Sorrel Tree (p.430): unrelated but remarkably alike out of flower, though with *finely serrated* leaves. Sassafras (p.276); Willow-leaved Magnolia (p.268). Snowdrop Tree (p.434): also with similar bark. Date-plum (p.432): glossier, longer-pointed leaves. Sweet Gum (p.278): browner and less rugged bark in winter.
OTHER TREES Water Tupelo, *N. aquatica* (Cotton Gum; marshy parts of SE USA), is in a few big gardens: leaves often coarsely toothed and downier beneath; buds tiny and *round*.
Chinese Tupelo, *N. sinensis*, a bushier plant in larger gardens since 1902, has hairier shoots; leaves (often longer) dark, red-flushing and never very glossy, and downy at their margins, with red-brown hairs persisting under the midrib; bark smooth for some years then more finely fissured.

Dove Tree *Davidia involucrata*

(Handkerchief Tree; Ghost Tree) One endangered species from W China. 1901. Occasional – larger gardens and some town parks in warmer areas; sometimes seeding. In its own family, Davidiaceae.
APPEARANCE Shape Conic, broad-domed, or multi-stemmed, to 24 m. **Bark** Orange-brown; scaly flakes. **Shoots** Dark brown: like Black Mulberry's (p.258) but with *shiny crimson/green* buds. **Leaves** Elegant, heart-shaped; sunken veins and *big sharp teeth*; readily frost-blighted or hanging from drought-stress. *White-downy* beneath in var. *involucrata*; glaucous but *hairless* in var. *vilmoriniana* (*D. vilmoriniana*) – which is confined in the wild to high-altitude cloud-forests but is hardier and commoner in Europe, and otherwise alike. **Flowers** The 'doves', 'handkerchiefs' or 'ghosts' unfold yellow-green in late spring then hang in spectacular white tiers.
COMPARE Idesia (p.408): larger, opposite leaves.

DOVE TREE

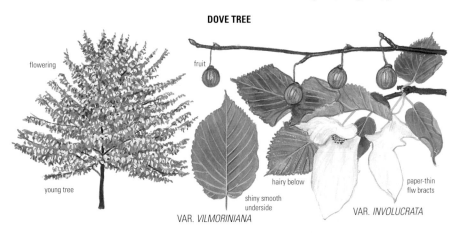

flowering · young tree · fruit · hairy below · shiny smooth underside · VAR. *VILMORINIANA* · paper-thin flw bracts · VAR. *INVOLUCRATA*

OLEASTER TO DOVE TREE **413**

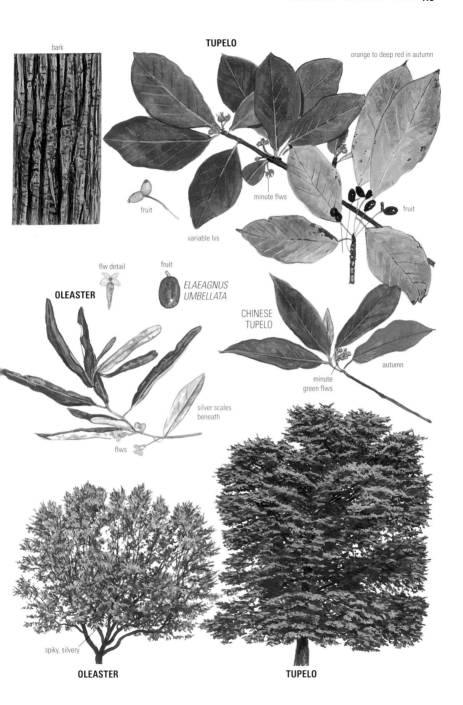

CHILEAN MYRTLE & EUCALYPTS

Myrtaceae is a big family of aromatic, evergreen trees and shrubs.

Chilean Myrtle *Luma apiculata*

(Orange-barked Myrtle; Luma; *Myrtus luma*) S central Andes. 1843. Occasional in very mild areas; naturalizing freely.
Appearance Shape Bushily domed/conic, to 20 m; almost *black evergreen foliage* sets off the bark. **Bark** Beautiful: long, thin flakes, *bright orange and white*. **Leaves** In opposite pairs; untoothed, to 25 × 15 mm; hairy under the sunken midrib; sweet, spicy aroma. **Flowers** White masses among black leaves in late summer. **Fruit** Small berries, ripening dark purple.

Eucalypts (500 Australasian species) include the world's tallest broadleaves and fastest-growing trees. Many mountain forms are still being tried out in N Europe, so that trees outside the scope of this book can be found in unexpected places. The aromatic oils can be irritant. The leaves are usually asymmetrical (curved) but the same on each 'side'; typically untoothed and hanging vertically to minimize transpiration under a high sun. Like many Australian plants, the foliage alters as the plant matures: 'juvenile leaves' are relatively broad, seldom stalked and in opposite pairs; a few species retain this foliage and some lose it quickly, while many retain odd sprouts. Adult leaves are narrower, stalked and appear alternate. All are evergreen; growth continues year-round. Flower-buds are usually carried in consistently sized clusters and are visible all year long; they have 'lids' (a mass of fused petals) generally defined by a scar when an outer cap has been shed. The bud's 'lid' is pushed off by a fuzz of stamens. The fruit is a woody capsule containing fine seeds, dispersed through valves in the lid. Identification can be tricky: flower-bud clusters are the best feature to look for. (Family: Myrtaceae.)

Things to Look for: Eucalypts

- Bark: How rough is it (at the base)?
- Juvenile leaves: What shape are they? How glacous? Are they toothed?
- Adult (alternate) leaves: Are they present? Are the main veins parallel to the midrib? Are they toothed? How long are they?
- Flower buds: How many are there together? What shape are they? Are they stalked? In single or paired clusters? With distinct lids (what shape?) Green or glaucous?

Key Species

Blue Gum (p.418): flowers usually borne singly.
Cider Gum (below): short leaves; flowers in 3s.
Broad-leaved Kindling Bark (p.416): long leaves; flowers in 3s. **Shining Gum** (p.418): long leaves; flowers in 4s–7s. **Snow Gum** (p.420): main veins parallel to the midrib; flowers in 7s–12s. **Tasmanian Yellow Gum** (p.418): finely round-toothed adult leaves. **Silver-leaved Mountain Gum** (p.422): largely retains opposite, juvenile leaves.

Cider Gum *Eucalyptus gunnii*

Marshy parts of the Tasmanian plateau and of the SE Australian Alps; its sap was once tapped for 'cider'. 1846. For long the most planted eucalypt in the UK (though scarcely the most ornamental) and abundant in warmer areas; often naturalizing freely.
Appearance Shape Billowing: irregular, narrow branch-domes from an often long but *sinuous* trunk, or on several stems; rarely spire-like for long. To 35 m; potentially long-lived, but easily blown down at any age. **Bark** *Orange-grey*, or salmon-pink; sometimes white/grey. Soon rough with fine vertical flakes at the base; limbs smooth between long, hanging scrolls. **Shoots** Pale green, then often vividly silver. **Juvenile leaves** Round, silvery, 3–6 cm, finely *round-toothed* (cf. all the leaves of Silver Gum, p.422). **Adult leaves** (Soon very predominant) *oval*/willow-like, 4–15 cm, dark dull grey (rarely shining green); *seldom hanging vertically*. *Less scent* than most eucalypts'. Specimens with narrow leaves (12 × 3 cm) are sometimes 'Whittinge-hamensis', descendants of a famous old East Lothian tree. **Flower-buds** In 3s, variably *bloomed silver*; ice-cream-cone-shaped, the lids with sharp central knobs; fruit usually silver-bloomed.
Compare Other gums with rather short, oval, dark-grey leaves: Mount Wellington Peppermint (p.422) and Small-leaved Gum (p.420). Narrow-leaved Black Peppermint (p.420), Spinning Gum (p.416), and Tasmanian Yellow Gum (p.418): can also have pink/orange barks. Tingiringi Gum (p.416): closely related; usually with longer, darker leaves. Silver-leaved Mountain Gum (p.422): similar jizz.
Other trees Alpine Cider Gum, *E. archeri*, a smaller high-altitude tree (rare to date) has usually greener, slender leaves; shoots, fruits *and flower-buds all apple-green, unbloomed*.

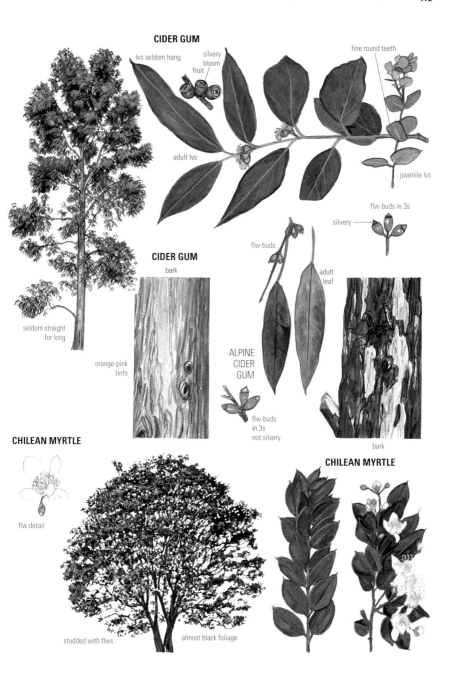

EUCALYPTS

Tingiringi Gum *Eucalyptus glaucescens*

A scarce mountain tree (New South Wales; Victoria); very rare to date.
Appearance Shape Generally stunted in the wild; very fast and straight in the UK so far. **Bark** Usually smooth white/blue-grey, with long shedding strips; rougher at the base. **Shoots** Red, bloomed white. **Juvenile leaves** Like Cider Gum's (p.414) but *untoothed*; intensely silver. **Adult leaves** Usually blackish grey-green, hanging on pink stalks; generally quite long (to 20 × 4 cm). **Flower-buds** In 3s, usually glaucous; lids *very flattened* but with a central point.
Compare Other gums with long, dark, hanging leaves: Snow Gum (p.420), Shining Gum (p.418) and Alpine Ash (p.422) have flowers in clusters of 4+; Blue Gum (p.418) has single flowers. Spinning Gum (below) can safely be separated only by fused juvenile leaves, and Urn Gum and Ribbon Gum (below) by flower-buds/fruit.

Spinning Gum *Eucalyptus perriniana*

Mountains of Tasmania, Victoria and New South Wales. Rather occasional: in large and small gardens for its unique juvenile foliage.
Appearance Shape Spire-shaped at first, but rounded often by 16 m, on light, ascending branches. **Bark** Orange-brown, or white/grey; long flaking strips. **Juvenile leaves** Vividly silver and untoothed, the pairs often fused to make a *disc pierced by the shoot*. After 'falling' this can spin round the twig for years. **Adult leaves** Flushing purplish, then dark grey-green/silvery; hanging; often narrower and more feathery than most gums' (12 × 2 cm). **Flower-buds** In 3s; silvery, with *smoothly pointed* caps.
Compare Narrow-leaved Black Peppermint (p.420): even slenderer leaves. Broad-leaved Kindling Bark (below): bigger, undulant leaves. Cider Gum (p.414); Tingiringi Gum (above). Only juvenile foliage identifies positively; but adult eucalypts in the UK with narrow, hanging leaves but no flowers or juvenile sprouts often seem to be this species.

Urn Gum *Eucalyptus urnigera*

Tasmania. By 1860. Rare; milder areas.
Appearance Shape Towering; to 38 m. **Bark** Mostly smooth white and pink, with long strips. **Juvenile leaves** Round, glaucous; untoothed. **Adult leaves** Hanging, dark grey to *vivid green*, to 15 × 3 cm. **Flower-buds** In 3s, *on individual 8 mm stalks; urn-shaped* (with a 'waist'). **Fruit** Carpeting the ground; also *urn-shaped*.
Compare Spinning Gum and Tingiringi Gum (above). The flower-buds and fruit are unique.

Broad-leaved Kindling Bark *Eucalyptus dalrympleana*

Tasmania, Victoria and New South Wales. Now locally frequent.
Appearance Shape A towering dome, often on a straight bole; rarely a giant bush with sinuous stems. **Bark** Pale orange-brown (rarely cream); great shed lengths lodge in the branch-forks. **Juvenile leaves** *Pale green*, untoothed; rather *heart-shaped*, to 6 cm. **Adult leaves** Long (to 20 × 4 cm), *undulant* and hanging; flushing purplish, then dark matt grey. **Flower-buds** In 3s, green, their caps rather long-pointed. **Fruit** To 9 mm.
Compare Narrow-leaved Black Peppermint (p.420); Shining Gum and Blue Gum (p.418).
Other trees Ribbon Gum, *E. viminalis* (Manna Gum), is rarer, and has *boat-shaped* juvenile leaves (10 × 2 cm). Bark often *whiter*, with finer hanging strips, but rough near the base of old trees; leaves often slenderer, and a darker grey-green.

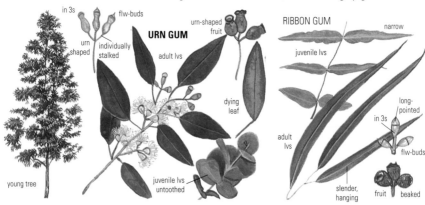

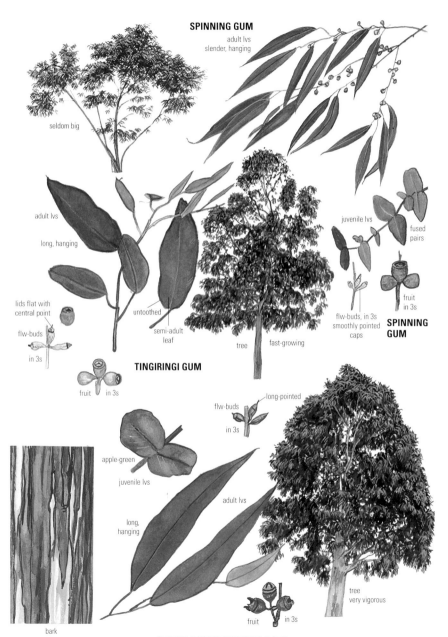

EUCALYPTS

Varnished Gum *Eucalyptus vernicosa*

High mountains in Tasmania. In a few big gardens. **APPEARANCE Shape** A densely rounded, *shining green bush* on twisting, smooth, red-brown stems. May not be immediately recognized as a eucalypt. **Leaves** Leathery, oval, to *5 cm only*; they remain largely *opposite* (cf. the very glaucous foliage of Silver-leaved Mountain Gum, p.422). **Flower-buds** Stalkless and often carried *singly* (cf. only Blue Gum); caps with sharp beaks.

Tasmanian Yellow Gum
Eucalyptus vernicosa ssp. *johnstonii*

(*E. johnstonii*; *E. muelleri*) Tasmania: effectively a lower-altitude form of Varnished Gum, with a very different habit and more 'adult' foliage. Rare – big gardens in milder areas. **APPEARANCE Shape** A *dense, vivid green spire* to 40 m, nearly always on a straight trunk. **Bark** *Pinkish* grey; smooth except at the base. **Juvenile leaves** *Glossy green*, round (to 6 cm) and thick; shallowly round-toothed. **Adult leaves** Similar but longer, slenderer (to 12 cm) and alternate; the *faint rounded teeth* persist (cf. Silver Gum, p.422). **Flower-buds** In 3s; big – like short tubes with *flattened, beaked caps*. **Fruit** *Big*: to 13 × 8 mm. **OTHER TREES** Tasmanian Alpine Yellow Gum, *E. subcrenulata*, is rarer but similar (usually less tall and straight); its flower-buds have conic caps and its fruits are smaller (to 9 × 6 mm).

Blue Gum *Eucalyptus globulus*

Tasmania (and rarely on the mainland). 1829. One of the world's most planted timber trees, but tender and confined in N Europe (as an old plant) to Ireland and the Isle of Man, where it seeds freely: younger ones are very occasionally seen elsewhere. **APPEARANCE Shape** Conic, often on a straight bole; then a towering, dense dome (to 44 m in Ireland). **Bark** Mostly smooth white and grey, with long spiralling strips. **Juvenile leaves** *Large* (to 15 × 4 cm), close-set and *floppy*, and an intense silvery-white; predominant on trees until 5 years old and 10 m tall (cf. Silver-leaved Mountain Gum, p.422). **Adult leaves** Very long (to 30 × 5 cm), hanging and *dark* grey-green. **Flower-buds** Usually carried *singly*: huge, warty, and silvery. **Fruits** Large: *to 3 cm long*; carpeting the ground under old trees. **COMPARE** Broad-leaved Kindling Bark (p.416); Shining Gum (below); Alpine Ash (p.422).

Shining Gum *Eucalyptus nitens*

(Silver Top) Mountains of New South Wales and Victoria. Still rare, but one of the most magnificent of hardy eucalypts (and a fine timber tree); has grown 20 m in 9 years in Argyll. **APPEARANCE Shape** Rather *densely* conic so far; likely to reach 40 m. **Bark** Mostly smooth white and grey; hanging strips lodge in the branch-forks. **Juvenile leaves** Big and floppy (like Blue Gum's; to 17 × 8 cm); rather glaucous. **Adult leaves** Hanging, long (to 25 × 4 cm), *very crescent-shaped*; fresh green to brilliantly blackish. **Flower-buds** In 4s–7s; small, *densely clustered*, with long conic caps. **Fruit** Small (6 mm), stalkless and shiny. **COMPARE** Blue Gum (above). Small-leaved Gum (p.420): similar flower-buds; much smaller leaves. Alpine Ash (p.422): flower-buds lack distinct 'caps'. Mountain Ash (p.422): paired flower-heads. Snow Gum (p.420).

VARNISHED GUM — bushy bright green; smooth brown bark

SHINING GUM — in 4s–7s; flw-buds; no stalks (to each flw-bud); adult lvs; long, hanging crescent-shaped; juvenile lvs big

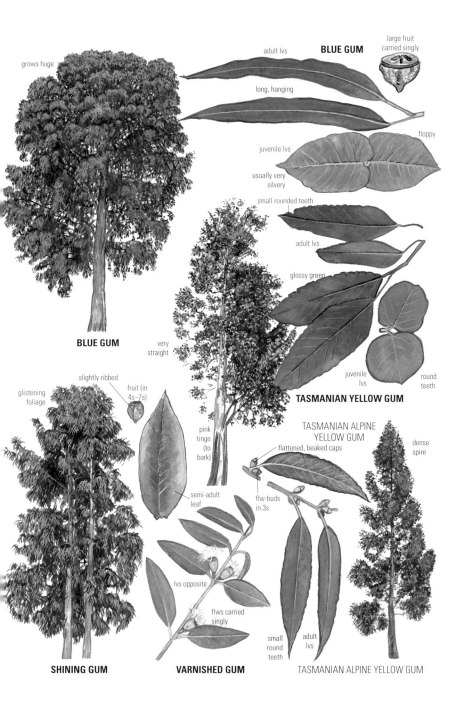

EUCALYPTS

Snow Gum
Eucalyptus pauciflora ssp. *niphophila*

(*E. niphophila*) High mountains of Victoria and New South Wales. Occasional, but now more grown: perfect for small gardens.
APPEARANCE Shape A *dark, glistening dome* (cf. Shining Gum, p.418), on steep, thin limbs from a *short* bole; exceptionally to 25 m. **Bark** Sometimes white and satiny-smooth, or with vertical, *tight* grey strips; sometimes a multi-coloured jigsaw. **Shoots** Red; silver-bloomed. **Juvenile leaves** Rarely seen; oval, *not glaucous*. **Adult leaves** Carried by seedlings after 4 pairs; flushing red-brown, then glossy dark grey-green; 7–14 cm, oblong (broad or narrow); *plane and scarcely curved*, with *parallel main veins running right up them* (cf. Blackwood, p.346) and an abrupt, hooked tip. **Flowers** In *big clusters* (7s–12s); buds glaucous, *smoothly club-shaped* and upcurving.
OTHER TREES Close relatives with the same *parallel, longitudinal* leaf-veins include:
Jounama Snow Gum, ssp. *debeuzevillei*, still rare; *angular* flower-buds (like an azalea's); white bark may develop more loosely spiralling strips.
Cabbage Gum, ssp. *pauciflora* (*E. coriacea*), rare; often untidily bushy, with *glaucous* juvenile leaves and more curved adult ones; its flower-buds (smoothly club-shaped) are *not* always glaucous.
Weeping Sally, *E. mitchelliana* (Mt Buffalo, Victoria): very rare but beautiful: smooth white bark; *weeping* branches. Leaves always narrow (to 15 × 2 cm); flower-buds in particularly *dense, star-like clusters*; narrow, smooth and *long-pointed*.
Black Sally, *E. stellulata*: very rare, differs from Weeping Sally in its *shorter, oval leaves* (to 8 × 3 cm), rough lower bark, and bushier habit.

Small-leaved Gum
Eucalyptus parvifolia

New South Wales; a rare species in the wild. Very occasional in the UK, in gardens of all sizes.
APPEARANCE Shape Usually low (to 20 m), spreading and quite *dense*, on rather twisting limbs. **Bark** Mostly smooth and grey; rarely more orange. **Juvenile leaves** Tiny (3 × 1 cm); dark green or greyish; mature crowns tend to retain many sprouts. **Adult leaves** Small (60 × 15 mm; sometimes longer and narrower) and *occasionally in opposite pairs* (cf. Varnished Gum, p.418); dark grey-green; often *spreading stiffly* at all angles. **Flower-buds** Tiny (4 mm); in *stalkless* groups of 4–7 (cf. Narrow-leaved Black Peppermint, below). **Fruit** Only 5 × 4 mm.
COMPARE Mount Wellington Peppermint (p.422): spiralled bark and flowers in 3s on many trees. Cider Gum (p.414).

Narrow-leaved Black Peppermint
Eucalyptus nicholii

New South Wales; rather rare as a young foliage tree.
APPEARANCE Shape A narrow, graceful dome. **Bark** Orange-brown; quickly with close *fibrous ridges*. **Shoots** Purple-bloomed. **Juvenile leaves** Very narrow (50 × 5 mm): delicately feathery. **Adult leaves** Also very narrow (12 × 1 cm) and hanging; they flush red-purple (cf. Spinning Gum, and Broad-leaved Kindling Bark, p.416), and the bloom patchily wears off to leave them dark grey-green. **Flower-buds, Fruit** Like Small-leaved Gum's (above).
OTHER TREES Black Gum, *E. aggregata* (swamps in New South Wales), shares the fibrous brown bark and tiny, clustered flower-buds but has slightly broader, duller green leaves; very rare.

BLACK SALLY

fruit big clusters short lvs parallel veins flw-buds often bushy star-like cluster

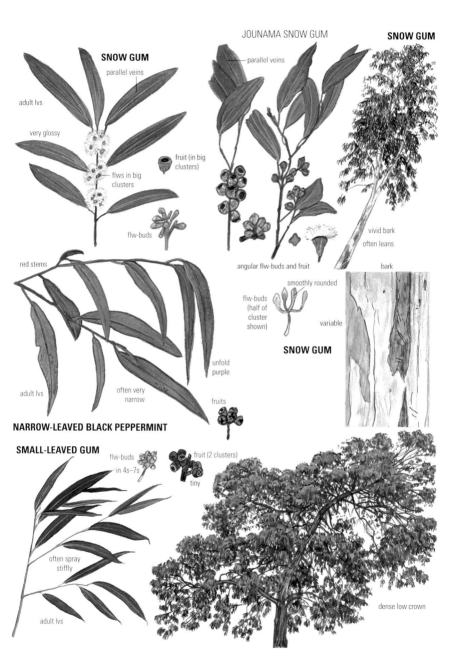

EUCALYPTS

Alpine Ash — *Eucalyptus delegatensis*

(*E. gigantea*) Mountains of Tasmania and Victoria. By 1905. Rare: milder areas.
Appearance Shape Towering, usually on a straight trunk (cf. Tasmanian Yellow Gum, p.418); to 42 m. **Bark** Smoothly white and grey on the limbs and upper trunk; *abruptly brown and fibrous near the base*. **Shoots** Red. **Juvenile leaves** (*Alternate* after 4 pairs) flushing red then dark grey-green; rather narrowly *oblong* (to 20 × 10 cm), hanging, *stalked*; they *mutate imperceptibly* into the similar but narrower adult leaves (*dull* green/grey; conspicuously veined). **Flower-buds** In 7s–15s, club-shaped; *no line around them* to define the usual distinct 'cap'.
Compare Broad-leaved Kindling Bark (p.416).

Mount Wellington Peppermint — *Eucalyptus coccifera*

(Tasmanian Snow Gum) Tasmania. 1840. Rare: big gardens.
Appearance Shape Wide-domed on twisting limbs; to 28 m. **Bark** White/grey; narrow, smoothly *spiralling* strips (cf. Jounama Snow Gum, p.420). **Juvenile leaves** Heart-shaped, pale green/silvery. **Adult leaves** *Small* (to 10 × 2 cm), green or often very silvery, and held *at all angles*; peppermint-scented. **Flower-buds** In 3s on some trees but 4s–7s on others; with *flat, warty caps*. **Fruit** Large (to 11 × 13 mm) and usually silvery.
Compare Small-leaved Gum (p.420): tiny, pointed flower-buds. Cider Gum (p.414): smooth bud-caps; different bark. Spinning Gum and Urn Gum (p.416): hanging leaves and differently shaped flower-buds.

Mountain Ash — *Eucalyptus regnans*

The tallest eucalypt in the wild (Victoria), and once perhaps exceeding even the American W coast conifers. Rare in the UK, and one of the less hardy species (its common name is shared by the Rowan, p.300).
Appearance Shape A towering dome. **Bark** Smooth grey/white ribbons, but brown and fibrous near the base. **Juvenile leaves** Green. **Adult leaves** *Dull* green; long (to 20 cm) and generally (like elms') *starting higher on one side of the stalk than the other* (cf. occasional leaves of Shining Gum, p.418). **Flower-buds** Slender, long-pointed; often in *a pair of clusters of 4–10 on long stalks from 1 leaf-base*.

Silver-leaved Mountain Gum — *Eucalyptus pulverulenta*

Mountains of New South Wales. Rare, but beginning to be grown for its intensely silvery foliage.
Appearance Shape A spiky, often leaning tree (to 35 m), which normally carries *only juvenile leaves*. **Bark** Smooth; white and orange. **Shoots** Smoothly rounded. **Leaves** Stalkless, heart-shaped, pinkish white then grey; to 5 cm long. **Flower-buds** In 3s, with long-conic caps.
Compare Juvenile Blue Gums (p.418).
Other trees Argyle Apple, *E. cinerea* (*E. pulverulenta* var. *lanceolata*), soon grows a closely *fibrous red-brown bark*. Its more 'adult' leaves may become narrower and longer (to 10 cm). Very rare.
Silver Gum, *E. cordata* (SE Tasmania, by 1850), differs from Silver-leaved Mountain Gum in its often longer leaves which have tiny, distant *round teeth* (cf. the alternate-leaved Tasmanian Yellow Gum, p.418, and juvenile leaves of Cider Gum, p.414), and are on warty, *squared* shoots. Very rare.

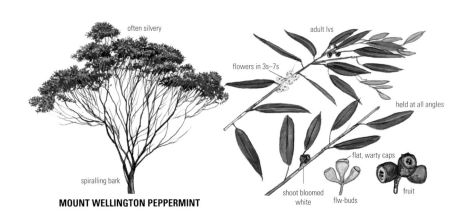

MOUNT WELLINGTON PEPPERMINT

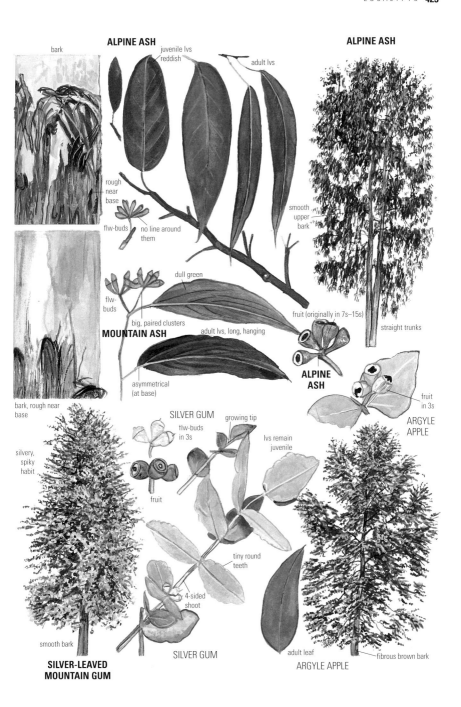

CASTOR ARALIA TO DOGWOODS

Castor Aralia *Kalopanax septemlobus*

(*K. pictus*) E Asia. 1864. Rare; big gardens; sometimes seeding. (Family: Araliaceae.)
APPEARANCE Shape Often gaunt; to 20 m. **Bark** Dark grey; *spiny warts*, then rugged criss-cross ridges. **Shoots** Thick, green (spined from 3rd year). **Buds** Conic, *to 4 cm*. **Leaves** Maple-like (but alternate), to 20 cm; hard, thick and shiny with tiny hooked teeth; they flush dark red but fall with little colour. Lobes shallowly triangular (cf. Sweet Gum, p.278) in the type, but *deep and narrowing to the base* in the commoner var. *maximowiczii* and recalling *Fatsia japonica* (a frequent bush, from Japan, with 9-lobed *evergreen* leaves); intermediate forms exist. **Flowers** Ivy-like; huge (50 cm), open heads late in summer. **Fruit** Rings of black berries showy until winter.

Papauma *Griselinia littoralis*

(Broad-leaf) New Zealand. *c*.1850. Frequent in mild parts, naturalizing: thrives on chalk and in salt spray. (Family: Griseliniaceae.)
APPEARANCE Shape Densely bushy (but to 20 m), on twisting, sprouty stems. **Bark** Dark brown; shaggy *curling scales*. **Leaves** Broad, *blunt*, to 11 × 8 cm; hairless, leathery. Usually matt and *pale apple-green*; some brightly variegated forms are now grown. **Flowers** Dioecious: little yellowy tails in spring. **Fruit** Blue-black berries, seldom seen in the UK.
COMPARE Ulmo (p.406): opposite leaves.

Dogwoods (40 species) have flowers and fruit that are very diverse. Their leaves are untoothed, with elegantly incurved main veins, and usually opposite; if torn carefully in half they can remain held together by threads (cf. only Hardy Rubber Tree, p.278). (Family: Cornaceae.)

Things to Look for: Dogwoods

- Bark: What is it like?
- Leaves: Are they alternate? Evergreen? How many vein-pairs? How downy?
- Flowers, Fruit: What are they like?

Common Dogwood *Cornus sanguinea*

(Cornel; *Thelycrania sanguinea*; *Swida sanguinea*) Europe, including England, Wales and Ireland. Abundant in downland scrub; frequent on heavy clays; now much planted. Skewers ('dogs') were made from the hard, straight twigs.
APPEARANCE Shape Broadly bushy, to 10 m; the leaves hanging elegantly, sometimes in tiers. **Bark** Grey; smooth, then with shallow, rounded ridges. **Shoots** Slim, straight and shiny (with some hairs at first): clear crimson in sun, lime-green in shade. **Buds** Scale-less: like *black bristles*. **Leaves** In opposite pairs; about 6 × 3 cm, with scattered stiff hairs on both sides; 3 or 4 vein-pairs. Rich *purple* in autumn. **Flowers** In early summer: dull white, heavily scented heads. **Fruit** Purple-black 7 mm berries, their oil once used for soap and in oil-lamps.
COMPARE Cornelian Cherry (below); Flowering Dogwood (p.426). Purging Buckthorn (p.398): toothed leaves, scaly bark.

Cornelian Cherry *Cornus mas*

S Europe, W Asia. Long grown in the UK, but now rather occasional.
APPEARANCE Shape Usually a stiff, twiggy bush; to 13 m. **Bark** *Orange-brown, closely scaly*. **Shoots** Green/brown, grey-hairy. **Leaves** Dull above, shinier below, with flattened hairs; 3–5 vein-pairs. **Flowers** In *mustard-yellow clusters in late winter* – almost as showy as Witch-Hazel's. **Fruit** Red, *cherry-like* (edible, but see below).
OTHER TREES *Lonicera maackii*, a tree-sized honeysuckle in some big gardens, has a very similar jizz (its leaves are also *opposite*, but downier). The fruits (slightly smaller, but showy; toxic) come *in pairs* and the branches in the late spring months are strung with small honeysuckle flowers (white then yellow).

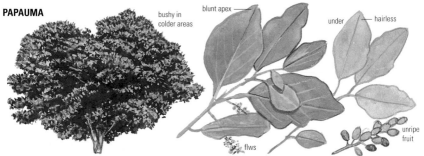

PAPAUMA — bushy in colder areas — blunt apex — under — hairless — flws — unripe fruit

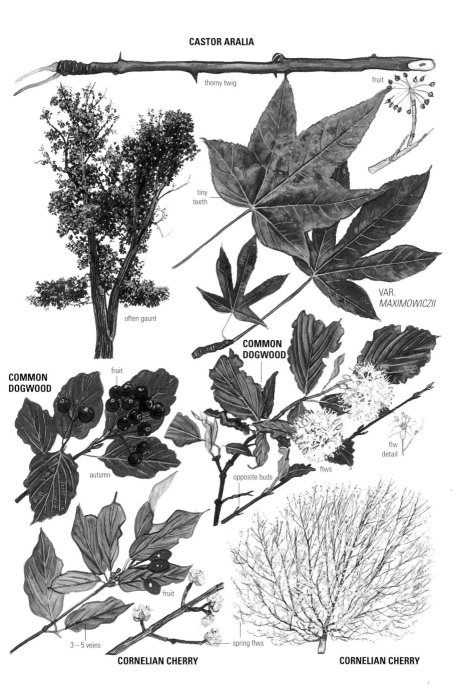

DOGWOODS

Table Dogwood — *Cornus controversa*

(*Swida controversa*) Japan; China; the Himalayas. By 1880. Rather rare: bigger gardens.
Appearance Shape Exquisitely poised *tiers of short, curved crimson shoots* and big, hanging leaves (cf. Emmenopterys, p.448). To 18 m. **Bark** Pale grey; developing shallow criss-cross ridges. **Leaves** Glossy, *alternate on extension-shoots*, to 15 cm; *6–9 vein-pairs*. **Flowers** In *flat white heads* in mid-summer, to 18 cm wide. **Fruit** *Blue*-black, 6 mm.
Variants Wedding-cake Tree, 'Variegata', has smaller, twisted, white-margined leaves, in tiers. Now occasional: the most desirable of variegated trees.
Other trees Large-leaved Cornel, *C. macrophylla* (*Swida macrophylla*), from the same habitats, is rarer and differs in its *opposite* leaves.

Flowering Dogwood — *Cornus florida*

(American Boxwood; *Benthamidia florida*) Ontario to NE Mexico; one of America's most loved flowering trees. c.1730. In the UK, confined to large gardens, and needing summer warmth.
Appearance Shape Bushy; to 8 m. **Bark** Grey-brown; *rugged, scaly ridges*. **Leaves** 7–13 cm, with 5–7 vein-pairs; elegantly hanging and glaucous beneath; microscopically serrated. Brilliant autumn colours. **Flowers** *Before the leaves*, in tight green heads, but backed by *4 big flame-shaped bracts* – white, or streaked a strangely intense smoky-pink in most garden trees (f. *rubra*).

Pacific Dogwood — *Cornus nuttallii*

(*Benthamidia nuttallii*) W North America. 1835. Rare: large gardens in warmer areas.
Appearance Shape *Tall and open*; light, rather level branches; to 17 m. **Bark** Purple-brown; *square-cracking*. **Leaves** Long (to 18 cm), with 5 or 6 vein-pairs; elegantly hanging. Brilliant autumn colours. **Flowers** Carrying *4–8 big showy bracts* (creamy, then pink-flushed) in late spring, sometimes with more in autumn.
Other trees *C.* 'Eddie's White Wonder' is a selection of the hybrid with Flowering Dogwood; rare. Flowers with 4 almost *circular*, overlapping, 5 cm creamy-white bracts.

Strawberry Dogwood — *Cornus kousa*

(Japanese Cornel; *Benthamidia kousa*) Japan, Korea and (var. *chinensis*) central China. 1875. One of the most admired of all garden trees year-round, yet still very occasional.
Appearance Shape Broad; rather tiered; to 10 m. **Bark** Thin *flakes of cream, orange and grey* (cf. Persian Ironwood, p.278, and Deciduous Camellia, p.408). **Leaves** 4–8 cm, with 3–4 vein-pairs, elegantly hanging. Brilliant autumn colours. **Flowers** Through early summer, backed by 4 big creamy-white bracts. **Fruit** Edible: *magenta-red 'strawberries'* (cf. Bentham's Cornel).
Variants 'Snowboy' has boldly white-margined leaves; 'Gold Star' has leaves centrally blotched in yellow (red towards autumn); both very rare to date.

Bentham's Cornel — *Cornus capitata*

(Himalayan Cornel; *Benthamidia capitata*) SW China, the Himalayas. 1825. Rare and confined to mild areas (where, oddly, it tolerates salt spray well).
Appearance Shape Conic, then rounded; to 18 m. **Bark** Grey-brown; shallowly scaly. **Leaves** Leathery and more or less *evergreen*, to 12 cm; dense, flattened hairs on both sides. **Flowers** With 4–6 big, *sulphur-yellow* bracts, in mid-summer. **Fruit** Edible: *big magenta 'strawberries'* (cf. Strawberry Dogwood).

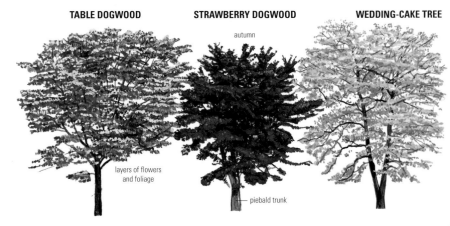

TABLE DOGWOOD — layers of flowers and foliage

STRAWBERRY DOGWOOD — autumn; piebald trunk

WEDDING-CAKE TREE

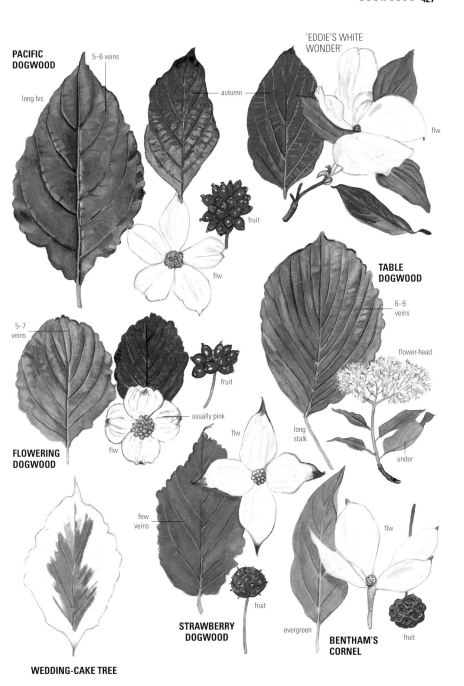

RHODODENDRONS

Few of the 1000 Rhododendrons reach tree size. They need (in cultivation) an acid soil; their leaves are untoothed. (Family: Ericaceae.)

Gurass *Rhododendron arboreum*

(Himalayan Tree Rhododendron) The Himalayas, to SW China; Nilgari Hills in S India; Sri Lanka. c.1810. Occasional – the commonest tree-rhododendron in large gardens in milder areas.
Appearance Shape Densely domed, on a bare interior of twisty limbs, to 16 m; often multi-stemmed. **Bark** Closely scaling ridges, dull red and grey. **Leaves** Leathery, to 20 cm; matt, blackish and soon smooth above, with a dense *brownish* (rarely silvery) *felt underneath*. **Flowers** Huge domes of 10–20 waxy trumpets in mid-spring; typically blood-red, but magenta, pink or white in some hardier, high-altitude forms (ssp. *campbelliae*, with a thick brown leaf-felt; subsp *cinnamomeum*, with a rufous leaf-felt): often most conspicuous when they fall and carpet the bare ground under the tree.
Compare Southern Evergreen Magnolia and Michelia (p.260).
Other trees *R. protistum* (in its var. *giganteum*) is perhaps the largest rhododendron in the wild, reaching 25 m on the China/Burma border. c.1930. In a few big gardens in mild areas. Leaves to 35 × 12 cm, with slight *auricles*; the felt under them buff/brown; flowers magenta-purple.
'Hardy Hybrid' rhododendrons were bred in great variety in the 19th century from *R. ponticum*, from Gurass and from American species: bushier, with often hairless leaves, and frost-proof flowers (of many colours) at the end of spring. Many can no longer be named with confidence. They were usually grafted on *R. ponticum*, which soon takes over the garden and then the neighbourhood.

Korlinga *Rhododendron falconeri*

The Himalayas. 1850. Very occasional. Large gardens in milder, wet areas.
Appearance Shape Densely domed; to 11 m. **Bark** A jigsaw of papery flakes: purples, mauves, and yellows. **Leaves** Like leather-pattern plastic: strongly *veined and wrinkled; to 35 cm*; a rich, rusty felt underneath. **Flowers** Domes of 15–20 pale creamy yellow (or pallid pink) trumpets, in mid-spring.
Compare Loquat (p.282); Chinese Evergreen Magnolia (p.260).
Other trees *R. sinogrande* is grown in mild areas for its *enormous leaves* (to 50 × 30 cm), glossy but wrinkled above and densely *silver-scurfy* underneath; flowers huge and pale cream.

Fortune's Rhododendron *Rhododendron fortunei*

E China. 1859. Rare: big gardens.
Appearance Shape Often irregular, on a slanted bole. **Bark** Finely flaking; orange-grey. **Leaves** Rather oblong, with tapered bases and sudden sharp tips, to 20 × 8 cm; *hairless* beneath and silvery between a network of green veins. **Flowers** Very scented heads of 8–12 big, soft-pink, frilly 7 cm trumpets, in late spring.
Compare Tarajo (p.366); Winter's Bark (p.274). The big, hairless leaves of other widely grown evergreens (eg. Cherry Laurel, p.344) are serrated.
Other trees 'Loderi Rhododendrons' – hybrids with the huge-flowered, tenderer Himalayan *R. griffithianum*, bred by Sir Edmund Loder at Leonardslee (Sussex) – are in many big gardens, to 12 m: trees/bushes with larger, headily scented white/pink blossoms. The blush-pink 'King George' is probably the mostly widely planted clone.
The aggressively naturalizing *R. ponticum* (1763) has narrower, finely tapered, hairless leaves and smaller, hard mauve flowers with a yellow blaze, at the end of spring: a sprawling bush (to 10 m).

LODERI RHODODENDRON ('KING GEORGE')
enormous very pale pink flws

pink flws

RHODODENDRON PONTICUM

FORTUNE'S RHODODENDRON

shrubby

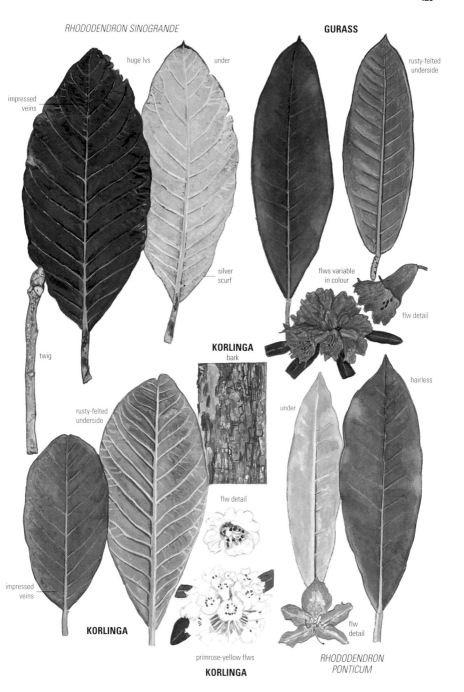

SORREL TREE & STRAWBERRY TREES

Sorrel Tree *Oxydendrum arboreum*

(Sour Gum; *Andromeda arborea*) E USA. 1752. Rather rare: large gardens in warm areas; it demands a moist, acidic soil.
Appearance Shape A high, thin dome, on sinuous branches; to 20 m. **Bark** Grey; very rugged crisscross ridges. **Leaves** Deep, rather glossy green; 10–20 cm, usually *finely serrated*; nearly hairless (sometimes bristly under the *conspicuous white midrib*); with the herb Sorrel's lemony taste. Bright red in autumn. **Flowers** *Great off-white drooping plumes* of heather-bells festoon the crowns even of saplings, from high summer to leaf-fall.
Compare Tupelo (p.412): unrelated but remarkably similar out of flower (its leaves *untoothed*). Black Cherry (p.344); Hardy Rubber Tree (p.278).

Strawberry Tree *Arbutus unedo*

Mediterranean region and SW Ireland, where it forms dense, low woods. Rather occasional in garden shrubberies in warmer parts; sometimes seeding.
Appearance Shape Dense, low and rounded, to 15 m, on twisted limbs; a *more vivid* green than most evergreens. **Bark** Closely, shallowly scaly; *dull* red-grey. **Leaves** Small (8 × 3 cm), usually *serrated*, hairless but on downy stalks/young shoots. **Flowers**, **Fruit** Ivory-white bells in sprays *in autumn*, as last year's yellow 'strawberries' ripen scarlet. These tempt but disappoint ('unedo', the Roman name, means 'I eat one – only'), but make a good jam, and (in Portugal) the alcoholic drink *Medronho*.
Compare Hybrid Strawberry Tree (below); Phillyrea (p.442).
Variants f. *rubra* has pink flowers; rare and rather ineffective.

Madrona *Arbutus menziesii*

W North America. 1827. Rather rare: big gardens. **Appearance Shape** A tall, open dome; *to 26 m*. **Bark** Smoothly peeling: *brick-red*, yellow-pink, or with green lights; finally grey and square-cracked at the base. **Leaves** *Large* (to 14 × 7 cm), *rather oblong* and (except on saplings) more or less *untoothed*; glaucous beneath. **Flowers** In *late spring*, on erect spires. **Fruit** 'Strawberries' small (1 cm), orange.
Compare Cyprus Strawberry Tree (below): a smaller tree with small leaves. Madrona's big foliage may suggest a tree rhododendron.

Cyprus Strawberry Tree
Arbutus andrachne

(Grecian Strawberry Tree) Albania E to the Caucasus and Palestine. 1724. Very rare: warm areas.
Appearance Shape Stunted, to 10m. **Bark** Smooth, creamy-yellow; or red and green peeling scales. **Leaves** *Small* (to 8 × 4 cm), and (except saplings) *untoothed*. **Flowers** White sprays in *spring*. **Fruit** 'Strawberries' smaller (12 mm) and smoother than Strawberry Tree's.

Hybrid Strawberry Tree
Arbutus × andrachnoides

Natural hybrid (Greece; Cyprus) of Strawberry and Cyprus Strawberry Trees. Very occasional.
Appearance Shape Irregularly domed, on twisting branches; taller than either parent. **Bark** Usually smooth and peeling or finely scaling in *ruby reds* and creams; or sometimes dull, like Strawberry Tree's (above). **Leaves** (Usually) *serrated*; slightly larger and paler underneath than Strawberry Tree's. **Flowers** White sprays in autumn *or early spring*.
Compare Strawberry Tree: bark *and/or* spring flowering distinguish.

SORREL TREE
flower-head
flws still open in autumn

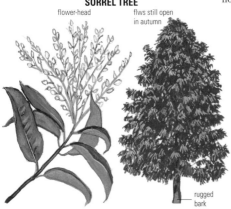

rugged bark

HYBRID STRAWBERRY TREE
vigorous

often red bark

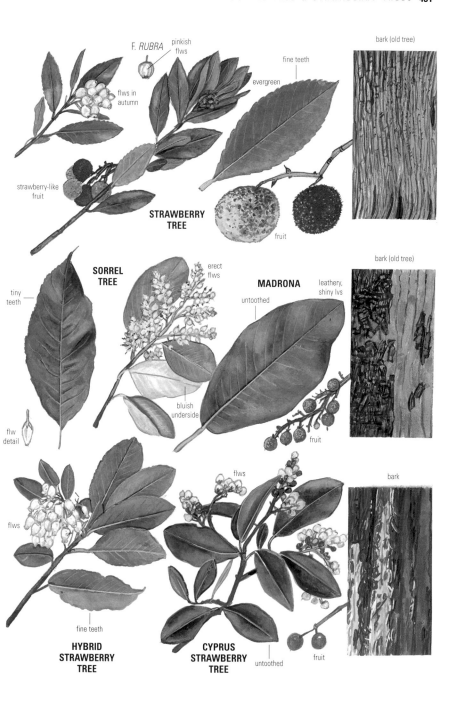

DATE-PLUM TO EPAULETTE TREE

Date-plum
Diospyros lotus

Cultivated across Asia and widely naturalized; long grown in the UK but now very occasional in old gardens. A tree in the Ebony family (Ebenaceae). **APPEARANCE Shape** Often rather elegantly and narrowly domed, to 16 m, or on big, low stems; *very glossy, hanging leaves*. **Bark** *Blackish, with small oblongs* – sometimes like a bar of plain chocolate (cf. Common Pear, p.316). **Shoots** Green/brown, downy at first, with flattened, conic, 6 mm buds but no big end-bud. **Leaves** Deciduous and thin but looking rather evergreen, to 18 × 5 cm (though some are much smaller); untoothed; strange, heavy smell in dank weather (hints of lentil stew and wet dog). There are golden hairs *above the midrib*, scattered beneath, and on the *6–12 mm stalk*. **Flowers** Dioecious; 5–8 mm red/green urn-shaped bells in mid-summer (singly on female trees; up to 3 together on males). **Fruit** The 1–2 cm date-plums are purple or yellow, but in the UK never ripen sufficiently to become edible.
COMPARE Tupelo (p.412); Bay Willow (p.166); Emmenopterys (p.448); Bamboo Oak (p.230): evergreen. Black Cherry (p.344), Hardy Rubber Tree (p.278) and Almond (p.338): with serrated leaves. A puzzling plant, best told by its bark and very glossy, hanging foliage.
OTHER TREES Persimmon, *D. virginiana* (SE USA), is much rarer in the UK but very similar: leaves on *longer stalks* (10–25 mm) and flowers slightly longer (10–15 mm). The edible persimmons (4 cm; yellow flushed scarlet) again fail to ripen in the UK.

Kaki
Diospyros kaki

(Chinese Persimmon) China; long grown in Japan for its fruit (the European 'persimmon' or 'Sharon fruit'). 1796. In a few collections, needing maximum summer warmth.
APPEARANCE Shape Low, sturdy. **Bark** May be paler and craggier than Date-plum's. **Leaves** *Large and oval*, to 20 × 9 cm; hairless above, downy beneath. **Fruit** To 8 cm, yellow/orange, ripening in the UK only against a warm wall.

Epaulette Tree
Pterostyrax hispida

China, Japan. 1875. Rare: larger gardens. (Family: Styracaceae.)
APPEARANCE Shape Elegant and rather conic (to 22 m) or bushy, on sinuous stems. May sucker profusely. **Bark** Grey-brown; flat-topped or finely scaly interlacing ridges. **Shoots** *Hairless*. **Buds** Slender, purple – generally 2 or 3 above each other (as in *Halesia* and *Styrax*, p.434). **Leaves** Big (to 20 × 10 cm) and matt; tiny distant teeth, and long white hairs under the veins. **Flowers** In densely hairy, hanging plumes in mid-summer; white and lemon-scented. **Fruit** 1 cm, very slender and covered in long pale-brown hairs, so that the whole *light-trapping hanging plume* remains showy into autumn.
COMPARE Snowdrop Tree (p.434): different flowers and fruits; slightly craggier bark. Kaki (above): untoothed leaves. Hemsley's Storax (p.434): smooth bark. Cucumber Tree (p.260): leaves untoothed; can look similar out of flower.
OTHER TREES *P. corymbosa* (similar habitats) has smaller, quickly hairless leaves and slightly larger fruit with *5 shallow wings*. Bushier; very rare.

EPAULETTE TREE

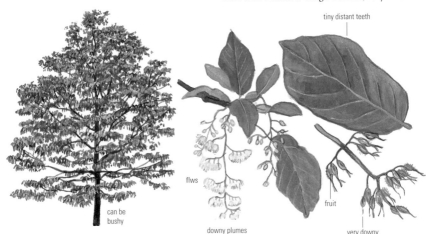

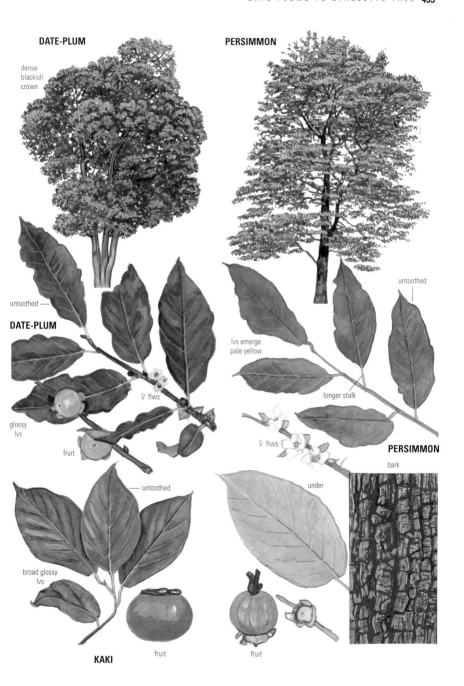

SNOWDROP TREE & STORAXES

Snowdrop Tree *Halesia monticola*

(Mountain Silverbell) SE USA: N Carolina to Arkansas. *c.*1897. Rather rare – larger gardens.
Appearance Shape Often untidily and openly conic; to 16 m. **Bark** Dark grey; *deep*, criss-cross, flat-topped ridges. **Shoots** Downy at first. Cut twigs, unlike Epaulette Tree's (p.432), show chambered pith (see Wingnuts, p.172, and get permission first!). **Leaves** Large (12–22 cm), dull green, with impressed veins and *tiny distant teeth*; there are scattered hairs underneath (but dense white down in the western var. *vestita*). **Flowers** Massed ivory snowdrops in late spring hang under the flushing leaves (pinkish in f. *rosea*, a variant of var. *vestita*). **Fruit** 5 cm, woody, green ripening brown, with *4 deep wings* rather like a dart's flights (cf. the small, shallowly 5-winged fruit of Pterostyrax corymbosa, p.432).
Compare Epaulette Tree (p.432); Tupelo (p.412). Snowbell Tree (below), with similar but showier and later flowers, has much smaller leaves.
Other trees *H. carolina*, from lower elevations (1756), is less rare in the UK but normally a coarsely spreading bush. Bark scaly but not deeply fissured; leaves smaller (5–12 cm), always grey-downy beneath; flowers slightly smaller, fruits 35–40 mm.

Snowbell Tree *Styrax japonicus*

Japan; Korea; central China. 1862; occasional in larger gardens.
Appearance Shape Short-boled; dainty, with dense rather tiered foliage, to 12 m. **Bark** Grey-brown; developing close, knobbly/scaly ridges, between orange fissures. **Shoots** Slender, orange/purple. **Buds** With 2 pale-brown main scales, like *furry mittens*. **Leaves** *Small (2–12 cm);* plane, slightly glossy; rather rhombic, with *distant, tiny teeth* and scattered hairs. **Flowers** Snow-white, dangling prettily (late frosts permitting) in early summer under even the inmost leafy shoots. **Fruit** Smoothly *globular*, 14 mm.
Compare Sugarberry (p.254); Osage Orange (p.256). Snowdrop Tree (above): similar flowers, though in spring; a much coarser, duller plant.
Other trees Storax, *S. officinalis* (E Mediterranean to W Asia, but also California and W Mexico) has *untoothed* white-downy leaves, especially underneath, and smooth grey bark. Cultivated for its fragrant resin; a rare garden bush in the UK. (Liquid storax is derived from Sweet Gum, p.278.)

Big-leaved Storax *Styrax obassia*

NE China; Japan; Korea. 1879. Rare: big gardens.
Appearance Shape Slender and graceful, with tiered foliage; to 14 m. **Bark** Grey; mostly *smooth*. **Buds** Orange-woolly, to 1 cm, but *hidden* by the base of the leaf-stalks in summer. **Leaves** Large (to 20 cm) and hanging; almost round with a sudden point and odd *big* teeth; deep green; *velvety beneath* (cf. Quince, p.290). **Flowers** Snow-white though not showy; hyacinth-scented, hanging from *15 cm tails* in early summer. **Fruit** Green, velvety, egg-shaped, 2 cm.
Other trees Hemsley's Storax, *S. hemsleyanus* (central and W China, 1900), is as graceful but rarer. *Bright orange* woolly buds are visible in summer; leaves slenderer (to 16 × 9 cm), with *tapering bases*, distant peg-like teeth, and only *sparse tufts of down* underneath – darker and thicker than Epaulette Tree's (p.432) or Snowdrop Tree's.

SNOWBELL TREE
flowering

BIG-LEAVED STORAX
velvety beneath
lvs hide flws
flws
hidden buds
fruit

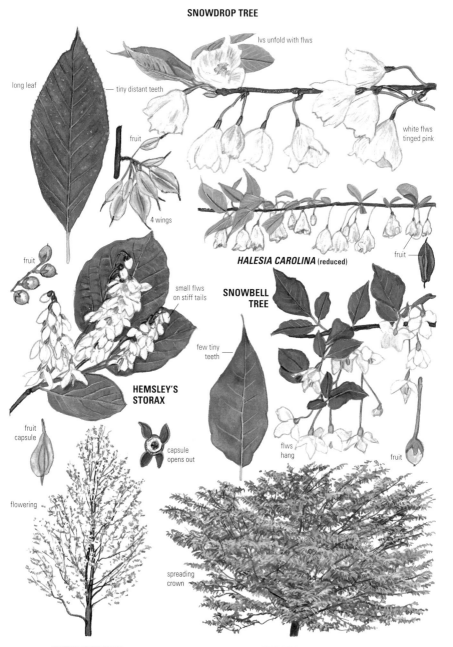

ASHES

Grouped in the same family as lilacs, jasmines and forsythias, ashes (60 species) have aerodynamic 'keys' like maples' (but symmetrical, and not paired) and opposite, usually compound leaves (in 3s on some strong shoots); they need a rich soil. (Family: Oleaceae.)

Things to Look for: Ashes

- Bark: What is it like?
- Buds: Are they black when mature?
- Leaves: Is the leaf-stalk hairy? How many leaflets? (Are they hairy? Individually stalked?)

Common Ash *Fraxinus excelsior*

Europe (including Britain and Ireland); the Caucasus. Abundant/dominant except on light sands; planted everywhere. The flexible 'ash-blonde' timber (used for tool-handles) burns even when green.
APPEARANCE Shape *Very open*; slender, cleanly curving limbs on an often long bole. The silvery shoots may droop then curl up like *branches of a chandelier*. Festooned frequently in ivy, which the airy crown fails to suppress, and affected by 'ash dieback' (probably due to a combination of environmental stresses): once known to 45 m; now rarely to 30 m. **Bark** Pale grey, developing a usually *regular* network of shallow, criss-cross ridges; rarely more like English Oak's. Erupting black bacterial cankers disfigure many trees. **Shoots** Grey. **Buds** Mitre-shaped, soon *sooty-black* – other ashes have brown buds. **Leaves** In opposite pairs; 9–13 irregularly serrated leaflets (the side ones *stalkless*), dull above and *white-downy under the lower midrib*, on a slightly *downy main stalk*. The last wild tree into leaf, and one of the first to go bare, fleetingly pale yellow. **Flowers** Nominally dioecious. Some trees change sex yearly, some carry branches of the wrong sex, some are hermaphrodite, and some produce dual-sex ('perfect') flowers. **Fruit** Bunched keys ripen biscuit-brown.
COMPARE Narrow-leaved Ash (p.438): slender, *hairless* leaflets. Claret Ash (p.438): slender leaflets on *hairless main stalks*. Manna Ash (p.438): smooth bark; 5–9 leaflets. Red Ash (p.440): 7–9 glossy leaflets. Oregon Ash (p.440): 5–9 often woolly leaflets. Elder (p.448): 5–7 leaflets.
VARIANTS Weeping Ash, 'Pendula', has straight shoots, like a shower of sleet, from angular branches; frequent, to 17 m. Wentworth Ash, 'Pendula Wentworthii', is rare but taller: the long-hanging shoots twist then *curl up at their tips*.
Golden Ash, 'Jaspidea', has *gold* not silver shoots for up to 5 years and brighter yellow autumn colours; a *densely leafy*, very slightly yellowish dome in summer. Occasional and vigorous; old trees (with little shoot-growth) often only noticeable from being grafted. 'Aurea' is much rarer: *stunted, twiggy growth*; smaller (green) leaves in summer.
Weeping Golden Ash, 'Aurea Pendula', is rare: a slow-growing, flat, gaunt plant with stiffly and shortly hanging golden shoots.
Single-leaved Ash, f. *diversifolia*, is occasional and has *1 big undivided leaf* (rarely 3 leaflets; cf. 'Veltheimii', p.438, and Box Elder, p.390), *jaggedly toothed/lobed* and to 20 cm. A puzzling graft (rarely found wild), but with typical bark *and black buds*. The weeping 'Diversifolia Pendula' is very rare.

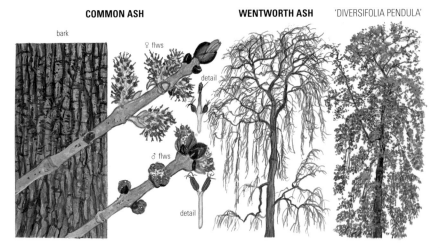

COMMON ASH — bark; ♀ flws; detail; ♂ flws; detail. **WENTWORTH ASH**. 'DIVERSIFOLIA PENDULA'

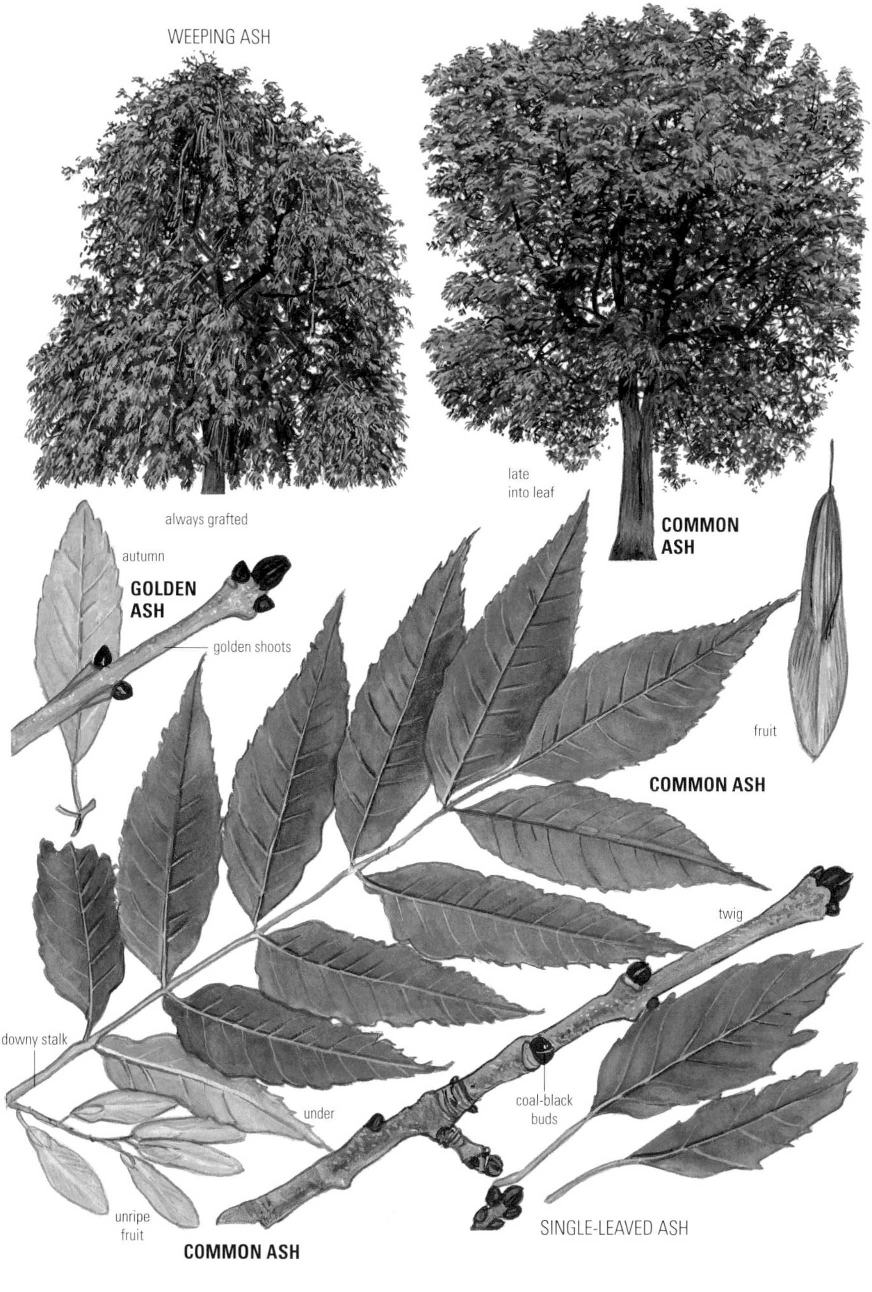

ASHES **437**

WEEPING ASH

always grafted

autumn
GOLDEN ASH
golden shoots

late into leaf
COMMON ASH

fruit

COMMON ASH

twig

downy stalk

under

coal-black buds

unripe fruit
COMMON ASH

SINGLE-LEAVED ASH

ASHES (EUROPEAN)

Narrow-leaved Ash *Fraxinus angustifolia*

W Mediterranean, N Africa. 1800. Occasional.
APPEARANCE Shape Denser than Common Ash (p.436), to 30 m; upswept or slightly weeping; ethereal, *feathery clouds* of shiny foliage. Old trees are often grafts on a stock of Common Ash – which they outgrow bizarrely. **Bark** Dark grey, soon with quite *deep, rugged ridges*. **Buds** Brown, finely grey-woolly. **Leaves** With 7–13 slender (60 × 15 mm) leaflets, shiny, stalkless and *completely hairless* on *hairless* main stalks. **Flowers** As Common Ash.
COMPARE Red Ash (p.440); Claret Ash (below).
VARIANTS 'Veltheimii' is this species' (very rare) answer to Single-leaved Ash (p.436): the 1 (*hairless*) leaflet is jaggedly toothed and to 12 cm long; its shoots are particularly densely upswept. 'Pendula' (less spectacular than Weeping Ash, p.436) has *long-hanging* shoots; very rare.

Claret Ash
Fraxinus angustifolia ssp. *oxycarpa* 'Raywood'

An Australian selection. By 1925. Now abundant as a park and street tree: breathtaking in autumn as the shoot-tips turn *royal purple* and the interior follows with orange, pink and gold.
APPEARANCE Shape Openly globular, to 25 m, with narrow forks which are the tree's main fault – the limbs shear off easily. **Bark** Dark grey; *smooth for many years* then finely ridged. **Leaves** With usually 7 *slender*, feathery, glossy, rich-green, stalkless leaflets with a *band of hairs* under the base of the midrib, on *hairless* common stalks.
OTHER TREES The wild Caucasian Ash (ssp. *oxycarpa* (*F. oxycarpa*); Romania to Iran) is rare: there may be only 3 leaflets (rarely jaggedly lobed; cf. Red Ash, p.440), usually turning *gold* in autumn.
Pallis' Ash, *F. pallisiae* (SE Europe; in some collections), has a similar aspect but *downy shoots* and 5–13 almost stalkless leaflets with stiff hairs above and *dense down beneath* (cf. Oregon Ash, p.440); *F. holotricha* (similar distribution) has less downy leaflets each on *8 mm stalks* (cf. Red Ash, p.440).

Manna Ash *Fraxinus ornus*

S Europe; Asia Minor. Long grown in the UK: very frequent in parks and streets.
APPEARANCE Shape A *dense, leafy dome* on rather twisting branches; rarely to 27 m. **Bark** Dark grey, usually *very smooth*; it has been tapped for a sugary 'manna'. **Buds** Grey-brown, woolly. **Leaves** With 5–9 very dark, broad, rather glossy leaflets, hairy under the lower midrib and on their *individual 1 cm stalks*; common stalk *kinky*, with tufts of hair at each junction. Muted yellow and *deep pink* autumn colours. (Var. *rotundifolia*, from S Italy and the Balkans, has leaflets only 3 cm long.) **Flowers** Insect- not wind-pollinated: *dense, fluffy plumes* of very narrow 1 cm creamy petals, *showy* in late spring.
COMPARE Euodia (p.356): a similar but later-flowering tree with rather larger, glossier leaves.
OTHER TREES Maries' Ash, *F. mariesii* (central China; 1878), is very rare and very slow (to 10 m) but has equally bold scented flowers. Leaflets 3–7, scarcely stalked.
Chinese Manna Ash, *F. chinensis* var. *rhyncophylla* (also from Korea; 1892), is in some collections and has *5 rather oval leaflets*, almost stalkless; insect-pollinated flowers in bold sprays but *lacking petals*.

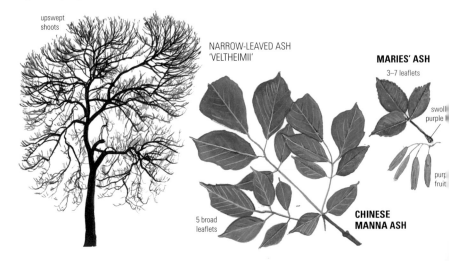

upswept shoots

NARROW-LEAVED ASH 'VELTHEIMII'

MARIES' ASH
3–7 leaflets

swoll purple

purp fruit

5 broad leaflets

CHINESE MANNA ASH

ASHES

White Ash — *Fraxinus americana*

E North America. 1724. Occasional in warmer areas.
APPEARANCE Shape Often gaunt: steep branches, straight shoots. To 30 m. **Bark** Pale grey; *shallow* ridges. **Shoots** Hairless; leaf-scars have *concave* upper edges. **Buds** Brown-woolly, the end ones *blunt*. **Leaves** With 7 or 9 *large* (to 15 × 7 cm) leaflets with distant teeth and *5–10 mm individual stalks*; often hairless but *whitish beneath with papillae* (waxy warts about the size of the ink-dots on this page); main stalk *hairless*. Butter-yellow in autumn. **Flowers**, **Fruit** As Common Ash (p.436). **COMPARE** Red Ash (below): differences emphasized. Bitternut (p.176): similar jizz but *alternate* leaves.

Red Ash — *Fraxinus pennsylvanica*

(Green Ash) E North America. 1783. Locally occasional (town parks, streets).
APPEARANCE Shape Often domed (to 23 m); like a *glossy*, well-groomed Common Ash (straighter shoots; brighter autumn golds). **Bark** Usually grey-brown with *close, sharp ridges* (like Norway Maple's, p.372), which may become shaggy, like some hickories' (pp.174–6). **Shoots** Sometimes palely velvety at first; the leaf-scars have *level* upper edges. **Buds** Brown-woolly, the end ones *pointed*. **Leaves** With 3–9 leaflets, large (to 14 × 6 cm) or slender and long-pointed with few jagged teeth or none (or in odd trees forking/jaggedly lobed); the lower leaflets, at least, have *5–8 mm individual stalks*; the grey-green under-leaf *lacks papillae* but has hairs (reduced, in the glossy-green widespread form distinguished as 'Green Ash', and once as var. *lanceolata*/*F. subintegerrima*, to tufts in the lower vein-joints). The main stalk may be downy.

COMPARE White Ash (above): differences emphasized. Caucasian Ash (p.438); Oregon Ash (below). A variable tree.
VARIANTS 'Variegata' is richly white-variegated; very rare and apt to revert.

Oregon Ash — *Fraxinus latifolia*

(*F. oregona*) W USA. *c.*1870. Very rare (but in a few town parks).
APPEARANCE Shape Domed, to 20 m; short, thick, *clumsy*, upswept shoots. **Bark** Like Red Ash's (above). **Shoots** Finely hairy. **Buds** Brown-woolly. **Leaves** With 5–9 *stalkless* leaflets, often large (each to 12 × 6 cm) and largely untoothed; usually finely downy above and *densely so* on underside, which can also be whitish with papillae; main stalks hairy.
COMPARE Red Ash, whose rare very downy forms have 2–8 mm leaflet-stalks. Pallis' Ash (p.438).
OTHER TREES Arizona Ash, *F. velutina* (SW USA; N Mexico), is also very rare and has typically velvety foliage, but *3–5 rather thick leaflets* (sometimes stalked; occasionally just 1 leaflet).

Japanese Big-leaved Ash — *Fraxinus spaethiana*

Japan. *c.*1873. A striking foliage tree in some collections.
APPEARANCE Shape Domed, to 15 m; often stunted in the UK. **Bark** Grey; rather shaggy vertical ridges. **Leaves** Flushing early: vivid green, with 7 or 9 *long-oblong* leaflets (cf. Sargent's Rowan, p.302), each to 20 cm and *crowded* so that the whole leaf can be as *broad as long*. The main stalk has a markedly *swollen bulbous base*.

EVERGREEN OLIVE FAMILY TREES

Chinese Tree Privet
Ligustrum lucidum

Central China; long cultivated as the food-plant of a wax-producing aphid. 1794. Locally occasional.
Appearance Shape Neatly domed, often on many closely forking stems; to 18 m; crown sparser than most evergreens'. **Bark** Grey-buff; fluted and sprouty; distant fissures. **Leaves** In opposite pairs; to 15 cm; thick, long-pointed, hairless and *brilliantly glossy* (matt beneath). **Flowers** Yellow-white spires decking the tree from mid-summer into winter, when next year's green heads develop. **Fruit** Blue-black, 1 cm berries (poisonous, but with medicinal uses); rare in the UK.
Compare Peruvian Nutmeg (p.276); Japanese Evergreen Oak (p.230). Bay (p.276): duller, alternate leaves.
Variants 'Excelsum Superbum' is as magnificent as its name: bright double variegation of *gold and cream* so the crown glows pale yellow. Very occasional. 'Tricolor' (rare) has narrow silver leaf-margins, *pink* at first. 'Aureovariegatum' (very rare) has a *dull, greeny-gold* variegation.

Olive
Olea europaea

Probably originating in SW Asia and Saudi Arabia; long grown around the Mediterranean. Now a fashionable pot-plant, but scarcely surviving in the UK to make a tree: demands long, hot summers and mild winters.
Appearance Shape Low and spiky, on a short, slanting trunk; steely grey. Wild plants (var. *sylvestris*) are spiny. **Bark** Grey; very gnarled ridges. **Shoots** Silver-scaled. **Leaves** In opposite pairs, to 8 × 2 cm; leathery and evergreen; silver-scaled beneath (cf. Oleaster, p.412). **Flowers** White, scented, in short hanging tails. **Fruit** Olives 10–35 mm, green then black/brown (or white) in their 2nd summer.

Phillyrea
Phillyrea latifolia

(Incorporating *P. media*) Around the Mediterranean; long grown in warmer parts of the UK but now very occasional.
Appearance Shape A low, dense, *blackish* evergreen dome; to 11 m. **Bark** Grey-black; closely but roughly square-cracked. **Shoots** With tiny hairs at first. **Leaves** In opposite pairs; hairless and glossy above; typically 5 × 2 cm, with neat little rounded teeth. **Flowers** Dioecious: small green-white clusters early in summer. **Fruit** Females carry blue-black, 1 cm berries.
Compare Holm Oak (p.222): Phillyreas, when noticed at all, are regularly passed over for this. Strawberry Tree (p.430); Maiten (p.448).
Variants f. *spinosa* has strongly toothed leaves (cf. Holly 'Myrtifolia', p.364); very rare.
f. *buxifolia* has small, untoothed leaves – very like Box's (p.358), but with the type's dark bark; also very rare.

Giant Osmanthus
Osmanthus yunnanensis

Central China. 1923. A handsome, puzzling tree in big gardens.
Appearance Shape A broad dome of *matt, pale, hanging foliage*; to 16 m. **Bark** Grey, finely roughened. **Leaves** *To 20 cm*, sometimes spinily toothed; leathery, undulant; both surfaces scattered with small black dots. **Flowers** Little massed clusters in spring, deliciously scented. **Fruit** Purple-bloomed berries.
Compare Tarajo (p.366): *alternate* leaves, like all other widely grown evergreens with foliage this big.
Other trees Holly-leaved Osmanthus, *O. heterophyllus* (1856), carries its scented flowers in autumn and is generally bushy. Leaves *very holly-like*, unspined with age – but *in opposite pairs.*

PHILLYREA — HOLLY-LEAVED OSMANTHUS — GIANT OSMANTHUS
fruit; spines; flws; small round teeth; lvs hang

EVERGREEN OLIVE FAMILY TREES

FOXGLOVE & INDIAN BEAN TREES

Things to Look for: Foxglove & Bean Trees

- Bark: How fissured is it?
- Leaves: What shape are they (especially at the tip and the base)? How hairy beneath? Is the leaf-stalk downy? What colour is it?

Foxglove Tree — *Paulownia tomentosa*

(Empress Tree; *P. imperialis*) N China – where only an empress could have one on her grave. 1838. Occasional in warmer areas. Glorious in flower (weather permitting), but sometimes cut annually as a foliage plant, regrowing 3 m shoots – hollow, like postage tubes – and elephantine leaves. One of a single small genus of trees among the many herbs of the Foxglove/Figwort family (Scrophulariaceae).
Appearance Shape Wide-spreading then upcurving branches from a straight trunk; to 26 m, but rarely even 15 m. Fragile, open and short-lived; may sucker. **Bark** Grey, finely roughened; very *shallow, wide, rounded* ridges in age. **Shoots** Stout, pink-brown, finely scurfy. Winter shoots end with 3 tiny purplish buds round a dead tip (which has failed to ripen) or – older plants – in candelabra of huge brown-woolly flower-buds. **Leaves** Gigantic (to 35 cm) and flimsy, heart-shaped and sometimes lobed/coarsely toothed; *softly woolly on both sides*, on a very *woolly stalk*. **Flowers** In late spring, but before the leaves: tall heads of deep mauve, highly fragrant bells. The buds – developing the summer before – are vulnerable both to winter cold and late frosts.
Compare Hybrid Bean Tree (p.446). Chinese Necklace Poplar (p.162): *alternate* leaves.
Other trees *P. fargesii* (W China, *c.*1896), is very rare: *almost hairless leaves* (cf. bean trees, with rugged barks) and slightly paler flowers.
P. fortunei (S China, Taiwan, E Himalayas; by 1940) is also very rare: slenderer, darker leaves, never lobed; flowers lilac outside and *cream inside* with purple splashes.

Indian Bean — *Catalpa bignonioides*

(Southern Catalpa) Georgia, Florida, Alabama and Mississippi. 1726. Quite frequent in hot areas. (Family: Bignoniaceae.)
Appearance Shape Typically very broad and open: level, twisting branches on a short, slanted bole; to 18 m. **Bark** *Shallowly scaly/square-cracked; orange to pink-brown* (rarely greyish). **Shoots** Stout, grey-brown, soon hairless, with 1.5 mm orange buds – often in (approximate) 3s, the end ones surrounding the scar of a dead shoot-tip (cf. Foxglove Tree), and helping to make the tree look dead for 8 months of the year. **Leaves** Big (15–30 cm) and foul-smelling when bruised; rather *pale fresh green; rounded* or very shallowly heart-shaped at the base and *tapering quickly to a sudden 1 cm point*; on saplings frequently but on mature trees *rarely* with small side-lobes. Closely hairy beneath; stalk green and *hairless*. They emerge briefly purplish and fall without autumn colour. **Flowers** In big candles at the growth-tips in late summer: white with yellow and purple splashes. **Fruit** 'Bean pods', to 40 cm, dangle after hot summers.
Compare Other Catalpas (p.446); Idesia (p.408).
Variants Golden Bean Tree, 'Aurea' (locally occasional) has *rich yellow* foliage, fading to soft green by autumn; usually conspicuously grafted. Shy-flowering.
'Nana' is rare and is grafted on to a stem of the type, producing a frizz of short, weak stems, with smaller leaves. (Compare, out of leaf, Mop-head Robinia, p.354.)

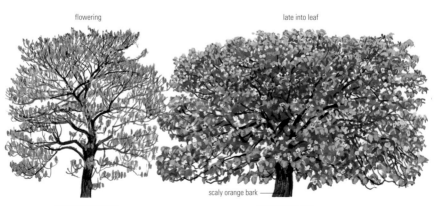

flowering — late into leaf — scaly orange bark

FOXGLOVE TREE   **INDIAN BEAN**

CATALPAS

Western Catalpa — *Catalpa speciosa*

(Northern Catalpa) Central USA: Arkansas to SW Indiana. 1880. Occasional in warm areas; less vulnerable to frost damage than Indian Bean (p.444) but still needs plenty of summer heat to ripen its wood.
APPEARANCE Shape A *tall, densely leafy* dome of short, twisting branches, to a narrow top (20 m); bole often straight. **Bark** *Grey*; quite *deep*, scaly vertical ridges. **Leaves** Longer and darker than Indian Bean's, and scentless; usually *deeply heart-shaped at the base* and tapering *narrowly to their long apex; rarely* on mature trees with small side-lobes. Closely brown-hairy beneath; stalk finely hairy at first. **Flowers** A little ahead of Indian Bean's, in large but *open* heads.
COMPARE Indian Bean; Hybrid Bean Tree (below).

Hybrid Bean Tree — *Catalpa × erubescens* 'J. C. Teas'

(Indian Bean × Yellow Catalpa; raised by J. C. Teas in Indiana in 1874; introduced 1891.) Occasional in warm areas.
APPEARANCE Shape Wide, tall (to 20 m) and rather *open*; trunk often sinuous. **Bark** Grey-brown, with quite deep scaly ridges (like Western Catalpa's (above) – but less predominantly vertical). **Leaves** Very big (to 35 cm), deep green; heart-shaped at the base, *often as broad as long; 20–50%* (on older trees) with small side-lobes. **Flowers** A little later than Indian Bean's; in slightly less dense candles, but richly fragrant. **Fruit** 40 cm 'bean-pods' (filled only with fluff) hang on profusely in warm areas.
COMPARE Western Catalpa (above); Yellow Catalpa (below).
VARIANTS 'Purpurea' has strongly purple-splashed flowers (creating a *lilac* tinge), and smaller, duller, darker leaves which, for several days as they open, are *more intensely black-purple* than other bean trees', and then retain some purplish veins and *purple stalks*. Rare.

Yellow Catalpa — *Catalpa ovata*

China (and Japan?). 1849. Very occasional in warm areas.
APPEARANCE Shape Rounded, or slender on a *long* sinuous bole. To 22 m. **Bark** Grey-brown; scaly ridges. **Leaves** Matt, *dark yellowish green*, smaller than Indian Bean's (but to 25 cm); *usually with big lobes* and at least as broad as long; on *dark red* stalks with short, stiff hairs. Purple glands under the leaf-base ooze nectar (cf. Idesia, p.408). **Flowers** Appearing with Indian Bean's; white, with yellow staining and red spots – appearing dull *creamy yellow* at a distance. In small, sparse heads but deliciously scented (strawberries, with a hint of soap flakes – like Silver Lime, p.404).

Farges' Catalpa — *Catalpa fargesii*

W China. 1900. Rare – big gardens.
APPEARANCE Shape *Upright*, on rather gaunt branching; to 20 m. **Bark** Grey-brown; broad *shallowly but shaggily scaling* ridges. **Shoots** Velvety at first (but hairless in f. *duclouxii*). **Leaves** (Relatively) *small* – about 12 × 10 cm – and rather shiny above; densely white-hairy underneath in the type, on velvety stalks. **Flowers** Pinkish, in *mid-summer*. One of the most spectacular of flowering trees – if conditions suit it.
OTHER TREES Bunge's Catalpa, *C. bungei* (China, 1905), a sturdier tree in a few collections, has almost hairless small leaves often with *1–6 scalloped teeth* each side, and smaller showier flowers.

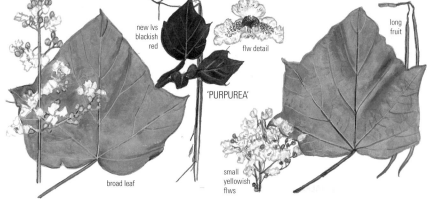

HYBRID BEAN TREE — new lvs blackish red; flw detail; 'PURPUREA'; broad leaf

YELLOW CATALPA — long fruit; small yellowish flws

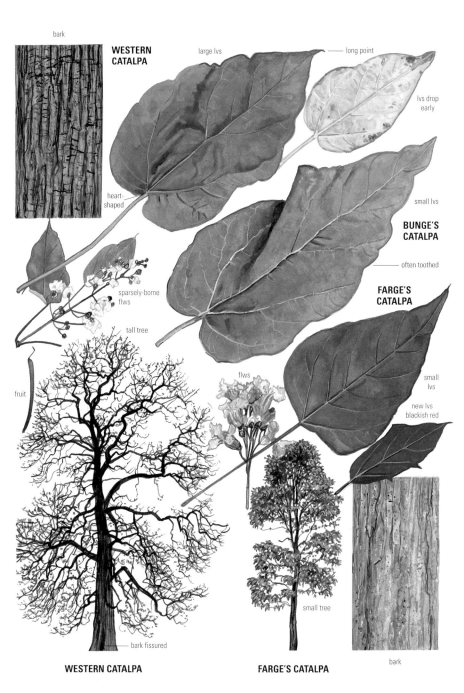

Elder — Emmenopterys

Elder — *Sambucus nigra*

Europe (including Britain and Ireland); N Africa; SW Asia. Ubiquitous except on sands; scarcely planted. (Family: Caprifoliaceae.)
Appearance Shape Branches arch over with vigorous erect sprouts; to 10 m. **Bark** *Creamy-grey*: corky criss-cross ridges. **Shoots** Cream-grey, with *raised warts and often wandering tips*. **Buds** Purplish; *spiky-scaled*, like pineapples – expanding in mid-winter, soon after the last lemon-yellow or pink leaves drop. **Leaves** In opposite pairs; *5 or 7* (rarely 3 or 9) leaflets, dull, with some stiff hairs. **Flowers** In *creamy plates* in early summer, heavily scented; sometimes made into 'champagne'. **Fruit** Elderberry wine comes from the black berries – 'the Englishman's grape' – which ripen in early autumn and are poisonous raw (like the leaves), but have many medicinal uses.
Variants Parsley-leaved Elder, f. *laciniata*, is very feathery and rare; less cut-leaved sports are occasional. Golden Elder, 'Aurea', is much grown. (The gold cut-leaved 'Plumosa Aurea' belongs to a Continental scarlet-berried bush, *S. racemosa*.)

Spindle — *Euonymus europaeus*

Europe (including Britain and Ireland) to the Caucasus. Abundant on richer soils. The straight, hard twigs were used for skewers and spindles. (Family: Celastraceae.)
Appearance Shape Generally bushy; to 9 m. **Bark** Deep green, with faint fawn 'snakes'; then grey-brown, closely grooved. **Shoots** Thin, straight, slightly squared and *deep green*. **Buds** Greenish, short-conic. **Leaves** Blackthorn-shaped (but *opposite*), to 8 cm, with *minute sharp teeth*; a little shiny and waxy, hairless. **Flowers** In small creamy heads in early summer. **Fruit** Poisonous *orange berries, from 4 magenta lobes*, can be delicately spectacular among leaves of the same colours.
Compare Purging Buckthorn (p.398); Snowbell Tree (p.434).
Other trees *E. bungeanus* (NE Asia, 1883; very rare) is perhaps the handsomest *tree-sized* spindle (domed, to 10 m): broader hairless leaves (to 10 × 6 cm) hang gracefully. Fruit-lobes pale creamy-pink.

Maiten — *Maytenus boaria*

S Andes to Brazil. 1822. Some big gardens in mild areas.
Appearance Shape An open evergreen dome; *cascading* outer shoots. **Bark** Grey; short criss-cross grooves. **Leaves** 1–6 cm; dark, glossy, hairless; closely saw-toothed; stalks 3–5 mm.
Compare Phillyrea (p.442): opposite leaves. Coigüe (p.202); Small-leaved Azara (p.408).

Emmenopterys — *Emmenopterys henryi*

Central and SW China. 1907. Rare: large gardens. (Family: Rubiaceae.)
Appearance Shape Narrow, on straight stems, to 17 m; tiers of hanging foliage (cf. Table Dogwood, p.426). **Bark** Dark grey, finely roughened. **Shoots** Smooth, glossy. **Buds** With *2 crimson 2 cm scales with long fine tips*. **Leaves** In opposite pairs; to 22 cm, soft and thick; with odd small teeth; dull but often *flushed red* on crimson stalks; downy under the sunken veins. **Flowers** White, 3 cm: over some weeks in airy plumes which sport 2 or 3 8-cm greeny-white bracts. Sadly only one tree in the UK (at Wakehurst Place, W Sussex) has so far produced them.

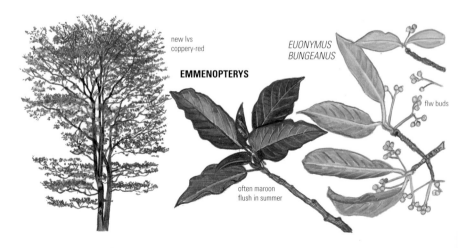

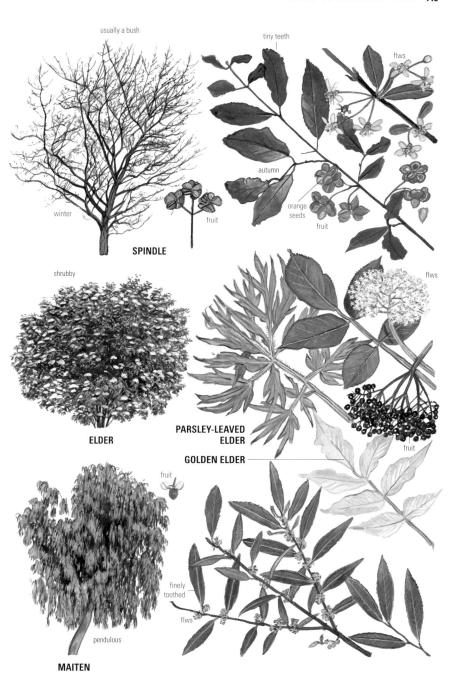

CABBAGE PALM TO CHUSAN PALM

Cabbage Palm — *Cordyline australis*

(Cabbage Tree; *Dracaena australis*) New Zealand. 1823. Abundant in milder, coastal areas; seeding freely and suckering. (Family: Liliaceae.)
Appearance Shape Branching after flowering (typically at 5 m and about 8 years); ultimately densely funnel-shaped, to 16 m: one of the few monocotyledons (plants, like grasses and lilies, which on germinating grow 1 'seed-leaf' not 2, and have parallel, not networked, leaf-veins) that steadily builds a thick, branching bole. **Bark** Creamy-grey; rather corkily square-cracked; whiskery at first from shed leaf-bases. **Leaves** Linear, to 90 × 8 cm; very tough and can be made into paper, though young sprouts are eaten in New Zealand; dying ones may phosphoresce in the wild. (cf. Yuccas – bushier plants, with towers of creamy-white big bell-flowers – which have shorter, thicker, greyer leaves.) **Flowers** Fragrant, in huge airy white plumes through early summer. **Fruit** Blue-white 6 mm berries.
Variants Many purple, brownish and prettily variegated clones are now in commerce.

In the palm family (Palmae), there are about 3000 species. Their leaves and flowers usually form from a single bud at the top of a stem that increases in height but not in diameter.

Canary Palm — *Phoenix canariensis*

Endemic to the Canary Islands but widely cultivated; scarcely hardy in the UK, but now frequent as a young plant in warm areas.
Appearance Shape An amazing star-burst of arching and hanging blue-green leaves, on a bole sturdier than Chusan Palm's (below). **Bark** Tiled with clean, bright brown leaf-stumps. **Leaves** To 7 m (on saplings necessarily much smaller), the *parallel leaflets* folded upwards at their bases.
Other trees Date Palm, *P. dactylifera*, has similar leaves (less drooping). It is easy to raise from date-seeds and even grows up in rubbish heaps, but fails to survive N European winters.
P. theophrastii (with a suckering habit, like Date Palm's) is endemic to Crete; not hardy in the UK.
Chilean Wine Palm, *Jubaea chilensis*, an ally of Coconut Palm, was represented in the 19th century at Kew Gardens by a large tree near the Main Gates, but is scarcely grown now and considered tender; rare in Chile, as tapping the sugary sap involved felling the tree. The trunk, with *smoothly wrinkled, pale grey bark*, can be 1 m thick; leaves very like Canary Palm's but on thicker, greyer central stems.

Chusan Palm — *Trachycarpus fortunei*

(Windmill Palm) China; N Vietnam; Upper Burma. 1836. The hardiest palm, occasional in milder areas, where it seeds freely.
Appearance Shape To 15 m. **Bark** Covered in the matted fibres of old leaf-bases, like an ancient hot-water pipe (and for much the same reason: the lagging helps keep frosts out); highly inflammable and something of a liability in public parks. Rarely smooth and green near the base. **Leaves** *Fan-like*, the tips usually shredded and hanging. The 1 m stalk has tiny spines. **Flowers** In huge (to 60 cm) egg-yolk-yellow, drooping plumes. **Fruit** Blue-black, 1 cm berries; rare in the UK.
Other trees Dwarf Fan Palm, *Chamaerops humilis* (W Mediterranean, 1731), is rare: to 5 m, suckering and usually on *several stems*. Leaves smaller, stiffer, greyer, on *viciously spined* stalks; flower-heads (dioecious) only 15 cm high.

CHUSAN PALM — brown fibres

DWARF FAN PALM

CABBAGE PALM — may form a clump — branches after flowering

GLOSSARY

Alternate: Not opposite each other.

Anther: Pollen-bearing male part of flower, tipping a stamen.

Appressed: Lying (almost) flat against.

Auricle: Backward-pointing lobe at base of leaf (or petal).

Back-cross: Offspring of a hybrid crossed with one of its parents.

Bloom: Waxy surface layer, easily rubbing off.

Bract: Modified leaf, behind a flower/flower-head/seed.

Bud: Embryo leaf, flower and/or shoot, plus any protective covering.

Canker: Fungal/bacterial infection, causing bark lesions.

Catkin: Tail-like single-sex flower-head, with scale-like bracts and stalkless flowers.

Chimaera: Plant originating in the non-sexual fusion of two species' tissues.

Chlorosis: Sickness due to a failure to obtain iron from alkaline soils, leading to yellow foliage.

Clone: All the plants propagated vegetatively from an individual.

Columnar: Narrow, erect, with parallel sides.

Compound: Made up of several parts (e.g. leaflets).

Conic: Narrowly pointed from a broad base.

Coppice: To fell a tree so that the base sprouts to make new trees; a stump treated in this way; to regrow like this.

Crown: A tree's whole structure above the trunk.

Cultivar: A '**cultiv**ated **var**iety' or mutant clone, nursery-distributed.

DED: Dutch Elm Disease: fungal pathogen of plants in the elm family (researched largely in Holland).

Dioecious: With male and female flowers on separate plants.

Distant: Widely separated.

Double (flowers): With some or all stamens mutated into extra petals.

Doubly compound: Compound, with each leaflet itself compound.

Doubly toothed: With serrations that are themselves toothed.

Down: Covering of tiny, soft hairs.

Endemic: Native to one small area.

Family: Scientific grouping of plants, divided into genera and into which orders are divided.

Fastigiate: Growing nearly vertical branches and/or shoots.

Flush: To come into leaf.

Forma (plural **formae**): Scientifically described sport or minor variant of a species.

Free: Not attached; projecting.

Genus (plural **genera**): Scientific grouping of species, into which families are divided.

Gland: Secreting organ: a tiny, hard, often sticky swelling.

Glaucous: Pale bluish grey (usually due to a bloom).

Graft: To fuse a shoot of one kind of tree on to roots of another; a tree fused in this way.

Habit: Manner of growth.

Hybrid: The offspring of different species, sub-species or varieties.

Impressed: Sunk below the surface.

Layer: A branch touching the ground and rooting; the secondary trunk that results.

Leader: Top shoot of a tree adding height.

Leaflet: Discrete part of a compound leaf.

Lenticel: Raised breathing-pore on a shoot, trunk or fruit.

Lobe: Promontory on a leaf (or petal, etc.), larger than a tooth.

Microspecies: A species which is usually local and produces fertile seed without pollination.

Midrib: A leaf's central main vein.

Needle: Linear leaf of a conifer.

Opposite: In a pair, one on either side of a shoot.

Original: The first tree of a taxon in cultivation in a country.

Petal: Segment of a flower's inner whorl.

Plane: Flat-surfaced.

Pollard: To cut a tree at 2–4 m so that it grows new limbs; a tree cut in this way.

Scale: A flattened appendage, not leaf-like or petaloid.

Scion: Shoot grafted on to a stock to make a new tree.

Semi-double (flower): With a few but not many stamens mutated into extra petals.

Sepal: Segment of a flower's outer whorl (outside/behind the petals and persisting, withered, on top of some fruits).

Serrated: With many teeth (which point forwards).

Shoot: Young twig.

Simple: Not double or not compound.

Sinus: Recess between two lobes.

Sport: A genetic freak.

Spur: Short side-twig, growing very slowly each year.

Stamen: Male organ in a flower: an anther and its stalk.

Stipule: Appendage, usually leafy, at the base of a leaf- or flower-stalk.

Stock: Roots of a plant on to which a scion is grafted.

Style: Female organ in a flower: a stalk for the stigma which receives the pollen.

Subspecies: A scientifically recognized (regional) variant of a species.

Sucker: A sprout from a root, which may grow into a new tree.

Taxon (plural **Taxa**): A single discrete nomenclatural unit.

Tepal: Undifferentiated sepal/petal in some flowers (eg. Magnolias, Tulips).

Tooth: Promontory on the margin of a leaf or petal, smaller than a lobe.

Trifoliate: With leaves divided into three leaflets.

Type: The usual form of a species (as opposed to a variety or clone); the individual on which a taxon's scientific description was based.

Variegated: With leaves consistently lacking green coloration in patches.

Variety: A scientifically recognized (morphological) variant of a species.

Whorl: Circle of branches, leaves, flowers or other organs.

Wing: Flat membranous extension.

Woolly: Covered in long, soft hairs

INDEX

Subspecies, varieties and cultivars are separately indexed only when their coverage extends across two or more spreads.

Abele 150
Abies alba 68
  amabilis 74
  bifida 84
  bifolia 78
  borisii-regis 68
  bornmuelleriana 70
  brachyphylla 86
  bracteata 88
  brevifolia 80
  cephalonica 76
  chensiensis 84
  cilicica 70
  concolor 76
  delavayi 82
  equi-trojani 70
  fabri 82
  fargesii 82
  faxoniana 82
  firma 84
  forrestii 82
  gamblei 80
  georgii 82
  grandis 74
  holophylla 84
  homolepis 86
  kawakamii 78
  koreana 86
  lasiocarpa 78
  lowiana 76
  magnifica 88
  mariesii 74
  minensis 82
  nebrodensis 68
  nephrolepis 72
  nobilis 88
  nordmanniana 70
  numidica 80
  pindrow 80
  pinsapo 78
  procera 88
  recurvata 84
  sachalinensis 72
  sibirica 72
  spectabilis 80
  squamata 86
  sutchuenensis 82
  veitchii 72
  venusta 88
  × *vilmorinii* 78
  webbiana 80
Acacia, False 354
Acacia baileyana 346
  dealbata 346
  decurrens 346
  melanoxylon 346

'Acer' 384
Acer acuminatum 386
  argutum 386
  barbatum 372
  buergerianum 374
  campestre 368
  capillipes 380
  cappadocicum 374
  carpinifolium 384
  circinatum 386
  cissifolium 390
  × *conspicuum* 380
  crataegifolium 382
  creticum 376
  dasycarpum 378
  davidii 380
  diabolicum 370
  distylum 382
  forrestii 380
  ginnala 376
  granatense 368
  griseum 388
  grosseri 382
  heldreichii 388
  henryi 390
  hersii 382
  hyrcanum 376
  japonicum 386
  laetum 374
  lobelii 374
  macrophyllum 378
  maximowiczianum 390
  miyabei 386
  mono 374
  monspessulanum 376
  negundo 390
  nigrum 372
  nikoense 390
  obtusifolium 376
  olivierianum 384
  opalus 376
  orientale 376
  palmatum 384
  pensylvanicum 382
  pictum 374
  platanoides 372
  pseudoplatanus 370
  pseudosieboldianum 386
  pycnanthum 378
  rubrum 378
  rufinerve 380
  saccharinum 378
  saccharum 372
  sempervirens 376
  shirasawanum 386
  sieboldianum 386
  spicatum 386
  stachyophyllum 382
  syriacum 376
  tataricum 376
  tetramerum 382

  trautvetteri 388
  triflorum 390
  truncatum 374
  ukurunduense 384
  velutinum 388
  × *zoeschense* 376
Aceraceae 368
Aesculus californica 396
  × *carnea* 394
  + *dallimorei* 394
  flava 396
  glabra 396
  hippocastanum 392
  × *hybrida* 396
  indica 394
  × *neglecta* 396
  octandra 396
  pavia 396
  × *plantierensis* 394
  turbinata 392
Afghan Hound Tree 42
Agathis australis 28
Ailanthus altissima 358
  vilmoriniana 358
Akagashi 230
Aka-shide 196
Akiraho 410
Albizia distachya 346
  julibrissin 346
  lophantha 346
Alder 190
  Black 190
  Caucasian 192
  Common 190
  Downy 190
  Green 192
  Grey 192
  Italian 192
  Japanese 192
  Oregon 190
  Oriental 192
  Red 190
  Spaeth's 192
Alder Buckthorn 398
Alerce 52
Almond 338
  Double 338
  Sweet 338
Alnus cordata 192
  firma 192
  glutinosa 190
  hirsuta 190
  × *hybrida* 192
  incana 192
  japonica 192
  oregona 190
  orientalis 192
  × *pubescens* 192
  rubra 190
  × *spaethii* 192
  subcordata 192
  viridis 192

Amelanchier × *grandiflora* 320
  laevis 320
  lamarckii 320
  ovalis 320
  rotundifolia 320
Amygdalus dulcis 338
  persica 338
Anacardiaceae 360
Andromeda arborea 430
Angelica Tree, Japanese 350
Apple 306
  Argyle 422
  Danube 306
  Orchard 306
  Pillar 308
Apricot 338
  Briançon 338
  Japanese 338
Aquifoliaceae 362
Aralia, Castor 424
Aralia elata 350
Araliaceae 350, 424
Araucaria angustifolia 28
  araucana 28
  bidwillii 28
  heterophylla 28
  imbricata 28
Araucariaceae 28
Arbor-vitae 32
Arbutus andrachne 430
  × *andrachnoides* 430
  menziesii 430
  unedo 430
Arceuthos drupaceae 54
Armeniaca vulgaris 338
Ash 436
  Alpine 422
  Arizona 440
  Caucasian 438
  Chinese Manna 438
  Claret 438
  Common 436
  Golden 436
  Green 440
  Japanese Big-leaved 440
  Manna 438
  Maries' 438
  Mountain 300, 422
  Narrow-leaved 438
  Oregon 440
  Pallis' 438
  Red 440
  Weeping 436
  Weeping Golden 436
  White 440
Aspen 152
  Weeping 152
Atherospermataceae 276
Athrotaxis cupressoides 58

INDEX **455**

laxifolia 58
selaginoides 58
Austrocedrus chilensis 30
Azara, Small-leaved 408
Azara microphylla 408
Azarole 286

Balm of Gilead 160
Basswood 402
   White 404
Bay 276
   Bull 260
   Narrow-leaved 276
   Sweet 276
Bean, Golden 444
   Hybrid 446
   Indian 444
Beech 204
   American 210
   Antarctic 200
   Black 202
   Blue 196
   Chinese 210
   Cockscomb 206
   Common 204
   Contorted 208
   Copper 204
   Dawyck 204
   Dombey's 202
   Fern-leaved 206
   Golden 204
   Golden Weeping 208
   Japanese 210
   Large-leaved 206
   Menzies' 202
   Mountain 202
   Myrtle 202
   Oak-leaved 208
   Oriental 210
   Oval-leaved Southern 202
   Purple 204
   Red 202
   Roblé 200
   Silver 202
   Small-leaved 206
   Southern 200
   Weeping 208
   Weeping Copper 208
Benthamidia capitata 426
   florida 426
   kousa 426
   nuttalli 426
Betula alba 182
   albo-sinensis 188
   alleghaniensis 184
   × *aurata* 182
   celtiberica 182
   coerulea-grandis 184
   costata 186
   davurica 184
   ermanii 186

   grossa 188
   jacquemontii 186
   × *koehnii* 184
   lenta 184
   lutea 184
   mandshurica 188
   maximowicziana 186
   medwediewii 188
   nigra 184
   obscura 182
   papyrifera 184
   pendula 182
   platyphylla 188
   populifolia 184
   pubescens 182
   resinifera 188
   szechuanica 188
   utilis 186
   verrucosa 182
   viridis 192
Betulaceae 182
Big Tree 62
Biota 34
Birch 182
   Black 184
   Blue 184
   Brown 182
   Canoe 184
   Cherry 184
   Chinese Red-barked 188
   Downy 182
   Erman's 186
   Grey 184
   Himalayan 186
   Japanese Cherry 188
   Japanese White 188
   Manchurian 188
   Maximowicz's 186
   Monarch 186
   Paper-bark 184
   River 184
   Sichuan 188
   Silver 182
   Swedish 182
   Sweet 184
   Transcaucasian 188
   Warty 182
   White 182
   Yellow 184
   Young's Weeping 182
Bitternut 176
Blackthorn 340
Blackwood 346
Box 358
   Balearic 358
Box Elder 390
Boxwood, American 426
Broad-leaf 424
Broom, Mount Etna 348
Broussonetia papyrifera 256

Buckeye, Californian 396
   Hybrid 396
   Red 396
   Sweet 396
   Yellow 396
Buckthorn, Alder 398
   Purging 398
   Sea 410
Bullace 340
   Black 340
   Shepherd's 340
Bunya-bunya 28
Butter-nut 180
Buttonwood 280
Buxaceae 358
Buxus balearica 358
   sempervirens 358

Cabbage Tree 450
Calocedrus decurrens 30
Camellia, Deciduous 408
Caprifoliaceae 448
Carpinus betulus 194
   caroliniana 196
   cordata 196
   japonica 196
   laxiflora 196
   orientalis 196
   turczaninowii 196
   viminea 196
Carya alba K. Koch 174
   alba Nutt. 176
   amara 176
   cordiformis 176
   glabra 176
   illinoinensis 174
   laciniosa 174
   ovalis 176
   ovata 176
   porcina 176
   sulcata 174
   tomentosa 174
Castanea dentata 212
   sativa 212
Castanopsis chrysophylla 212
   cuspidata 230
Catalpa, Bunge's 446
   Farges' 446
   Northern 446
   Southern 444
   Western 446
   Yellow 446
Catalpa bignonioides 444
   bungei 446
   × *erubescens* 446
   fargesii 446
   ovata 446
   speciosa 446
Cedar 90
   Atlas 92
   Blue Atlas 92

   Chilean Incense 30
   Chinese 358
   Chinese Red 60
   Cyprus 90
   Eastern Red 56
   Eastern White 34
   of Goa 48
   Incense 30
   Japanese Red 60
   of Lebanon 90
   Northern White 34
   Pencil 56
   Port Orford 36
   Smooth Tasmanian 58
   Summit 58
   Western Red 32
Cedrela sinensis 358
Cedrus atlantica 92
   brevifolia 90
   deodara 92
   libani 90
Celastraceae 448
Celtis australis 254
   biondii 254
   bungeana 254
   caucasica 254
   laevigata 254
   occidentalis 254
Cephalotaxaceae 24
Cephalotaxus fortunei 24
   harringtonii 24
Cerasus avium 322
   mahaleb 342
   vulgaris 324
Cercidiphyllaceae 274
Cercidiphyllum japonicum 274
Cercis canadensis 348
   siliquastrum 348
Chain Tree, Golden 350
Chamaecyparis formosensis 44
   lawsoniana 36
      'Albomaculata' 36
      'Albospica' 36
      'Allumii' 36
      'Columnaris' 36
      'Elegantissima' 38
      'Ellwoodii' 38
      'Ellwood's Gold' 38
      'Ellwood's White' 38
      'Erecta Viridis' 36
      'Filifera' 36
      'Filiformis' 36
      'Fletcheri' 38
      'Fraseri' 36
      'Golden King' 38
      'Grayswood Gold' 38
      'Grayswood Pillar' 36
      'Green Pillar' 36
      'Green Spire' 36

INDEX

'Hillieri' 38
'Intertexta' 36
'Kilmacurragh' 36
'Lane' 38
'Lemon Queen' 38
'Lutea' 38
'Naberi' 38
'Pembury Blue' 38
'Pottenii' 38
'Silver Queen' 38
'Smithii' 38
'Stewartii' 38
'Triomf van Boskoop' 38
'Versicolor' 36
'Westermannii' 38
'Winston Churchill' 38
'Wisselii' 38
'Youngii' 36
nootkatensis 44
obtusa 40
pisifera 42
thyoides 44
Chamaerops humilis 450
Checkers Tree 296
Cherry 322
  All Saints' 324
  Bird 342
  Black 344
  Canada Red 342
  Cheal's Weeping 332
  Chinese Flowering 330
  Choke 342
  Cornelian 424
  Dawyck 324
  Fuji 332
  Great White 328
  Greyleaf 324
  Hubei 334
  Japanese 326
  Judd's 330
  Korean Hill 330
  Lombardy Poplar 326
  Manchurian 324
  Oshima 336
  Rhex's 324
  Rose-bud 334
  Rum 344
  St Lucie 342
  Sargent's 330
  Schmitt's 324
  Seagull 336
  Sour 324
  Swallow 332
  Tibetan 324
  Weeping Spring 332
  Wild 322
  Wild Hill 330
  Winter 334
  Yoshino 332
Cherry-Crab 312

Cherry Laurel 344
Cherry-Plum 340
  Double 340
Chestnut, American 212
  Cut-leaved 212
  Dallimore's 394
  Golden 212
  Horse 392
  Indian Horse 394
  Japanese Horse 392
  Red Horse 394
  Sunrise Horse 396
  Sweet 212
  Variegated 212
Chinkapin, Giant 212
Chittam Wood 360
Christmas Tree 102
Chrysolarix amabilis 98
Chrysolepis chrysophylla 212
Citrus 356
Cladrastis kentukea 352
  lutea 352
  sinensis 352
  tinctoria 352
Coffee Tree, Kentucky 350
Coigue 202
  de Magellanes 202
Compositae 410
Cordyline australis 450
Cork Tree 356
  Amur 356
  Japanese 356
Cormus domestica 298
Cornaceae 424
Cornel 424
  Bentham's 426
  Himalayan 426
  Japanese 426
  Large-leaved 426
Cornus capitata 426
  controversa 426
  'Eddie's White Wonder' 426
  florida 426
  kousa 426
  macrophylla 426
  mas 424
  nuttallii 426
  sanguinea 424
Corylus avellana 198
  colurna 198
  maxima 198
Cotinus americanus 360
  coggygria 360
  obovatus 360
Cotoneaster, Himalayan Tree 282
Cotoneaster 'Cornubia' 282
  'Exburiensis' 282

  frigidus 282
  'Rothschildianus' 282
  × watereri 282
Cottonwood, Eastern 156
Coyan 200
Crab, Chinese 310
  Chonosuki's 308
  Cutleaf 308
  Hall's 314
  Hubei 310
  Italian 312
  Japanese 308
  Magdeburg 310
  Maple-leaved 312
  Plum-leaved 310
  Prairie 308
  Pratt's 308
  Purple 314
  Siberian 310
  Siebold's 308
  Sweet 308
  Weeping Purple 314
  Wild 306
  Yunnan 308
+ Crataegomespilus dardarii 290
Crataegus azarolus 286
  calycina 284
  × carrierei 290
  chrysocarpa 288
  coccinioides 288
  crus-galli 288
  × dippeliana 286
  ellwangeriana 288
  × grignonensis 290
  heldreichii 286
  laciniata 286
  laevigata 284
  × lavallei 290
  × media 284
  mollis 288
  monogyna 284
  nigra 286
  oxycantha 284
  oxycanthoides 284
  pedicellata 288
  pentagyna 286
  persimilis 288
  pinnatifida 286
  prunifolia 288
  punctata 286
  sanguinea 286
  schraderiana 286
  stipulacea 290
  submollis 288
  tanacetifolia 286
× Crataemespilus grandiflora 290
Cryptomeria fortunei 60
  japonica 60
Cucumber Tree 260
  Yellow 260

Cunninghamia konishii 66
  lanceolata 66
  sinensis 66
Cupressaceae 30
× Cupressocyparis leylandii 45
  notabilis 46
  ovensii 46
Cupressus arizonica 50
  benthamii 48
  cashmeriana 44
  corneyana 52
  forbesii 52
  glabra 50
  goveniana 52
  guadalupensis 52
  × leylandii 46
  lindleyi 48
  lusitanica 48
  macrocarpa 48
  nootkatensis 44
  × notabilis 46
  × ovensii 46
  sempervirens 50
  stephensonii 52
  torulosa 52
Cyathea dealbata 20
Cyatheaceae 20
Cydonia oblonga 290
Cypress, Alice Holt 46
  Arizona 50
  Bald 64
  Bentham's 48
  Bhutan 52
  Cuyamaca 52
  Deciduous 64
  False 36
  Gowen 52
  Guadalupe 52
  Hinoki 40
  Hybrid 46
  Italian 50
  Kashmir 44
  Lawson 36
  Leyland 46
  Mexican 48
  Monterey 48
  Nootka 44
  Ovens 46
  Patagonian 52
  Pond 64
  Rough Arizona 50
  Sawara 42
  Smooth Arizona 50
  Swamp 64
  Taiwan 44
  Tecate 52
  True 48
  West Himalayan 52
  White 44
Cytisus purpureus 350

Dacrydium franklinii 26
Damson 340
Date-plum 432
Davidia involucrata 412
 vilmoriniana 412
Davidiaceae 412
Deodar 92
Dicksonia antarctica 20
Diospyros kaki 432
 lotus 432
 virginiana 432
Dogwood 424
 Common 424
 Flowering 426
 Pacific 426
 Strawberry 426
 Table 426
Doronoki 162
Dove Tree 412
Dracaena australis 450
Drimys winteri 274

Ebenaceae 432
Elaeagnaceae 410
Elaeagnus angustifolia 412
 umbellata 412
Elder 448
 Box 390
 Golden 448
 Parsley-leaved 448
Elm 240
 American White 248
 Belgian 246
 Boxworth 242
 Caucasian 252
 Chichester 246
 Chinese 250
 Coritanian 244
 Cornish 244
 Dengie 242
 Dickson's Golden 246
 Dutch 248
 English 244
 European White 248
 Exeter 242
 Field 242
 Golden Wych 242
 Goodyer's 244
 Guernsey 244
 Huntingdon 246
 Hybrid 242
 Japanese 250
 Jersey 244
 Lace-bark 250
 Lock 244
 Plot 244
 Scots 242
 Siberian 250
 Smooth-leaved 246
 Wheatley 244
 Wych 242

Embothrium coccineum 274
Emmenopterys 448
Emmenopterys henryi 448
Empress Tree 444
Encina 222
Epaulette Tree 432
Ericaceae 428
Eriobotrya japonica 282
Eriolobus trilobatus 312
 tschonoskii 308
Eucalypt 414
Eucalyptus aggregata 420
 archeri 414
 cinerea 422
 coccifera 422
 cordata 422
 coriacea 420
 dalrympleana 416
 delegatensis 422
 gigantea 422
 glaucescens 416
 globulus 418
 gunnii 414
 johnstonii 418
 mitchelliana 420
 muelleri 418
 nicholii 420
 niphophila 420
 nitens 418
 parvifolia 420
 pauciflora 420
 perriniana 416
 pulverulenta 422
 regnans 422
 stellulata 420
 subcrenulata 418
 urnigera 416
 vernicosa 418
 viminalis 416
Eucommia ulmoides 278
Eucommiaceae 278
Eucryphia 406
 Nymans 406
 Rose-leaved 406
 Rostrevor 406
Eucryphia cordifolia 406
 glutinosa 406
 × *intermedia* 406
 lucida 406
 milliganii 406
 × *nymansensis* 406
 'Penwith' 406
 pinnatifolia 406
Eucryphiaceae 406
Euodia 356
Euodia hupehensis 356
 velutina 356
Euonymus bungeanus 448
 europaeus 448

Fagaceae 200

Fagus crenata 210
 engleriana 210
 grandifolia 210
 lucida 210
 moesiaca 210
 orientalis 210
 sieboldii 210
 sylvatica 204
  'Albomarginata' 208
  'Albo-variegata' 208
  'Aspleniifolia' 206
  'Aurea Pendula' 208
  'Bornyensis' 208
  'Brathay Purple' 206
  'Cockleshell' 206
  'Cristata' 206
  'Dawyck' 204
  'Dawyck Gold' 208
  'Dawyck Purple' 208
  'Grandidentata' 206
  'Heterophylla' 206
  f. *laciniata* 208
  f. *latifolia* 206
  'Luteovariegata' 208
  'Miltonensis' 206
  'Pagnyensis' 208
  f. *pendula* 208
  'Prince George of Crete' 206
  'Purple Fountain' 208
  f. *purpurea* 204
  'Purpurea Pendula' 208
  'Purpurea Tricolor' 208
  'Quercifolia' 208
  'Remillyensis' 208
  'River's Purple' 204
  'Rohan Gold' 208
  'Rohanii' 208
  'Roseomarginata' 208
  'Rotundifolia' 206
  f. *tortuosa* 208
  'Tricolor' 208
  'Zlatia' 204
 taurica 210
Fatsia japonica 424
Ficus carica 256
Fig 256
Filbert 198
 Purple 198
Fir, Algerian 80
 Apollo 76
 Beautiful 74
 Blue Douglas 120
 Bornmüller's 70
 Bristlecone 88
 Caucasian 70
 Chinese 66
 Cilician 70

 Colorado White 76
 Common Silver 68
 Cork 78
 Douglas 120
 East Siberian 72
 European Silver 68
 Faber's 82
 Farges' 82
 Flaky 86
 Forrest's 82
 Gamble's 80
 Grand 74
 Grecian 76
 Himalayan 80
 Japanese Douglas 120
 King Boris' 68
 Korean 86
 Large-coned Douglas 120
 Low's 76
 Manchurian 84
 Maries' 74
 Min 84
 Momi 84
 Needle 84
 Nikko 86
 Noble 88
 Nordmann 70
 Pacific Silver 74
 Pindrow 80
 Red 88
 Sakhalin 72
 Salween 84
 Santa Lucia 88
 Siberian 72
 Sicilian 68
 Silver 66
 Smith's 82
 Spanish 78
 Spruce 102
 Subalpine 78
 Taiwan 78
 Trojan 70
 Veitch's 72
 Vilmorin's 78
 Water 64
 Webb 80
Firebush, Chilean 274
Fitzroya cupressoides 52
 patagonica 52
Flacourtiaceae 408
Foxglove Tree 444
Frangula alnus 398
Fraxinus americana 440
 angustifolia 438
 chinensis 438
 excelsior 436
 holotricha 438
 latifolia 440
 mariesii 438
 ornus 438
 oxycarpa 438

INDEX

pallisiae 438
pennsylvanica 440
spaethiana 440
subintegerrima 440
velutina 440

Garland Tree 308
Gean 322
  Double 322
  Weeping 322
Genista aetnensis 348
Ghost Tree 412
Ginkgo biloba 20
Ginkgoaceae 20
Gleditsia sinensis 348
  triacanthos 348
Grain Tree 222
Grannies Ringlets 60
Greengage 340
Griselinia littoralis 424
Griseliniaceae 424
Guindo 202
Gum, Alpine Cider 414
  Black 412, 420
  Blue 418
  Cabbage 420
  Cider 414
  Cotton 412
  Jounana Snow 420
  Manna 416
  Ribbon 416
  Shining 418
  Silver 422
  Silver-leaved Mountain 422
  Small-leaved 420
  Snow 420
  Sour 430
  Spinning 416
  Sweet 278
  Tasmanian Alpine Yellow 418
  Tasmanian Snow 422
  Tasmanian Yellow 418
  Tingiringi 416
  Urn 416
  Varnished 418
Gurass 428
Gutta-Percha Tree 278
Gymnocladus canadensis 350
  dioica 350

Hackberry 254
  Mississippi 254
Hagberry 342
Hales, French 296
Halesia carolina 434
  monticola 434
Hamamelidaceae 278
Handkerchief Tree 412
Hawkberry 342

Hawthorn, Common 284
  Midland 284
Hazel 198
  Common 198
  Corkscrew 198
  Cut-leaved 198
  Golden 198
  Turkish 198
  Weeping 198
Headache Tree 276
Hemlock 116
  Carolina 118
  Chinese 118
  Eastern 116
  Himalayan 118
  Jeffrey's 118
  Mountain 118
  Northern Japanese 118
  Southern Japanese 118
  Western 116
  Yunnan 118
Hesperopeuce mertensiana 118
Hiba 32
Hickory 174
  Big-bud 174
  Big Shellbark 174
  Red 176
  Shagbark 176
  Smoothbark 176
Hippocastanaceae 392
Hippophae rhamnoides 410
  salicifolia 410
Holly 362
  American 362
  Common 362
  Egham 362
  Golden 362
  Golden Hedgehog 364
  Hedgehog 364
  Henderson's 366
  Highclere 366
  Himalayan 366
  Hodgins' 366
  Laurel-leaved 364
  Leather-leaf 364
  Madeira 362
  Moonlight 364
  Myrtle-leaved 364
  Perny's 366
  Perny's Weeping 364
  Pyramid 364
  Silver 364
  Silver Hedgehog 364
  Weeping 364
  Yellow-berried 364
Honey Locust 348
  Chinese 348
Hop-hornbeam, European 196
  Japanese 196
Hornbeam 194

  American 196
  Common 194
  Japanese 196
  Oriental 196
  Sawa 196
Hualle 200
Hualo 200

Idesia 408
Idesia polycarpa 408
Ilex × *altaclarensis* 366
  aquifolium 362
  'Argentea Marginata' 364
  f. *argenteomarginata* 364
  'Argenteomarginata Pendula' 364
  'Aurea Pendula' 364
  f. *aureomarginata* 362
  f. *bacciflava* 364
  'Crassifolia' 364
  'Crispa' 364
  'Crispa Picta' 364
  'Ferox' 364
  'Ferox Argentea' 364
  'Ferox Aurea' 364
  'Flavescens' 364
  'Fructu-Luteo' 364
  'Golden Milkboy' 362
  'Golden Milkmaid' 362
  'Golden Queen' 362
  'Handsworth New Silver' 364
  f. *heterophylla* 364
  'Heterophylla Aureomarginata' 362
  'Laurifolia' 364
  'Laurifolia Aurea' 364
  'Madame Briot' 362
  'Myrtifolia' 364
  'Myrtifolia Aurea' 364
  'Pendula' 364
  'Pyramidalis' 364
  'Silver Milkboy' 364
  'Silver Queen' 364
  'Watereriana' 362
  × *beanii* 366
  dipyrena 366
  latifolia 366
  opaca 362
  perado 362
  pernyi 366
Ironwood 196
Persian 278
Ivy, Poison 360

Jubaea chilensis 450
Judas Tree 348
Juglandaceae 172
Juglans ailanthifolia 170

  cathayensis 180
  cinerea 170
  × *intermedia* 178
  mandshurica 180
  nigra 178
  regia 178
  sieboldiana 180
Juneberry 320
Juniper 54
  Chinese 56
  Coffin 56
  Common 54
  Drooping 56
  Golden Chinese 56
  Grecian 58
  Himalayan 56
  Irish 54
  Meyer's 56
  Phoenician 58
  Prickly 54
  Rocky Mountain 58
  Spanish 58
  Stinking 58
  Swedish 54
  Syrian 54
  Temple 54
Juniperus chinensis 56
  communis 54
  drupacea 54
  foetidissima 58
  oxycedrus 54
  phoenicea 58
  recurva 56
  rigida 54
  scopulorum 58
  squamata 56
  thurifera 58
  virginiana 56
    'Canaertii' 56
    'Glauca' 58

Kaki 432
Kalopanax pictus 424
  septemlobus 424
Katsura 274
Keaki 252
Kindling Bark, Broad-leaved 416
Kingnut 174
Kobushi 268
Koelreuteria paniculata 398
Kohuhu 410
Korlinga 428
Kowhai 352
Kuma-shide 196
Kuro-matsu 142

+ *Laburnocytisus adamii* 350
Laburnum, Adam's 350
  Common 350
  Golden-leaved 350

Scots 350
Voss's 350
Laburnum alpinum 350
  anagyroides 350
  × *watereri* 350
Lagarostrobus franklinii 26
Larch 94
  Common 94
  Dahurian 96
  Dunkeld 96
  European 94
  Golden 98
  Japanese 96
  Polish 94
  Russian 94
  Siberian 94
  Sikkim 98
  Western 98
Larix dahurica 96
  decidua 94
  × *eurolepis* 96
  europaea 94
  gmelinii 96
  griffithiana 98
  griffithii 98
  kaempferi 96
  laricina 98
  leptolepis 96
  occidentalis 98
  × *pendula* 98
  potaninii 98
  russica 94
  sibirica 94
Lauraceae 276
Laurel, Californian 276
  Cherry 344
  Poet's 276
  Portugal 344
Laurelia sempervirens 276
  serrata 276
Laurus nobilis 276
Leatherwood, Tasmanian 406
Leguminosae 346
Lemon 356
Lenga 200
Leylandii 46
Libocedrus chilensis 30
  decurrens 30
Ligustrum lucidum 442
Liliaceae 450
Lime 400
  American 402
  Broad-leaved 400
  Caucasian Hybrid 402
  Common 402
  Crimean 402
  Cut-leaved 400
  Golden 402
  Mongolian 404
  Oliver's 404
  Silver 404

Silver Pendent 404
Small-leaved 400
Liquidambar formosana 278
  orientalis 278
  styraciflua 278
Liriodendron chinense 272
  tulipifera 272
Lithocarpus edulis 230
Locust, Black 354
  Chinese Honey 348
  Clammy 354
  Honey 348
Lonicera maackii 424
Loquat 282
Luma apiculata 414

Maackia 352
Maackia amurensis 352
Maclura aurantiaca 256
  pomifera 256
Madrona 430
Magnolia 258
  Big-leaf 262
  Campbell's 264
  Chinese Big-leaf 262
  Chinese Evergreen 260
  Dawson's 266
  Fishtail 262
  Fraser 262
  Glossy 260
  Japanese Big-leaf 262
  Lily-flowered 270
  Loebner's 268
  Sargent's 266
  Saucer 270
  Southern Evergreen 260
  Sprenger's 266
  Star 268
  Veitch's 264
  Watson's 262
  Willow-leaved 268
Magnolia acuminata 260
  'Butterflies' 270
  campbellii 264
  dawsoniana 266
  delavayi 260
  denudata 270
  'Elizabeth' 270
  fraseri 262
  'Galaxy' 266
  grandiflora 260
  'Heaven Scent' 270
  heptapeta 270
  hypoleuca 262
  'Iolanthe' 266
  × *kewensis* 268
  kobus 268
  liliiflora 270
  × *loebneri* 268
  macrophylla 262

nitida 260
obovata 262
officinalis 262
'Peppermint Stick' 270
× *proctoriana* 268
robusta 266
salicifolia 268
sargentiana 266
'Sayonara' 270
× *soulangiana* 270
sprengeri 266
'Star Wars' 266
stellata 268
tripetala 262
× *veitchii* 264
'Vulcan' 270
× *watsonii* 262
× *weiseneri* 262
Magnoliaceae 258
Mahogany, Chinese 358
Maidenhair Tree 20
Maiten 448
Malosorbus florentina 312
Malus baccata 310
  bhutanica 308
  'Butterball' 312
  coronaria 308
  'Crittenden' 312
  'Dartmouth' 312
  dasyphylla 306
  domestica 308
    'Elise Rathke' 314
  florentina 312
  floribunda 308
  × *gloriosa* 312
  halliana 314
  × *hartwigii* 314
  hupehensis 310
  ioensis 308
  'John Downie' 312
  × *magdeburgensis* 310
  × *moerlandsii* 314
  prattii 308
  prunifolia 310
  pumila 306
  × *purpurea* 314
  'Red Tip' 308
  × *robusta* 312
  'Royalty' 314
  × *scheideckeri* 314
  sieboldii 308
  'Snowcloud' 314
  sylvestris 306
  theifera 310
  toringo 308
  toringoides 308
  transitoria 308
  trilobata 312
  tschonoskii 308
  'Van Eseltine' 310
  'Winter Gold' 312
  'Wisley Crab' 314

yunnanensis 308
× *zumi* 312
Maple 368
  Amur 376
  Ash-leaved 390
  Balkan 376
  Bigleaf 378
  Birch-leaved 382
  Black 372
  Cappadocian 374
  Chosen 390
  Coliseum 374
  Coral-bark 384
  Cretan 376
  Cut-leaved Japanese 384
  Cut-leaved Silver 378
  Cyprus 376
  Deep-veined 386
  Downy 388
  Downy Japanese 386
  Eagle's Claw 372
  English 368
  Fastigiate Field 368
  Field 368
  Forrest's 380
  Full Moon 386
  Great 370
  Green Snake-bark 382
  Grey Snake-bark 380
  Hawthorn-leaved 382
  Heldreich's 388
  Hers' 382
  Honshu 380
  Hornbeam-leaved 384
  Horned 370
  Italian 376
  Keijo 386
  Kyushu 380
  Lime-leaved 382
  Lobel's 374
  Miyabe's 386
  Montpelier 376
  Mountain 386
  Nikko 390
  Norway 372
  Oregon 378
  Paper-bark 388
  Père David's 380
  Purple Japanese 384
  Red 378
  Red-bud 388
  Red Snake-bark 380
  Rough-barked 390
  Shandong 374
  Siebold's 386
  Silver 378
  Smooth Japanese 384
  Snake-bark 380
  Spanish 368
  Sugar 372
  Syrian 376
  Tang 374

INDEX

Tartar 376
Trautvetter's 388
Trident 374
Uri 382
Van Volxem's 388
Vine 386
Vine-leaved 390
Zoeschen 376
May 284
  Pink 284
Maytenus boaria 448
Mazzard 322
Medlar 290
  Bronvaux 290
  Japanese 282
Medlar-Thorn 290
Mei 338
Meliaceae 358
Mespil, Snowy 320
Mespilus germanica 290
Metasequoia glyptostroboides 64
Michelia 260
Michelia doltsopa 260
Mimosa 346
Mockernut 174
Monkey Puzzle 28
Moosewood 382
Moraceae 256
Morus alba 258
  nigra 258
Mulberry, Black 258
  Paper 256
  White 258
Myrtaceae 414
Myrtle, Chilean 414
  Orange-barked 414
Myrtus luma 414

Naden 334
Negundo aceroides 390
Nettle-tree 254
  Caucasian 254
  Southern 254
Nirre 200
Nothofagus antarctica 200
  betuloides 202
  cunninghamii 202
  dombeyi 202
  fusca 202
  glauca 200
  menziesii 202
  nervosa 200
  obliqua 200
  procera 200
  pumilio 200
  solanderi 200
Nutmeg, Californian 24
  Japanese 24
  Peruvian 276
Nyssa aquatica 412
  sinensis 412

sylvatica 412
Nyssaceae 412

Oak 214
  Armenian 224
  Bamboo 230
  Bartram's 238
  Basket 232
  Black 234
  Black-jack 240
  Burr 232
  Californian Black 234
  Californian Live 222
  Caucasian 228
  Champion 234
  Chestnut 232
  Chestnut-leaved 226
  Chinese Cork 230
  Chinkapin 232
  Common 216
  Cork 224
  Cut-leaved 216
  Cypress 216
  Daimyo 224
  Downy 218
  Durmast 214
  English 216
  Fulham 220
  Golden 216
  Golden, of Cyprus 212
  Green 218
  Holly 222
  Holm 222
  Hungarian 226
  Japanese Chestnut 230
  Japanese Evergreen 230
  Kermes 222
  Laurel 240
  Lebanon 224
  Lea's 240
  Lucombe 220
  Ludwig's 238
  Macedonian 222
  Mediterranean 222
  Medlar-leaved 214
  Mirbeck's 228
  Moscow 214
  Mossy-cup 232
  Northern Pin 236
  Pagoda 238
  Palestine 222
  Pedunculate 216
  Pin 236
  Pontine 224
  Portuguese 228
  Purple 216
  Pyrenean 228
  Quercitron 234
  Red 234
  Scarlet 236
  Sessile 214
  Shingle 240

Shumard 238
Sindian 222
Southern Red 238
Spanish 238
Swamp White 232
Thonp 230
Trojan 222
Turkey 218
Turner's 220
Ubame 222
Valonia 218
Water 240
White 232
Willow 238
Olea europaea 442
Oleaceae 436
Olearia paniculata 410
Oleaster 412
Olive 442
  Russian 412
Orange 356
  Osage 256
Osier, Common 170
  Purple 170
  Weeping 170
Osmanthus, Giant 442
  Holly-leaved 442
Osmanthus heterophyllus 442
  yunnanensis 442
Ostrya carpinifolia 196
  japonica 196
  virginiana 196
Oxydendrum arboreum 430

Padus racemosa 342
Pagoda Tree 352
Palm, Cabbage 450
  Canary 450
  Chilean Wine 450
  Chusan 450
  Date 450
  Dwarf Fan 450
  Windmill 450
Palmae 450
Paraserianthes lophantha 346
Parrotia persica 278
Paulownia fargesii 444
  fortunei 444
  tomentosa 444
Paupauma 424
Pear 316
  Almond-leaved 318
  Austrian 316
  Bollwyller 292
  Callery 318
  Chanticleer 318
  Choke 316
  Common 316
  Himalayan 318

Kieffer 318
Michaux's 318
Plymouth 316
Sand 318
Snow 320
Wild 316
Willow-leaved 320
Pecan 174
Peppermint, Mount Wellington 422
  Narrow-leaved Black 420
Persimmon 432
  Chinese 432
Phellodendron, Japanese 356
Phellodendron amurense 356
  chinense 356
  lavallei 356
  japonicum 356
Phillyrea 442
Phillyrea latifolia 442
  media 442
Phoenix canariensis 450
  dactylifera 450
  theophrastii 450
Photinia, Deciduous 282
  Giant 282
Photinia beauverdiana 282
  serratifolia 282
  serrulata 282
Picea abies 102
  alba 114
  alcoquiana 108
  alpestris 102
  asperata 112
  bicolor 108
  brachytyla 110
  breweriana 100
  engelmannii 114
  glauca 114
  glehnii 108
  × hurstii 114
  jezoensis 110
  koyamai 104
  likiangensis 110
  × lutzii 114
  mariana 104
  maximowiczii 108
  morinda 100
  morindoides 112
  obovata 102
  omorika 106
  orientalis 108
  polita 112
  pungens 114
  purpurea 110
  rubens 104
  schrenkiana 100
  sitchensis 106

smithiana 100
spinulosa 112
torano 112
watsoniana 106
wilsonii 106
Pignut 176
Pinaceae 66
Pine 122
  Aleppo 130
  Apache 146
  Armand's 142
  Arolla 134
  Austrian 124
  Bhutan 138
  Big-cone 148
  Bishop 128
  Black 124
  Blue 138
  Bosnian 134
  Bristlecone 136
  Calabrian 130
  Chile 28
  Chinese Red 142
  Chinese White 142
  Cluster 132
  Corsican 124
  Coulter 148
  Crimean 124
  David's 142
  Digger 148
  Durango 132
  Endlicher 132
  Foxtail 136
  Gregg 144
  Hartweg's 132
  Holford's 140
  Huon 26
  Jack 128
  Japanese Black 142
  Japanese Red 142
  Japanese Umbrella 66
  Japanese White 142
  Jeffrey's 146
  Jelecote 148
  Kauri 28
  King William 58
  Knobcone 136
  Korean 134
  Lacebark 148
  Limber 146
  Loblolly 144
  Lodgepole 126
  Macedonian 138
  Maritime 132
  Mexican Weeping 148
  Mexican White 140
  Monterey 144
  Montezuma 132
  Mountain 134
  Murray 126
  Norfolk Island 28
  Northern Pitch 144
  Norway 128
  Parana 28
  Ponderosa 146
  Pyrenean 124
  Red 128
  Rocky Mountain
    Bristlecone 136
  Royal 146
  Scots 122
  Scrub 128
  Shore 126
  Shortleaf 126
  Sierra Lodgepole 126
  Singleleaf Nut 144
  Soft 122
  Stone 130
  Sugar 140
  Swiss Stone 134
  Umbrella 130
  Western White 146
  Western Yellow 146
  Weymouth 138
  Whitebark 136
  Yunnan 142
Pinus albicaulis 136
  aristata 136
  armandii 142
  attenuata 136
  ayacahuite 140
  balfouriana 136
  banksiana 128
  brutia 130
  bungeana 148
  cembra 134
  contorta 126
  coulteri 148
  densiflora 142
  divaricata 128
  durangensis 132
  echinata 126
  engelmannii 146
  excelsa 138
  flexilis 146
  greggii 144
  griffithii 138
  halepensis 130
  hartwegii 132
  heldreichii 134
  × *holfordiana* 140
  jeffreyi 146
  koraiensis 134
  lambertiana 140
  leucodermis 134
  longaeva 136
  maritima 132
  monophylla 144
  montana 134
  montezumae 132
  monticola 146
  mughus 134
  mugo 134
  muricata 128
  nigra 124
  parviflora 142
  patula 148
  peuce 138
  pinaster 132
  pinea 130
  ponderosa 146
  pseudostrobus 132
  radiata 144
  reflexa 146
  remorata 128
  resinosa 128
  rigida 144
  rudis 132
  sabiniana 148
  sinensis 142
  strobus 138
  sylvestris 122
  tabuliformis 142
  taeda 144
  thunbergii 142
  tuberculata 136
  uncinata 134
  virginiana 128
  wallichiana 138
  yunnanensis 142
Pittosporaceae 410
Pittosporum tenuifolium 410
  tobira 406
Plane, Corstorphine 370
  London 280
  Oriental 280
  (*in Scotland*) 370
Planataceae 280
Platanus × *acerifolia* 280
  × *hispanica* 280
  occidentalis 280
  orientalis 280
Platycladus orientalis 34
Plum 340
  Cherry 340
  Mirabelle 340
  Myrobalan 340
  Naples 340
  Pissard's 340
Podocarp, Willow 26
Podocarpaceae 26
Podocarpus andinus 26
  nubigenus 26
  salignus 26
  totara 26
Poirier Sauger 316
Poplar 150
  Afghan 154
  Balsam 160
  Balsam Spire 160
  Berlin 152
  Black 150
  Black Italian 158
  Bolle's 150
  Chinese Necklace 162
  Downy Black 152
  Eastern Balsam 160
  Female Lombardy 154
  Golden 158
  Golden Lombardy 154
  Grey 150
  Hybrid Black 156
  Japanese Balsam 162
  Lombardy 154
  May 156
  Noble 158
  Picart's 150
  Plantier's 154
  Prince Eugene's 156
  Railway 156
  Simon's 162
  Variegated 160
  Western Balsam 160
  White 150
  Wild Black 152
  Yellow 272
Populus × *acuminata* 162
  alba 150
  'Androscoggin' 162
  'Balsam Spire' 160
  balsamifera 160
  × *berolinensis* 152
  × *canadensis* 156
    'Casale 78' 158
    'Eugenei' 156
    'Florence Biondi' 158
    'Gelrica' 156
    'Heidemij' 156
    'I-78' 158
    'Marilandica' 156
    'OP226' 158
    'Regenerata' 156
    'Robusta' 156
    'Serotina' 158
    'Serotina Aurea' 158
    'Serotina de Selys' 158
    'Serotina Erecta' 158
  candicans 160
  canescens 150
  deltoides 156
  × *euramericana* 156
  × *generosa* 158
  × *jackii* 160
  lasiocarpa 162
  laurifolia 152
  maximowiczii 162
  nigra 'Afghanica' 154
    ssp. *betulifolia* 152
    'Elegans' 154
    'Foemina' 154
    'Gigantea' 154
    'Italica' 154
    'Lombardy Gold' 154
    'Plantierensis' 154

INDEX

'Thevestina' 154
'Vereecken' 154
nivea 150
'Oxford' 162
przewalskii 162
'Rochester' 162
simonii 162
szechuanica 162
tacamahaca 160
tremula 152
trichocarpa 160
'TT32' 160
violascens 162
Pride of India 398
Privet, Chinese Tree 442
Protaceae 274
Prumnopitys andina 26
Prunus 'Accolade' 336
 'Amanogawa' 326
 'Amarmeniaca' 338
 'Asano' 332
 avium 322
 × *blireana* 340
 brigantina 338
 campanulata 336
 canescens 324
 cerasifera 340
 cerasus 324
 cocomila 340
 'Collingwood Ingram' 336
 communis 338
 conradinae 334
 × *dawyckensis* 324
 divaricata 340
 domestica 340
 dulcis 338
 'Fudan Zakura' 334
 'Fugenzo' 326
 'Geraldinae' 332
 'Hillieri' 336
 × *hillieri* 336
 'Hilling's Weeping' 332
 hirtipes 334
 'Hokuzai' 326
 'Ichiyo' 328
 incisa 332
 insititia 340
 'Ito-kukuri' 326
 jamasakura 330
 'James Veitch' 326
 'Jo-Nioi' 328
 × *juddii* 330
 'Kanzan' 326
 'Kiku-shidare-zakura' 332
 'Kursar' 336
 laurocerasus 344
 'Longipes' 328
 lusitanica 344
 maackii 324
 mahaleb 342
 'Mount Fuji' 328
 mume 338
 'Ojochin' 328
 'Okame' 336
 padus 342
 'Pandora' 336
 pendula 332
   var. *ascendens* 334
   'Beni Higan Sakura' 334
   'Fukubana' 334
   'Pendula' 332
   'Pendula Plena Rosea' 322
   'Pendula Rosea' 332
   'Pendula Rubra' 322
   'Pink Star' 334
   'Stellata' 334
 persica 338
 'Pink Perfection' 326
 'Royal Burgundy' 326
 sargentii 330
 × *schmittii* 324
 serotina 344
 serrula 324
 serrulata 330
 'Shimidsu Zakura' 328
 'Shirofugen' 328
 'Shirotae' 328
 'Shogetsu' 328
 'Shosar' 336
 × *sieboldii* 334
 'Snow Goose' 336
 speciosa 336
 spinosa 340
 'Spire' 336
 × *subhirtella* 334
 'Tai Haku' 328
 'Taoyame' 328
 'Ukon' 328
 'Umineko' 336
 'Uzuzakura' 326
 × *verecunda* 336
 virginiana 342
 webbii 338
 'Yae-murasaki-zakura' 326
 × *yedoensis* 332
Pry Tree 400
Pseudolarix amabilis 98
Pseudotsuga douglasii 120
 japonica 120
 macrocarpa 120
 menziesii 120
 taxifolia 120
Pterocarya fraxinifolia 172
 × *rehderiana* 172
 rhoifolia 172
 stenoptera 172
Pterostyrax corymbosa 432
 hispica 432
Pyrus amygdaliformis 318
 austriaca 316
 bourgaeana 318
 calleryana 318
 caucasica 318
 communis 316
 cordata 316
 cydonia 290
 elaeagrifolia 320
 × *lecontii* 318
 magyarica 316
 × *michauxii* 318
 nivalis 320
 pashia 318
 pyraster 316
 pyrifolia 318
 rossica 316
 salicifolia 320
 salvifolia 316
 ussuriensis 318

Quercus acuta 230
 acutissima 230
 aegilops 218
 agrifolia 222
 alba 232
 alnifolia 212
 bicolor 232
 borealis 234
 brachyphylla 218
 brutia 216
 calliprinos 222
 canariensis 228
 castaneifolia 226
 cerris 218
 coccifera 222
 coccinea 236
 conferta 226
 congesta 218
 × *crenata* 220
 daimio 224
 dalechampii 214
 dentata 224
 ellipsoidalis 236
 faginea 228
 falcata 238
 frainetto 226
 glandulifera 232
 glauca 230
 hartwissiana 216
 × *heterophylla* 238
 × *hispanica* 220
 iberica 214
 ilex 222
 imbricaria 240
 ithaburensis 218
 kelloggii 234
 laevigata 230
 lanuginosa 218
 laurifolia 240
 × *leana* 240
 libani 224
 lucombeana 220
 × *ludoviciana* 238
 lusitanica 228
 macedonica 222
 macranthera 228
 macrocarpa 232
 macrolepis 218
 marilandica 240
 mas 214
 maxima 234
 mirbeckii 228
 muehlenbergii 232
 myrsinifolia 230
 nigra 240
 pagoda 238
 palustris 238
 pedunculata 216
 pedunculiflora 216
 petraea 214
 phellos 238
 phillyreoides 222
 polycarpa 214
 pontica 224
 prinus 232
 pubescens 218
 pyrenaica 228
 robur 216
 × *rosacea* 216
 rubra 234
 × *schochiana* 238
 serrata 240
 shumardii 238
 suber 224
 thomasii 216
 toza 228
 trojana 222
 × *turneri* 220
 variabilis 230
 velutina 234
 virgiliana 218
Quickthorn 284

Rain Tree, Golden 398
Rauli 200
Redbud 348
Redwood, Californian 62
 Coast 62
 Dawn 64
 Giant 62
 Sierra 62
Rhamnaceae 398
Rhamnus cathartica 398
 frangula 398
Rhododendron 428
 Fortune's 428
 Hardy Hybrid 428
 Himalayan Tree 428
 Loderi 428
Rhododendron arboreum 428
 falconeri 428
 fortunei 428

griffithianum 428
'King George' 428
ponticum 428
protistum 428
sinogrande 428
Rhus cotinus 360
 potaninii 360
 radicans 360
 typhina 360
 vernicifulua 360
Robinia, Golden 354
 Mop-head 354
 Single-leaved 354
Robinia × ambigua 354
 kelseyi 354
 pseudoacacia 354
 × slavinii 354
 viscosa 354
Rosaceae 282
Rowan, American 300
 Common 300
 Cut-leaved 300
 Edible 300
 Ghose's 302
 Golden 300
 Golden-fruited 300
 Hubei 304
 Japanese 302
 Joseph Rock's 304
 Kashmir 304
 Kew Hybrid 302
 Ladder-leaf 304
 Sargent's 302
 Vilmorin's 304
Rubber Tree, Hardy 278
Rubiaceae 448
Rutaceae 356

Salicaceae 150
Salix alba 164
 amygdalina 170
 atrocinerea 168
 babylonica 166
 × blanda 170
 caprea 168
 cinerea 168
 coaetanea 168
 daphnoides 168
 × ehrhartiana 166
 elaeagnos 170
 × elegantissima 170
 fragilis 166
 lucida 166
 matsudana 166
 × meyeriana 166
 pedicellata 168
 × pendulina 170
 pentandra 166
 purpurea 170
 × reichardtii 168
 × rubens 166
 × sepulcralis

'Chrysocoma' 168
'Salamonii' 170
× sericans 170
× smithiana 170
triandra 170
viminalis 170
Sallow, Common 168
 Great 168
 Grey 168
 Rusty 168
 Weeping 168
Sally, Black 420
 Weeping 420
Sambucus nigra 448
 racemosa 448
Sapindaceae 398
Sassafras 276
Sassafras albidum 276
 officinale 276
Sato-zakura 326
Saxegothaea conspicua 26
Scholar's Tree 352
Sciadopitaceae 66
Sciadopitys verticillata 66
Scrophulariaceae 444
Sequoia, Giant 62
 Weeping Giant 62
Sequoia gigantea 62
 sempervirens 62
Sequoiadendron giganteum 62
Service, Arran 298
 Bastard 298
 Bristol 296
 Exmoor 296
 Tree of Fontainebleau 296
 True 298
 Wild 296
Serviceberry 320
Shadblow 320
Sharon Fruit 432
Silk Tree 346
Silver Top 418
Silverbell, Mountain 434
Simaroubaceae 358
Siris, Pink 346
Sloe 340
Smoke-bush 360
Snowbell Tree 434
Snowdrop Tree 434
Sophora japonica 352
 tetraptera 352
× Sorbopyrus auricularis 292
Sorbus alnifolia 294
 americana 300
 anglica 298
 aria 292
 arranensis 298
 aucuparia 300

 austriaca 298
 bristoliensis 296
 cashmiriana 304
 commixta 302
 croceocarpa 296
 cuspidata 294
 decipiens 296
 devoniensis 296
 domestica 298
 eminens 292
 esserteauana 302
 fennica 298
 folgneri 294
 'Ghose' 302
 glabrescens 304
 hedlundii 294
 hibernica 292
 hupehensis 304
 hybrida 298
 intermedia 298
 'Joseph Rock' 304
 × kewensis 302
 koehneana 304
 KW 7746 302
 lancastriensis 292
 latifolia 296
 meinichii 298
 'Mitchellii' 294
 mougeotii 298
 oligodonta 304
 pallescens 294
 pohuashanensis 302
 porrigentiformis 292
 prattii 304
 pseudofennica 298
 pseudohupehensis 304
 Rock 23657 304
 rockii 302
 rupicola 292
 sargentiana 302
 scalaris 304
 scandica 298
 subcuneata 296
 'Theophrasta' 296
 thibetica 294
 thuringiaca 298
 torminalis 296
 umbellata 292
 × vagensis 296
 vestita 294
 vexans 292
 vilmoriniana 304
 vilmorinii 304
 'Wilfrid Fox' 294
 wilmottiana 292
Sorrel Tree 430
Spindle 448
Spruce 100
 Alcock's 108
 Black 104
 Blue Colorado 114
 Blue Engelmann 114

 Brewer 100
 Colorado 114
 Dragon 112
 East Himalayan 112
 Engelmann 114
 Hondo 110
 Koyama's 104
 Lijiang 110
 Maximowicz's 108
 Morinda 100
 Norway 102
 Oriental 108
 Purple-coned 110
 Red 104
 Sakhalin 108
 Sargent's 110
 Schrenk's 100
 Serbian 106
 Siberian 102
 Sikkim 112
 Sitka 106
 Snake-bark 102
 Tiger-tail 112
 White 114
 Wilson's 106
Spur-Leaf 272
Stewartia koreana 408
 monadelpha 408
 pseudocamellia 408
 sinensis 408
Storax 434
 Big-leaved 434
 Hemsley's 434
Strawberry Tree 430
 Cyprus 430
 Grecian 430
 Hybrid 430
Stryphnolobium japonicum 352
Stuartia, Chinese 408
Styracaceae 432
Styrax hemsleyanus 434
 japonicus 434
 obassia 434
 officinalis 434
Sugarberry 254
Sumach, Cut-leaved 360
 Stag's-horn 360
 Venetian 360
Sweet Gum 278
 Chinese 278
 Oriental 278
Swida controversa 426
 macrophylla 426
 sanguinea 424
Sycamore 370
 American 278, 280
 Golden 370
 Purple 370
 Variegated 370

Taiwania 60

INDEX

Taiwania cryptomerioides 60
 flousiana 60
Tamarack 98
Tamaricaceae 410
Tamarisk 410
Tamarix africana 410
 anglica 410
 canariensis 410
 dalmatica 410
 gallica 410
 parviflora 410
 ramosissima 410
 smyrnensis 410
 tetrandra 410
Tarajo 366
Taxaceae 22
Taxodiaceae 58
Taxodium ascendens 64
 distichum 64
 sempervirens 62
Taxus baccata 22
 celebica 22
 cuspidata 22
 mairei 22
 × *media* 22
Tetracentron sinense 272
Tetracentronaceae 272
Tetradium daniellii 356
Theaceae 408
Thelycrania sanguinea 424
Thorn 284
 Broad-leaved Cockspur 288
 Cockspur 288
 Glastonbury 284
 Grignon's 290
 Hungarian 286
 Hybrid Cockspur 290
 Large-leaved 286
 Oriental 286
 Red 286
 Scarlet 288
 Spotted 286
 Tansy-leaved 286
Thuja 32
 Japanese 34
 Korean 34
 Oriental 34
Thuja gigantea 32
 koraiensis 34
 lobbii 32
 occidentalis 34
 orientalis 34
 plicata 32
 standishii 34
Thujopsis dolabrata 32
Tilia americana 402
 cordata 400
 dasystyla 402
 × *euchlora* 402

 × *europaea* 402
 heterophylla 404
 'Moltkei' 404
 mongolica 404
 oliveri 404
 × *petiolaris* 404
 platyphyllos 400
 'Spectabilis' 404
 tomentosa 404
Tobira 406
Toona sinensis 358
Torreya californica 24
 grandis 24
 nucifera 24
Totara 26
 Chilean 26
Trachycarpus fortunei 450
Tree Fern, Common 20
 Silver 20
Tree of Heaven 358
 Downy 358
Trochodendraceae 274
Trochodendron aralioides 274
Tsuga canadensis 116
 caroliniana 118
 chinensis 118
 diversifolia 118
 dumosa 118
 heterophylla 116
 × *jeffreyi* 118
 mertensiana 118
 sieboldii 118
 yunnanensis 118
Tulip Tree 272
 Chinese 272
 Pink 264
Tupelo 412
 Chinese 412
 Water 412

Ulmaceae 240
Ulmo 406
Ulmus americana 248
 campestris 244
 canescens 246
 carpinifolia 246
 chinensis 250
 'Commelin' 248
 coritana 246
 × *diversifolia* 242
 'Dodoens' 248
 effusa 248
 × *elegantissima* 242
 glabra 242
 × *hollandica* 242
 'Belgica' 246
 'Dampieri' 246
 'Dampieri Aurea' 246
 'Groeneveld' 248
 'Hollandica' 248

 'Major' 248
 'Vegeta' 246
 'Wredei' 246
 japonica 250
 laevis 248
 'Lobel' 248
 microphylla 250
 minor 242
  var. *angustifolia* 244
  ssp. *canescens* 246
  var. *cornubiensis* 244
  'Dicksonii' 246
  var. *lockii* 244
  'Louis van Houtte' 244
  var. *minor* 246
  var. *plotii* 244
  'Sarniensis' 244
  'Viminalis' 246
  var. *vulgaris* 244
 montana 242
 parvifolia 250
 'Pinnato-Ramosa' 250
 'Plantijn' 248
 procera 244
 pumila 250
 'Regal' 250
 'Sapporo Autumn Gold' 250
 stricta 244
Umbrella Tree 262

Varnish Tree 360
 Chinese 360

Walnut 178
 Black 178
 Common 178
 Cut-leaved 178
 Japanese 180
 Manchurian 180
Wattle, Cape 346
 Cootamundra 346
 Silver 346
Wedding-cake Tree 426
Wellingtonia 62
Wheel Tree 274
Whitebeam, Alder-leaved 294
 Arran 298
 Cliff 292
 Common 292
 Finnish 298
 Folgner's 294
 Golden 292
 Himalayan 294
 Mitchell's 294
 Swedish 298
Whitethorn 284
Willow 164
 Almond 170
 Basford 166

 Bay 166
 Chinese Weeping 166
 Coral-bark 164
 Corkscrew 166
 Crack 166
 Cricket-bat 164
 Dragon's Claw 166
 Duke of Bedford's 166
 French 170
 Goat 168
 Golden 164
 Golden Weeping 168
 Hoary 170
 Pussy 168
 Salomon's Weeping 170
 Silver 164
 Smith's 170
 Thurlow Weeping 170
 Violet 168
 White 164
 Wisconsin Weeping 170
Wingnut 172
 Caucasian 172
 Chinese 172
 Hybrid 172
 Japanese 172
Winteraceae 274
Winter's Bark 274

Xanthocyparis nootkatensis 44

Yellow-wood 352
 Chinese 352
Yew 22
 Chilean Plum 26
 Chinese 22
 Chinese Plum 24
 Common 22
 Fastigiate Plum 24
 Golden 22
 Hicks' 22
 Irish 22
 Japanese 22
 Plum 24
 Prince Albert's 26
 Westfelton 22
Yulan 270

Zelkova 252
 Chinese 252
 Cretan 252
Zelkova abelicea 252
 acuminata 252
 carpinifolia 252
 crenata 252
 cretica 252
 serrata 252
 sicula 252
 sinica 252
 'Verschaeffeltii' 252